SUSTAINABLE DESALINATION HANDBOOK

SUSTAINABLE DESALINATION HANDBOOK

Plant Selection, Design and Implementation

VEERA GNANESWAR GUDE

ELSEVIER

Butterworth-Heinemann
An imprint of Elsevier

Butterworth-Heinemann is an imprint of Elsevier
The Boulevard, Langford Lane, Kidlington, Oxford OX5 1GB, United Kingdom
50 Hampshire Street, 5th Floor, Cambridge, MA 02139, United States

Library of Congress Cataloging-in-Publication Data
A catalog record for this book is available from the Library of Congress

British Library Cataloguing-in-Publication Data
A catalogue record for this book is available from the British Library

ISBN: 978-0-12-809240-8

For information on all Butterworth-Heinemann publications
visit our website at https://www.elsevier.com/books-and-journals

Publisher: Mathew Deans
Acquisition Editor: Ken McCombs
Editorial Project Manager: Peter Jardim
Production Project Manager: Surya Narayanan Jayachandran
Cover Designer: Victoria Pearson

Typeset by SPi Global, India

DEDICATION

To my late father Gude Bhaskar Rao, to my mother Gude Fathima, to my sister Kalyani, and to my brothers Kiran and Karunakar, and to Preeti Muire.

CONTENTS

CONTRIBUTORS

Ozeair Abessi
Babol Noshirvani University of Technology, Babol, Iran

Aisha Al Bloushi
Khalifa University of Science and Technology, Abu Dhabi, United Arab Emirates

Edo Bar-Zeev
Ben-Gurion University of the Negev, Beersheba, Israel

Jamel Belhadj
Université de Tunis El Manar; Université de Tunis, Tunis, Tunisia

Natalia Belkin
Bar Ilan University, Ramat Gan, Israel

Ilana Berman-Frank
Bar Ilan University, Ramat Gan, Israel

Liat Birnhack
Faculty of Civil and Environmental Engineering, Technion, Haifa, Israel

Francesco Calise
University of Naples Federico II, Naples, Italy

David A. Caron
University of Southern California, Los Angeles, CA, United States

Habib Cherif
Université de Tunis El Manar, Tunis, Tunisia

Massimo Dentice d'Accadia
University of Naples Federico II, Naples, Italy

Hila Frank
Ben-Gurion University of the Negev, Beersheba; National Institute of Oceanography, Haifa, Israel

Adewale Giwa
Khalifa University of Science and Technology, Abu Dhabi, United Arab Emirates

Brent Haddad
University of California, Santa Cruz, CA, United States

Shadi W. Hasan
Khalifa University of Science and Technology, Abu Dhabi, United Arab Emirates

Nadine Heck
University of California, Santa Cruz, CA, United States

J. Jaime Sadhwani Alonso
University of Las Palmas de Gran Canaria, Las Palmas de Gran Canaria, Spain

Nurit Kress
The National Institute of Oceanography, Haifa, Israel

Ori Lahav
Faculty of Civil and Environmental Engineering, Technion, Haifa, Israel

Noemi Melián-Martel
University of Las Palmas de Gran Canaria, Las Palmas de Gran Canaria, Spain

Ângelo Paggi Matos
Federal University of Santa Catarina, Florianópolis, Brazil

Toufic Mezher
Khalifa University of Science and Technology, Abu Dhabi, United Arab Emirates

Adina Paytan
University of California, Santa Cruz, CA, United States

Karen L. Petersen
University of California, Santa Cruz, CA, United States

Antonio Piacentino
University of Palermo, Palermo, Italy

Donald Potts
University of California, Santa Cruz, CA, United States

Edoardo Quiriti
University of Naples Federico II, Naples, Italy

Antonio Santos Sánchez
Federal University of Ouro Preto, Ouro Preto, Brazil

Erica L. Seubert
University of Southern California, Los Angeles, CA, United States

Abhishek Shrivastava
The Dow Chemical Company, Edina, MN, United States

Derek Stevens
The Dow Chemical Company, Edina, MN, United States

Maria Vicidomini
University of Naples Federico II, Naples, Italy

Nikolay Voutchkov
Water Globe Consultants, LLC, Winter Springs, FL, United States

Dennis E. Williams
GEOSCIENCE Support Services, Inc., La Verne, CA, United States

BIOGRAPHY

Veera Gnaneswar Gude is an associate professor of Civil and Environmental Engineering at Mississippi State University (MSU). He has more than 15 years of academic, research and industrial experience in desalination, water-wastewater treatment, biofuel scientific, and technological areas. He received a BS degree in chemical engineering technology from Osmania University in 2000 and worked for Du Pont Singapore after his graduation from 2000 to 2004. He received an MS degree in environmental engineering from the National University of Singapore in 2004 and a PhD degree in environmental engineering from the New Mexico State University in 2007, under the direction of Prof. Nagamany Nirmalakhandan, for his research in low-temperature thermal desalination. He has expanded his research interests into water and wastewater treatment, microbial desalination, biofuel synthesis using sustainable chemistry principles and process intensification topics during his postdoctoral research, industrial and academic appointments. He is a licensed professional engineer and a board certified environmental engineer (also known as diplomate of environmental engineering) by the American Academy of Environmental Engineers and Scientists (AAEES).

Dr. Gude has published more than 70 scientific research articles on desalination (thermal, membrane, hybrid, and microbial), water-wastewater treatment, and on biofuels research in well-regarded journals. His research work is well utilized in the literature with more than 2500 citations by 2017. He has published 3 books on desalination research (Elsevier) and 2 books on biofuels research (CRC Press), 15 invited book chapters, 50 conference proceedings papers, 15 technical reports, several popular press articles and media releases, and 2 patents on low-temperature desalination and microalgae biofuels technologies. He has delivered 35 invited lectures including 6 plenary/keynote lectures and more than 140 scientific research and educational presentations. He has organized many workshops on water-energy-environment nexus topics at national and international conferences. He was the chair and board representative for clean energy and water division of ASES between 2011 and 2016. He serves on numerous scientific advisory boards and task committees across the world including ASCE-EWRI, ASEE, ASES, and AWWA. He is a member of several editorial boards and editor for many scientific journals including *ASCE Journal of*

Environmental Engineering, *Heliyon*, *Nature npj Clean Water*, *Renewable Energy*, *Resource-Efficient Technologies*, and *Water Environment Research Journal*. His research is supported by NSF, USEPA, USGS, USDA, and by many industrial and international agencies. He has received much recognition for his research, teaching, and service activities at regional, national, and international conferences and by professional societies (ASCE, ASEE, ASES, and Chi Epsilon National Civil Engineering Honor Society) including the 2017 Albert Nelson Marquis Lifetime Achievement Award and 2017 ASCE MS Engineer of the Year Award.

PREFACE

The future water supply system portfolio for many inland and coastal communities will include desalination of seawater or brackish water sources to meet the increasing demands for freshwater supplies. Population growth coupled with higher living standards, climate change, and dwindling groundwater reserves together have triggered this phenomenon. Desalination of brackish water or seawater is both energy- and cost-intensive. Numerous novel materials and processes have been developed in the past few decades to address these essential drawbacks that hinder successful implementation of desalination systems. The goal of these efforts is to achieve the minimum possible specific energy consumption closer to the theoretical thermodynamic minimum requirements. Though originally predominant in thermal desalination technologies, membrane technologies have outnumbered thermal desalination plants in the last decade mainly due to their elegant nature and significant improvements in specific energy consumption and water flux rates. Thermal desalination technologies still hold promise for potential applications, especially in cogeneration, trigeneration, and polygeneration schemes, where thermal energy sources are abundant or they can be derived from other processes mainly power generation plants. Hybridization of membrane and thermal processes along with colocation of power generation and water desalination plants have proven in terms of higher productivities, economic benefits, and lower environmental footprint. Yet, successful implementation of desalination plants and their sustainability depends on such factors, which are often categorized as environmental, economic, and social. Desalination byproduct is another significant environmental consideration, which has been found to be useful for many beneficial processes involving agriculture, fish and shrimp farming, high-value product, and metal recovery, and as feedstock to other chemical processes.

As desalination technology is here to stay, it is critical to understand and develop knowledge on various elements of process implementation including process selection, design, energy utilization and recovery, process integration, product diversification and resource utilization, seawater intake and discharge systems, concentrate management, and upgrading of desalinated water for potable use and other environmental impacts and regulations, economic and social issues related to its implementation.

This book is intended to serve as a resource and guidance manual for understanding the principles, mechanisms, design, and successful implementation of desalination technologies for undergraduate and graduate researchers, science and engineering professionals, and industrial practitioners. This book is a critical resource for researchers from chemical engineering and science, environmental science and engineering, mechanical and electrical engineering, and industrial and process engineering and management disciplines with a desire to understand the science, engineering, and process management related to sustainable desalination.

This book is presented in two parts and 15 chapters. The first part presents the design and process performance of desalination plants, while the second part discusses issues related to sustainable implementation of desalination plants. Details of the chapters are presented as follows.

Chapter no.	Title
1	Membrane Desalination—Process Selection, Design, and Implementation
2	Energy Efficiency in Reverse Osmosis
3	Environmental Impact and Technoeconomic Analysis of Hybrid MSF/RO Desalination: The Case Study of Al Taweelah A2 Plant
4	Trigeneration and Polygeneration Configurations for Desalination and Other Beneficial Processes
5	Design and Construction of Open Intakes
6	Design and Construction of Subsurface Intakes
7	Brine Disposal and Management—Planning, Design, and Implementation
8	Posttreatment of Desalinated Water
9	Desalination Concentrate Management and Valorization Methods
10	Environmental Regulations—Inland and Coastal Desalination Case Studies
11	Impact of Seawater Desalination on Coastal Environments
12	Microbial Communities in the Process and Effluents of Seawater Desalination Plants
13	Impact of Algal Blooms and Their Toxins on RO Desalination Plant Operations
14	Social Issues and Public Acceptance of Seawater
15	Environmental Life-Cycle Analysis of Desalination Processes

PART 1—DESALINATION PROCESS SELECTION AND IMPLEMENTATION

Chapter 1 presents various types and applications of membrane desalination processes based on the source water salinity, whether brackish water or seawater. Single- or multiple-pass and single- or multiple-stage and hybrid systems with multiple passes and stages reverse osmosis processes have been discussed. Process configurations for nanofiltration, reverse osmosis, and brackish water reverse osmosis processes have been presented with illustrations and case studies. Water quality issues and life-cycle costs for different saline water sources are discussed in detail along with issues to be considered in practical implementation of membrane processes.

Chapter 2 elaborates on the energy efficiency of reverse osmosis process. Theoretical development of thermodynamics and fundamentals of energetics from first and second laws of thermodynamics are discussed in depth. Models related to estimation of energy consumption along with membrane resistance and resistance due to membrane fouling, kinetics, flux, and rejection tradeoff are discussed. Energy efficiencies of different flow module configurations are elaborated, followed by systemic evaluation of energy consumption including power source and finally future opportunities for improvements.

Chapter 3 explores various possibilities for hybrid desalination processes through an extensive literature review and critical analysis. Hybrid configurations of membrane and thermal processes are discussed both in exclusive and integrated concepts. Environmental impact and technoeconomic analyses for a hybrid MSF/RO desalination plant are presented in detail.

Chapter 4 provides a detailed account on investigations related to polygeneration schemes for energy- and resource-efficient desalination process. Integration of renewable energy sources like geothermal, solar, and biomass is considered for polygeneration schemes. Solar desalination including photovoltaic and concentrated solar collectors for multieffect distillation desalination process is used along with heating and cooling applications. Thermoeconomic, exergic, and an exergoeconomic analyses aiming at defining the best process design variables are presented with two case studies for the city of Naples and the Pantelleria island, both characterized by solar and geothermal resources and scarcity of fossil fuels and freshwater.

Chapter 5 discusses different types and key features of open intake systems, both onshore and offshore, used in desalination plants with several examples. Selection, implementation, and challenges involved in open intake systems are discussed in detail. Benefits of colocation, and selection and design criteria are also presented for different scenarios.

Chapter 6 presents various options for cost-effective and efficient subsurface water intake systems for desalination. Vertical beach wells, Ranney collector wells, horizontal directional drilling wells, slant wells, and subsurface infiltration galleries are discussed with case studies around the world. A detailed discussion on their applicability based on the benefits and drawbacks is included followed by brief discussions on recent developments and field experiences and findings and future directions.

Chapter 7 explores brine disposal and management topics including planning, design, and implementation to minimize the adverse environmental impacts on coastal ecosystems. Environmental considerations, siting of desalination systems, brine discharge numerical and experimental models, and design details for single-port discharges and multiport diffusers with operational variable for surface and subsurface discharges.

Chapter 8 deals with the final water quality of the freshwater produced by desalination plants. Posttreatment required to address issues related to slightly acidic, low-buffering capacity soft waters. Basics and fundamental chemistry of aqueous, gaseous, and solid phases relevant to posttreatment processes are covered in detail. Rationale for the desired water quality criteria, advantages and drawbacks of current posttreatment processes are detailed, engineering and cost considerations are discussed, state-of-the-art posttreatment alternatives are covered, and contemporary research trends are listed.

Chapter 9 discusses two alternatives for desalination concentrate management, which are volume minimization or valuable resource recovery. Volume minimization methods such as zero liquid discharge configurations are presented. A detailed account on the beneficial use of desalination concentrates for renewable energy recovery, salts recovery, acid/alkali production, irrigation and fertigation, aquaculture, microalgae cultivation, and integrated production alternatives is discussed with extensive literature references.

PART 2—ISSUES IN SUSTAINABLE DESALINATION

Chapter 10 presents the environmental impacts associated with desalination plants, which include sensitive effects attributed to the rejected brine discharge into the receiving water bodies, the air pollutant emissions attributed to the energy demand of the processes, the noise pollution, and the use of land and others. Case studies are provided to illustrate the environmental impacts of desalination plants.

Chapter 11 provides a comprehensive overview highlighting the environmental effects of brine effluent on various coastal species, including bacteria, zooplankton, seagrass, fish larvae, and corals. Effects of desalination brine effluent on the activity and diversity of bacteria and microalgae, the abundance of meiofauna, and the physiology of seagrass meadows around the outfall site are discussed. Alternatives and measures to monitor and minimize the interface of brine effluent with marine coastal biota are provided.

Chapter 12 presents an overview of the impacts of desalination intakes and discharges on marine microorganisms (heterotrophic bacteria, cyanobacteria, and unicellular eukaryotes), which represent the "unseen majority" of the marine environment. A specific focus on the effects of desalination on the microbial community alterations throughout the seawater desalination process from intake to discharge is included. An assessment of the effects of desalination discharges on coastal microbial communities based on the data from laboratory and in situ experiments is provided. Recommendations are made for plant operations that could mitigate the effects on the marine microbial community.

Chapter 13 discloses the global increase in occurrence and incidence of harmful algae blooms and potential threat for RO desalination plants due to the high biomass and a variety of substances released by them. Noxious substances and powerful neurotoxins and their effect on RO membrane performances (chemical consumption and fouling rates) are discussed. Algal blooms can cause significant operational issues that result in increased chemical consumption, increased membrane fouling rates, and in extreme cases, a plant to be taken offline. Methods for identifying the toxic substances, their known producers are discussed through a case study and bench-top SWRO experiments modeled the worst-case scenario.

Chapter 14 examines public perception issues surrounding proposed and existing coastal desalination facilities with respect to regional water security, costs, impacts on quality of life, and demographic drivers shaping public attitudes and acceptance of seawater desalination plants.

Chapter 15 presents the environmental impact assessment for desalination plants using Life-Cycle Assessment (LCA) technique with a special focus on reverse osmosis process. Discussion on various software programs used to study LCA and a brief review of renewable energy powered desalination are included. Based on LCA, greenhouse gases emission for different desalination technologies are compared with reverse osmosis desalination coupled to renewable energies and fossil fuels. In addition, global embodied energy of a typical reverse osmosis desalination plant is investigated during a life-cycle period of 20 years, and finally it provides some guidelines for minimizing the environmental impact in desalination plants.

ACKNOWLEDGMENTS

I would like to express my appreciation to the colleagues in the Civil and Environmental Engineering department, the Bagley College of Engineering, and the Office of Research and Economic Development at Mississippi State University (MSU) for their support. My sincere appreciation and gratitude are extended to my doctoral and postdoctoral advisors Dr. Nagamany Nirmala Khandan, Dr. Ricardo Jacquez, Dr. Shuguang Deng, and Dr. Adrian Hanson of New Mexico State University (NMSU) for their continuous support in my professional career. Prof. Khandan is gratefully acknowledged for introducing me to desalination research during my doctoral research studies at NMSU. His encouragement and guidance had tremendous impact on my professional development. Colleagues at the Southwest Technology Development Institute of NMSU, especially Corey Asbill and Andrew Rosenthal of this institute and Robert Keith Payne of Engineering Technology department, are thanked for their support in my research while at NMSU. Dr. Craig Ricketts is thanked for sharing his work place at the engineering research facility. Industrial collaborations with Thomas Leggiore, Charles Oborne, and George Forbes are also graciously acknowledged. My graduate research students at Mississippi State University and New Mexico State University are appreciated for their diligence and energy in work environment and research contributions, especially Dr. Bahareh Kokabian and Dr. Edith Martinez-Guerra of MSU for their work on microbial desalination and solar desalination research projects respectively. Finally, I would like to thank Ms. Preeti Muire for her support and care.

Veera Gnaneswar Gude

PART 1

Desalination Process Selection and Implementation

CHAPTER 1

Membrane Desalination—Process Selection, Design, and Implementation

Nikolay Voutchkov
Water Globe Consultants, LLC, Winter Springs, FL, United States

1.1. OVERVIEW

Based on the source water salinity they process, desalination plants can be divided into three main groups: nanofiltration (softening) plants, brackish water desalination plants, and seawater desalination plants. In addition, depending on the number of sequential RO systems for treatment of permeate and concentrate, RO system configurations could be divided into two main categories: (1) single- and multiple-pass RO systems; (2) single- and multiple-stage RO systems. In all types of desalination plants, multiple-pass and multiple-stage RO systems could also be combined into configurations that allow achieving target RO system recovery and product water quality at optimum life-cycle cost of water production. The various types of systems and their practical application are discussed in the following sections.

1.1.1. Single- and Multiple-Pass RO Systems

An NF or RO system where the saline source water is desalinated (i.e., treated by reverse osmosis) only once is referenced as a single-pass RO system (see Fig. 1.1). RO systems that are designed to re-treat permeate from a single-pass (also referred to as first-pass) RO system by reverse osmosis multiple times are termed multiple-pass RO systems.

Fig. 1.1 depicts a general schematic of single-pass, two-pass, and three-pass RO systems. Since each RO pass provides additional treatment of permeate produced by the previous RO pass, the overall system permeate water quality improves with each pass. Therefore, multiple-pass RO systems are applied when source water salinity is relatively high and the target product water quality cannot be achieved by treating the saline source water by reverse osmosis only once.

Sustainable Desalination Handbook
https://doi.org/10.1016/B978-0-12-809240-8.00001-0

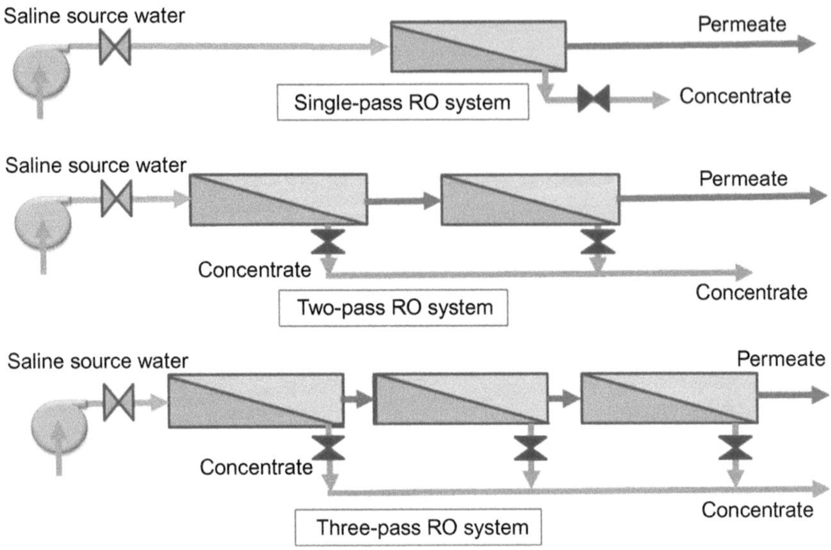

Fig. 1.1 Single- and multiple-pass RO systems.

While multiple RO pass systems allow to generate product water of very high quality, such systems are also more costly and produce less water than a single-pass RO system processing the same source water flow. Therefore, multiple-pass RO systems are typically applied mainly when a single-pass RO system cannot produce source water of desired quality. Another reason when multiple-pass (usually two-pass) RO systems are preferred is when the source water has a very high fouling potential.

The rate of RO membrane fouling is exponentially proportional to the membrane flux, which in turn (for a given RO system) is proportional to the difference between RO system feed and permeate pressures. By using a two-pass vs single-pass RO system, the feed pressure to the first pass could be reduced significantly (typically down to 65%–75% of the total RO feed pressure), whose reduction could result in a measurable improvement of RO system operations (reduced RO membrane cleaning frequency and lower overall differential/feed RO pressure). In this case, the use of a two-pass RO system, however, would require the installation of interpass RO pump to deliver the remaining (25%–35%) of the total RO feed pressure needed for production of the target RO system water quality. The pump, which boosts the permeate pressure from the first pass to the second pass, is referred to as booster (or second-pass) RO feed pump.

1.1.2. Single and Multiple RO Stage Systems

A key challenge associated with the use of multiple-pass RO systems is that the overall recovery of such systems decreases with the number of the installed RO passes, because a portion of the saline source water is converted into concentrate at each pass (see Fig. 1.2).

In order to reduce the total volume of concentrate (i.e., increase the overall recovery/produce more fresh water) from the same volume of source water, the concentrate generated by the individual RO passes can be treated by a separate RO system, referenced as "stage." Usually, the membrane vessels of the BWRO system stages are grouped in parallel and are referred to as "arrays." The stages are arranged in series such that adequate crossflow and minimum concentrate flow are maintained in each stage.

Typically, the ratio between the number of vessels in the first stage and second stage of NF and BWRO systems is selected at 2:1 and such configuration is referenced to as "2:1 Array." In three-stage NF and BWRO systems, the typical array configuration is—3:2:1—i.e., the number of vessels in the first stage is three times higher than those in the third stage. The second stage has two times higher number of vessels than the third stage.

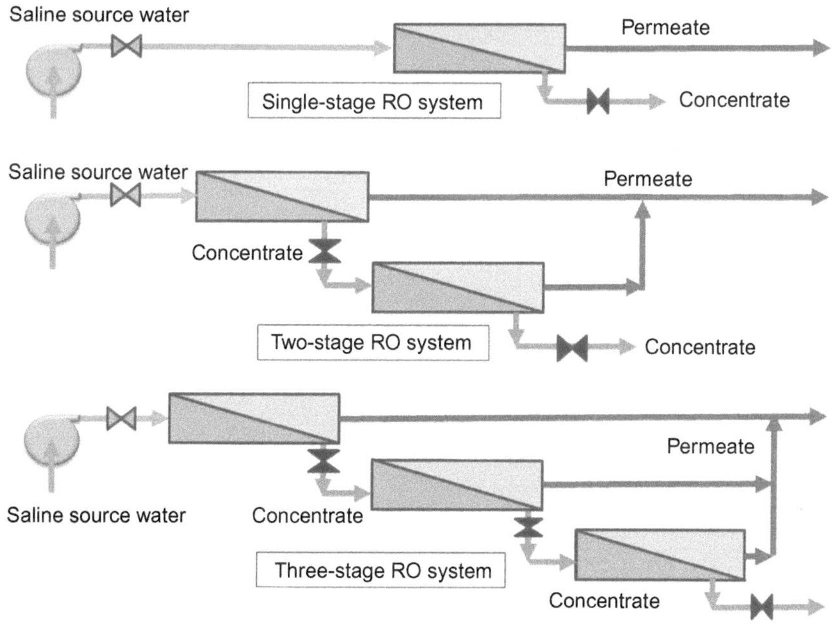

Fig. 1.2 Single and multiple-stage RO systems.

Fig. 1.2 depicts single-stage, two-stage, and three-stage RO systems. Use of multiple stages allows improving the overall recovery of the entire RO system. However, the additional concentrate treatment steps also increase RO system costs. The optimum number of stages and passes for a given RO system depends on many factors, such as source water quality, target permeate water quality and fresh-water production flow, and cost of equipment and energy, and should be determined based on the site-specific conditions of each project.

It is important to note that the single-pass and single-stage RO systems shown in Figs. 1.1 and 1.2 are identical in configuration—i.e., a RO system in which the source water is treated only once can be referenced both as single pass and single stage. The difference of nomenclature can be explained by the main difference in configuration of SWRO and BWRO systems—i.e., in SWRO desalination plants, it is more common that the source water undergoes RO treatment multiple times; while in NF and BWRO plants the concentrate rather than the permeate generated by the first RO treatment undergoes additional processing by reverse osmosis. As a result, an SWRO system, where the source water and permeate are processed through reverse osmosis in sequence is referred to as two-pass SWRO system; while a BWRO system where the source water and the concentrate are processed through reverse osmosis in sequence is referred to as two-stage BWRO system.

1.2. NF SYSTEM CONFIGURATIONS

Nanofiltration plants usually treat relatively low-salinity water (typically less than 1000 mg/L), which has high level of hardness, organic content, color, iron, or other bivalent ions such as nitrates. Typically, NF plants are equipped with two-stage RO systems and are designed to operate at 80%–90% recovery. The most commonly used NF membrane configuration includes pressure vessels with seven elements per vessel and a 2:1 membrane array.

Fig. 1.3 depicts the general process schematic of the largest operational NF plant in the United States—the 151,000 m^3/day (40 MGD) Boca Raton plant is Florida. This plant processes well water of relatively low salinity (466 mg/L) but has high hardness (265 mg/L), alkalinity (265 mg/L), iron (0.3 mg/L), and NOM content (TOC = 12 mg/L). The plant is designed to operate for total recovery of 85%, and it has 10 two-stage NF trains with 2:1 arrays (72 vessels in the first stage and 36 vessels in the second stage). The plant uses Hydranautics ESNA1-LF2 elements operating at an average flux of 20.7 Lmh (12.2 gfd) and producing and average of

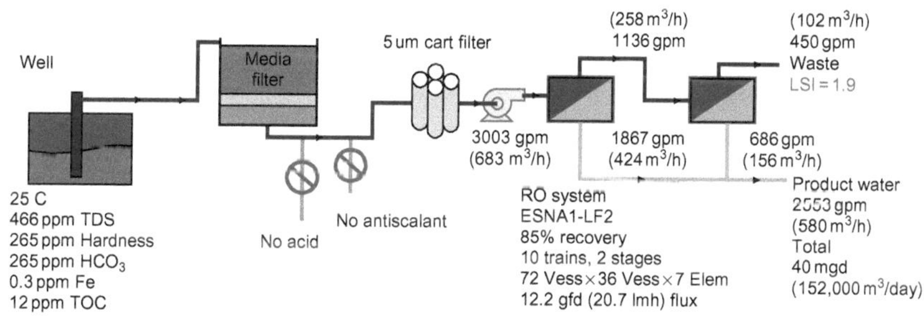

Fig. 1.3 Schematic of Boca Raton NF Plant, Florida. *(Courtesy: Hydranautics.)*

20.1 m^3/day (5300 gpd) per element. The plant permeate has total hardness of 50–80 mg/L, iron content below 0.04 mg/L, and TOC of less than 1 mg/L [1].

1.3. BWRO SYSTEM CONFIGURATIONS

Source water TDS concentration of BWRO plants typically ranges between 500 and 10,000 mg/L. Plants processing source water with salinity between 500 and 2500 mg/L and in a range of 2500–10,000 mg/L (or above) are referred to as low-salinity and high-salinity brackish water reverse osmosis (BWRO) desalination facilities, respectively.

Fig. 1.4 illustrates a typical schematic of a low-salinity BWRO desalination plant. For such plants blending a portion (5%–30%) of the source water flow (indicated on Fig. 1.4 as "Bypass of brackish source water") with RO permeate is very common practice for remineralization of the desalinated water. Low-salinity BWRO plants often process the source water through a

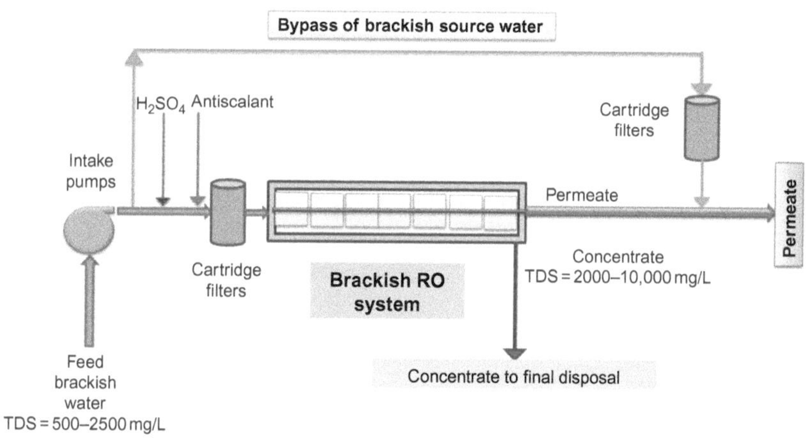

Fig. 1.4 Schematic of typical low-salinity BWRO plant.

single RO stage (pass) only. However, two-stage BWRO plants configured with 2:1 arrays are also very common.

Table 1.1 provides an illustrative hypothetical example of the permeate water quality produced by low-salinity BWRO plant operating at blending ratio of 28.6% and permeate recovery of 85% [2]. In this specific example, the TDS of the source seawater and RO permeate are 647.3 and 215 mg/L, respectively.

Fig. 1.5 depicts a typical schematic of high-salinity BWRO desalination plant. As indicated in this figure, brackish source water in this case is

Table 1.1 Example of product water quality generated by low-salinity BWRO plant

Water quality parameter	Source water quality	Blended permeate water quality
Temperature (°C)	25	25
pH	7.0	6.6
Ca^{2+} (mg/L)	96	29
Mg^{2+} (mg/L)	11.7	3.5
Na^+ (mg/L)	90	32.1
K^+ (mg/L)	6.5	2.4
HCO_3^- (mg/L)	72.6	30.4
SO_4^{2-} (mg/L)	158.4	47.2
Cl^- (mg/L)	190.7	61
F^- (mg/L)	0.2	0.1
SiO_2 (mg/L)	24.3	9.3
TDS (mg/L)	647.3	215

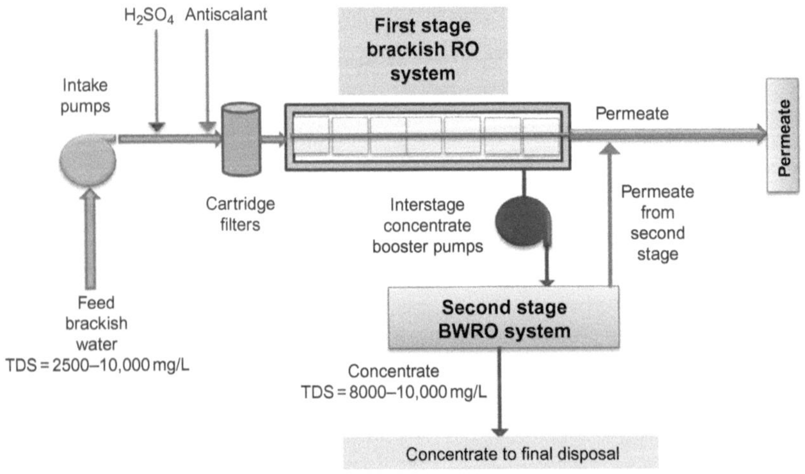

Fig. 1.5 Typical schematic of high-salinity Brackish water RO plant.

usually treated by a two-stage RO system where concentrate generated by the first RO pass is processed through a second RO system after increasing (boosting) the concentrate pressure via an interstage booster pumps. Under this configuration the first-stage permeate usually contributes 75%–85% of the total permeate flow while the RO system, which processes concentrate from the first RO pass (typically referred to as "second stage"), produces the remaining 15%–25% of the total RO system permeate flow.

A hypothetical example of the permeate water quality produced by a high-salinity BWRO plant is presented in Table 1.2 [2]. In this example, the desalination plant uses brackish well water source of salinity of 5881 mg/L and processes this source through a two-stage BWRO system similar to that illustrated in Fig. 1.5. The overall plant recovery is 80% (i.e., 80% of the source water is converted into low-salinity permeate). The system is equipped with an interstage booster pump, which directs the concentrate generated by the first SWRO stage to the second RO stage.

The recovery of the first RO stage for the example presented in Table 1.2 is 64%. The second stage recovers approximately 44% of the concentrate produced by the first-stage RO system.

1.4. SEAWATER SYSTEM CONFIGURATIONS

The SWRO system configurations, which are most widely applied at present, include single-pass treatment, where the source water is processed by reverse osmosis only once (Fig. 1.6); and two-pass RO treatment, where

Table 1.2 Example of high-salinity BWRO plant permeate water quality

Water quality parameter	Source water quality	Blended permeate water quality
Temperature (°C)	28	28
pH	7.0	6.0
Ca^{2+} (mg/L)	105	1.0
Mg^{2+} (mg/L)	130	1.3
Na^{+} (mg/L)	1837	84
K^{+} (mg/L)	85	4.8
CO_3^{2-} (mg/L)	0.3	0.0
HCO_3^{-} (mg/L)	250	18.4
SO_4^{2-} (mg/L)	479	5.1
Cl^{-} (mg/L)	2970	123
F^{-} (mg/L)	1.4	0.1
NO_3^{-} (mg/L)	5.0	1.4
SiO_2 (mg/L)	17.0	0.5
TDS (mg/L)	5881	240

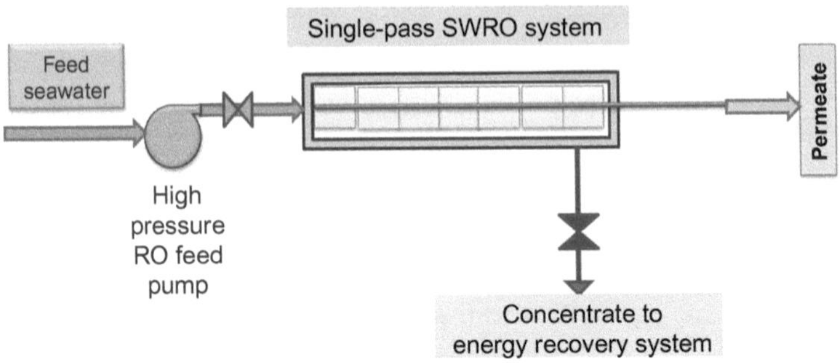

Fig. 1.6 Single-pass SWRO system.

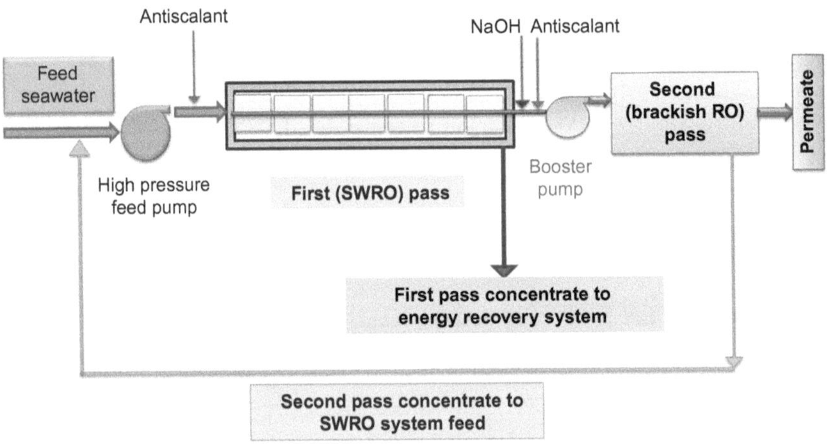

Fig. 1.7 Conventional full two-pass SWRO system.

the seawater is first processed through an SWRO system and then permeate produced by this system is reprocessed by brackish RO membranes (see Figs. 1.7 and 1.8).

1.4.1. Single-Pass SWRO Systems

Single-stage SWRO systems are designed to produce desalinated seawater (permeate) in one step using only a single set of RO trains operating in parallel. In general, between 800 and 900 SWRO membrane elements installed in 100–150 vessels are needed to produce 10,000 m^3/day (2.6 MGD) of permeate suitable for potable use in a single-stage SWRO system.

Under a typical single-stage SWRO system configuration, each RO train has a dedicated system of transfer pump for pretreated seawater followed by a high-pressure RO feed pump. The high-pressure feed pump motor/operation is coupled with that of energy recovery equipment (see Fig. 1.6).

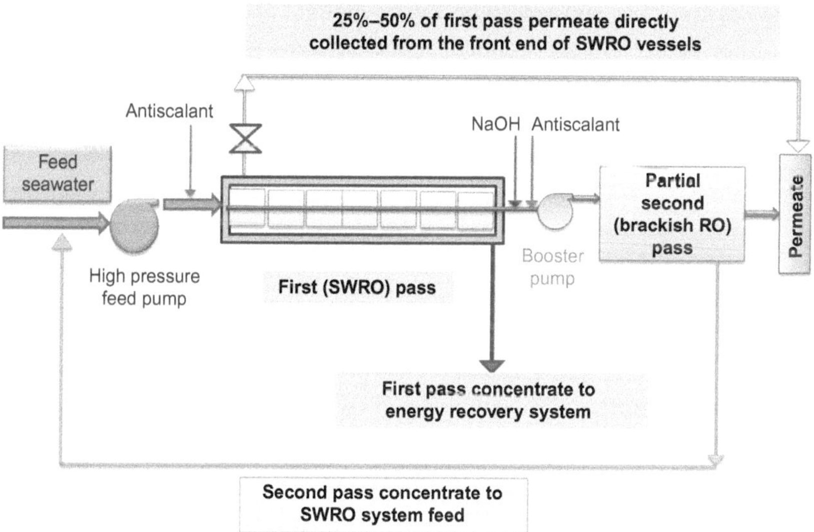

Fig. 1.8 Split-partial two-pass SWRO system.

Single-stage SWRO systems are widely used for production of drinking water. However, these systems have found limited industrial application mainly because of the water quality limitations of the produced permeate. Even if using the highest-rejection RO membrane elements commercially available today (nominal minimum rejection of 99.85%), the single-stage SWRO desalination systems typically cannot consistently yield permeate with TDS concentration lower than 200 mg/L, chloride level of less than 100 mg/L, and boron concentration lower than 0.5 mg/L, especially when source water temperatures exceed 18–20°C (64–68°F).

If enhanced boron removal is needed in such systems, high boron rejection membranes are used, and/or sodium hydroxide and antiscalant is added to the RO system feed water to increase pH to 8.8 or more, which in turn improves boron rejection.

1.4.2. Two-Pass SWRO Systems

Two-pass SWRO systems are typically used when either the source seawater salinity is relatively high (i.e., higher than 35,000 mg/L) and/or the product water quality requirements are very stringent. For example, if high-salinity/high-temperature source water (such as Red Sea and Persian Gulf seawater for example) is used in combination with standard-rejection (99.6%) SWRO membranes, then single-stage SWRO systems may not be able to produce permeate suitable for drinking water use. In this case, two-pass

SWRO systems have proven to be a very efficient and cost-effective configuration for potable water production. RO systems with two or more passes are also widely used for production of high-purity industrial water.

The two-pass SWRO systems typically consist of a combination of a single-pass SWRO system and a single- or multiple-pass brackish water RO (BWRO) system connected in series (Figs. 1.7 and 1.8). Permeate from the SWRO system (i.e., first pass) is directed for further treatment to the BWRO system (i.e., second pass) to produce a high-quality TDS permeate. The concentrate from the pass-two BWRO system is returned to the feed of the pass-one SWRO system to maximize the overall desalination system production capacity and efficiency. Two-pass SWRO systems are classified into two main groups: conventional full two-pass systems and split-partial two-pass systems.

1.4.2.1. Conventional Full-Two-Pass SWRO Systems

In conventional full two-pass SWRO membrane systems (see Fig. 1.7), the source seawater is first treated by a set of SWRO membrane trains (referred to as first RO pass) and then the entire volume of desalinated water from the first pass is processed through a second set of brackish water desalination membrane trains [3]. If enhanced boron removal is needed, sodium hydroxide and antiscalant are added to the feed permeate of the second RO pass to increase pH and improve boron rejection.

1.4.2.2. Split-Partial Two-Pass SWRO Systems

In split-partial two-pass systems, the second RO pass typically processes only a portion (50%–75%) of permeate generated by the first pass. The rest of the low-salinity permeate is produced by the front (feed) SWRO elements of the first pass. This low-salinity permeate is collected and without additional desalination it is directly blended with permeate produced by the second RO pass (see Fig. 1.8).

As indicated in Fig. 1.8, the second-pass concentrate is returned to the feed of the first RO system pass. When the desalination system is designed for enhanced boron removal, this concentrate will have pH of 9.5–11 and potentially could cause precipitation of calcium carbonate on the membranes. In order to avoid this challenge, typically antiscalant is added to the feed to the partial second-pass (brackish RO) system. Long-term experience with such configuration indicates that this solution is very effective in preventing scaling of the first-pass RO membranes by the recycled second-pass concentrate. Because boron level in the front permeate stream is usually between 0.25 and 0.5 mg/L, no additional treatment of this stream is needed even if the plant is designed and operated for enhanced boron removal.

While the recycling of the second-pass concentrate returns a small portion of the source water salinity and, therefore, it slightly increases the salinity of the seawater fed to the first RO pass, this the energy use associated with this incremental salinity increase is significantly smaller than the energy savings of processing the entire volume of the first-pass permeate through the second pass. Under the split-partial two-pass configuration, the volume of permeate pumped to the second RO pass and the size of this pass are typically 25%–50% smaller than the volume pumped to the second RO pass under conventional once-trough operation. Since pumping energy is directly proportional to flow, the energy costs for the second-pass feed pumps are reduced proportionally—i.e., with 25%–50%. At SWRO system operating at 45% recovery, such savings will amount to 14%–22% of the energy of the first-pass RO pump.

The concentrate returned from the second pass carries only 1%–2% of additional salinity to the first-pass RO feed, which reduces the energy benefit from such recovery proportionally—i.e., by 1%–2% only. As a result, the overall energy savings of the use of split-partial two-pass RO system as compared to conventional two-pass RO system are between 12% and 20%. Practical experience with large SWRO desalination plants indicates that the average total RO system life-cycle cost savings associated with applying such SWRO system configuration are typically between 14% and 16%.

At present, most new SWRO desalination systems are designed with split-partial second-pass configuration because this configuration allows reducing the size of the second-pass RO system and the overall fresh-water production costs. It should be pointed out that split-partial second RO pass systems can be configured in several alternatives, which may involve the use of the same or different type of membrane elements within the first-pass SWRO system.

Example of plant with partial-second-pass configuration is the 95,000 m^3/day (25 MGD) Tampa Bay seawater desalination facility in Florida. The second pass at this facility is designed to treat up to 30% of the permeate produced by the first-pass SWRO system as needed in order to maintain the concentration of chlorides in the plant product water always below 100 mg/L.

The partial second pass at the Tampa Bay water seawater desalination plant was installed to provide operational flexibility and to accommodate the wide fluctuations of source water salinity (16,000–32,000 mg/L) and temperature (18–40°C/64–104°F). Typically, the product water quality target chloride concentration of 100 mg/L at this plant is achieved by only operating the first pass of the system. However, when source water TDS concentration exceeds 28,000 mg/L and/or the source water temperature exceeds 35°C (95°F), the second pass is activated to maintain adequate product water quality. The percent of first-pass permeate directed for additional treatment through the second pass is a function of the actual combination of source water TDS and temperature and is adjusted based on the plant product water chloride level.

1.4.3. Product Water Quality of Single- and Two-Pass SWRO Systems

Tables 1.3–1.7 present summary of the range of permeate water quality produced by typical single-pass and partial two-pass seawater desalination systems processing five different seawater sources: Pacific Ocean water, Atlantic Ocean water, Mediterranean seawater, Persian Gulf seawater, and Red Sea water.

1.4.4. Four-Stage SWRO Systems

In addition to the single- and two-pass SWRO systems described earlier, another reverse osmosis configuration that has found practical application in a number of large SWRO desalination plants that have to meet a very stringent

Table 1.3 Reverse osmosis permeate water quality seawater source—Pacific Ocean

Water quality parameter	Pacific Ocean Source seawater Quality	Permeate water quality	
		Single-pass SWRO system	Split-partial two-pass RO system
Temperature (°C)	14–28	15–29	16–30
pH	8.0	6.3–7.2	7.6–7.8
Ca^{2+} (mg/L)	358	0.6–1.1	0.2–0.5
Mg^{2+} (mg/L)	1720	1.8–2.8	0.07–0.10
Na^{+} (mg/L)	9900	78–134	9–20
K^{+} (mg/L)	600	3.0–6.0	0.43–0.60
CO_3^{2-} (mg/L)	2.0	0.0	0.0
HCO_3^{-} (mg/L)	170	1.8–2.5	0.4–0.7
SO_4^{2-} (mg/L)	2570	2.6–5.3	0.7–1.3
Cl^{-} (mg/L)	18,100	130–195	13–20
F^{-} (mg/L)	2.1	0.9–1.2	0.7–0.9
NO_3^{-} (mg/L)	0.00	0.00	0.00
B^{-} (mg/L)	4.5	0.7–1.2	0.3–0.5
Br^{-} (mg/L)	73	0.6–0.9	0.2–0.4
TDS (mg/L)	33,500	220–350	25–45

Table 1.4 Reverse osmosis permeate water quality seawater source—Atlantic Ocean

Water quality parameter	Atlantic Ocean Source seawater Quality	Permeate water quality	
		Single-pass SWRO system	Split-partial two-pass RO system
Temperature (°C)	16–30	17–31	18–32
pH	8.0	6.3–7.2	7.7–8.0
Ca^{2+} (mg/L)	410	0.7–1.4	0.3–0.5
Mg^{2+} (mg/L)	1302	1.6–2.4	0.35–0.8
Na^{+} (mg/L)	10,506	83–160	11–25
K^{+} (mg/L)	390	2.4–4.5	0.35–0.50
CO_3^{2-} (mg/L)	2.0	0.0	0.0
HCO_3^{-} (mg/L)	145	1.4–2.0	0.5–0.8
SO_4^{2} (mg/L)	2720	2.4–5.8	1.1–1.2
Cl^{-} (mg/L)	19,440	146–220	20–34
F^{-} (mg/L)	2.5	1.0–1.6	0.8–1.2
NO_3^{-} (mg/L)	0.00	0.00	0.00
B^{-} (mg/L)	4.5	0.7–1.2	0.3–0.5
Br^{-} (mg/L)	78	0.8–1.1	0.3–0.5
TDS (mg/L)	35,000	240–400	35–65

Table 1.5 Reverse osmosis permeate water quality seawater source—Mediterranean Sea

Water quality parameter	Mediterranean Source seawater Quality	Permeate water quality	
		Single-pass SWRO system	Split-partial two-pass RO system
Temperature (°C)	16–28	17–29	18–30
pH	8.1	6.3–7.2	7.9–8.1
Ca^{2+} (mg/L)	480	1.0–2.0	0.35–0.45
Mg^{2+} (mg/L)	1558	1.9–2.8	0.5–1.0
Na^{+} (mg/L)	12,200	98–196	15–34
K^{+} (mg/L)	480	3.0–5.5	0.8–1.8
CO_3^{2-} (mg/L)	5.6	0.0	0.0
HCO_3^{-} (mg/L)	160	1.7–2.4	0.5–0.8
SO_4^{2-} (mg/L)	3190	2.9–6.3	1.4–2.95
Cl^{-} (mg/L)	22,340	169–260	25–52
F^{-} (mg/L)	1.4	0.7–1.1	0.5–0.8
NO_3^{-} (mg/L)	0.00	0.00	0.00
B^{-} (mg/L)	5.0	0.9–1.5	0.4–0.6
Br^{-} (mg/L)	80	0.9–1.3	0.35–0.6
TDS (mg/L)	40,500	280–480	45–95

Table 1.6 Reverse osmosis permeate water quality seawater source—Red Sea

Water quality parameter	Red Sea Source seawater Quality	Permeate water quality	
		Single-pass SWRO system	Split-partial two-pass RO system
Temperature (°C)	22–33	23–34	24–35
pH	7.0–8.0	6.8–7.8	7.6–8.0
Ca^{2+} (mg/L)	500	1.1–2.1	0.5–0.7
Mg^{2+} (mg/L)	1540	1.8–3.4	0.7–1.0
Na^{+} (mg/L)	13,300	142–220	20–38
K^{+} (mg/L)	489	3.2–6.5	1.2–1.8
CO_3^{2-} (mg/L)	2.4	0.0	0.0
HCO_3^{-} (mg/L)	142.4	1.4–2.0	0.5–1.0
SO_4^{2-} (mg/L)	3100	2.8–6.2	1.9–2.6
Cl^{-} (mg/L)	22,840	195–276	29–58
F^{-} (mg/L)	0.9	0.5–0.7	0.3–0.5
NO_3^{-} (mg/L)	0.00	0.00	0.00
B^{-} (mg/L)	5.3	1.2–1.7	0.45–0.80
Br^{-} (mg/L)	80	1.0–1.4	0.45–0.60
TDS (mg/L)	42,000	350–520	55–105

Table 1.7 Reverse osmosis permeate water quality seawater source—Persian Gulf

Water quality parameter	Persian Gulf Source seawater Quality	Permeate water quality	
		Single-pass SWRO system	Split-partial two-pass RO system
Temperature (°C)	18–35	19–36	20–37
pH	6.0–7.0	5.1–6.0	5.1–6.0
Ca^{2+} (mg/L)	570	1.4–2.6	0.6–0.8
Mg^{2+} (mg/L)	1600	2.0–3.6	0.9–1.3
Na^{+} (mg/L)	14,100	142–228	25–45
K^{+} (mg/L)	530	4.3–6.8	1.5–2.2
CO_3^{2-} (mg/L)	4.2	0.0	0.0
HCO_3^{-} (mg/L)	155	1.8–2.3	0.6–0.9
SO_4^{2-} (mg/L)	3300	3.1–6.5	2.1–3.2
Cl^{-} (mg/L)	24,650	222–305	37.5–64
F^{-} (mg/L)	1.5	0.9–1.2	0.5–0.8
NO_3^{-} (mg/L)	0.00	0.00	0.00
B^{-} (mg/L)	6.3	1.3–2.5	0.7–1.0
Br^{-} (mg/L)	83	1.2–1.5	0.60–0.80
TDS (mg/L)	45,000	380–520	70–120

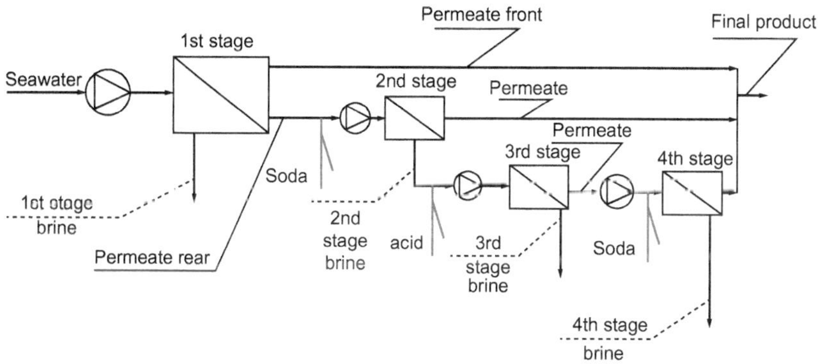

Fig. 1.9 Schematic of Ashkelon four-stage SWRO system [4].

Table 1.8 Reverse osmosis permeate water quality Ashkelon seawater desalination plant

Water quality parameter	Mediterranean Source seawater Quality	Fresh-water quality	
		Permeate	Finished water
Temperature (°C)	16–28	17–29	18–30
pH	8.1	7.5–8.5	8–8.5
Ca^{2+} (mg/L)	483	0.2–0.5	90–110
Mg^{2+} (mg/L)	1557	0.5–0.8	0.5–0.8
Na^{+} (mg/L)	12,200	6–30	30–39
K^{+} (mg/L)	481	1.5–1.8	1.5–1.8
HCO_3^- (mg/L)	162	0.6–0.8	45–50
SO_4^{2-} (mg/L)	3186	1.0–1.4	1.0–1.8
Cl^- (mg/L)	22,599	10–15	10–15
F^- (mg/L)	1.4	0.1–0.2	0.8–1.0
NO_3^- (mg/L)	0.00	0.00	0.00
B^- (mg/L)	5.3	0.2–0.3	0.2–0.3
Br^- (mg/L)	80	0.2–0.3	0.2–0.3
TDS (mg/L)	40,679	20–40	180–220

boron limit of 0.4mg/L or less (i.e., Ashkelon, Hadera, and Sorek SWRO plants in Israel) is the four-stage SWRO system depicted in Fig. 1.9 [4].

Table 1.8 presents a summary of the source water quality and the design and actual product water quality of the Ashkelon SWRO plant in Israel, which was the first facility where such reverse osmosis configuration was used and therefore has the longest track record of successful performance [5].

1.4.5. Two-Stage SWRO Systems

Two-stage SWRO membrane systems are mainly used to maximize the overall desalination plant recovery and reduce the volume of concentrate discharged by the desalination plant. A general schematic of a two-stage RO system is shown in Fig. 1.10. In these SWRO systems, typically the entire volume of the concentrate generated by the first-stage SWRO system is directed to a second-stage SWRO system for further treatment and enhanced recovery. Permeate from both systems is blended prior to final use.

The main advantage of such SWRO system configuration is that it allows achieving very high level of use (recovery) of the available source seawater and the energy used by the first-stage RO system. For example, while a single-stage SWRO system configuration typically allows turning 35%–50% of the source seawater to potable water, the two-stage SWRO system recovery may reach 60%–65%. Designing the SWRO plant around higher recovery allows minimizing the size of the plant intake and pretreatment facilities, and the capital expenditures for their construction and operation.

However, because of the high salinity of the first-stage concentrate (typically above 55,000 mg/L), the practical implementation of two-stage SWRO systems requires use of SWRO membranes, membrane vessels, piping, and auxiliary equipment that can withstand and perform well at very high pressures (up to 98 bars/1400 psi). The cost of this equipment is usually higher than the cost of the same size and type of equipment built to operate at more "standard" pressures (i.e., below 70 bars/1000 psi). Therefore, the viability of using two-stage SWRO system has to be carefully assessed based on a comprehensive life-cycle cost analysis for the site-specific conditions of a given project. To date, two-stage systems have been applied mainly to upgrade the capacity and improve energy use of existing conventional single-stage SWRO plants.

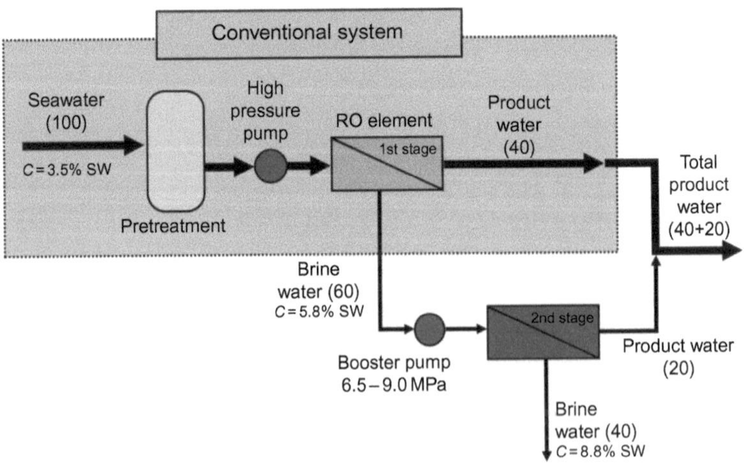

Fig. 1.10 Two-stage SWRO system.

Sometimes, two-stage SWRO systems are designed to be operated with lower feed pressure to the first stage, because of the highly fouling nature of the feed water. Since operation at lower feed pressure allows to reduce fouling of the first stage/pass elements, this configuration could be beneficial if the pretreatment system is not very robust and the water quality from the SWRO system is of high alluvial organic or particular content.

1.4.6. Hybrid SWRO Systems With Multiple Passes and Stages

The two-pass and two-stage RO system configurations may be combined to achieve an optimum plant design and tailor desalination plant operation to the site-specific water source water quality conditions and product water quality goals.

An example of a full-scale two-pass, two-stage SWRO system application is the 170,500 m^3/day Fujairah seawater desalination plant [6]. A general treatment process schematic of this plant is depicted on Fig. 1.11. The plant uses Gulf of Oman source seawater.

The first pass of the Fujairah plant consists of 17 duty and 1 standby RO trains using standard rejection SWRO membranes producing permeate of TDS concentration of 400–500 mg/L. The overall recovery of the first-pass SWRO system is 43%. The second pass has eight BWRO trains with a total recovery rate of 90%. The second-pass BWRO trains have two stages and treat approximately 80% of the first-pass permeate to TDS concentration of 10–20 mg/L. The rest (i.e., 20%) of the first-pass permeate is blended with

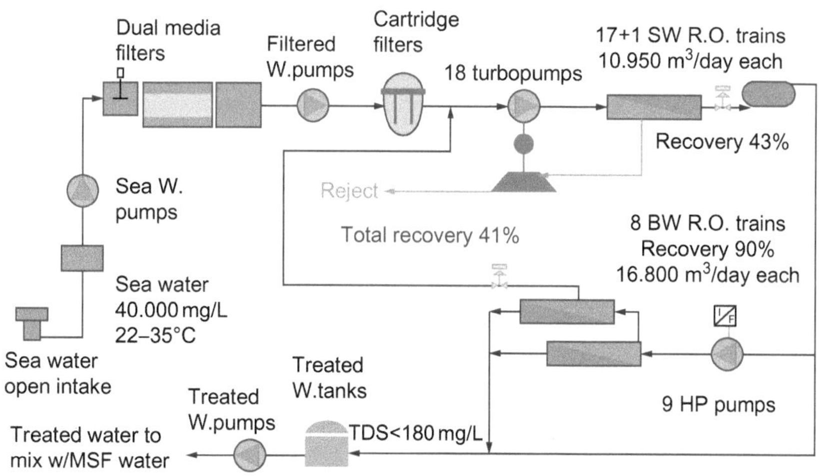

Fig. 1.11 Schematic of Fujairah seawater desalination plant.

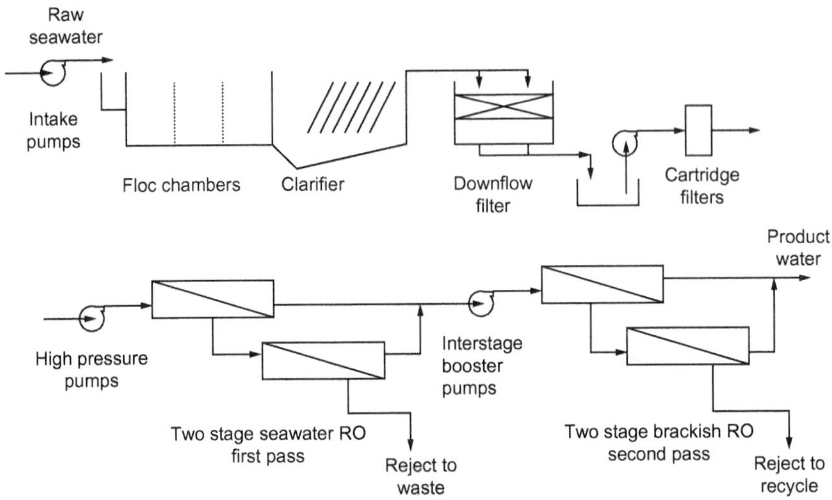

Fig. 1.12 Schematic of Point Lisas seawater desalination plant.

the second-pass permeate to produce finished water of TDS concentration of less than 120 mg/L. Concentrate produced by the second-pass RO system has salinity lower than that of the source seawater and is recycled to the feed of the first-pass RO system (see Fig. 1.11). Despite its complexity, the two-pass, two-stage RO system in Fujairah performs very reliably.

An example of a two-pass/two-stage SWRO plant is the 146,000 m^3/day (38 MGD) Point Lisas facility in Trinidad (Fig. 1.12). This plant produces high-quality desalinated seawater of TDS concentration of 85 mg/L or less, which is predominantly used for industrial applications. The first pass of the Point Lisas SWRO system consists of six two-stage RO units. Each of the first-stage RO trains uses SWRO membranes and is coupled with Pelton wheel energy recovery device. The second-stage trains of the first pass are equipped with brackish water RO elements.

The entire volume of permeate from the first pass of the Point Lisas SWRO system is further treated in a second-pass RO system to meet the final product water quality specifications. The second-pass system also consists of two stages—each equipped with BWRO membranes. The Point Lisas seawater desalination plant has the same number of first-pass and second-pass RO membrane trains.

Because of the high fouling potential of its source seawater, this two-pass-two-stage configuration has proven to be beneficial. The high-pressure pumps of the first stage are designed for 80% lower feed pressure than these of typical single-stage system, which allows to significantly reduce the fouling on the RO elements of this pass. The permeate from the first pass is treated through a second-pass system and an interstage booster pump is installed to deliver the remaining 20% of the feed pressure needed to produce water similar to that in a single-pass RO. Overall, this configuration produces better water quality, higher recovery and results in less membrane fouling than a typical single-pass system for the same source water quality.

1.4.7. Three-Center RO System Configuration

As indicated previously, a typical conventional SWRO system is configured in individual equally sized RO trains each of which is serviced by a separate transfer pump, cartridge filter vessel, high-pressure feed pump, and energy recovery equipment dedicated to this RO train (see Fig. 1.13). The size of the individual RO trains depends on the overall production capacity of the SWRO plant and typically varies between 2000 m^3/day (0.5 MGD) and 21,000 m^3/day (5.5 MGD). The main advantage of this RO-train-based configuration is that it is modular and allows for relatively easy flow distribution and service of the individual trains. Since typically the size of the individual RO trains does not exceed 10%–20% of the total plant production capacity, train shut down for maintenance (membrane cleaning and replacement, and equipment service) is handled either by using a standby RO train or by temporary increase of the production capacity of the RO trains remaining in service.

The RO-train-based configuration is very suitable when the SWRO plant is designed and intended to operate at a constant production output.

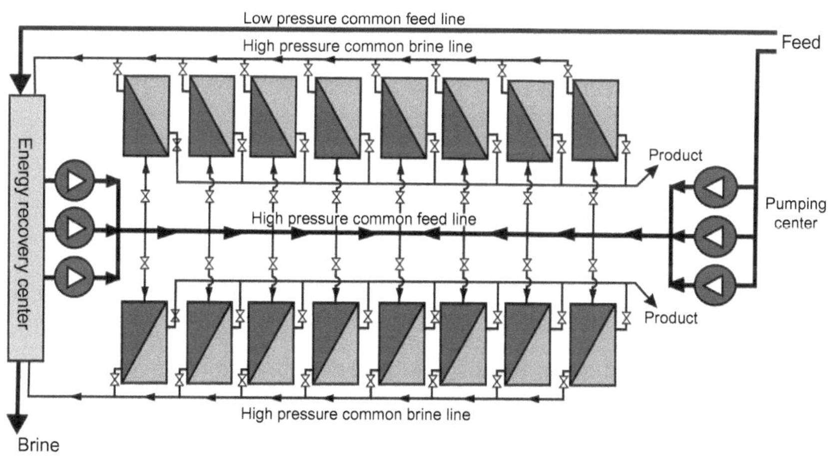

Fig. 1.13 Three-center SWRO system configuration.

At present, most of the existing large municipal SWRO plants worldwide are designed to supplement existing conventional water supply sources rather than to be the primary or the only source of water supply for a given area. Therefore, the operation of these SWRO plants does not need to have the flexibility to follow the actual diurnal and monthly fluctuations in product water demand and most of the existing plants are designed to operate at constant production capacity.

In the future, SWRO desalination is likely to become a prime rather than a supplemental source of water supply for many coastal communities pressured by population growth, especially for large urbanized or industrial areas with limited traditional local sources of fresh-water supply (i.e., groundwater or river or lake water). The SWRO systems servicing such areas have to be designed to have the operational flexibility of matching desalination plant production with the product water demand fluctuations.

Shift of the SWRO plant operational paradigm from constant to variable production flow requires a change of the typical SWRO configuration from one that is most suitable for constant production output to one that is most cost effective for delivery of varying permeate production. A response to such shift of the desalination plant operational paradigm is the three-center RO system configuration in Ashkelon, Israel (see Fig. 1.13). Under this configuration, the RO membrane vessels, high-pressure pumps, and energy recovery equipment are no longer separated in individual RO trains, but are rather combined in three functional centers—a high-pressure RO feed pumping center, a membrane center, and an energy recovery center [7]. The three functional centers are interconnected via service piping.

The RO feed pumping center includes only a few large-capacity high-pressure pumps that deliver seawater to the RO membrane center. The main benefit of using few large-capacity high-pressure pumps rather than a large number of small-capacity units is the gain in overall pumping efficiency.

Typically, the smaller the ratio between the pressure and the flow delivered by a given pump, the better the pump efficiency and the "flatter" the pump curve (i.e., the pump efficiency is less dependent on the variation of the delivered flow). Therefore, pump efficiency can be improved by either reducing the pressure delivered by the pump or by increasing pump flow. Since the pump operating pressure decrease is limited by the RO system target salt separation performance, the main approach to improve pump efficiency is to increase unit pump flow. While a conventional size high-pressure RO feed pump of small capacity would typically have maximum total energy use efficiency of 80%–85%, the use of 10 times larger size pump may allow to increase the pump efficiency to 88%–92%, especially for large SWRO plants. This beneficial feature of the three-center design is very valuable in the case of systems delivering varying flow.

While in a conventional RO train design the membrane vessels are typically grouped in 100–200 units per train and in 2–20 RO trains, the membrane center configuration contains two to four times larger number of RO vessel groups (banks) and a smaller number of membrane vessels per bank. Under this configuration, the individual vessel banks are directly connected to the high-pressure pump feed lines and can be taken off service one at a time for membrane replacement and cleaning.

Although the feed water distribution piping for such membrane center configuration is more elaborate and costly than the use of individual RO trains that contain two to three times more vessels per train, what is lost in capital expenditure is gained in overall system performance reliability and availability. A reliability analysis completed for a 95,000 m^3/day (25 MGD) SWRO plant [8] indicates that the optimum number of vessels per bank for this scenario is 54 and number of RO banks per plant is 20. A typical RO-train-based configuration would include two to four times more (108–216) vessels per RO train and two to four times less (5–10) RO trains. According to this analysis, the use of the three-center configuration instead of the conventional RO-train-based approach allows to improve RO system availability from 92%–96% (avg. 95%) to 98%, which is a significant benefit in terms of additional amount of water delivered to the customers and improvement in water supply reliability.

The centralized energy recovery system included in the three-center configuration (Fig. 1.13) uses high-efficiency pressure-exchanger-based energy recovery technology. The proposed configuration allows to improve the overall energy recovery efficiency of the RO system and to reduce system power, equipment, and construction costs. While typically the energy recovery of the conventional Pelton wheel systems drops significantly when there is reduction of the overall SWRO plant recovery, the energy recovery efficiency of the pressure-exchanger systems improves with lowering the plant recovery rate. This allows operating the SWRO plant cost effectively while delivering variable product water flow. For example, if the SWRO plant output has to be reduced by 40% to accommodate low diurnal demand, an SWRO system with RO-train-based configuration has to shut down 40% of its trains and if this low demand persists, it has to flush these trains to prepare them for the next start up.

An RO system with three-center configuration would only need to lower its overall recovery to achieve the same reduction of the diurnal demand. Although temporary operation at lower recovery would result in elevated costs for pumping and pretreatment of larger volumes of source water, these extra operations expenses are typically compensated for by the improved energy recovery efficiency that results from operating the SWRO system at lower water recovery ratio. In addition to providing flexibility in operating cost effectively the desalination plant at varying production flows, the three-center design yields low-fouling RO plant operations of very high availability factor.

REFERENCES

[1] Bartels G, Lai K, Wilf M. In: New generation low fouling nanofiltration membranes. Proceedings of April 2007 conference of the European Desalination Society, Halkidiki, Greece; 2007.

[2] Wilf M, Awerbuch L, Bartels C, Mickley M, Pearce G, Voutchkov N. The guidebook to membrane desalination technology: reverse osmosis, nanofiltration and hybrid systems process design. In: Applications and economics. Rehovot: Balaban Publishers; 2007.

[3] Greenlee LF, Lawler DF, Freeman BD, Marrot B, Moulin P. Reverse osmosis desalination: water sources, technology, and today's challenges. Water Res 2009;43:2317–48.

[4] Gorenflo A, Brusilovsky M, Faigon M, Liberman B. High pH operation in seawater reverse osmosis permeate: first results from the world's largest SWRO plant in Ashkelon. Desalination 2007;203:82–90.

[5] Dreizin Y, Tenne A, Hoffman D. Integrating large scale seawater desalination plants within Israel's water supply system. Desalination 2008;220:132–49.

[6] Sanz MA, Bonnelye V, Cremer G. Fujairah reverse osmosis plant: 2 years of operation. Desalination 2007;203:91–9.

[7] Liberman B. The importance of energy recovery devices in reverse osmosis desalination. In: The future of desalination in Texas—Volume 2, technical papers, case studies desalination technology resources (report # 363). Texas Water Development Board; 2002. December.

[8] Liberman B, Wilf M. In: Evolution of configuration of RO seawater desalination elements. Proceedings of IDA World Congress—Singapore 2005; 2005. Ref: SP05-059.

CHAPTER 2

Energy Efficiency of Reverse Osmosis

Abhishek Shrivastava, Derek Stevens
The Dow Chemical Company, Edina, MN, United States

2.1 INTRODUCTION

Water-energy nexus has been studied significantly over the last couple of decades [1–4]. The underlying idea behind water-energy nexus is that water is needed for energy generation, e.g., water is the working fluid in power plants where it is used as a heat transfer fluid in power cycles to generate electricity from fossil fuels. Water treatment is needed during exploration and production of oil and gas, e.g., removal of particles that can foul up a production well or removal of sulfate that can sour production wells. Water is needed for harvesting biofuels. At the same time, energy is needed for water treatment and purification, e.g., energy is required for waste water treatment in terms of aeration to support biological activity to reduce chemical and biological oxygen demand, nutrient levels and subsequent eutrophication of natural resources where the waste water is discharged, energy is also required for thermal, electrical, or pressure-based water purification, e.g., ultrafiltration, nanofiltration, reverse osmosis, ion exchange, forward osmosis, thermal distillation processes, electrodeionization, electrodialysis reversal, and other hybrid processes [5,6] that might use a combination of thermal, membrane, or electrical desalination techniques.

The mega trends in energy [7] and water consumption [8] are quite disconcerting. With population increase from approximately 7.5 billion in 2017 to approximately 9 billion by 2050 accompanied by rapid industrialization and urbanization, it is predicted that by 2035 world energy demand will increase by approximately 50% to 18 billion tonne oil equivalents as compared to 2009 energy demand levels. With rising population, energy and food demands, climate change, it is anticipated that by 2025 about 3.5 billion people will be living in water-stressed regions as compared to

Sustainable Desalination Handbook
https://doi.org/10.1016/B978-0-12-809240-8.00002-2

2.3 billion in 2011 [9] and by 2030 world is projected to face a 40% global water deficit if current trends continue [8].

In order to avert this severe crisis in one of the key resources for human sustenance—water, it is imperative we look at technologies that are energy and cost efficient at water treatment. Such technologies not only allow us to treat and make use of existing water resources like seawater and brackish water but also enable reuse and recycling of water for industries, agricultural, and municipal applications.

Desalination technologies can be broadly classified into two main categories thermal and nonthermal technologies. Nonthermal technologies generally refer to membrane-based technologies that don't employ thermal methods for separation, e.g., reverse osmosis. Compared to thermal method, membrane-based desalination technologies are relatively new as they came to commercialization only in the second half of 20th century [10–12] with the last few decades seeing rapid development and wide scale adoption [13]. By mid 2000s, thermal technologies accounted for around 60% of installed global desalination capacity. After year 2000, with continued advancements in membrane technology, module design, system design, high-pressure pumps, and energy recovery devices, >70% of new installations for desalination were membrane desalination plants. This trend resulted in membrane desalination overtaking thermal desalination and today >60% of installed desalination capacity worldwide is accounted for by membrane-based desalination systems.

The main reason for wide scale adoption of membrane-based desalination, e.g., reverse osmosis is its inherent lower energy consumption because of no phase change involved during the process of separation of salt from water. Thermal processes generally involve some phase change and the energy penalty even after energy recovery using heat exchangers is too high. Hybrid processes and utilization of waste heat by collocating thermal plants does make them favorable; however, thermodynamically net energy needed tends to be higher. When overall cost of water is considered energy consumption accounts for a significant portion of cost of water [14,15] and that where for most systems membrane-based desalination systems result in lower cost as compared to thermal-based systems. For example, current specific energy consumption (SEC) (energy required to produce a unit amount of permeate water from a given feed water) for large-scale commercial thermal systems is reported between 5 and 16 kWh/m^3, whereas for large-scale commercial RO (reverse osmosis) systems the reported SEC is between 2 and 4 kWh/m^3 [15–17]. For thermal desalination systems, the SEC can be broken into thermal and electrical components with electrical component being more expensive than thermal component due to power cycle efficiency and for multistage flash (MSF) systems the auxiliary electrical SEC to circulate brine via brine pumps is often comparable to the total energy required for RO desalination plants [18]. This SEC manifests itself in total cost of water. Reported desalination costs [15] for large-scale desalination plants range between US \$0.2 m^{-3} and US \$1.2 m^{-3} for reverse osmosis desalination and between US \$0.5 m^{-3} and US \$1.5 m^{-3} for thermal desalination plants.

As energy consumption plays a significant role in the overall cost of water, the focus of the rest of the chapter will be on discussing the energetics of reverse osmosis. In the following sections, two major topics will be discussed. First—fundamental energetics of desalination process from first and second laws of thermodynamics (Section 2.2) and second—a breakdown of energy efficiency of reverse osmosis (Section 2.3). Section 2.4 will discuss future opportunities as we plan on a future with sustainable water treatment technologies from both cost and energy perspective.

2.2 ENERGETICS OF DESALINATION

2.2.1 Thermodynamic Minimum Energy

To understand the energy consumption of any desalination process, it is important to first understand the theoretical energy needed for salt-water separation. This theoretical energy can be understood as the minimum energy needed to separate salt and water from each other via some ideal process. It is important to note that this energy requirement is independent of the process or the technology used because by definition this does not account for any losses that might be incurred during the separation or unmixing process.

From solution thermodynamics [19–22], for spontaneous processes like salt and water mixing (or mixing of solutions of salt and water), the change in Gibbs free energy of mixing is always negative for constant temperature mixing. In other words, the free energy of the salt-water mixture is lower than the total free energy of the components (salt, water, or salt-water mixtures) used to make the final mixture. This is primarily because entropy of the system increases with spontaneous mixing process. Salt-water mixing is a spontaneous process and hence when salt and water are mixed Gibbs free energy of released during mixing can be used to calculate the minimum energy that will be needed to reverse the change. We can call this the thermodynamic minimum energy needed for unmixing of separation and can be calculated from the following equation:

$$= d\left(\Delta G_{\text{unmixing}}\right) = -d\left(\Delta G_{\text{mixing}}\right) = \pi_s \bar{V}_w dn_w \tag{2.1}$$

where

$\Delta G_{unmixing}$ = Gibbs free energy of unmixing (units: joules or kWh/m^3 (when expressed as specific energy for reverse osmosis)).

ΔG_{mixing} = Gibbs free energy of mixing.

πs = osmotic pressure of salt-water mixture (units: pascal).

$\bar{V}_w$ = molar volume of water (units: m^3/mol).

nw = moles of water.

In context of a separation process, unmixing or separation of salt and water is specified in terms of recovery (R) of water from the initial salt-water mixture. Recovery can be defined as volume of pure water recovered from the initial volume of salt-water mixture. Higher the recovery, larger is the thermodynamic minimum energy needed for separation. This is because as a small volume of pure water is recovered from a given salt-water mixture, the resultant mixture is of higher concentration and hence of higher osmotic pressure πs. Hence, higher energy is needed to recover additional amount of water.

Eq. (2.1) applies to an infinitesimally small recovery, i.e., recovery of small volume ($\bar{V}_w dnw$) of water from a salt-water mixture with osmotic pressure πs. Eq. (2.1) can be integrated over a desired water recovery R [23,24], normalized to the desired recovery and can be rewritten as Eq. (2.2).

$$E_{\text{thermodynamic,min}} = \frac{1}{R}\int_0^R \pi \, dR \tag{2.2}$$

where $E_{thermodynamic,min}$ is the thermodynamic minimum specific energy of separation with units of kWh/m^3 and π is the osmotic pressure at a recovery R. In practice, if the starting feed concentration, ion composition, and temperature are known, osmotic pressure can be calculated as a function of different recoveries to build osmotic curves as shown in Fig. 2.1. To calculate $E_{thermodynamic, min}$, area under these curves can be calculated by numerical integration and divided by the recovery R [17].

It is important to note that $E_{thermodynamic, min}$ is the minimum energy needed for such separation and this minimum cannot be achieved due to the second law of thermodynamics. Based on second law of thermodynamics, as we are reversing a spontaneous process the actual energy needed for separation would be more than the Gibbs free energy of mixing. One way

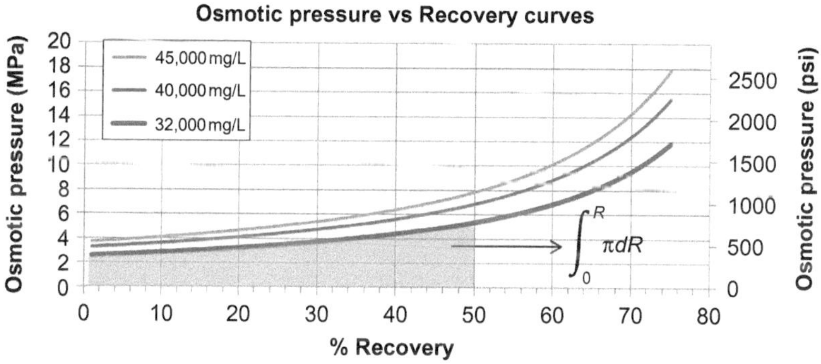

Fig. 2.1 Osmotic pressure vs recovery for different starting feed concentrations typical of a seawater feed. OLI Analyzer 3.1 was used to generate osmotic pressure data at 25°C. For typical seawater desalination, at a feed concentration of 32,000 ppm NaCl, operated at a recovery of 50% at 25°C, this energy was calculated as 1 kWh/m^3, which agrees with the numbers reported in the literature [16]. *(Reproduced with permission from Shrivastava A, Rosenberg S, Peery M. Energy efficiency breakdown of reverse osmosis and its implications on future innovation roadmap for desalination. Desalination 2015;368:181–92.)*

of understanding the application of second law of thermodynamics in the unmixing of salt-water mixture would be that the technology used to effect this separation would require additional energy due to losses like frictional losses, heat losses, and other system inefficiencies. Hence, for example, if we could design a reverse osmosis system where the high-pressure feed pump were 100% efficient, energy recovery devices had 100% efficiency, membranes had infinite water permeability and zero salt permeability, modules had no frictional losses and concentration polarization perhaps $E_{\text{thermodynamic, min}}$ could be achieved. An ideal reverse osmosis system that might be able to achieve this thermodynamic minimum can be conceptualized as a frictionless piston driving flow through an ideal semipermeable membrane where there is no concentration polarization and system losses [25].

2.2.2 Energy Consumption in Reverse Osmosis With Second Law of Thermodynamics Framework

As we start building framework for the second law of thermodynamics for reverse osmosis or for desalination in general, we can start by defining an efficiency for a specific desalination technology that is used for salt-water separation. This efficiency can be defined as:

$$\eta_E = \frac{E_{\text{thermodynamic,min}}}{E_{\text{thermodynamic,min}} + W} \tag{2.3}$$

where

ηE = energy efficiency.

W = work (units: joules or kWh/m^3 (when expressed as specific energy for reverse osmosis)).

This definition of efficiency is similar to a Carnot efficiency [26]. As defined in Eq. (2.3), the term W involves any losses and work done by the system to supply the thermodynamic minimum separation energy to the feed mixture, losses involved due to friction or heat or other losses specific to the technology, and any work needed to bring the streams back to their original state.

The technology that separates the salt from water is a semipermeable membrane that works on solution-diffusion mechanism [27–34]. An ideal semipermeable membrane would have infinite water permeability (A-value, gfd/psi) and zero salt permeability (B-value, gfd), the fact that actual membranes are not ideally semipermeable results in kinetic losses due to the need for high rejection and hence high flux [17]. Membrane fouling results in additional resistance to flux through the membranes and hence in additional energy input. Membranes are packed into modules, which have frictional losses and losses due to concentration polarization on the membrane surface. For reverse osmosis, the source of energy comes from a pump and an energy recovery device is used to recover the pressure energy from the concentrate. Both these technologies have efficiency associated with them which causes overall decrease in efficiency or increase in work input. And finally systems have frictional losses and design constraints that result in some practical losses. All these losses can be included in the W (work) term in Eq. (2.3) to define an overall efficiency for reverse osmosis (ηE) and will be discussed in detail in the subsequent sections.

2.3 REVERSE OSMOSIS ENERGY CONSUMPTION

2.3.1 Membrane Energy Consumption

2.3.1.1 The Solution-Diffusion Model

In the most fundamental terms, water and salt move through a membrane due to spatial gradients in their chemical potential, with the amount of transport, or flux, determined by the magnitude of the gradient. The solution-diffusion model, the derivation of which can be found in several places [28,35], is most often used to describe pressure driven osmotic transport of water and dissolved solutes through membranes. In its simplified form, the volumetric flux of water, Jw, is driven by pressure drop, while the mass flux of dissolved solute, *js*, is driven by concentration gradient. Pressure is assumed to be constant and equal to the feed through the entire thickness of the active layer, and interfacial chemical potentials are at equilibrium.

$$J_w = A\left(\Delta P_{f,p} - \Delta\pi\right) \tag{2.4}$$

$$j_s = B\left(C_{s,0} - C_{s,l}\right) \tag{2.5}$$

where

$$A = \frac{D_{wM} K_w c_{w,0} v_w}{lRT} \tag{2.6}$$

$$B = \frac{D_{sM} K_s}{l} \tag{2.7}$$

The net driving pressure, $\Delta Pfp - \Delta\pi$, is the difference between transmembrane hydraulic pressure ($\Delta Pfp = P_{\text{feed}} - P_{\text{permeate}}$) and trans-membrane osmotic pressure ($\Delta\pi = \pi s_{,0} - \pi s.l$). The membrane water permeability, *A*, and salt permeability, *B*, are each a consolidation of membrane properties. *DwM* and *DsM* are the diffusivity of water and salt in the membrane, respectively. *Kw* and *Ks* are sorption coefficients of water and salt into the membrane; $cw_{,0}$ is the molar concentration of water at the feed interface adjacent to the membrane; *vw* is the molar volume of water; *l* is the membrane thickness; *R* is the ideal gas constant; and *T* is the temperature. The concentration driving force is the difference in interfacial, or surface solute concentration at the feed side ($Cs_{,0}$) and permeate solute concentration ($Cs,l = Cs,p$). Rather than SI units, *A* values are most commonly expressed in practice through units of gal/ft^2/day/psi (gfd/psi) or alternatively liters/m^2/h/bar (lmh/bar). *B* values, in turn, are typically given either in gfd or lmh. This chapter uses lmh and bar as conventional units of flux and pressure. The total permeate mass flux, *jp*, or volumetric flux, *Jp*, is simply the sum of the solute and water flux. For a single solute:

$$j_p = J_w \rho_P + j_s \tag{2.8}$$

$$J_p = J_w + \frac{j_s}{\rho_P} \tag{2.9}$$

with ρP representing the permeate density. Fig. 2.2 provides a schematic illustration of the pressure and concentration profiles through a thin film composite membrane in the solution diffusion model.

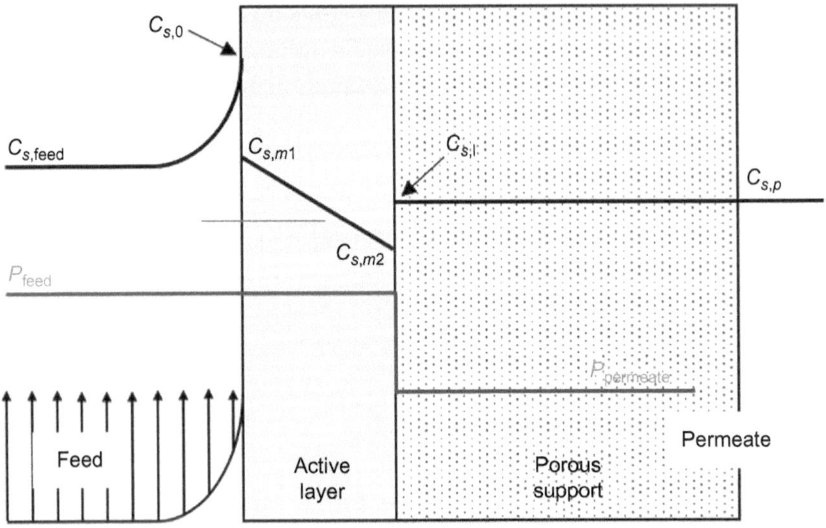

Fig. 2.2 Solution-diffusion model: pressure and concentration profiles in a thin film composite reverse osmosis membrane, consisting of a semipermeable active layer and porous support.

Concentration polarization occurs as solute moves through the mass transfer boundary layer from the bulk feed to the membrane surface. Polarization causes $Cs_{,0}$ to be greater than the bulk feed solute concentration, $Cs_{,\mathrm{feed}}$, by an amount, which we call Γ.

$$\Gamma = \frac{C_{s,0}}{C_{s,\mathrm{feed}}} \tag{2.10}$$

The size of Γ depends on transport characteristics within the feed channel and properties of the solute in question and will be explored further in Section 2.3.2. Reflecting polarization, salt passage is calculated as the ratio of permeate to feed solute concentration.

$$\text{Salt passage} = \frac{C_{s,\mathrm{permeate}}}{C_{s,\mathrm{feed}}} = \frac{j_s / J_w}{C_s} = \frac{1}{\Gamma} \cdot \frac{B}{B + J_w} \tag{2.11}$$

The solution-diffusion model may not be the most accurate depiction of membrane transport in all cases. For example, the solution-diffusion-imperfection model adds convective terms to reflect transport through imperfections in the membrane [36]. While other modifications have been proposed [37], in its basic form, the solution-diffusion-imperfection model assumes no change in solute concentration through the pores. The convective terms are thus driven entirely by transmembrane pressure drop (ΔP), proportional to a coefficient, L, which characterizes the convective permeability through the imperfections.

$$J_w = A(\Delta P - \Delta\pi) + L\Delta P \tag{2.12}$$

$$j_s = B(C_{s,0} - C_{s,p}) + LC_{s,0}\Delta P \tag{2.13}$$

The relatively tight polyamide reverse osmosis membranes sold commercially, however, usually tend to be modeled quite well with the basic solution-diffusion expressions.

2.3.1.2 Resistance in the Membrane

Resistance to water flow through a thin film composite membrane is essentially the inverse of its water permeability, or $1/A$. In practice, this resistance is dominated by, but not exclusive to, the semipermeable active layer. Resistance also comes from the porous support structure, which for polyamide RO membranes consists of a porous ultrafiltration membrane on top of a nonwoven web. Resistance in the support actually will manifest itself as additional pressure drop, which is not reflected by the idealized solution-diffusion schematic in Fig. 2.2. The resistance, or inverse permeability, of a composite with n layers can be broken down at a high level by the sum of resistances ($1/Ai$) in each layer:

$$\frac{1}{A_{\text{composite}}} = \sum_{1}^{n}\left(\frac{1}{A_i} + \frac{1}{A_{i,i-1}}\right) \tag{2.14}$$

For the second term in this equation, Ai,i_{-1}, we must recognize that additional resistance may arise at the interfaces between these layers, as the flow of water out of the active layer seeks out the finitely distributed surface pores of the support [38,39]. From a close inspection of commercial offerings, one will find most of today's polyamide thin film composite RO membranes have a water permeability in the range of 1–10 lmh/bar, while nanofiltration membranes can be as much as double this, 10–20 lmh/bar. Water permeabilities of UF membranes suitable as RO/NF supports can reach upward of 750–800 lmh/bar [40]. Under such cases, as seen in Fig. 2.3, the support will generally only make up <1% of the total resistance. However, ultrafiltration supports are susceptible to mechanical compaction [41,42], and desalination applications requiring high feed pressure and/or temperature can see significant effects. One must also recognize the growing contribution of the support with any hypothetical new membrane possessing high permeability (≫20 lmh/bar).

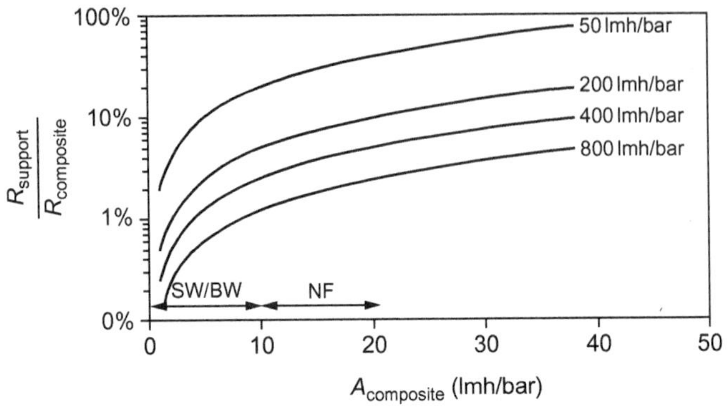

Fig. 2.3 Contribution of support film to total membrane resistance as a function of composite and support permeabilities, with the latter ranging from 50 to 800 lmh/bar.

2.3.1.3 *Resistance From Fouling*

Fouling is arguably the number one challenge faced by operators of reverse osmosis and nanofiltration systems, and can arise due to particulate matter, inorganic scale, organic fouling and/or biological fouling. A wide array of pretreatment operations are performed to minimize RO/NF membrane fouling, including antiscalant dosing, chlorination (which in most cases must subsequently be removed before the water stream contacts the membrane, as polyamides are not oxidatively stable), coarse upstream filtration, coagulation-flocculation, and softening, to name a few. The argument of series resistance introduced by Eq. (2.9) can be expanded one step further to include the effects of such foulant layers. Fig. 2.4 uses this simple series resistance argument to illustrate how sensitivity to fouling grows with membrane permeability. Assuming as a thought exercise that membranes of varying permeability but the same propensity to fouling are operated identically in the same feed waters, they should each build up the same amount of fouling on their surface. With identical loadings of foulant ($1/A_{\text{foulant}}$), a higher permeability membrane will see a larger percent increase in total resistance, giving it, all else equal, a larger percent drop in water permeability. The same foulant load, which reduces the permeability of a SW (1 lmh/bar) membrane by 10% reduces a BW (3.5 lmh/bar) membrane by nearly 30% and an NF (10 lmh/bar) membrane by >50%. This simplified relationship is instructive as a ballpark demonstration but might not hold true in practice for several reasons. Among them, membranes are known to vary in surface charge/hydrophilicity and roughness, all of which can affect their propensities to foul [43–47].

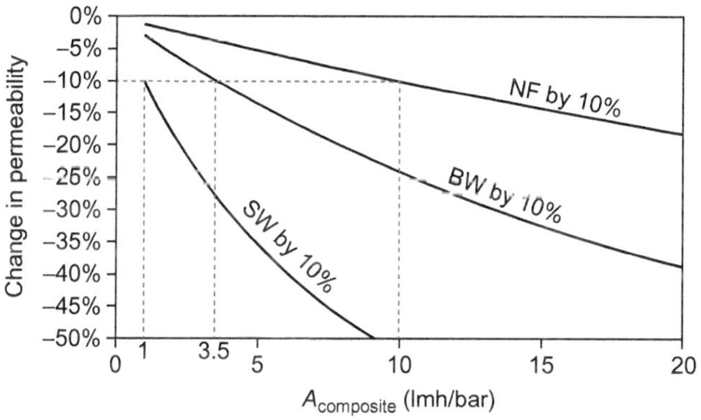

Fig. 2.4 Impact of initial composite water permeability on the response to fouling. Three different fouling loadings were chosen corresponding to a 10% drop in permeability for either SW (1 lmh/bar), BW (3.5 lmh/bar), or NF (10 lmh/bar) membranes.

2.3.1.4 Kinetics and the Flux/Rejection Trade-off

While osmotic pressure provides a baseline lower limit for energy consumption, the membrane's water permeability (composed of all its resistance terms) dictates the amount of net driving pressure needed to obtain a desired flux, as seen by Eq. (2.1). The term "kinetic energy consumption" or "configurational energy" cited in many publications [17,25] is a reflection of optimization in this area: choosing what net driving pressure or flux is used and how it is distributed across a system. Several cost-benefit analyses are at stake in the selection of an operating flux or net driving pressure. A feed pressure just slightly above the feed osmotic pressure is the thermodynamic minimum required to produce any water flux, but is unable to account for recovery and requires infinite active area. The choice of net driving pressure thus reflects a balance between the cost of energy and cost of capital required to install more elements and pressure vessels. At a vessel or system level, recovery and pressure drops (see Fig. 2.9 later in this chapter) lead to flux imbalances and inefficiencies in net driving pressure down the length of a vessel or series of vessels. Interstage designs, interstage boosting pumps, and, to an extreme, the concept of closed-circuit desalination (CCD) are all attempts to better balance flux out in such systems to avoid unnecessarily large net driving pressures that waste energy [48–51] (Fig. 2.5).

Again, here, there are economic tradeoffs. Single element stages with booster pumps ahead of each can be used to maintain a constant, idealized net driving pressure across a series of elements, but come at a higher installation cost than a system that has only one or two stages. CCD potentially

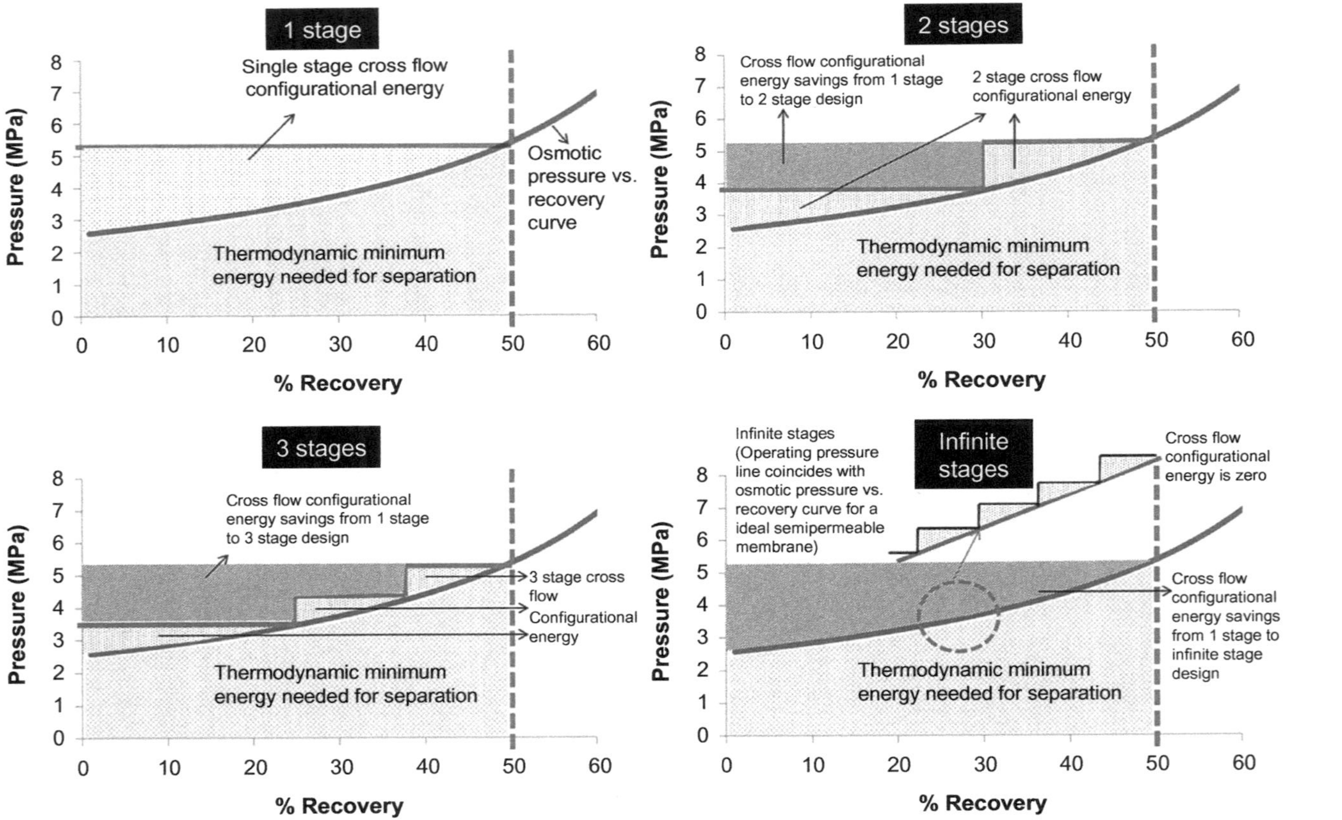

Fig. 2.5 Illustration of energy savings *(gray areas)* in going from single-stage operation to infinite stages. These savings come at the expense of capital. In addition, other factors like operating flux and pretreatment come into consideration. *(Reproduced with permission from Shrivastava A, Rosenberg S, Peery M. Energy efficiency breakdown of reverse osmosis and its implications on future innovation roadmap for desalination. Desalination 2015;368:181–92.)*

addresses this problem by operating in semibatch fashion, recirculating brine back to the feed and continuously increasing feed pressure as needed to maintain net driving pressure.

System size, however, is not the only cost of operating at small net driving pressures. According to the solution-diffusion model, RO membranes have constant salt permeability, and passage of salt through them varies inversely with operating flux (see Eq. 2.6). Fig. 2.6 plots the rejection of typical BW and SW membranes as a function of flux, assuming no effects from polarization or module friction described later. Rejection in both cases drops precipitously below 10 lmh. In applications with stringent permeate TDS requirements, operation at high net flux, and thus high net driving pressure, may be needed to achieve desired rejection rather than reduce system size, despite the implications on energy requirements.

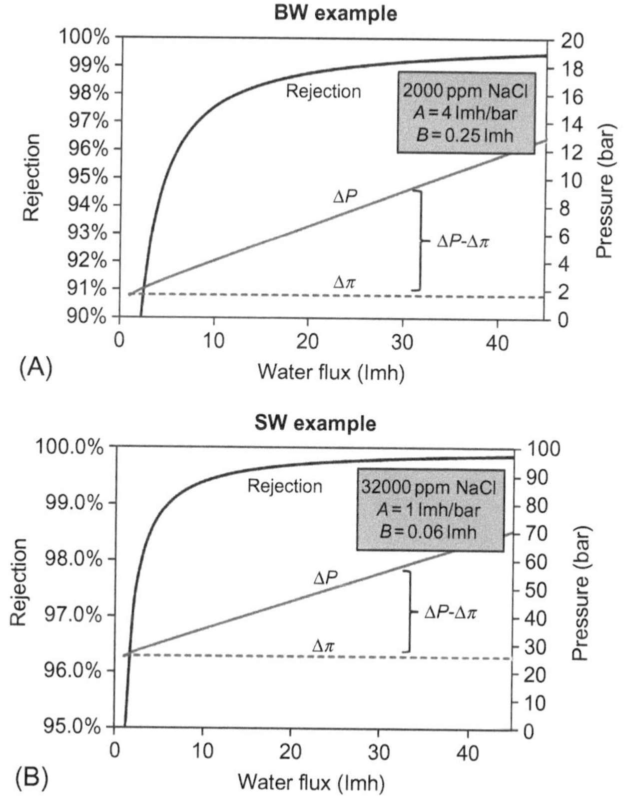

Fig. 2.6 BW (A) and SW (B) examples illustrating the impact of operating flux, driven by transmembrane net driving pressure ($\Delta P - \Delta\pi$) on salt rejection.

2.3.2 Module Energy Consumption

2.3.2.1 Spiral Wound Modules

Spiral wound modules are the predominant construction used for reverse osmosis membranes today [52]. One or more sheets of membrane are folded in half over a feed spacer and inserted between sheets of permeate carrier, called a tricot. After applying the necessary glue lines, this entire assembly of layers is wound around a perforated tube forming the central axis of the module. The diameter and length of the module varies with application, as small as 1.8″ diameter by 12″ length in residential filters and up to most recently 16″ diameter by 40″ length in large-scale desalination plants [53]. 8″ diameter, 40″ length elements are by far the most common construction used for industrial and municipal applications today. Feed water enters one end of the module, flowing through gaps provided by the feed spacer, permeates across the membrane and into the tricot channel, and eventually enters the permeate tube and exits the module through one or both ends. Water, which does not permeate, is rejected out of the feed channels on the other side of the module. After the membrane sheet, energy consumption in a module occurs in three main areas: (1) the feed channel, (2) the permeate channel, and (3) at entry and exit points associated with couplings from module to module and module to vessel (Fig. 2.7).

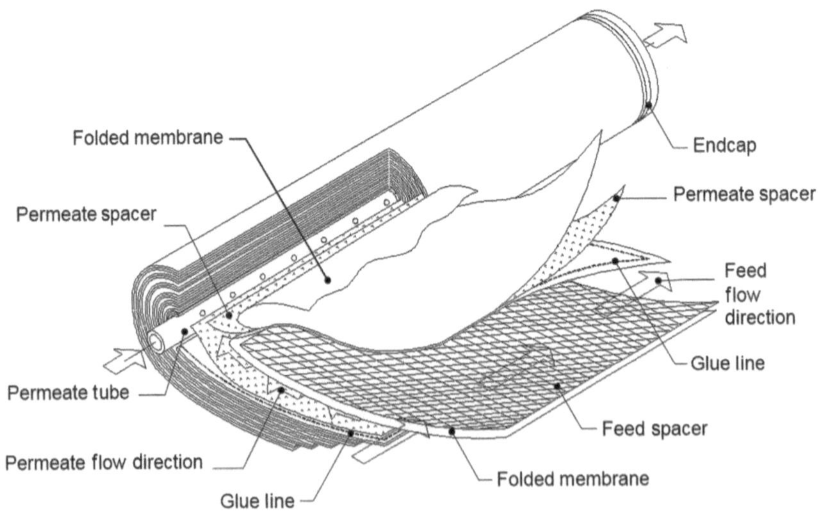

Fig. 2.7 RO spiral wound module construction. *(Reproduced from Johnson J, Busch M. Engineering aspects of reverse osmosis module design. Desalin Water Treat 2010;15(1–3):236–48.)*

Feed Channel Pressure Drop

Flow and permeation in the feed channel give rise to pressure drop (ΔP) and concentration polarization (Γ), which consume usable energy in a module. The feed spacer adds complexity to flow in the feed channel, and its engineered design offers a way to tune ΔP and Γ within a module to achieve desired results. Though complex, empirical fits to a few rather simple equations can be used to help describe these properties.

Pressure drop in the feed channel from feed to brine, $\Delta Pf,b$, can be simplified using the well-known Darcy-Weisbach equation for incompressible flow [54]. ΔP is proportional to the length of the channel, L, and the square of the velocity, v, by a friction factor *ff*. In this case, the friction factor can be fit well using the apparent Reynolds number (Re) in the channel in combination with two constants, c_1 and c_2, which are unique to the spacer design:

$$\Delta P_{f,b} = f_f\left(\frac{L\rho}{d_h}\frac{v^2}{2}\right) = c_1 \,\mathrm{Re}^{c_2}\left(\frac{L\rho}{d_h}\frac{v^2}{2}\right) = c_1\left(\frac{\rho v d_h}{\mu}\right)^{c_2}\left(\frac{L\rho}{d_h}\frac{v^2}{2}\right) \quad (2.15)$$

Here ρ is the density of the fluid, v is the apparent velocity in the channel, μ is the fluid viscosity, and dh is the hydrodynamic diameter. For the case of this flow profile, dh is equal to twice the channel height, and the apparent velocity accounts for channel height only and does not include the complicating effects of the spacer.

Feed Channel Concentration Polarization

The concentration polarization of solutes in the channel is a consequence of the balance between permeate flux, JP, the mass transfer coefficient of the solute, ks, and the operational salt passage SP [55].

$$\Gamma = \exp\left(\frac{J_P}{k_s}\right)(1 - SP) + SP \quad (2.16)$$

Higher flux pushes more solute to the surface, increasing the boundary layer and increasing Γ. A higher mass transfer coefficient reduces the boundary layer and decreases Γ. The mass transfer coefficient is affected by mixing in the channel as well as properties of the solute in question: [55,56].

$$k_s = (Sh)\left(\frac{D_s}{d_h}\right) = c_3 \,\mathrm{Re}^{c_4}\, Sc^{1/3}\left(\frac{D_s}{d_h}\right) = c_3\left(\frac{\rho v d_h}{\mu}\right)^{c_4}\left(\frac{\mu}{\rho D_s}\right)^{1/3}\left(\frac{D_s}{d_h}\right) \quad (2.17)$$

where *Sh* and *Sc* are the dimensionless Sherwood and Schmidt numbers, respectively, *Ds* is the diffusivity of the solute, and c_3 and c_4 again are parameters unique to the feed spacer being used. As can be seen from Eq. (2.11), for a given solute, a higher mass transfer coefficient can be achieved: (1) through operation by using high feed flow rates (high v), or (2) through design by using spacers, which promote mixing (high c_3, c_4 or small *dh*). The typical tradeoff, which exists here, is that channels that are designed for low polarization/high *ks* will have high-pressure drop (small *dh*, large c_1, c_2), and vice versa [57,58].

Permeate Channel Pressure Drop

Pressure drop in the permeate channel, or tricot, occurs due to frictional losses as water travels around the spiral on its way to the permeate tube. This pressure drop can be expressed through a friction parameter, *k*, which is a property of the tricot being used [52].

$$\frac{dP_p}{dx} = -k\frac{q(x)}{w} \tag{2.18}$$

Pp is the permeate pressure at a given location in the leaf, *q*(*x*) is the net flow rate of permeate at a given location in the leaf, *w* is the leaf width, and *x* is the coordinate position along the leaf length with origin at the outer edge of the leaf and terminus at the product water tube for a total length of l. The flow rate *q*(*x*) can be calculated by integrating the flux from both sides of the leaf to a given position, *x*:

$$q(x) = 2w\int_0^x J_P(x)\,dx \approx 2w\int_0^x A\left(\text{NDP}^{\text{o}} - P_p(x)\right)dx \tag{2.19}$$

To simplify the analysis, here we have made the assumption that *js* is small enough such that *JP* is approximately equal to *Jw*. NDP° is the net driving pressure for reverse osmosis with the exception of permeate pressure (i.e., $P_{\text{feed}} - \pi_{\text{feed}} + \pi_{\text{permeate}}$. Though permeate osmotic pressure may vary slightly with *x*, the entire term NDP° is assumed constant to simplify the solution. The relevant boundary conditions for this problem require zero permeate pressure at the tube and zero water flow at the edge of the leaf:

$$BC\#1:\ P_p(x=l)=0 \tag{2.20}$$

$$BC\#2:\ q(x=0)=0 \tag{2.21}$$

A permeate backpressure intentionally applied to the permeate tube could change the first boundary condition, but such cases could also be accounted for by making a corresponding reduction to NDP^o (i.e., let NDP^o = $P_{feed} - \pi_{feed} + \pi_{permeate} - P_{tube}$ and let $P_{permeate}$ be a relative pressure in the tricot of $P_{absolute} - P_{tube}$. With these boundary conditions, the second-order differential equation posed by the combination of Eqs. (2.12), (2.13) provides a convenient analytical solution. Fig. 2.8 shows the pressure and flow profiles calculated with these equations across a 30-in. leaf using membranes of low (SW) and high (BW) permeability, both operating at an average flux of 20 lmh. The permeate pressure profiles are fairly similar between the two membrane types, as they are operating at the same flux and are assumed to have the same tricot. This pressure drop, however, takes up a larger fraction of the overall net driving pressure in the BW module, leading to a greater flux imbalance in the high A-value membrane (Fig. 2.8B).

After solving for $Pp(x)$, Eq. (2.16) can be integrated to show the aggregate flow of water exiting each leaf at $x = l$ is:

$$q(x=l)=2Aw\cdot \mathrm{NDP}^o\,\frac{\tanh\left(\sqrt{2kA}\cdot l\right)}{\sqrt{2kA}} \tag{2.22}$$

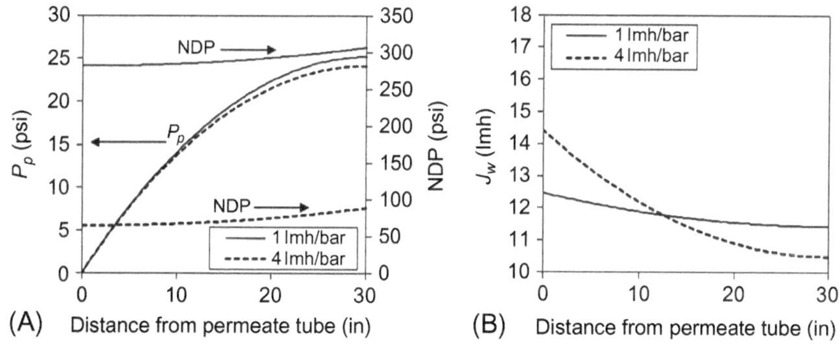

Fig. 2.8 Simulation of (A) permeate pressure (P_p), net driving pressure (NDP) and (B) local water flux (J_w) for a 30-in. leaf in a spiral wound module constructed with a seawater (A = 1 lmh/bar) or brackish water (A = 4 lmh/bar) membrane. Both membranes were simulated to run at an average flux over the length of the leaf of 20 lmh and use the same tricot channel with k = 130 psi s/in.^{3} friction coefficient. Effects from concentration polarization, feed pressure drop and recovery are not considered.

The element efficiency, ε, is the ratio of this actual permeate flow (q at $x = l$) to the amount one would expect in the absence of any permeate pressure drop:

$$\varepsilon = \frac{q(x=l)}{2wlA \cdot \text{NDP}^{\circ}} = \frac{\tanh\left(\sqrt{2kA} \cdot l\right)}{\sqrt{2kA} \cdot l} \tag{2.23}$$

As should be expected, efficiency depends inversely on leaf length and the tricot friction parameter. The value of k can be reduced by using thicker, more open tricots, but a balance must be struck against the active area, which can be achieved with given diameter constraints, as well as the potential for pressure driven intrusion of membrane into the tricot channel. Membrane forced into the tricot at high applied feed pressures can choke off the permeate channel and serve to increase k. Given these considerations, certain degrees of structural stability, minimum thickness, and maximum thickness are desired. Similarly, the improvement in leaf efficiency with shorter leaves must be balanced against the higher manufacturing cost and dead area introduced by glue lines when the requisite larger number of shorter leaves is used to maintain desired active area. As we saw in Fig. 2.8, efficiency also scales inversely with membrane A-value, as channel pressure drops make up a larger fraction of total resistance with higher permeability membranes.

Armed with this foundation, the average flux and net salt passage in a full spiral wound RO module can be expressed in two simplified equations in which net driving pressure is calculated at the midpoint of each element. Use of an average midpoint rather than a complete axial integration can lead to some error in the result, but is generally good enough of an approximation to well describe true behavior, particularly considering we are modeling here just a single element rather than a full vessel, which often contains six or more elements in series.

$$\bar{J}_w = \frac{A\varepsilon\left[P_f - \frac{1}{2}\Delta P_{f,b} - \pi\left(\bar{\Gamma}\bar{C}_{f,b}\right) + \pi\left(\bar{C}_p\right)\right]}{2nlw} \tag{2.24}$$

$$SP = \frac{\bar{\Gamma} B}{B + \bar{J}_w} \tag{2.25}$$

The net impact of losses due to ε, $\Delta Pf,b$ and Γ on an operational vessel containing spiral wound modules is illustrated in the examples of Fig. 2.9. The two panels on the left represent a seven-element vessel

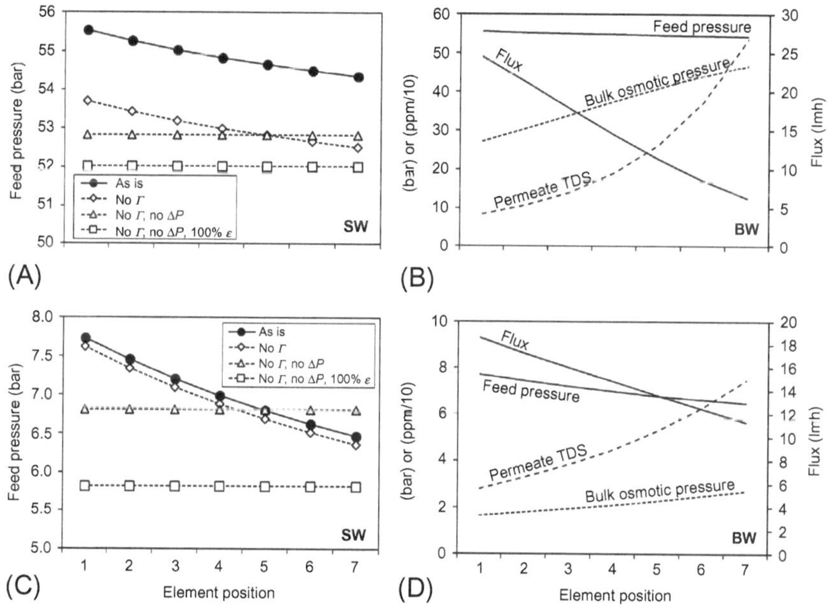

Fig. 2.9 Simple examples illustrating the impact of polarization and frictional losses on RO system performance. *(Left)* SW vessel with seven elements operating at 45% recovery with a 9.5 m^3/h feed of 32,000 ppm NaCl; *(Right)* BW vessel with seven elements also operating at 45% recovery but with a 9.5 m^3/h feed of 2000 ppm NaCl. All elements are 440 ft^2 modules constructed with the same assumed feed spacer, permeate tricot, and leaf lengths. The membranes have assumed performance parameters $A = 1$ lmh/bar, $B = 0.06$ lmh for SW and $A = 4$ lmh/bar, $B = 0.25$ lmh for BW. (A,B) Feed pressure requirement to achieve 45% recovery with and without losses from polarization and feed/permeate pressure drops; (C, D) Variation in feed pressure, flux, permeate TDS and bulk feed osmotic pressure with vessel position, accounting for all losses in polarization and feed/permeate pressure drops.

containing SW membranes ($A = 1$ lmh/bar, $B = 0.06$ lmh) operating at 45% recovery with 9.5 m^3/h feed of 32,000 ppm NaCl, while the two on the right represent a seven-element vessel containing BW membranes ($A = 4$ lmh/bar, $B = 0.25$ lmh) operating at 45% recovery with 9.5 m^3/h feed of 2000 ppm NaCl. These operating conditions are close to what a real SW plant might run at. While most BW installations operate at higher recoveries, the BW recovery in this case is kept constant at 45% to provide an easier comparison to the SW case. Elements in both cases are assumed to have the same construction consisting of 440-ft^2 active area, 33-in. leaf lengths, standard 28-mil feed spacers, and a "typical" permeate tricot friction parameter of $k = 130$ psi s/in.^{3} referenced in an earlier report from Johnson and Busch [52].

As seen by Fig. 2.9, polarization makes up the largest component of module losses in an SW system, owing to the large osmotic pressure of the bulk solution. Feed and permeate channel pressure drops are less important, due to the combination of large axial increases in bulk osmotic pressure and high required net driving pressure. In the BW example, on the other hand, polarization has only a minor effect, while feed and permeate pressure drops each contribute sizably to the overall losses. The membrane A-value has a stark impact on leaf efficiency, with ε of 93.5% and 79% calculated for the SW and BW examples, respectively. SW and BW elements in reality may have differences in their element constructions to best take advantage of their unique characteristics, but we have left these parameters fixed in this example to offer a uniform comparison.

2.3.2.2 Hollow Fiber Modules

Hollow fiber RO modules sold commercially today are predominately made from cellulose triacetate membranes. The modules used in RO contain thousands of small fibers, each around 160 μm in diameter, wound around a central, perforated feed tube [59]. Permeation is outside in, with water traveling through the membrane to the interior bore, or lumen, of the fibers. Each end of the fiber is potted with resin, with at least one end sliced open to provide an exit for permeate at the end of the module (Fig. 2.10).

The flow geometry in a hollow fiber module is more complex than in a spiral, and consequently the governing equations modeling their behavior have additional complexity. Sekino and others have published a series of reports outlining these models, with their 1995 publication providing a nice

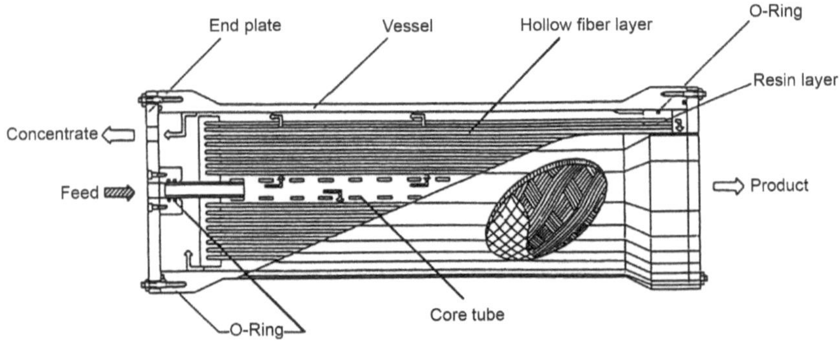

Fig. 2.10 Construction of a hollow fiber RO module. *(Reproduced with permission from Kumano A, et al. Study of mass transfer characteristics for a hollow fiber reverse osmosis module. J Membr Sci 2008:324(1–2);136–41. Copyright 2008, with permission from Elsevier.)*

overview of historical modeling efforts by other groups [59–61]. For the sake of brevity, only key equations from these models are reported herein as a reference, and the reader may refer to these cited publications for more in-depth analyses. Pressure drop in the permeate channel, or bore of the fiber, is given by the Hagen-Pouiseuille equation:

$$\frac{dP_p}{dz} = \frac{128\mu q_p(z)}{\pi d_I^4} \tag{2.26}$$

where z is the distance in the longitudinal direction down the length of the fiber, $qp(z)$ is the permeate flow at position z, dI is the inner diameter of the bore, and μ is again fluid viscosity. The permeate flow, in turn, is similar and integration of flux over the longitudinal coordinate across the outer diameter, dO of the fiber:

$$q(z) = \pi d_O \int_0^z J_P(z)\,dz \tag{2.27}$$

As the feed enters the module through a central tube, pressure drop on the feed side varies with radial coordinate r following the Ergun equation:

$$\frac{dP_{f,b}}{dr} = \frac{150\sigma^2\mu}{(1-\sigma)^3 d_P^2} v_s + \frac{1.75\sigma\rho}{(1-\sigma)^3 d_p} v_s^2 \tag{2.28}$$

Here r is the packing density of fibers, vs is the superficial velocity in the feed, and $dP = 1.5dO$ is the specific surface diameter of the fibers.

Polarization is calculated from the same relationship given in Eqs. (2.10), (2.11) for spiral wound modules, with values of c_3 and c_4 reported by Sekino of 0.048 and 0.6, respectively [59].

Hollow fiber modules used in RO tend to employ small fiber diameters that allow for large surface areas, as much as 10 times the surface area-to-volume ratio of spiral RO [60]. The CTA membranes have smaller A-values than polyamides and are operated at lower fluxes. Consequently, concentration polarization is much less severe, but the small interior diameters of the fibers (on the order of 100 μm [62] or less) can present problems with pressure drop [63–65]. While the permeation distance in spiral modules is dictated by radial leaf length, which is often in the neighborhood of 30″, the permeate in hollow fibers travels at least the entire length of the module, more if the fibers are cross wound. Today's commercially available, large-scale hollow fiber RO modules can be as much as 1–1.5 m (40–60″) long. Fujiwara and Mtsuyama have published an analysis on the benefits of extracting permeate from both ends of a hollow fiber module (a design termed both open-ended, or BOE) rather than just one end (single open-ended, or SOE). Fig. 2.11 shows the profile of permeate pressure, flux, and permeate concentration down the length of 1.61 m long module containing 870 m^2 of fiber active area. Operating at 55 bar and 30% recovery, permeate pressure reaches as high as 10 bar on the closed end of the SOE design. The BOE design has only half the distance to travel and reaches a peak pressure of <5 bar in the center of the module. Operating flux and salt passage are both consequently much improved in the BOE design.

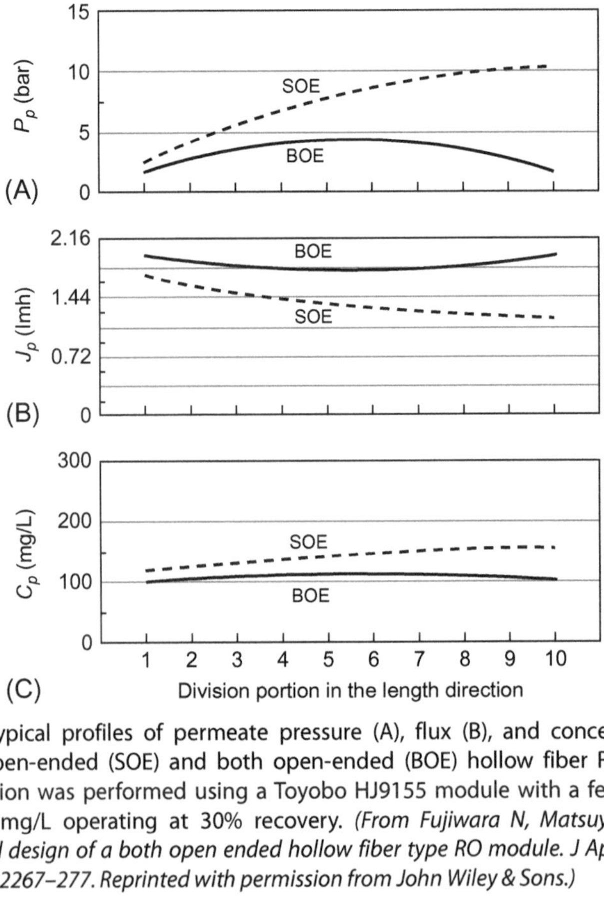

Fig. 2.11 Typical profiles of permeate pressure (A), flux (B), and concentration (C) in single open-ended (SOE) and both open-ended (BOE) hollow fiber RO modules. The simulation was performed using a Toyobo HJ9155 module with a feed of 55 bar and 35,000 mg/L operating at 30% recovery. *(From Fujiwara N, Matsuyama H. The analysis and design of a both open ended hollow fiber type RO module. J Appl Polym Sci 2008;110(4):2267–277. Reprinted with permission from John Wiley & Sons.)*

2.3.3 System-Level Energy Consumption and Efficiency of Power Source

Considering Eq. (2.3), and all the energy terms discussed in the preceding sections—the thermodynamic minimum energy, energy required to generate flux and rejection through a membrane, energy consumed in the modules in accounting for concentration polarization, and pressure drop—has to be supplied by a power source. In case of reverse osmosis, this source is a pressure pump. Pumps have evolved since their early discovery. With continuous improvements in the efficiency of electric motors, variable frequency drives (VFDs), impeller designs, the overall efficiency of pumps has in some cases exceeded 95% [66]. VFDs have particularly been very impactful in terms of energy efficiency as they eliminate the need for throttling valves for process flow control.

With seawater desalination, typical recoveries are around 50% and initial feed pressures can be in excess of 800 psi (55 bars). The high-pressure concentrate represents significant amount of energy that should be recovered prior to its disposal. This energy is recovered using energy recovery devices with the preferred and most efficient being pressure or work exchangers [67,68]. Overall system operation has to be synchronized considering appropriate operating ranges so that pump VFDs and energy recovery devices (energy or pressure exchangers) can work without impeding each other's optimum hydraulic operating conditions [69,70]. Reported efficiencies of pressure exchangers are in the range of >95% without any adverse effects on feed and concentrate mixing due to the need for contact between the high-pressure concentrate and low-pressure feed as the contact tie is very small minimizing any mixing between the feed and the concentrate.

Finally, the electrical energy to drive the pumps in reverse osmosis plants is the output of a power plant. Mostly power plants utilize energy sources in form of fuel or chemical energy. In United States, >60% of the energy sources for power plants is constituted by coal and natural gas with coal accounting for >40% followed by natural gas, nuclear power, oil, hydro, and renewables [7,71]. Power plants use power cycles to convert the chemical energy to hear and then to electricity, which is then used to drive the pumps in RO plants. There is an efficiency associated with these power cycles, which can range from 27% to 42% for coal-fired power plants, 33%–55% for natural-gas-fired power plants, and 24%–44% for oil-fired power plants. Hence, when comparing reverse osmosis energy efficiency with thermal desalination, this efficiency should be taken into account as thermal plants would also use heat as a source of energy in addition to electricity [72].

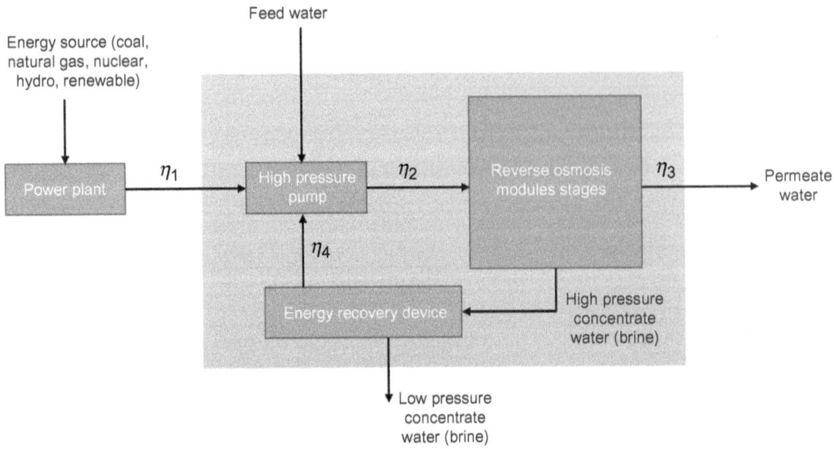

Fig. 2.12 Energy consumption for each major step in a RO operation. ηi represents the energy efficiency pertaining to the preceding step/unit operation. In order to get the total efficiency, each ηi has to be multiplied by the appropriate factor like recovery and location in the process flow.

Diagrammatically the energy consumption flow can be represented by Fig. 2.12.

Sections 2.3.1–2.3.3 pertain to the breakdown of reverse osmosis energy consumption, which is the box represented by "Reverse Osmosis Modules Stages" in Fig. 2.12. Optimized seawater desalination plants have reported energy consumption of 2 kWh/m^3 [16,17,72]. This number refers to the shaded control volume in Fig. 2.12. Analysis has been done on real-world optimized seawater desalination and brackish water desalination plants from first principles on real operating conditions and data [17] and is seen in Fig. 2.13.

The analysis shows that the energy efficiency of reverse osmosis as calculated by Eq. (2.29), which is basically Eq. (2.3) renamed for reverse osmosis, is approximately 53% as the other 47% of energy consumption is accounted for by thermodynamic minimum energy. Rest of the energy is accounted for by the work term *WRO*, which can be written as the sum of work done to generate membrane flux, overcome module configuration, friction and concentration polarization losses, and overcome pump losses, energy recovery device losses, and system friction losses as described in Sections 2.3.1–2.3.3.

$$\eta_{E,RO} = \frac{E_{\text{thermodynamic,min}}}{E_{\text{thermodynamic,min}} + W_{RO}} \tag{2.29}$$

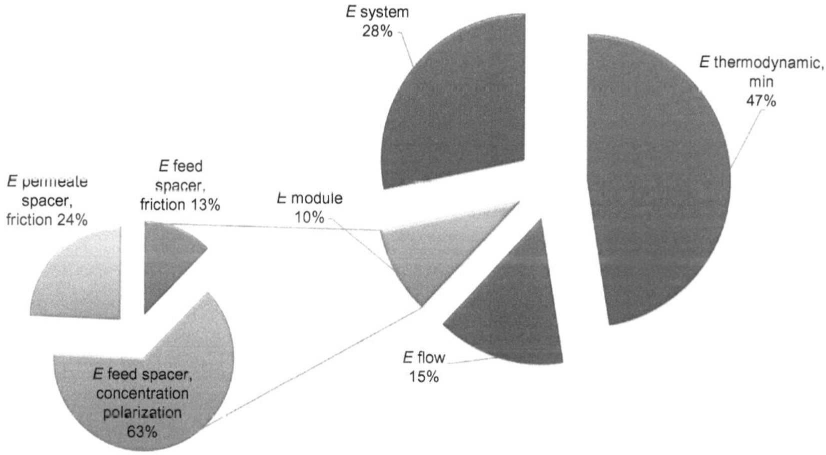

Fig. 2.13 Seawater desalination specific energy breakdown. Previously published in [24]. Total energy consumption = 2.12 kWh/m^3. *E* represents thermodynamic, min represents thermodynamic minimum energy, *E* flow represents energy needed to drive flux through the state-of-the-art commercial membrane, *E* module represents energy required to drive flow through the module (*E* feed spacer friction, *E* permeate spacer friction) and overcome concentration polarization (*E* feed spacer concentration polarization), *E* system represents losses in high-pressure pump, energy recovery devices, and elsewhere in the system.

2.4 CONCLUSION AND FUTURE OPPORTUNITIES

Efficiency of ~50% for the reverse osmosis unit operation (Fig. 2.12) with electrical energy as an input is quite good. Half of the energy consumption is accounted for by thermodynamics, which cuts any further improvements in half right at the outset. Pumps and energy recovery devices are quite efficient already as discussed in Section 2.3.3 and account for the next 28% of the energy losses. Hence, 75% of the total energy consumption cannot be optimized any further. Remaining 25% comprises membranes and modules with membranes accounting for 15% and modules 10%. Innovation in membranes will provide some energy savings but at a system scale it will not be that impactful, and multiple analyses suggest that we are already in an area of diminishing returns for energy savings with high permeability membranes for both seawater and brackish water reverse osmosis desalination [16,17,26,51,72].

This begs the question as to what are the future opportunities for energy and more importantly cost-savings opportunities for desalination. Here are the key areas for future opportunities also outlined in other references [13,73–75] in addition to those cited in the last paragraph:

- *Antifouling membranes and module designs*: One of the major limitations of membrane-based separations is the propensity of membranes to foul (Section 2.3.1.3). As discussed, fouling can be biological, organic, colloidal, or inorganic scaling related, which ultimately leads to lower membrane *A*-values and at times higher *B*-values. Hence, as we look for membranes with high permeabilities in the quest of further achieving the remaining energy efficiency, bigger impact will be made if these membranes can be fouling resistant so as to enable higher flux operation without the need for significant pretreatment. This can result in operational savings in terms of energy, chemical pretreatment, and capital savings. Capital savings will come from lower pretreatment capability needs and less membrane area for the same productivity. Materials and coatings that can impart inherent antifouling properties will be key.
- *Advanced materials for membranes chemical resistance and mechanical robustness*: Enhanced chemical resistance, e.g., chlorine tolerance for RO membranes can result in significant savings in replacement costs for membrane. Also chemical and solvent membranes can enable membranes application into difficult separations [76] that are quite energy intensive currently as these rely on thermal separation pathways, which are inherently inefficient due to need for phase change to effect a separation. Mechanically robust and chemical-resistant membranes and their commercial enablement with scalable and cost-effective manufacturing techniques will be key in reducing the overall cost of high-salinity water separations needed in brine treatment and concentrate management, which will enable minimal liquid discharge operations at low cost.
- *Pretreatment for water purification and waste water treatment technologies*: Significant amount of energy and capital is invested in pretreatment technologies for brackish and seawater plants depending upon the feed water quality [14]. These technologies can range from chemical pretreatment, membrane pretreatment using ultrafiltration and nanofiltration membranes to biological treatment using membrane bioreactors and disinfection strategies. New technologies that can reduce energy and cost of pretreatment for seawater and brackish water desalination, waste water treatment via bio digestion and nutrient removal will be quite impactful in overall cost and energy reduction for water purification and treatment.

- *Sensors and system-level analytics*: For a full-scale desalination system including pretreatment, concentrate treatment, and management, one of the key opportunity areas for cost and efficiency optimization is system-level data analytics leading to predicative maintenance and insight into operational efficiency. With the computing power at our disposal and the capability of running on demand online analytics, this can reduce response times and increase productivity of water treatment plants. The key areas of research here will be nonfouling sensors that can provide reliable data to act on using advanced analytics.
- *Specific ion/high rejection membranes*: Membranes designed at rejection specific solutes of interest like Boron, heavy metals, arsenic, and micropollutants can result in reduced capital costs by eliminating second pass operations or post treatment of permeate to achieve desired quality.

As we plan on a sustainable future, our thinking of water has to be more holistic and along with advancement of desalination and water treatment technologies, efficient use and consumption of water in agriculture and industries, use of renewable energy sources, and water harvesting also has to be a priority.

Overall, the opportunity for further enhancing energy efficiency and cost reduction is fragmented and application specific. Only efforts on many fronts as outlined earlier will provide any significant improvements.

REFERENCES

[1] Pate R., et al. Overview of energy-water interdependencies and the emerging energy demands on water resources. Report SAND, 2007.
[2] Younos T., Hill R., Poole H. Water dependency of energy production and power generation systems. Water Resour Impact 2012;14(1):9–12.
[3] Sinha S., Sharma B.R., Scott C.A. Understanding and managing the water-energy nexus: moving beyond the energy debate. Groundwater research and management: integrating science into management. Decision 2006;242–57.
[4] Schnoor J.L. Water–energy nexus. Environ Sci Technol 2011;45(12):5065.
[5] Thu K., et al. A hybrid multi-effect distillation and adsorption cycle. Appl Energy 2013;104:810–21.
[6] McGovern R.K., Zubair S.M., Lienhard V J.H. The benefits of hybridising electrodialysis with reverse osmosis. J Membr Sci 2014;469:326–35.
[7] Chu S., Majumdar A. Opportunities and challenges for a sustainable energy future. Nature 2012;488(7411):294–303.
[8] The United Nations World Water Development Report 2015; 2015.
[9] Service R.F. Desalination freshens up. Science 2006;313(5790):1088–90.
[10] Lonsdale H. The growth of membrane technology. J Membr Sci 1982;10(2):81–181.
[11] Bennett A. 50th anniversary: desalination: 50 years of progress. Filtration+Separation 2013;50(3):32–9.

[12] Glater J. The early history of reverse osmosis membrane development. Desalination 1998;117(1):297–309.
[13] Gude V.G. Desalination and sustainability—an appraisal and current perspective. Water Res 2016;89:87–106.
[14] Fritzmann C., et al. State-of-the-art of reverse osmosis desalination. Desalination 2007;216(1–3):1–76.
[15] Ghaffour N., Missimer T.M., Amy G.L. Technical review and evaluation of the economics of water desalination: current and future challenges for better water supply sustainability. Desalination 2013;309:197–207.
[16] Elimelech M., Phillip W.A. The future of seawater desalination: energy, technology, and the environment. Science 2011;333(6043):712–7.
[17] Shrivastava A., Rosenberg S., Peery M. Energy efficiency breakdown of reverse osmosis and its implications on future innovation roadmap for desalination. Desalination 2015;368:181–92.
[18] Ghiazza E., Borsani R., Alt F. Innovation in multistage flash evaporator design for reduced energy consumption and low installation cost. Tianjin: The International Desalination Association (IDA), World Congress on Desalination and Water Reuse; 2013.
[19] Sandler S.L. Chemical and engineering thermodynamics. 3rd ed. New York: John Wiley and Sons; 1999.
[20] Pitzer K.S. Activity coefficients in electrolyte solutions. 2nd ed. Boca Raton, FL: CRC Press; 1991.
[21] Robinson R.A., Stokes R.H. Electrolyte solutions. Revised ed. vol. 2. London: Dover Publications; 2002.
[22] Smith J.M., Van Ness H.C., Abbott M.M. Introduction to chemical engineering thermodynamics. 5th ed. New York: McGraw-Hill; 1996.
[23] Feinberg B.J., Ramon G.Z., Hoek E.M.V. Thermodynamic analysis of osmotic energy recovery at a reverse osmosis desalination plant. Environ Sci Technol 2013;47(6):2982–9.
[24] Shrivastava A., Rosenberg S., Peery M. In: Maximizing seawater reverse osmosis energy efficiency with high permeability membranes and system design. International Desalination Association World Congress on desalination and water reuse; Tianjin: IDA; 2013.
[25] Liu C., Rainwater K., Song L. Energy analysis and efficiency assessment of reverse osmosis desalination process. Desalination 2011;276(1–3):352–8.
[26] Cussler E.L., Dutta B.K. On separation efficiency. AIChE J 2012;58(12):3825–31.
[27] Paul D.R. Reformulation of the solution-diffusion theory of reverse osmosis. J Membr Sci 2004;241(2):371–86.
[28] Wijmans J., Baker R. The solution-diffusion model: a review. J Membr Sci 1995;107(1):1–21.
[29] Yaroshchuk A., Bruening M.L., Licón Bernal E.E. Solution-diffusion–electromigration model and its uses for analysis of nanofiltration, pressure-retarded osmosis and forward osmosis in multi-ionic solutions. J Membr Sci 2013;447:463–76.
[30] Van Gauwbergen D., Baeyens J. Modelling reverse osmosis by irreversible thermodynamics. Sep Purif Technol 1998;13(2):117–28.
[31] Voros N.G., Maroulis Z.B., Marinos-Kouris D. Salt and water permeability in reverse osmosis membranes. Desalination 1996;104(3):141–54.
[32] Heintz A., Stephan W. A generalized solution—diffusion model of the pervaporation process through composite membranes part II. Concentration polarization, coupled diffusion and the influence of the porous support layer. J Membr Sci 1994;89(1–2):153–69.
[33] Heintz A., Stephan W. A generalized solution—diffusion model of the pervaporation process through composite membranes part I. Prediction of mixture solubilities in the dense active layer using the UNIQUAC model. J Membr Sci 1994;89(1–2):143–51.

[34] Yaroshchuk A.E., Dukhin S.S. Phenomenological theory of reverse osmosis in macroscopically homogeneous membranes and its specification for the capillary space-charge model. J Membr Sci 1993;79(2–3):133–58.
[35] Rautenbach R., Albrecht R. Membrane processes. New York: John Wiley & Sons, 1989.
[36] Sherwood T.K., Brian P.L.T., Fisher R.E. Desalination by reverse osmosis. Ind Eng Chem Fundam 1967;6(1):2–12.
[37] Yaroshchuk A.E. Solution-diffusion-imperfection model revised. J Membr Sci 1995;101(1–2):83–7.
[38] Ramon G.Z., Wong M.C.Y., Hoek E.M.V. Transport through composite membrane, part 1: is there an optimal support membrane? J Membr Sci 2012;415–416:298–305.
[39] Wijmans J.G., Hao P. Influence of the porous support on diffusion in composite membranes. J Membr Sci 2015;494:78–85.
[40] Mehta A., Zydney A.L. Permeability and selectivity analysis for ultrafiltration membranes. J Membr Sci 2005;249(1–2):245–9.
[41] Mehdizadeh H., Dickson J.M., Eriksson P.K. Temperature effects on the performance of thin-film composite, aromatic polyamide membranes. Ind Eng Chem Res 1989;28(6):814–24.
[42] Pendergast M.T.M., et al. Using nanocomposite materials technology to understand and control reverse osmosis membrane compaction. Desalination 2010;261(3):255–63.
[43] Hoek E.M.V., et al. Modeling the effects of fouling on full-scale reverse osmosis processes. J Membr Sci 2008;314(1–2):33–49.
[44] Van Wagner E.M., et al. Surface modification of commercial polyamide desalination membranes using poly(ethylene glycol) diglycidyl ether to enhance membrane fouling resistance. J Membr Sci 2011;367(1–2):273–87.
[45] Sagle A.C., et al. PEG-coated reverse osmosis membranes: desalination properties and fouling resistance. J Membr Sci 2009;340(1–2):92–108.
[46] Lee S., Cho J., Elimelech M. Influence of colloidal fouling and feed water recovery on salt rejection of RO and NF membranes. Desalination 2004;160(1):1–12.
[47] Kang G.-D., Cao Y.-M. Development of antifouling reverse osmosis membranes for water treatment: a review. Water Res 2012;46(3):584–600.
[48] Efraty A., Barak R.N., Gal Z. Closed circuit desalination—a new low energy high recovery technology without energy recovery. Desalin Water Treat 2011;31(1–3):95–101.
[49] Efraty A. CCD series no-19: the lowest energy prospects for SWRO through single-element modules under plug-flow and closed-circuit desalination conditions. Desalin Water Treat 2016;57(46):21696–711.
[50] Efraty A. CCD series no-22: recent advances in RO, FO and PRO and their hybrid applications for high recovery desalination of treated sewage effluents. Desalination 2016;389:18–38.
[51] Lin S., Elimelech M. Kinetics and energetics trade-off in reverse osmosis desalination with different configurations. Desalination 2017;401:42–52.
[52] Johnson J., Busch M. Engineering aspects of reverse osmosis module design. Desalin Water Treat 2010;15(1–3):236–48.
[53] Bartels C., et al. Industry Consortium Analysis of large reverse osmosis/nanofiltration element diameters. In: Desalination and water purification research and development report no. 114. U.S. Department of the Interior Bureau of Reclamation; 2005.
[54] Streeter V.L., Wylie E.B. Fluid mechanics. New York: McGraw-Hill; 1979.
[55] Ho W.W., Sirkar K.K. Membrane handbook. New York: Springer; 1992.
[56] Cussler E.L. Diffusion: mass transfer in fluid systems. 3rd ed. Cambridge, UK: Cambridge University Press; 2009. p. 647.
[57] Zhou W., Song L., Guan T.K. A numerical study on concentration polarization and system performance of spiral wound RO membrane modules. J Membr Sci 2006;271(1–2):38–46.

[58] Kim S., Hoek E.M.V. Modeling concentration polarization in reverse osmosis processes. Desalination 2005;186(1–3):111–28.
[59] Sekino M. Precise analytical model of hollow fiber reverse osmosis modules. J Membr Sci 1993;85(3):241–52.
[60] Kumano A., et al. Study of mass transfer characteristics for a hollow fiber reverse osmosis module. J Membr Sci 2008;324(1–2):136–41.
[61] Sekino M. Study of an analytical model for hollow fiber reverse osmosis module systems. Desalination 1995;100(1):85–97.
[62] Kumano A., Matsuyama H. Analysis of hollow fiber reverse osmosis membrane module of axial flow type. J Appl Polym Sci 2012;123(1):463–71.
[63] Gill W.N., Bansal B. Hollow fiber reverse osmosis systems analysis and design. AIChE J 1973;19(4):823–31.
[64] Hanbury W.T., et al. Pressure drops along the bores of hollow fibre membranes—their measurement, prediction and effect on fibre bundle performance. Desalination 1981;38:301–18.
[65] Hanbury W.T., Yuceer A. Constriction of B10 hollow fibre bore diameters under operating conditions. Desalination 1985;54:27–44.
[66] Kaya D., et al. Energy efficiency in pumps. Energy Convers Manag 2008;49(6):1662–73.
[67] Stover R.L. Development of a fourth generation energy recovery device. A 'CTO's notebook. Desalination 2004;165:313–21.
[68] Harris C. Energy recovery for membrane desalination. Desalination 1999;125(1–3):173–80.
[69] Manth T., Gabor M., Oklejas Jr. E. Minimizing RO energy consumption under variable conditions of operation. Desalination 2003;157(1–3):9–21.
[70] Geisler P., Krumm W., Peters T.A. Reduction of the energy demand for seawater RO with the pressure exchange systems PES. Desalination 2001;135(1–3):205–10.
[71] Taylor P., et al. Energy efficiency indicators for public electricity production from fossil fuels. International Energy Agency; 2008. https://www.iea.org/publications/freepublications/publication/En_Efficiency_Indicators.pdf.
[72] Gordon J.M., Hui T.C. Thermodynamic perspective for the specific energy consumption of seawater desalination. Desalination 2016;386:13–8.
[73] Shannon M.A., et al. Science and technology for water purification in the coming decades. Nature 2008;452(7185):301–10.
[74] Semiat R. Energy issues in desalination processes. Environ Sci Technol 2008;42(22):8193–201.
[75] Peñate, B. and L. García-Rodríguez, Current trends and future prospects in the design of seawater reverse osmosis desalination technology. Desalination, 2012. 284(0): p. 1–8.
[76] Sholl D.S., Lively R.P. Seven chemical separations to change the world. Nature 2016;532:435–7.

FURTHER READING

[1] Fujiwara N., Matsuyama H. The analysis and design of a both open ended hollow fiber type RO module. J Appl Polym Sci 2008;110(4):2267–77.

CHAPTER 3

Environmental Impact and Technoeconomic Analysis of Hybrid MSF/RO Desalination: The Case Study of Al Taweelah A2 Plant

Aisha Al Bloushi, Adewale Giwa, Toufic Mezher, Shadi W. Hasan
Khalifa University of Science and Technology, Abu Dhabi, United Arab Emirates

NOMENCLATURE

A_b	brine heater area (m^2)
A_c	total condenser area (m^2)
A_j	rejection section area (m^2)
A_r	recovery section area (m^2)
A_S	salt permeability (m/s)
A_{total}	total heat transfer area (m^2)
A_W	pure water permeability (m/s Pa)
BPE	boiling point elevation
C_{civil}	civil cost estimation ($)
C_{rj}	concentration of reject flow rate for RO (kg/m^3)
Cp	heat capacity of the liquid streams liquid (kJ/kg °C)
C_p	final permeate concentration for RO (kg/m^3)
C_{cw}	sea water concentration at membrane wall (kg/m^3)
C_f	feed sea water concentration (kg/m^3)
$C_{membrane}$	membrane cost estimation ($)
C_{pump}	pump cost estimation ($)
C_w	new salt concentration value at membrane wall (kg/m^3)
DC	direct capital investment for MSF and RO plants ($/year)
F_f	RO feed (kg/h)
F_{rj}	reject flow rate for RO (kg/h)
F_p	total product flow rate for RO (kg/h)
Ks	mass transfer coefficient (m/s)
M_b	rejected brine flow rate (kg/h)
M_d	product flow rate for MSF (kg/h)
M_{cw}	cooling water for MSF (kg/h)
M_f	MSF feed (kg/h)

Sustainable Desalination Handbook
https://doi.org/10.1016/B978-0-12-809240-8.00003-4

M_{Hb}	total brine flow rate of hybrid MSF/RO (kg/h)
M_{Hd}	total permeate flow rate of hybrid MSF/RO (kg/h)
M_{Hf}	total hybrid seawater feed (kg/h)
M_r	brine recycle flow rate (kg/h)
M_s	steam flow rate (kg/h)
N	total number of heat recovery and rejection stages in MSF
Np	total product flux estimation (m^3/m^2s)
N_s	salt flux (kg/m^2s)
N_w	pure water flux (kg/m^2s)
NEA	nonequilibrium allowance
PR	performance ratio
P_1	atmospheric pressure (Pa)
P_2	operating pressure (Pa)
Q_y	yearly total RO plant capacity (m^3/year)
Q_{ref}	total production capacity for RO plant (m^3/day)
Rat	capacity ratio between production and reference
R_f	recovery fraction
sA	specific heat transfer area (m^2s/kg)
sM_{cw}	specific cooling water flow rate
T_o	top brine temperature (°C)
T_f	feed temperature (°C)
T_n	brine temperature in the last stage (°C)
T_s	steam temperature (°C)
T_v	condensing vapor temperature (°C)
U_b	overall heat transfer coefficient of brine heater (kW/m^2°C)
U_j	overall heat transfer coefficient of rejection section (kW/m^2°C)
U_r	overall heat transfer coefficient of recovery section (kW/m^2°C)
X_b	salinity of the rejected brine (ppm)
X_f	salinity of the intake seawater (ppm)
X_{Hb}	concentration of hybrid system (ppm)
X_{Hd}	concentration (salinity) of hybrid MSF/RO (ppm)
X_r	recycled brine concentration (ppm)
π_{cw}	osmotic pressure of feed seawater (Pa)
π_{cp}	osmotic pressure for permeate water (Pa)
ρ_{cw}	density of seawater (kg/m^3)
ρ_{cp}	density of permeate water (kg/m^3)
λ_{avg}	average latent heat (kJ/kg)

3.1 INTRODUCTION

Hybrid desalination systems are systems that combine two or more desalination technologies in order to achieve the comparative advantages of these technologies. Some hybrid systems combine systems that employ thermal distillation or membrane desalination, or both approaches. Some hybrid systems also produce electric power, in addition to pure water production [1]. Hybrid systems have been considered as an economic alternative to standalone systems in some studies because they are capable of reducing stress and pressure in terms of consumption of energy, scaling and fouling, and cost of desalinated water through the enhancements of the recovery rate and overall quality of water [2,3]. When hybrid systems were initially conceived and developed, they were only used for small-scale desalination. It took over two decades before they could be used in large-scale operations. The simplest application of the hybrid desalination systems is the elimination of the requirement for second passes in reverse osmosis (RO) process in such a way that brine management can be simplified by diluting the higher-salinity reject brine from the RO with a lower-salinity brine from another desalination technology, such as multistage flash (MSF) process.

Nowadays, the applications of hybrid systems have evolved and the concept of blending has allowed the emergence of simplified solutions that can ensure the production of pure water and rejection of byproducts that are able to meet local and international standard requirements. For example, the implementation of RO-MSF and RO-multieffect distillation (RO-MED) in desalination and power plants such as Ras Al-Khair in Saudi Arabia, Fujairah I and II in UAE, and Az-Zour in Kuwait has shown the feasibility and applicability of hybrid systems on a large scale [4]. The adoption of these systems has resulted in improving the efficiency of the desalination plants through the improvement of product water and brine quality. Moreover, the regions where these desalination plants are situated have recorded an increase in energy conservation and a reduction in the production costs incurred throughout the desalination processes [5]. Nair and Kumar [6] reported that the annual cost of producing 455,000 m^3/day of pure water and cogenerating 400-MW electric power using MSF technology stands at an excess of $735 million. Additionally, the amount of fuel consumption required to run the plant stands at an excess of 191 tons/h. However, a similar plant using RO-MSF hybrid system, where 40% of the feed is treated by RO and 60% by MSF, reduces the fuel consumption to 115 tons/h and the operational costs to $443 million annually [6]. The differences in these statistics show a reduction in the production cost and energy consumption when using hybrid systems. Also, hybrid systems can reduce the carbon footprint of desalination, by up to 40%, by further extending the life of RO membranes, thereby eradicating the occasional membrane replacement costs associated with high-salinity feeds to RO stages [7,8]. Therefore, recent studies have been able to show that hybrid systems can minimize wastage of available resources, as most of the byproducts can be recycled or reused, rather than disposing these products as wastes.

Desalination plays a very important role in the Middle East because drinking water is mainly sourced from saline water in this region [9]. In addition, most of the desalination plants in the region are cogeneration plants that produce both pure water and electric power. In the summer, water demand increases because of the arid ambient conditions and, consequently, the requirement of power also rises dramatically. However, as soon as the summer season is over, almost a drop of 30%–40% in the demand of power is observed as a result of the reduction in demand for drinking water but without a corresponding drop in total water demand. Therefore, as soon as summer is over, the overall water demand only reduces slightly because of other competing needs such as landscaping and irrigation. This implies that almost 50% of the generated power remains unutilized in the off-season, which is a drawback of cogeneration. But the silver lining is that, with hybrid systems, this excess generation of electricity in the off-season can be utilized for the production of water in a more sustainable way, since the overall water demand does not vary considerably [10].

During cooler seasons, a portion of the preheated (cooling) seawater leaving the heat rejection stage of the MSF distiller or the last effect of the MED can be used as feed water for RO plant. The remaining portion of the preheated seawater can be mixed with make-up water so as to reduce its temperature and recycled back to the heat rejection unit. The use of rejected cooling seawater from the MSF as RO seawater feed has been observed to increase the RO product water recovery by 40%–48% at a feed temperature of 33°C, when compared to an isolated RO plant using surface seawater as feed at a temperature of 15°C [11]. The higher temperature of the rejected cooling seawater, as compared with ambient seawater, reduces the viscosity of the feed water and increases the mass diffusivity of pure water across the RO membrane. This makes it possible for water flux enhancement across a RO membrane but, at the same, a slight decline in the rejection of salt by the membrane would be observed due to the slight enhancement of membrane salt permeability.

Therefore, the objective of this chapter is to discuss the recent advances in hybrid desalination technologies and the contribution of hybridization to sustainable desalination, cost-competitive, environmentally friendly and socially impactful desalination approaches. Recent progress has been achieved mostly in the aspects of RO-, forward osmosis- (FO-), MED-, and MSF-based hybrid systems. In addition, a case study has been presented and analyzed in this chapter to further confirm the observation of previous studies regarding the benefits of hybridization. In this case study, the technoeconomic and environmental impact assessments of hybridizing the MSF technology in an existing large-scale desalination plant—Al Taweelah A2 in Abu Dhabi Emirate of the UAE—with RO have been modeled. This special plant was chosen since its economic life is expected to end in 2021. The simulated MSF-RO hybrid plant is then compared with the existing MSF plant in terms of energy consumption, water cost, and CO_2 emission. The same input variables and parameters are used for the MSF technology in the existing plant and modeled case. This chapter attempts to fill in this gap by presenting a detailed model using Microsoft Excel and including detailed mass and energy balance and cost equations for standalone and hybrid systems.

3.2 HYBRID DESALINATION SYSTEMS

3.2.1 Hybrid RO Desalination Systems

There has been a surge in RO desalination recently because it is considered as an energy-efficient and cost-effective alternative to thermal distillation for large-scale desalination plant. Hybrid RO systems have also been explored to address some challenges that are still associated with RO, such as membrane fouling, brine treatment, and feed pretreatment. The most commonly studied hybrid RO systems are RO-pressure retarded osmosis (RO-PRO), RO-multistage flash (RO-MSF), RO-electrodialysis (RO-ED), RO-capacitive deionization (RO-CDI), and RO-reverse electrodialysis (RO-RED). RO-PRO hybrid combination for water desalination has shown good prospects for RO brine management. PRO separates water from a pressurized concentrated solution to a less concentrated solution through osmotic gradient and produces electric power from the hydraulic head obtained, as fresh water leaves the pressurized solution [12]. The cost of seawater desalination can be effectively reduced when the economic value of the energy produced is accounted for. In comparison to standalone PRO, energy production from the hybrid RO-PRO is enhanced because the denser brine generated from the RO process can be recycled back to the concentrated solution (i.e., seawater) to improve the salinity gradient across the PRO membrane [13]. In comparison to standalone RO, the RO-PRO system avails the opportunity of eliminating the adverse effect of brine disposal to the sea and natural habitat and removes the requirement of additional spending on excess brine management.

In a recent model proposed for the RO-PRO hybrid desalination, it was observed that the integrated approach can provide 50% water recovery rate and would help in reducing energy consumption by 65% when compared to RO standalone systems [14]. However, some recent studies have concluded that such PRO systems may not be economically viable because the energy required to pump the high-salinity feed to the PRO might be significantly higher than the energy gained from the hydraulic head and the energy credit allocated to less brine management. Membrane fouling resulting from the high salinity of the recirculated brine in RO-PRO hybrid systems is another barrier to the process economics. Another variation of RO-PRO hybrid desalination system aimed at the economic feasibility aspect of the system has been proposed by Kim et al. [12]. The specific energy consumption (SEC) and extent of fouling of the PRO membrane were evaluated through wastewater reclamation and seawater desalination tests. Water flux decline was observed after some time and it was observed that membrane fouling contributed significantly to the cost of water treatment. Antiscaling pretreatment was proposed as a membrane-fouling mitigation measure. This pretreatment was effective for the prevention of inorganic scaling [12]. Therefore, low-fouling, low-cost and energy-efficient pumping of brine might make the RO-PRO hybrid desalination system to be feasible.

RO-MSF has also been employed to manage the brine produced from RO. Helal et al. [15] carried out an optimization study for nine scenarios aimed at estimating the minimum cost of water that can be obtained from RO-MSF hybrid system, based on theoretical Modeling and experimental verification [15]. It was observed that RO is preferred for the desalination of feed containing low salt concentrations, especially for the desalination of brackish water to minimize fouling constraints and enhance energy efficiency. The RO-MSF system was found to be economically viable for the treatment of concentrate obtained from the RO desalination of low-salinity feed [16].

The interest in electrically induced desalination via ED has been growing steadily in recent times. RO-ED hybrid system has availed multiple benefits; chief among them is the ability to achieve high water recovery from ED through the RO treatment of ED concentrate. ED is normally used to desalinate dilute streams or feeds with low conductivity in order to minimize the huge electric power requirements of this technology [17]. An ED system that was counterflowed with RO has been used by McGovern et al. [18] to concentrate 120-ppt saline feed. Operating conditions such as current density was optimized so that the optimum values of output variables such as energy efficiency and cost of water (LCW) can be obtained. The current density of the hybrid system that is required to minimize LCW is greater than that required to maximize efficiency. This indicates that the influence of ED properties on the output variables is higher, relative to the properties of RO unit per unit membrane area [18]. At higher salinity, the dilute and concentrate resistances were found to be dominated by the membrane resistance. It was found that the performance is limited at high salinity because of higher concentration resistances, increase in water transport through osmosis, steady decline in efficiency, and increase in the LCW.

Hybrid RO-CDI desalination system has also been explored as a possible replacement for standalone RO desalination system. An integrated process that combines RO with constant-current-operated capacitive deionization (CCOCD) or constant-voltage-operated capacitive deionization (CVOCD) has been used recently to produce ultrapure water of high quality from seawater. At the same feed flow rates and concentration, the RO-CCOCD hybrid system has more advantages than RO-CVOCD, such as longer absorption times for the CDI cells when the spacer volume/dead volume and the capacitance is the same as that of CVOCD [19]. Meanwhile, the specific energy consumption is same for both RO-CVOCD and RO-CCOCD, when desalted water is produced at the same feed flow rate and concentration. Therefore, energy consumption is the same irrespective of whether a constant current or voltage is maintained. The combination of RO and CDI in RO-CCOCD has also been shown to increase pure water productivity. In addition to this, ultrapure water with a very low resistivity (18 MΩ cm) can be produced from the hybrid system due to the influence of the constant-current electric field; thus, the combination can be referred to as ultrapure water treatment (UPW) [19].

RO-RED is one of the newer concepts in hybrid desalination systems. The RED unit of the hybrid desalination system harvests energy from the salinity gradient that is present between a low-salinity solution (e.g., biologically treated secondary effluent or impaired water) and a highly concentrated solution (e.g., seawater or concentrated brine). The high-salinity solution, which has been RED-treated, serves as the RO unit's feed solution because the solution becomes diluted after RED treatment. By using the diluted stream as the RO feed, the pump work required to transport the stream to the RO unit becomes reduced. The concentrate from the RO unit can then serve as a source of energy recovery because it can be recirculated back to the RED unit to maintain high-salinity gradient. Additionally, the RED unit provides a platform for controlling the brine's concentration in such a way that water recovery can be optimized [20].

Some features of RO-based hybrid desalination systems such as membrane water flux, salt rejection, energy consumption, and cost, which have been reported in recent studies, are summarized in Table 3.1.

Table 3.1 RO hybrid systems

System	Feed water	Salt rejection	Flux
RO-membrane distillation (MD)-ED	Aqueous electrolyte solutions used in SGP-RE unit consisted of 0.1 $MK_3Fe(CN)_6$, 0.1 $MK_4Fe(CN)_6 \cdot 3H_2O$, NaCl (2.5 M) with conductivity 261.2 mS/cm	Water recovery of 92%	Transmembrane flux in the range of 1.2–2.4 $kg/m^2 h$
RO-MSF	The RO cost of water increases progressively according to feed concentration. It changes from \$0.56/m^3 at a feed desalination plants with common intake/outfall concentration of 20,000 ppm to \$0.91/m^3 at facilities 50,000 ppm	The reject brine concentration from the first stage in a two-stage RO plant should be <67,000 ppm	Not specified
RO-PRO	RO feed flow rates of 167, 195, 234, and 293 m^3/day were used for RO recoveries of 30%, 40%, 50% and 60%, respectively	Membrane salt rejection for PRO model results in 99.4%	Not specified, though formulae of salt and water flux are considered in the chapter
RO-RED	High-salinity feed solution is assumed to be seawater with a concentration of 0.57 mol/L in this chapter	Performance is limited at high recoveries because of differences in high stream-to-stream concentration, increase in transport of water through osmosis, steady decline in efficiency	Not specified

Performance			
Energy consumption	Cost	Remarks	Ref.
The demand of energy, theoretically, to produce 1 m^3 of potable water from 2 m^3 of seawater (50% recovery factor) is 1.1 kWh	Depending on the membrane could be higher than the current average cost of standalone systems or lower	The system currently shows overall consumption of 3–4 kWh/m^3 energy of desalted water which is lower in comparison to thermal systems	[21]
The energy required for pumping in the MSF plant is $0.03/kWh and energy in SWRO plant is $0.03/kWh	Not specified but theoretically costs are expected to exceed the standalone MSF process	Water cost can be reduced by 17%–24% of that of the sole MSF process. Meanwhile, water recovery can be doubled	[16]
Reduces the consumption of energy by 65% in compared to RO standalone systems. The RO-PRO system as a whole can achieve energy reduction of 40% in theory	Not specified but theoretically costs are expected to exceed the standalone RO and PRO processes and systems	Not specified but the study states that the brine would definitely be lower than either of the standalone system	[12]
Typical specific energy consumption of seawater reverse osmosis (SWRO) is 3.0 kWh/m^3	The system provided huge benefits by combining two simple process of RO and RED but the cost of the hybrid RO-RED desalination system is higher than both of the standalone systems		[20]

Continued

Table 3.1 RO hybrid systems—cont'd

System	Feed water	Salt rejection	Flux
RO-CDI	Not specified as the study considers different scenarios but feed water concentrations seawater salinity ranges from 25,000 to 45,000 ppm	Not specified but speculated to be lower that standalone systems of RO and CDI	Not specified
RO-ED	For the entire system the feed salinity is set at 120 ppt	The RO membrane is assumed to have perfect salt rejection and a permeability of 0.31 L/(m^2 h bar)	The flux of the RO system modeled in the study is an average permeate flux of 13 L/m^2 h

Performance			
Energy consumption	**Cost**	**Remarks**	**Ref.**
The energy consumption of RO-CCOCD appears to be similar to that of RO-CVOCD, despite the superior quality and amount of UPW produced	Not specified	The production of UPW from the total product water by both the RO-CDI hybrid desalination systems, RO-CCOD and RO-CVOD, is >38% and 5%, respectively. The RO-CCOCD system is superior in terms of using seawater to produce fresh/ UPW with a consumption of energy that is reasonable	[19]
The study considers three cases ED A, ED B and RO. The consumption of energy are 43, 21, and 6.1 kWh/m^3, respectively	The cost for LCW for ED A and ED B is $13 and 4.3/$m^3$	The performance is limited at high recoveries because of differences in high stream-to-stream concentration, increase in transport of water through osmosis, steady decline in efficiency	[18]

3.2.2 Hybrid FO Desalination Systems

The development of hybrid forward osmosis (FO) desalination systems began in the early 1960s [22]. The main purpose for the conception of these hybrid systems was the recovery of the draw solution. Studies in recent times have led to the integration of FO with other desalination technologies so that the efficiency of draw recovery and regeneration can be replaced [23]. Water is recovered from the diluted draw solution in the second process, such that the overall water recovery is enhanced. Hybrid FO desalination systems depend highly on the nature of the feed water and draw solution [24]. The recent FO-based hybrid desalination systems include FO-RO, FO-low pressure retarded osmosis (FO-LPRO), FO-ED, FO-membrane distillation (MD), FO-MSF, and FO-nanofiltration (NF).

It has been shown that higher water recovery can be achieved through FO-RO when compared with standalone RO process [25]. However, the energy consumption of a hybrid FO-RO process is higher than that of RO standalone process due to the energy requirements for draw solution regeneration, thereby leading to an increase in the overall cost, unless the environmental impacts of the recovery of the renewable osmotic energy from brine are internalized. Therefore, a crucial factor in FO-RO hybrid system is the selection of an appropriate draw solution that can be regenerated through energy-efficient means so that the FO-RO process can be competitive to the conventional RO process [26]. The productivity of the FO-RO hybrid system is higher at higher feed salinities due to higher osmotic energy. However, an increase in the concentration of the draw solution to reach a high recovery rate might lead to fouling and concentration polarization problems, depending on the nature of the FO membrane [26].

FO-LPRO hybrid desalination system combines FO with LPRO. FO-LPRO has been used to desalinate seawater of the red sea and it was found that the hybrid system consumed around 50% less energy than the conventional RO system, i.e., the FO-LPRO hybrid system consumed about 1.5 kWh/m^3, which was 50% of the energy used in seawater RO (SWRO) [27]. However, the total cost of the hybrid system was higher due to the high cost of the FO unit. In order to make the hybrid system become more profitable than the standard RO system with ultrafiltration pretreatment (UF), a specific flux of the FO module would be needed. After analyzing the life-cycle cost of the system, it was observed that FO-LPRO could provide a 16% cost reduction compared to the conventional SWRO if the cost of operating the FO module is reduced by maintaining high flux. The cost of the FO-LPRO hybrid system can be controlled, depending on the operating conditions. Meanwhile, a minimum average of $5.5\,L/m^2h$ FO flux would be required to desalinate seawater at a cost that is less than that of SWRO. Also, a minimum FO flux of $10.52\,L/m^2h$ would be required to ensure that the FO-LPRO becomes more economically viable than the UF-LPRO technology [27].

The concentrated brine fed to the FO system as draw solution can also be obtained from ED in the FO-ED hybrid system [28]. For this system, it is important to optimize the balance between the salinity of the draw solution and the diluted draw solution so that maximum volume of pure water can be obtained with minimum use of energy. FO-ED desalination is an energy-efficient process when ED is powered by solar energy. The cost of the system is about EUR3.32–4.92/m^3 for a small-sized potable water production system but the ED unit would be unsuitable to desalinate highly saline water [28].

FO-MD is another hybrid desalination system that has been tested for small-scale water production, e.g., small-scale decentralized sewer mining [29]. In the work of Wang et al. [29], a laboratory-scale FO-MD system was used for realizing a stable water flux by continuously feeding raw sewage to the feed side of the FO unit, such that water recovery of up to 80% was achieved. Excellent removal of trace organic contaminants (TrOCs), up to 98%, was achieved. It was further demonstrated that the accumulation of TrOCs in the draw solution can be prevented by carbon adsorption via granular-activated or ultraviolet oxidation so that a near-complete rejection (>99.5%) of TrOCs is achieved. In FO-MD system, MD can be used for the regeneration of the draw solution and production of pure water at the same time [29]. Some of the advantages of this hybrid include high salt rejection of MD without the requirement of high pressure or high temperatures, and high flux of FO because of the high temperature of the recycled draw solution. However, one of the main problems associated with this hybrid is the reverse salt flux, which reduces the osmotic driving force in the FO membrane and increases the rate of membrane scaling and the cost of the process. Meanwhile, hydroacid complex (Na_5Fe-CA) has been presented recently as a suitable draw solution because it has desirable properties such as thermal stability (suitable for MD) and low reverse flux at the high temperatures of MD [29]. These properties enhance the performance of FO for seawater or brackish water desalination, such that high flux and low fouling can be achieved.

The draw solution for the FO process can also be sourced from the reject brine from MSF. In the FO-MSF hybrid desalination system, FO is used as the pretreatment technique, i.e., through dilution, to reduce the concentration of the ionic species that can cause scaling in MSF. Therefore, FO-MSF can ensure the reduction of the cost of desalination. By comparing FO-MSF with NF-MSF system, it has been observed that less power consumption can be achieved by FO-MSF, making FO an attractive pretreatment system [25]. With FO as the pretreatment method, the use of antiscalants can be minimized and operations can be carried out at elevated brine temperature in order to increase membrane performance and recovery rate [25]. However, at high salinities, NF gives a higher recovery rate. When compared to RO as a pretreatment method, NF has also shown some comparative advantages over RO, such as lower hydraulic pressure, low fouling, high flux recovery, and less chemical cleaning requirements, thereby leading to a reduction in power consumption [30]. Table 3.2 shows the features and performance of recent FO-based hybrid desalination systems.

3.2.3 Hybrid MED Desalination Systems

Multieffect distillation (MED) is one of the well-established desalination technologies. There are many hybrid desalination systems that now use the MED process. A novel hybrid desalination system that combines the processes of MED and MSF has shown that this integrated system can reduce the pumping power required by 58% in comparison to MSF-BR and 16% in comparison to MED-TVC. MED-MSF is particularly suited for Gulf Cooperation Council (GCC) countries because a considerable number of plants that employ MED and MSF are still available in these countries [31]. The integration of MED and adsorption desalination (AD) cycle is another solution proposed to strengthen the thermally driven MED process. AD cycle is a low-cost desalination system that requires a low-temperature waste heat or solar energy as the source of thermal energy. In the conventional MED systems, the low brine temperature (LBT) is controlled by ambient conditions (40°C) and the operational temperature range (between 40°C and 65°C) limits the number of stages. The introduction of AD system allows the use of a larger number of stages. A larger number of stages mean that the system can achieve higher fresh water recovery and improved performance. A parallel-feed MED integrated with AD cycle has been able to increase the water production of a regular MED cycle by more than three times at the same top brine temperature [32]. Also, the MED-AD hybrid system was able to reduce corrosion associated with the regular MED.

In addition, life-cycle cost of the hybrid MED-AD cycle was found to be around \$0.485/m^3, which is even lower than the life-cycle cost of the RO process.

The viability of combining MED with two other desalination technologies, in such a way that the hybrid system consists of three technologies having mutual advantages, has also been studied recently. A mathematical model aimed at integrating three desalination techniques—MED, TVC, and RO—by using the laws of conservation of energy and concentration has shown the feasibility of this approach for enhanced productivity and cost reduction [33]. Different thermodynamic properties including the gained output ratio (GOR) were evaluated at various points of the modeled process. An exergy analysis was performed for the RO plant in order to investigate its contribution to the hybrid system. The study also performed a computational model based on the mechanisms of diffusion and convection transport along with the concept of concentration polarization in order to predict RO membrane's performance. The ACWA Power SASAKURA Company in collaboration with Saline Water Desalination Research Institute (SWDRI) and Water Reuse Promotion Center (WRPC) of Japan has developed a hybrid desalination system that combines three processes—MED, NF, and RO [34]. The NF process reduces the TDS of the feed and removes the scale components from the feed. The brine rejected from the RO process is reused as feed for the MED process of the system. The features and performance of recent hybrid MED-based systems are shown in Table 3.3.

3.2.4 Hybrid MSF Desalination Systems

Apart from RO-based, FO-based and MED-based systems, many of the hybrid desalination systems also use MSF combined with other processes. The hybrid systems that utilize MSF are mainly MSF-ED, MSF-NF, and MSF-RO systems. MSF-ED system has been explored for brine management in such a way that the seawater is initially treated via ED and then MSF and crystallization to ensure improved productivity of desalinated water and enhanced salt rejection [35]. A study in 2015 aimed at reviewing and analyzing all the technologies and processes that utilize the NF process has indicated that MSF can be integrated with NF or RO for higher yield of pure water [36]. Several studies have been carried out on MSF/RO hybrid systems with different design configurations [37–39]. Cardona et al. [40] proposed an optimization methodology for the design of MSF-RO plants in order to maximize overall profitability of plant operation. The optimization parameters are extraction pressure (or temperature) of steam

Table 3.2 FO hybrid systems

System	Draw solution	Feed waster	Salt rejection	Flux
FO-LPRO	Red Sea seawater (40,500 ppm as TDS)	Secondary wastewater (TDS: 2430 ppm)	98%	Certain minimum FO fluxes are needed to make it a competitive option
FO-MSF	Brine reject from MSF plant operated by BR mode (TDS 50% higher than seawater)	Seawater (32,000–50,000 ppm)	FO > 95%	FO flux increased with increasing salinity of seawater (8.5 L/m^2h for 50,000 ppm)
FO-RO	NaCl and $MgCl_2$	Seawater	Permeate TDS increased with increasing concentration of draw solution	Maximum recovery rate can be increased with increasing concentration of draw solution at the expense of higher power consumption. But in general, higher recovery rates (which will give higher water flux) can be reached in FO-RO than in RO process
FO-ED	NaCl	Secondary wastewater effluent or brackish water	High quality water meeting potable water standards. Low concentration of total organic carbon (TOC), carbonate, and cations from feed water	Not specified
FO-MD	Hydro-acid complex (Na_5Fe-CA)	Seawater and brackish water	99.99%	High water flux and negligible reverse salt flux
FO-NF	Divalent salt (Na_2SO_4 or $MgSO_4$) with high osmotic pressure	Brackish desalination	Up to 98%	>90%

Performance			
Energy consumption	**Cost**	**Remarks**	**Ref.**
Consumes 50% (approx. 1.5 kWh/m^3) of the energy used in SWRO	It is needed a minimum average of 5.5 L/m^{2}h FO flux to recover and desalinate water at less cost than SWRO and a minimum of 10.52 L/m^{2}h to replace the UF-LPRO technology	FO-LPRO has more benefits in terms of environmental impact	[27]
FO is more efficient than NF	Using the brine reject of MSF as draw solution in FO will reduce the operation cost of FO process	Potential of reducing antiscalant and operating at an elevated TBT. Increase in efficiency and recovery rate	[25]
Total power consumption is higher than a stand-alone RO process	Increasing the concentration of draw solution to reach a high recovery rate, increases the cost of desalination and power consumption	Efficiency was higher at higher seawater salinities and the selection of the draw solution is crucial. The simulated system did not take into consideration fouling problems	[25]
Energy efficient process when ED powered by solar energy	EUR3.32–4.92/m^3 for small size potable water production system	Unsuitable to desalinate high saline water. This system was driven by a renewable energy source (solar energy)	[28]
Not specified	Not specified	Low risk for membrane fouling. Not membrane wetting observed	[29]
It requires less hydraulic pressure than an RO system, causing a reduction in power consumption	High flux recovery is very high, which may reduce the chemical cleaning that is needed in the typical RO system	No pretreatment required	[30]

Table 3.3 MED hybrid systems

System	Feed water	Performance				Ref.
		Salt rejection	Energy consumption	Cost	Remarks	
MED-MSF	Not specified but brine and the same feed utilized for both systems	Speculated to be lower than both of the standalone systems	The power of pumping is 58% lesser in comparison to the MSF-BR and 16% lesser than that of MED-TVC	The cost of the water per unit is at its lowest value when the price of oil is lower than $40/bbl	Best suited for GCC nation	[31]
MED-AD	Sea water maintained at an ambient temperature to suit both the processes	Not specified	Ratios of power-to-water for exergy and energetic analyses are 95.7%:4.3% (exergy) and 72.2%:27.8% (energetic) methods, respectively	The projected water production LCC can be lowered to $0.485/m^3$		[32]
MED-TVC-RO	The feed water is not specified but the flow rate of the feed is taken as 31 kg/s	Rejection of salt not specified but the energy from the rejection of brine is speculated to be high and to have high exergy destruction	Not specified but the study mentions that the loss of energy will be lesser in comparison to standalone systems	Reduction of $0.5/m^3$ per unit during production because of the hybrid desalination system	Reduction of intake of seawater by 50%	[33]
MED-NF-RO	Brine rejected from the RO process is reused as feed for the MED process	Not specified but the brine will be reusable for other processes of the system	Not specified	Not specified	–	[34]

from turbine and the capacity ratio between the MSF and the RO sections. The optimization results concluded that model is suitable for desalination applications. In another study [41], MSF is fed by the brine of the RO and the permeate from RO is mixed with the distillate from MSF. The benefit of this hybridization is to reduce the cost of the product, i.e., fresh water. The cost reduction resulted from the decrease in the size of the pretreatment unit and the opportunity to use less expensive membranes because there was no need for the management of brine via membrane technology. The integrated configuration improves the thermal efficiency of the cogeneration plant.

The economic feasibility of MSF-RO plant has been further examined through the evaluation of seven different designs of hybrid desalination plants [15,16,42]. Initially, the cost of water production was estimated for standalone MSF and RO, including once-through MSF, MSF with brine recycling, and standalone two-stage SWRO. These standalone systems were compared with the different configurations of the hybrid systems. The lowest water production cost yields was achieved through the standalone two-stage RO plant. Meanwhile, the hybridization of the two-stage SWRO with MSF improved the economics and operation characteristics compared with standalone MSF processes. The properties and performance of recent MSF-based hybrid systems are summarized in Table 3.4.

3.3 CASE STUDY: TECHNOECONOMIC ASSESSMENT AND ENVIRONMENTAL IMPACTS OF HYBRIDIZATION FOR AL TAWEELAH A2 MSF PLANT IN ABU DHABI—UAE

Recently, the problem of water shortage reached the crisis level in the UAE due to the absence of surface water resources, regular rainfall, and renewable groundwater resources. The UAE is an arid land with an average summer temperature of 34.1°C and high humidity in coastal area [43]. As a result, there is an absence of permanently existing natural surface water such as rivers or lakes, limited groundwater resources, and water demand exceeding the availability from renewable resources [10]. Due to the economic development in UAE, significant changes were noticed in the country population, standard of living, and industrial development. Consequently, water consumption has increased dramatically to satisfy the requirement for irrigation, domestic, and industrial uses. Desalination is used to meet the UAE's water demand [11] through which 98% of the water supply comes from the desalination of seawater or brackish water [6]. UAE contributes around 14% to the global desalinated water capacity indicating that UAE is the world's second largest producer of desalinated water after Saudi Arabia.

Table 3.4 MSF hybrid systems

		Performance				
System	**Feed water**	**Salt rejection**	**Flux**	**Energy consumption**	**Cost**	**Ref.**
MSF-ED	Seawater	Water recovery 66.4% (73.8% for ED and 90% for the EDR stage)	Permeate with a TDS concentration around 10,000 ppm (380 ppm calcium, 1260 ppm magnesium, 4080 ppm chloride, and 2780 ppm sulfate)	The energy cost was assumed as \$0.06/kWh, the efficiency of pumps as 0.85, and the membranes' life as 10 years	Not specified but ED unit with 80% effective membrane surface MSF unit cost was equal to \$1.0/m^3	[35]
MSF-NF	Not specified, discharged concentrated brine was used as feed for crystallization	Fresh water production increased by 19%	Not specified	Not specified	Their water production costs were about \$0.71 and 0.43/m^3. Subtracting the salt price of \$30/ton, the cost of fresh water production could be further lowered to \$0.37/m^3	[36]
MSF-RO	MSF is fed by the brine of the RO	Salt passage decreased by an average of 2.2% Water recovery increased by 38% compared with the isolated systems	Not specified	The integrated configuration improves the thermal efficiency of the cogeneration plant The specific electrical energy consumption reduced by 25%	Reduction in both the capital and operating costs of the RO product water	[40]

The average water consumption in the UAE is as high as 1.16 m^3 a day per individual [44]. By region, Abu Dhabi city is responsible for 54.7% of the total domestic water consumption, followed by Al Ain and Al Gharbia regions with 33.6% and 11.7%, respectively [44]. MSF, RO, and MED processes are the most commonly used technologies in the UAE. Currently, RO technology represents the leading world market, since it has lower specific energy consumption, as compared with other technologies. However, MSF and MED are still being used in the UAE because of the poor quality and the high salinity of Gulf seawater. 63% of the desalinated water in the UAE is produced via MSF while MED accounts for about 6% of the total installed capacity. For instance, Jebel Dhana plant and Al Mirfa plant are MED-based plants with huge capacities of 18,189 and 9080 m^3/day, respectively [45]. Fig. 3.1 shows the distribution of different desalination technologies used in the UAE.

As shown in Fig. 3.1, RO accounts for 12% of the whole installed capacity. It is expected that RO technology will be the leading desalination technology in the near future due to the surging rate of the deployment of RO technology in the UAE. Hamriyah plant in Sharjah is an example of standalone RO plant with a production capacity of 90,920 m^3/day [46]. Also, RO plants are currently being integrated with other desalination technologies. For instance, a large-scale hybrid MSF-RO plant was installed in Fujairah (UAE), i.e., Fujairah I. It is one of the most important hybrid plants in the UAE and is one of the largest hybrid plants in the world with a water production of 454,600 m^3/day. It consists of five MSF units that contributes about 62.5% of the plant's production and the remaining 37.5%

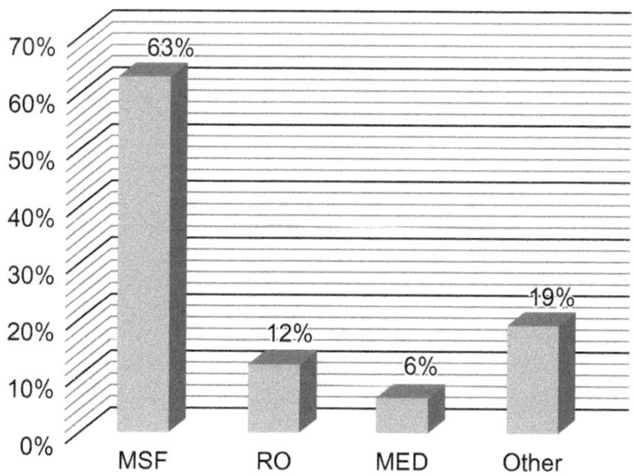

Fig. 3.1 The share (*y*-axis) of different desalination technologies (*x*-axis) implemented in UAE [45].

is contributed by seawater RO. Fujairah II is another large hybrid plant in the UAE, which comprises 100 MIGD MED and 30 MIGD SWRO [1,4]. For a typical unit size MSF plant with a production of 50,000–70,000 m^3/day, the total equivalent energy consumption ranges between 19.58 and 27.25 kWh/m^3. On the other hand, the energy consumption of the RO unit depends mainly on the salinity of the feed water and the recovery rate. The average reported energy consumption of RO ranges from 4 to 6 kWh/m^3 for a unit size with a production up to 128,000 m^3/day [47]. The estimated cost of desalinated water production can be determined by the plant size, site characteristics, and design features. Plant size depends on the process equipment, pumping units, and heating steam cost for thermal processes and the membrane area for membrane processes, while the pretreatment and posttreatment equipment and the chemical rates affect the site characteristics. The design features that influence the water production cost are electric power consumption, steam requirement, and chemicals [7]. The water production cost for MSF with production of 23,000–528,000 m^3/day ranges from \$0.56 to 1.75/m^3. RO plant produces water with less cost compared with MSF plant, it costs around \$0.45–0.66/m^3 for a production of 100,000–320,000 m^3/day [47].

Through economic necessity, a major driver of the planned and ongoing investments in desalination in Abu Dhabi is the development of less energy-intensive methods of desalinating water. The Middle East desalination industry is now focused on cheaper, simple and efficient, new technologies. Integrating MSF desalination plant with RO plant as a hybrid system is gaining importance for large-scale water production due to the flexibility of this integration, higher plant availability, and better match of power and water in cogeneration systems [48]. Therefore, the Al Taweelah A2 MSF plant in Abu Dhabi has been integrated with RO in this case study so that the technoeconomic and environmental impact of this integration can be assessed and evaluated.

3.3.1 Al Taweelah A2 Plant Case Study—Methodology

In this section, mass and energy balance equations were used for three different desalination plants configurations, MSF standalone, RO standalone and hybrid (MSF/RO). Also, economic analysis and environmental impact assessment were performed to predict the unit water production cost, energy consumption, and CO_2 emissions for the different desalination configurations. As mentioned earlier, the objective is first to evaluate the performance of Al Taweelah A2 MSF plant. Then, assume two alternative solutions for the existing plant after the end of its economic life, either to replace the MSF by RO or to integrate MSF with RO as a hybrid system. Al Taweelah A2 plant is located approximately 60 km northeast of Abu Dhabi city. The plant started to operate in 2001 with a concession period of 20 years. It contains a 720-MW combined cycle power plant and 50-MIGD MSF desalination plant. It consists of four Hanjung MSF distillers, where each has a net water production of 56,750 m^3/day, to fulfill the required 227,000 m^3/day net production from the station. The plant consists of 16 stages in the heat recovery section and 3 stages in the heat rejection section. The distiller type of MSF plants are conventional crosstube units with brine recirculation. It comprises 16 heat recovery and three heat rejection stages [49]. The subsequent sections detail the modeling equations, design input data, and assumptions of the three selected designs (i.e., MSF, RO, and hybrid MSF/RO).

3.3.1.1 Modeling, Economics, and Environmental Impact of MSF

The simplified model of MSF (Fig. 3.2) is used for obtaining quick design data and to evaluate the system performance. Although simplified, the model maintains the essential characteristics of accurate and more complex models. The model is based on several assumptions. These include linear temperature profiles of all streams flowing within the plant, constant value for heat transfer coefficient, heat transfer area, boiling point rise, and specific heat capacity of brine solution for each section in the plant, constant latent heat of vaporization of water (λ) and independent of temperature, specific heat capacity of brine solutions is independent of salt concentration, and distillate product is assumed to be salt free [50]. Table 3.5 summarizes

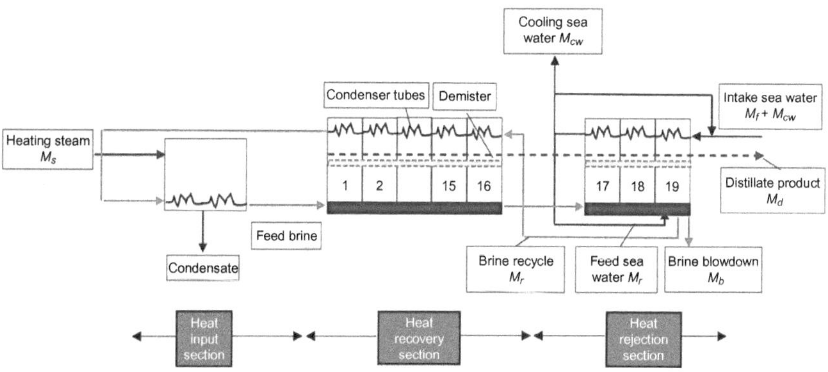

Fig. 3.2 MSF schematic diagram showing heat rejection, heat recovery, and heat input sections.

Table 3.5 Modeling equations of MSF process

Model	Equation	Notation and/or model assumption	Ref.
Overall material balance	$M_f = M_d + M_b$	M is mass flow rate, f, b, and d refer to feed, brine, and distillate, respectively	[50]
Overall salt balance	$X_f M_f = X_b M_b$	X is the salt concentration, and distillate is salt free (i.e., $X_d = 0$)	[50]
Temperature drop per stage	$\Delta T = \frac{T_o - T_n}{n}$	n is the number of recovery and rejection stages	[50]
Temperature of stage i	$T_i = T_o - i\Delta T$	–	[50]
Condenser temperature in stage i	$T_{ri} = T_n + (n - j)\Delta T - (i - 1)\Delta T$	–	[50]
Stage material and salt balance	$D_1 = \gamma M_r$ $\gamma = \frac{C_p \Delta T}{\lambda_{avg}}$ $T_{avg} = \frac{T_o - T_n}{2}$ $M_d = M_r(1 - (1 - \gamma)^n)$	D_1 is the amount of flashing vapor formed in the first stage, M_r is the recycle brine flow rate, C_p is the specific heat capacity, and γ is the specific ratio of sensible heat and latent heat (λ_{avg})	[50]
Steam flow from the brine heater energy balance	$M_s \lambda_s = M_r Cp(T_o - T_{f1})$	T_{f1} and M_s refer to temperature leaving the last condenser and entering the brine heater, and steam flow rate, respectively	[50]

Brine heat transfer area	$A_b = \frac{M_s \lambda_s}{U_b (LMTD)_b}$ $(LMTD)_b = \frac{(T_s - T_o) - (T_s - T_{r1})}{\ln \frac{(T_s - T_o)}{(T_s - T_{r1})}}$ $U_b = 1.7194 + 3.2063 \times 10^{-3} T_s + 1.5971 \times 10^{-5} T_s^2 - 1.9918 \times 10^{-7} T_s^3$	U is the overall heat transfer coefficient, λ_s is the steam latent heat, and A_b is the brine heat area	[50]
Condenser heat transfer area in the first stage	$A_r = \frac{M_r C_p (T_{r1} - T_{r2})}{U_r (LMTD)_r}$ $(LMTD)_r = \frac{(T_{v1} - T_{r1}) - (T_{v1} - T_{r2})}{\ln \frac{(T_{v1} - T_{r1})}{(T_{v1} - T_{r2})}}$ $U_r = 1.7194 + 3.2063 \times 10^{-3} T_{v1} + 1.5971 \times 10^{-5} T_{v12} - 1.9918 \times 10^{-7} T_{v13}$ $T_{v1} = T_1 - BPE_1 - NEA_1$ $BPE = X_r(B + (X_r C)) 10^{-3}$ $B = (6.71 + 6.34 \times 10^{-2} T_n + 9.74 \times 10^{-5} T_n^2) 10^{-3}$ $C = (22.238 + 9.59 \times 10^{-3} T_n + 9.42 \times 10^{-5} T_n^2) 10^{-8}$	BPE is the boiling point elevation, NEA is the nonequilibrium allowance ($NEA = 0.213$ and 1.217°C for heat recovery and heat rejection sections, respectively), T_v is the condensing vapor temperature, and U_r is the condenser overall heat transfer coefficient	[50]

Continued

Table 3.5 Modeling equations of MSF process—cont'd

Model	Equation	Notation and/or model assumption	Ref.
Condenser heat transfer area in heat rejection stages	$A_j = \frac{(M_f + M_r)C_p(T_{jn} - T_{cw})}{U_j(LMTD)_j}$ $(LMTD)_j = \frac{(T_{vn} - T_{cw}) - (T_{vn} - T_{jn})}{\ln\frac{(T_{vn} - T_{cw})}{(T_{vn} - T_{jn})}}$ $U_j = 1.7194 + 3.2063 \times 10^{-3}\, T_{vn} + 1.5971 \times 10^{-5}\, T_{vn}^2 - 1.9918 \times 10^{-7}\, T_{vn}^3$ $T_{vn} = T_n - (BPE)_n - (NEA)_n$	Heat transfer area for the condenser in each stage in the heat recovery section and heat rejection section are assumed equal	[50]
Total plant heat transfer area	$A_c = (n - j)A_r + (j)A_j$	–	[50]
Plant performance parameters	$PR = \frac{M_d}{M_s}$ $sA = \frac{A_b + A_c}{M_d}$ $sM_{cw} = \frac{M_{cw}}{M_d}$	PR, sA, and sM_{cw} are the plant performance ratio, the specific heat transfer area, and the specific cooling water flow rate, respectively	[50]
Annual fixed charges	$C_1 = a \times DC$ $DC = \frac{0.0963\psi\ A_{total}}{M_d^{0.27}}$	a is amortization factor, DC is the direct capital cost, ψ has value between 5000 and 9000 (in our cause 6000 was chosen), A_{total} is the total heat transfer area, and M_d is distillate flow rate	[50]
Annual heating steam cost	$C_2 = s\lambda f m 365/(1000\,PR)$	s is steam heating cost, λ is average latent heat of heating steam, f is plant availability, and m is plant capacity	[50]

Annual electric power cost	$C_3 = cwfm365$	c is electric cost and w is specific consumption of electric power	[50]
Annual chemicals cost	$C_4 = kfm365$	k is specific chemicals cost	[50]
Annual labor cost	$C_5 = \ell fm365$	ℓ is specific cost of operating labor	[50]
Total annual cost	$C_{A\text{-}MSF} = C_1 + C_2 + C_3 + C_4 + C_5$	–	[50]
Unit product cost	$C_{MSF} = \dfrac{C_{A\text{-}MSF}}{fm365}$	–	[50]
Total annual energy consumption	$E_{T\text{-}MSF} = E_{e\text{-}MSF} + E_{s\text{-}MSF}$ $E_{e\text{-}MSF} = \dfrac{C_3}{\text{Electricity tariff}}$ $E_{s\text{-}MSF} = \dfrac{C_2}{\text{Steam tariff}}$	$E_{e\text{-}MSF}$ *is* the annual electrical energy consumption and $E_{s\text{-}MSF}$ is annual equivalent electrical to thermal energy consumption	[51] [52]
Annual amount of CO_2 emissions	$A_{CO2\text{-}MSF} = E_{T\text{-}MSF}\,\text{GHG}$	GHG is greenhouse gas emission factor	[53]

MSF process model equations for material balance, salt balance, heat and energy, and cost and environmental analysis. Also, Tables 3.6 and 3.7 detail the design and economic data, which were used to model the MSF plant. Electricity tariff and energy cost were assumed as 0.08 $/kWh [51] and 0.023 $/kWh [52], respectively, which can also be subjected to variation as per the global energy price. The production capacity (as per the existing Al Taweelah A2 MSF plant) was set at 56,750 m^3/day.

Table 3.6 Design data model input of MSF plant

MSF design data	Symbol	Input	Unit
Number of stages	n	19	–
Number of heat recovery stages	i	16	–
Number of heat rejection stages	j	3	–
Seawater feed flow rate	M_f	1125	kg/s
Intake seawater temperature	T_f	25	°C
Steam temperature	T_s	116	°C
Top brine temperature	T_o	110	°C
Brine temperature in the last stage	T_n	35	°C
Heat capacity of liquid streams	C_p	4.18	kJ/kg
Salinity of intake seawater	X_f	30,000	ppm
Salinity of the brine blow down	X_b	70,000	ppm
Nonequilibrium allowance for heat recovery section	$(NEA)_r$	0.213	°C
Nonequilibrium allowance for heat rejection section	$(NEA)_j$	1.217	°C
Constant used to calculate direct capital cost	ψ	6000	–

Table 3.7 Economic data model input of MSF plant

MSF economic data	Input	Unit
Electric cost for MSF	0.05	$/m^3$
Average latent heat of heating steam	2222.3	kJ/kg
Steam heating cost	1.466×10^{-3}	$/kJ
Specific cost of operating labor for MSF	0.1	$/m^3$
Interest rate	5	%
Plant availability	0.9	–
Specific consumption of electric power for MSF	5	kWh/m^3
Specific chemicals cost for MSF	0.025	$/m^3$
Electricity tariff in Abu Dhabi in 2016	0.08	$/kWh
Steam tariff for desalination plants in Abu Dhabi	0.023	$/kWh
Greenhouse emission factor	0.0005	tCO_2/kWh

3.3.1.2 Modeling, Economics, and Environmental Impact of RO

The simplified RO (Fig. 3.3) model is based on several assumptions. These include constant mass transfer coefficient (k), constant water permeability (A_w), and independence of pressure in the RO stage; salt concentration at the membrane wall has a uniform bulk concentration and is equal to the feed seawater concentration (i.e., $C_w = C_f$). Table 3.8 summarizes RO process model equations for material balance, salt balance, heat and energy, and cost and environmental analysis. Also, Tables 3.9 and 3.10 detail the design and economic data, which were used to model the RO plant. Electricity tariff and energy cost were assumed as 0.08 $/kWh [51] and 0.023 $/kWh [52], respectively, which can also be subjected to variation as per the global energy price. The production capacity (as per the existing Al Taweelah A2 MSF plant) was set at 56,750 m^3/day.

3.3.1.3 Modeling, Economics, and Environmental Impact of Hybrid MSF/RO

The intake seawater in the proposed hybrid MSF/RO process (Fig. 3.4) is split between MSF and RO desalination plants at different split ratio (which is explained in the sensitivity analysis shown in Section 3.4.4). In this configuration, the permeates and brines from both plants were collected independently and mixed prior to collection or discharge, respectively. Cooling water (used in MSF plant) is rejected back to the sea. Table 3.11 summarizes hybrid MSF/RO process model equations for material balance, salt balance, heat and energy, and cost and environmental analysis. Also, the design and economic data, which were used to model the MSF and RO plants, were used in the case of hybrid MSF/RO process. Electricity tariff and energy cost were assumed as 0.08 $/kWh [51] and 0.023 $/kWh [52], respectively, which can also be subjected to variation as per the global energy price. The production capacity (as per the existing Al Taweelah A2 MSF plant) was set at 56,750 m^3/day.

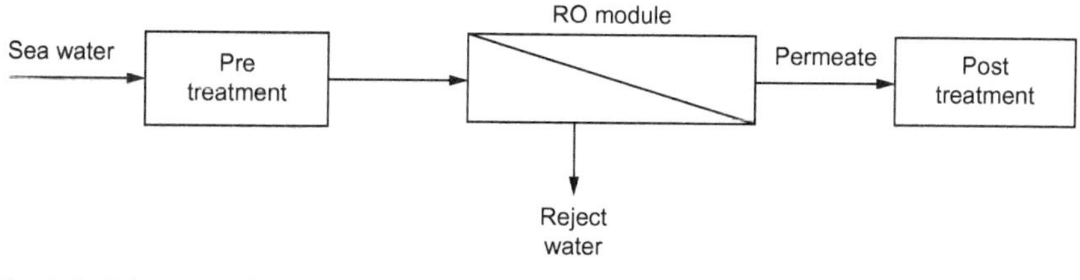

Fig. 3.3 Schematic diagram of RO process.

Table 3.8 Modeling equations of RO process

Model	Equation	Notation and/or model assumption	Ref.
Applied pressure	$\Delta P = P_2 - P_1$	P_1 defines the atmospheric pressure and P_2 defines the operated pressure	[54]
Osmotic pressure at feed and permeate sides	$\pi_{cw} = (0.6955 + 0.0025t)10^8 \frac{c_{cw}}{\rho_{cw}}$ $\rho_{cw} = 498.4\,m + (248{,}400\,m^2 + 752.4\,m\,c_{cw})^{½}$ $\pi_{cp} = (0.6955 + 0.0025t)10^8 \frac{c_{p1}}{\rho_{cp}}$ $\rho_{cp} = 498.4\,m + (248{,}400\,m^2 + 752.4\,m\,c_{p1})^{½}$ $m = 1.0069 - 2.757 \times 10^{-4}\,t$ $\Delta\pi = \pi_{cw} - \pi_{cp}$	π_{cw} is the osmotic pressure of feed seawater, t defines temperature, C_{cw} is the sea water concentration at membrane wall, C_f is the feed sea water concentration, ρ_{cw} is seawater density, π_{cp} is the osmotic pressure for permeate water, C_{p1} is permeate concentration as first assumption $C_{p1} = 1000\,\text{ppm}$, ρ_{cp} is permeate water density, m is constant, $\Delta\pi$ is the osmotic pressure difference	[54]
Water and salt flux	$N_w = A_w(\Delta P - \Delta\pi)$ $N_s = A_s(C_{cw} - C_{p1})$ $C_w = C_{p1} + (C_f - C_{p1})\exp^{\left[\frac{n_w}{1000k_s}\right]}$ $n_w = N_w \times \rho_{cp}$	N_w is pure water flux, N_s is the salt flux, A_w is the water permeability coefficient, A_s is the salt permeability coefficient, ΔP is the net applied pressure, $\Delta\pi$ is the net osmotic pressure Test if absolute $(C_{cw} - C_f) < 10^{-3}$, it means our assumptions are right ($C_{cw} = C_f$, $Cp_1 = 100\,\text{ppm}$), C_w is the new Salt concentration value at membrane wall, k_s is mass transfer coefficient n_w is the pure water flux	[54]
Permeate water flow rate	$F_p = A_{total} N_p 3600$ A_{total} = number of trains × number of modules × module area $N_p = N_w + N_S$	A_{total} is the total area of the membrane, N_p is the total product flux estimation	[54]
Reject brine flow rate	$F_{rj} = F_f - F_p$	F_f is the feed flow rate	[54]
Reject flow rate concentration	$C_{rj} = \frac{(F_f \times C_f) - (F_p \times C_p)}{F_{rj}} 1000$ $C_p = \frac{N_W}{N_S}$	C_p is the new permeate concentration	[54]
Yearly total plant capacity	$Q_y = F_p \frac{1}{\rho_{cp}} 24 \times 365$	–	[54]

Annual fixed charges	$C_1 = a \times DC$ $DC = 0.0963[C_{\text{membrane}} + (C_{\text{Civil}}(Rat)^{0.9}) + (C_{\text{Pump}}(Rat)^{0.7})$ $C_{\text{membrane}} = \text{cost of one module} \times \frac{\text{area of one module}}{\text{total membrane area}}$ $C_{\text{Civil}} = 2390(Q_{\text{ref}})^{0.8}$ $C_{\text{Pump}} = 0.0141 \frac{Q_{ref}\, 101.32P}{R_f}$ $Rat = \frac{24F_p}{\rho_{cp} Q_{ref}}$ $R_f = \frac{F_p}{F_f}$	a is amortization factor, DC is the direct capital cost, Rat is the capacity ratio between production and reference, Q_{ref} defines total production capacity	[50] [15]
Annual electric power cost	$C_2 = cwfm365$	c is the electric cost, w is the specific consumption of electric power, f defines plant availability, m is the plant capacity	[50]
Annual chemicals cost	$C_3 = kfm365$	k is specific chemicals cost	[50]
Annual membrane replacement cost	$C_4 = 10\%$ membrane purchase cost	Membrane purchase cost = 60% DC	[50]
Annual labor cost	$C_5 = \ell fm365$	ℓ is the specific cost of operating labor	[50]
Total annual cost	$C_{\text{A-RO}} = C_1 + C_2 + C_3 + C_4 + C_5$	–	[50]
Unit product cost	$C_{RO} = \frac{C_{A\ RO}}{fm365}$	–	[50]
Total annual energy consumption	$E_{\text{T-RO}} = E_{\text{e-RO}}$ $E_{e\ RO} = \frac{C_2}{\text{Electricity tariff}}$	$E_{\text{e-RO}}$ is the annual electrical energy consumption	[51]
Annual amount of CO_2 emissions	$A_{\text{CO2-RO}} = E_{\text{T-RO}}\,\text{GHG}$	GHG is the greenhouse emission factor	[53]

Table 3.9 Design data model input of RO plant

RO design data	Symbol	Input	Unit
Pure water permeability	A_W	2.06×10^{-12}	m/s Pa
Salt permeability	A_S	0.000000443	m/s Pa
Mass transfer coefficient	K_s	0.000059154	m/s
Feed temperature	T_{cw}	25	°C
Salinity of RO feed	C_f	30	kg/m^3
Feed flow rate	F_f	4.05×10^6	kg/h
Operating pressure	P_1	7.0×10^6	Pa
Atmospheric pressure	P_2	101,325	Pa
Number of trains	–	4	–
# of modules in each train	–	480	–
Area of one module	–	34.5	m^2

Table 3.10 Economic data model input of RO plant

RO economic data	Input	Unit
Electric cost for RO	0.04	\$/m^3
Specific cost of operating labor for RO	0.06	\$/m^3
Interest rate	5	%
Plant availability	0.9	–
Specific consumption of electric power for RO	3	kWh/m^3
Specific chemicals cost for RO	0.033	\$/m^3
Electricity tariff in Abu Dhabi in 2016	0.08	\$/kWh
Steam tariff for desalination plants in Abu Dhabi	0.023	\$/kWh
Greenhouse emission factor	0.0005	tCO_2/kWh

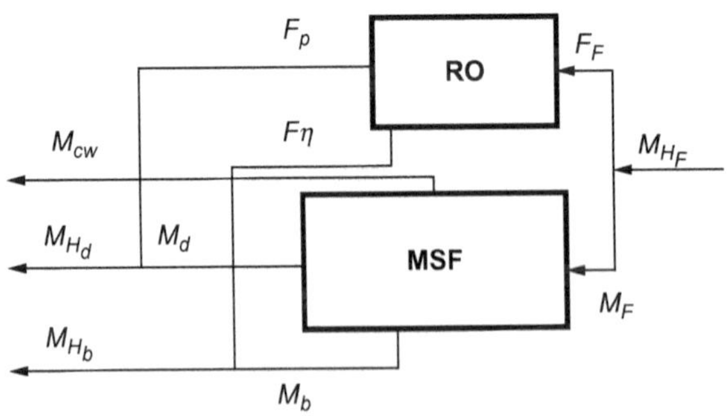

Fig. 3.4 Hybrid MSF/RO proposed in this study.

Table 3.11 Modeling equations of hybrid MSF/RO process

Model	Equation	Notation and/or model assumption	Ref.
Overall material balance	$MH_f = MH_d + M_{Hb} + M_{cw}$	M is the mass flow rate, H_f defines the total feed, H_d defines the total distillate, H_b defines the total brine	[15]
Mass balance for hybrid permeate	$M_{Hd} = M_d + F_p$	–	[15]
Salt balance	$X_{Hd} M_{Hd} = c_p F_P$	–	[15]
Overall salt balance	$X_{Hf} M_{Hf} = X_{Hd} M_{Hd} + X_{Hb} M_{Hb} + X_f M_{cw}$	–	[15]
Total brine of hybrid system	$M_{Hb} = M_b + F_{rj}$	–	[15]
Salinity of total rejected brine	$X_{H_b} = \frac{M_b X_b + F_{rj} c_{rj}}{M_{H_b}}$	–	[15]
Electromechanical equipment cost	$C_{H1} = 44{,}000 + 70\,F_{ref}$	F_{ref} is reference feed flow rate	[15]
Civil work cost	$C_{H2} = 150{,}000 + 375\,F_{ref}$	F_{ref} is reference feed flow rate	[15]
Electrochlorination	$C_{H3} = 22{,}000 + 12{,}000\,F_{ref}$	F_{ref} is reference feed flow rate	[15]
Brine disposal cost	$C_{H4} = 25{,}000 + 10\,B_{ref}$	B_{ref} is reference blow-down (brine) flow rate	[15]
Annual hybrid plant intake-outfall cost	$C_{A,Hybrid} = 0.0963\,C_{H\,total}(\text{feed ratio})^{0.9}$	–	[15]
Total annual cost of simple hybrid	$C_{total,Hybrid} = (C_{A,MSF} + C_{A,RO} + 1.1\,C_{A,Hybrid})$	–	[15]
Unit product cost	$C_{Hybrid} = \frac{C_{total,Hybrid}}{M_{H_d}}$	–	[15]
Total annual energy consumption for hybrid plant	$E_{Total\text{-}Hybrid} = E_{e\text{-}Hybrid} + E_{e\text{-}RO} + E_{e\text{-}MSF} + E_{s\text{-}Hybrid}$ $E_{e\ Hybrid} = \frac{C_{A,Hybrid}}{\text{Electricity tariff}}$ $E_{s\ Hybrid} = E_{s\ MSF} = \frac{C_2}{\text{Steam tariff}}$	$E_{e\text{-}Hybrid}$ is the annual electrical energy consumption for the hybrid and $E_{s\text{-}Hybrid}$ is annual equivalent electrical to thermal energy consumption	[51] [52]
Annual amount of CO_2 emissions	$A_{CO2\text{-}Hybrid} = E_{Total\text{-}Hybrid}\,GHG$	GHG is greenhouse emission factor	[53]

3.4 RESULTS AND DISCUSSION

The modeling results of the MSF, RO, and hybrid MSF/RO processes are presented in this section in terms of unit water cost, annual energy consumption, and annual CO_2 emission. Also, the developed models were able to estimate the blow-down brine flow rate and salinity, brine recycle flow rate and salinity, cooling water flow rate, condensers and brine heater heat transfer areas, and process performance ratio. Tables 3.12–3.16 show the model output, energy consumption, and economic analysis of the MSF, RO, and hybrid MSF/RO processes.

3.4.1 Energy Consumption of MSF, RO, and Hybrid MSF/RO

In general, desalination processes are energy intensive, yet they all share a common minimum energy requirement for the separation of water from salt regardless of the deployed technology or number of stages or membrane units. The concept of minimal energy is well established in thermodynamics

Table 3.12 MSF plant design model output

Model output	Symbol	Value	Unit
Blow-down brine flow rate	M_b	482.14	kg/s
Product water flow rate	M_d	642.86	kg/s
Brine recycle flow rate	M_r	5084.68	kg/s
Steam flow rate	M_s	113.26	kg/s
Cooling water flow rate	M_{cw}	4896.33	kg/s
Overall heat transfer coefficient for the brine heater	U_b	1.99	$kW/m^2 °C$
Brine heater area	A_b	11,608.36	m^2
Recycle brine concentration	X_r	61,149.88	ppm
Overall heat transfer coefficient for the heat recovery	U_r	2.00	$kW/m^2 °C$
Condenser area in the heat recovery section	A_r	5178.68	m^2
Overall heat transfer coefficient for the heat rejection	U_j	1.83	$kW/m^2 °C$
Condenser area in the heat rejection section	A_j	8816.91	m^2
Total condenser area	A_c	109,309.55	m^2
Total heat transfer area	A_{total}	12,017.90	m^2
Yearly total plant capacity	Q_y	20,336,587.7	m^3/year
Specific heat transfer area	sA	107.48	m^2 s/kg
Performance ratio	PR	5.68	–
Specific cooling water flow rate	sM_{cw}	7.61	–

Table 3.13 Economics, energy consumption, and gas emission model output of MSF rocess

Model output	Symbol	Value	Unit
Annual fixed charges	C_1	79,679.00	\$/year
Annual heating steam cost	C_2	10,472,528.75	\$/year
Annual electric power cost	C_3	3,649,165.71	\$/year
Annual chemical cost	C_4	456,145.71	\$/year
Annual labor cost	C_5	1,824,582.85	\$/year
Total annual cost	$C_{A\text{-}MSF}$	16,482,102.04	\$/year
Unit water production cost	C_{MSF}	0.90	\$/m^3
Electrical energy consumption	$E_{e\text{-}MSF}$	45,614,571.43	kWh/year
Equivalent electrical to thermal consumption	$E_{s\text{-}MSF}$	455,327,336.9	kWh/year
Total energy consumption	$E_{T\text{-}MSF}$	500,941,908.4	kWh/year
CO_2 emissions	$A_{CO2\text{-}MSF}$	250,470.9	tCO_2/year

Table 3.14 RO plant design model output

Model output	Symbol	Value	Unit
Permeate concentration	C_p	726.19	ppm
Final salt concentration at membrane wall	C_{cw}	35,261.75	ppm
Total product flux estimation	N_p	0.00960	m^3/m^2 s
Total membrane area	A_{total}	66,949.42	m^2
Permeate water flow rate	F_p	2,314,286	kg/h
Reject brine flow rate	F_{rj}	1,735,714	kg/h
Reject flow rate concentration	C_{rj}	69,031.75	ppm
Yearly total plant capacity	Q_Y	20,336,587.7	m^3/year

Table 3.15 Economics, energy consumption, and gas emission model output of RO process

Model output	Symbol	Value	Unit
Annual fixed charge	C_1	500,288.81	\$/year
Annual membrane replacement cost	C_2	461,439.91	\$/year
Annual electric power cost	C_3	3,649,166.16	\$/year
Annual chemical cost	C_4	602,112.42	\$/year
Annual labor cost	C_5	912,291.54	\$/year
Total annual cost	$C_{A\text{-}RO}$	6,125,298.85	\$/year
Unit water production cost	C_{RO}	0.34	\$/m^3
Electrical energy consumption	$E_{e\text{-}RO}$	45,614,577.06	kWh/year
CO_2 emissions	$A_{CO2\text{-}RO}$	22,807.3	tCO_2/year

Table 3.16 Plant design, economics, energy consumption, and gas emission model output of hybrid MSF/RO process

Model output	Symbol	Value	Unit
Total permeate flow rate	M_{Hd}	$20.3\times10^{+6}$	m^3/year
Salinity of permeate	X_{Hd}	580.95	ppm
Total brine flow rate	M_{Hb}	$1.7\times10^{+6}$	kg/h
Brine salinity	X_{Hb}	69,225.40	ppm
Reference feed flow rate	F_{ref}	1000	m^3/h
Reference brine flow rate	B_{ref}	750	m^3/h
Hybrid electromechanical equipment cost	CH_1	114,000	$
Hybrid civil work cost	CH_2	525,000	$
Hybrid electrochlorination	CH_3	12,022,000	$
Hybrid brine disposal cost	CH_4	32,500	$
Annual plant intake-outfall cost	$C_{Total\text{-}Hybrid}$	351,037.63	$/year
Total annual cost	C_{Hybrid}	8,608,929.874	$/year

through which the movement of solute is determined by fluctuations of thermal collisions with solvent molecules. The minimum work needed is equal to the difference in free energy between the feed seawater and both streams (i.e., permeate and concentrated brine). Using the van't Hoff equation for a typical seawater having salinity of 33,000 ppm at 25°C, the minimum work is 0.77 kWh/m^3 [55]. The actual work requirement is likely to be much higher than the minimum obtained from van's Hoff equation. This is attributed to the additional work required to maintain the process at a finite rate, rather than to achieve the separation. Presently, desalination plants account for 5–26 times as much work as the theoretical minimum. That is why there is a huge need to make desalination processes as energy efficient as possible by improving the technology and economies of scale. Different factors are impacting the energy requirements in desalination processes. These include plant capacity and plant design, feed seawater quality, and materials used. It should be noted that the energy consumption in thermal-based desalination processes such as MSF is not affected by the feed salinity (i.e., salt concentration in feed seawater), while salinity presents a major challenge in membrane-based desalination process such as RO (i.e., higher the feed salinity is, the higher the energy requirements are due to higher osmotic pressure). RO systems use electricity (which can be generated from fossil fuel (coal, oil, and gas), nuclear sources, and renewable energy) as a primary source of energy, while MSF systems use thermal energy (which can be produced from power-plant waste heat, fossil-fuel-fired boilers, industrial-waste heat sources, and renewable energy sources) as a primary source and electricity to drive associated pumps as a secondary source. It was reported in literature that MSF requires higher energy when compared to RO process (i.e., 19.58–27.25 kWh_e/m^3 of water for MSF contrary to 4–6 kWh_e/m^3 of water for the SWRO process with an energy recovery system) [47]. The reason for that is attributed to the high energy needed for water vaporization, and the continuous improvement in the membrane technology, which resulted in less energy requirements membrane with longer lifetime. Tables 3.13, 3.15, and 3.16 show that the annual energy consumption of MSF, RO, and hybrid MSF/RO plants was $50.0 \times 10^{+7}$ kWh (24.2 kWh/m^3), $4.5 \times 10^{+7}$ kWh (2.2 kWh/m^3), and $14.1 \times 10^{+7}$ (6.6 kWh/m^3) kWh, respectively. It is clear that the impact of MSF hybridization in Al Taweelah A2 MSF plant has resulted in significant reduction of annual energy consumption that is equivalent to 71.9% maintaining the same water production capacity of 56,750 m^3/day (20,713,750 m^3/year).

3.4.2 Economics and Water Cost of MSF, RO, and Hybrid MSF/RO

Feed seawater quality, energy and operation-, plant capital-, auxiliaries-, disposal-, and maintenance cost are key factors in the economics of desalination. The energy cost of thermal-based desalination plants such as MSF is approximately 60% of water production costs while it presents 44% in membrane-based desalination plants such as RO [7]. For example, it was reported that the water cost for MSF plants varied in the range of 0.52 and 1.75 $/m^3$, whereas RO plants reported a water production cost between 0.45 and 0.66, 0.48–1.62, and 0.7–1.72 $/m^3$ for large-, medium-, and small-sized plants, respectively [7]. Tables 3.13, 3.15, and 3.16 show that the unit water production of MSF, RO, and hybrid MSF/RO plants was 0.90, 0.34, and 0.42 $/m^3$, respectively. That said, the impact of MSF hybridization in Al Taweelah A2 MSF plant has resulted in significant reduction of the unit water cost from 0.90 to 0.42 $/m^3$ that is equivalent to 53.3% reduction. This could be explained by the fact that although thermal-based desalination processes such as MSF generate permeate with almost salt free compared to 400–500 ppm in RO systems, improved technology has led to developing membranes with longer lifetime, require less energy, and thus reduced cost [7].

3.4.3 Environmental Impacts of MSF, RO, and Hybrid MSF/RO

Three environmental impacts are associated with seawater desalination. These include the release of significant amount of greenhouse gases such as CO_2, the discharge of chemicals used in pretreatment, and the high temperature and large volumes of discharged concentrated brine. It was estimated that 25 kg of CO_2 is generated per cubic meter of product water [56]. It was also reported that the salinity of the discharged brine is 100% higher, at ambient temperature, than the feed seawater in RO plants, while 15% increase in the brine salinity was reported in MSF plants both at a temperature of 5–10°C higher than the ambient temperature [56]. Consequently, thermal desalination plants such as MSF have a greater adverse impact on the environment, mainly on the marine life and ecosystems through which the increase in temperature of the surface water would reduce the dissolved oxygen that is required for the growth of habitants. Moreover, the amounts of pretreatment chemicals are considered relatively low in RO when compared to MSF. Tables 3.13, 3.15, and 3.16 show that the annual CO_2 emission of MSF, RO, and hybrid MSF/RO plants was 250,470.9, 22,807.3, and 70,636.8 tons, respectively. Also, the concentrated brine was discharged at a flow rate and salinity of 482.14 kg/s (61,149.88 ppm), 482.12 kg/s (69,031.75 ppm), and 472.2 kg/s (69,225.40 ppm) for MSF, RO, and MSF/RO, respectively. That said, the impact of MSF hybridization in Al Taweelah A2 MSF plant has resulted in significant reduction of greenhouse gas emissions that are equivalent to 71.8%. Table 3.17 summarizes the results obtained for MSF, RO, and hybrid MSF/RO systems.

3.4.4 Sensitivity Analysis in Hybrid MSF/RO—Impact of Feed Seawater Flow Rate

As explained in the methodology, the split ratio in the hybrid MSF/RO was set at three different MSF: RO levels: 20%:80%, 40%:60%, and 50%:50%. The objective was to determine the optimal design (i.e., flow rate) of the hybrid MSF/RO process by varying the flow rate of the feed seawater that is pumped and treated in the RO process. Optimization was indicated by the minimum unit water production cost, annual energy consumption, and

Table 3.17 Summary of water production cost, annual energy consumption, and annual gas emission of MSF, RO, and hybrid MSF/RO systems

Desalination system	Unit water cost ($/m^3)	Total energy consumption (kWh/year)	Gas emission (tCO_2/year)
MSF	0.90	$500.9\times10^{+6}$	$250.5\times10^{+3}$
RO	0.34	$45.6\times10^{+6}$	$22.8\times10^{+3}$
Hybrid MSF/RO	0.42	$141.2\times10^{+6}$	$70.6\times10^{+3}$

Table 3.18 Sensitivity analysis showing the impact of RO feed flow rate

Feed MSF:RO ratio	Unit water cost ($/m^3)	Total energy consumption (kWh/year)	Gas emission (tCO_2/year)	No. of RO trains
20%:80%	0.42	$141.2\times10^{+6}$	70,636.8	3
40%:60%	0.53	$232.4\times10^{+6}$	116,200.9	2
50%:50%	0.58	$277.9\times10^{+6}$	138,983.1	2

gas emission. Table 3.18 shows that the unit water cost was estimated at 0.42, 0.53, and 0.58 $/m^3 at 20%:80%, 40%:60%, and 50%:50%, respectively. Similarly, the annual energy consumption and greenhouse gas emissions were $141.2\times10^{+6}$ kWh (70,636.8 tCO_2), $232.4\times10^{+6}$ kWh (116,200.9 tCO_2), and $277.9\times10^{+6}$ (138,983.1 tCO_2), at 20%:80%, 40%:60%, and 50%:50%, respectively. As discussed in Sections 3.4.1–3.4.3, it is deduced that the performance of the hybrid MSF/RO system was improved when the flow rate of feed seawater was increased. This is also supported by the fact that membrane-based desalination processes such as RO require less energy due to the rapid developments of long lifetime and efficient membranes, in addition to the low chemical requirements in the pretreatment stage in RO when compared to MSF.

3.5 FUTURE PROSPECTS OF DESALINATION TECHNOLOGIES

Today, hybrid desalination systems that combine both thermal and membrane desalination processes are becoming more economical because of their flexibility in operation, low energy consumption, low construction cost, high plant availability, and better power and water matching [1,58]. Today more concerns are being raised regarding the sustainability of desalination in terms of energy use and environmental impacts of different processes. Globally, desalination processes are producing around 76 million tons of carbon emissions. They have an efficiency rate of 10%–15%, yet are not considered sustainable. Thermally driven hybridization can achieve 20%–25% of process efficiency. Innovation in membrane technologies will still need another 5–10 years of intensive research to reach the desired level of sustainability. In addition, the increase in the use of renewable energy in desalination will also contribute to the mitigation effect of CO_2 emissions and to achieve the goals of COP of 21 [57].

3.6 CONCLUSION

Desalination continues to grow due to the increase in water scarcity, population growth, and industrial development. Recent desalination plants have improved in both system design and operation, which led to huge savings in power consumption and water production cost. Hybridization between desalination processes appeared to have a great impact and provide cost-effective, environmentally friendly, and innovative solutions. This chapter reviewed existing literature on different hybrid desalination technologies and also presented a modeling study via selecting Al Taweelah A2 MSF plant in Abu Dhabi (UAE) as a case study. Two possible scenarios were investigated: (1) replacement of thermal MSF plant with membrane RO plant, and (2) integration of RO process within MSF (i.e., hybrid MSF/RO). The results revealed significant reduction in energy consumption (71.9%), water production cost (53.3%), and annual gas emission (71.8%) (i.e., tCO_2/year).

ACKNOWLEDGMENT

The authors are grateful to Masdar Institute of Science and Technology, Abu Dhabi, United Arab Emirates, for the financial support (Grant No. SSG2015-000068) and for providing the platform where research activities on water desalination were collected.

REFERENCES

[1] Hamed OA. Overview of hybrid desalination systems—current status and future prospects. Desalination 2005;186:207–14. https://doi.org/10.1016/j.desal.2005.03.095.
[2] Fath H, Sadik A, Mezher T. Present and future trend in the production and energy consumption of desalinated water in GCC countries. Int J Therm Environ Eng 2013;5:155–65. https://doi.org/10.5383/ijtee.05.02.008.
[3] Mezher T, Fath H, Abbas Z, Khaled A. Techno-economic assessment and environmental impacts of desalination technologies. Desalination 2011;266:263–73. https://doi.org/10.1016/j.desal.2010.08.035.
[4] Almulla A, Hamad A, Gadalla M. Integrating hybrid systems with existing thermal desalination plants. Desalination 2005;174:171–92. https://doi.org/10.1016/j.desal.2004.08.041.
[5] Khawaji AD, Kutubkhanah IK, Wie JM. Advances in seawater desalination technologies. Desalination 2008;221:47–69. https://doi.org/10.1016/j.desal.2007.01.067.
[6] Nair M, Kumar D. Water desalination and challenges: the Middle East perspective: a review. Desalin Water Treat 2013;51:2030–40. https://doi.org/10.1080/19443994.2013.734483.
[7] Ghaffour N, Missimer TM, Amy GL. Technical review and evaluation of the economics of water desalination: current and future challenges for better water supply sustainability. Desalination 2013;309:197–207. https://doi.org/10.1016/j.desal.2012.10.015.
[8] Borsani R, Rebagliati S. Fundamentals and costing of MSF desalination plants and comparison with other technologies. Desalination 2005;182:29–37. https://doi.org/10.1016/j.desal.2005.03.007.

[9] Study C. Achieving sustainable watersheds. Innovation 2009;https://doi.org/10.1016/0304-3924(85)90001-2.
[10] Sommariva C, Syambabu VSN. Increase in water production in UAE. Desalination 2001;138:173–9. https://doi.org/10.1016/S0011-9164(01)00261-2.
[11] Saif O. The future outlook of desalination in the Gulf, UAE. 2012. 1–63. http://inweh.unu.edu/wp-content/uploads/2013/11/The-Future-Outlook-of-Desalination-in-the-Gulf.pdf.
[12] Kim DI, Kim J, Shon HK, Hong S. Pressure retarded osmosis (PRO) for integrating seawater desalination and wastewater reclamation: energy consumption and fouling. J Membr Sci 2015;483:34–41. https://doi.org/10.1016/j.memsci.2015.02.025.
[13] Kim J, Park M, Snyder SA, Kim JH. Reverse osmosis (RO) and pressure retarded osmosis (PRO) hybrid processes: model-based scenario study. Desalination 2013;322:121–30. https://doi.org/10.1016/j.desal.2013.05.010.
[14] Achilli A, Cath TY, Childress AE. Power generation with pressure retarded osmosis: an experimental and theoretical investigation. J Membr Sci 2009;343:42–52. https://doi.org/10.1016/j.memsci.2009.07.006.
[15] Helal AM, El-Nashar AM, Al-Katheeri E, Al-Malek S. Optimal design of hybrid RO/MSF desalination plants part I: modelling and algorithms. Desalination 2003;154:43–66. https://doi.org/10.1016/S0011-9164(03)00207-8.
[16] Helal AM. Optimal design of hybrid RO/MSF desalination plants part II: results and discussion. Desalination 2004;160:13–27. https://doi.org/10.1016/S0011-9164(04)90014-8.
[17] Pellegrino J, Gorman C, Richards L. A speculative hybrid reverse osmosis/electrodialysis unit operation. Desalination 2007;214:11–30. https://doi.org/10.1016/j.desal.2006.09.024.
[18] McGovern RK, Zubair SM, Lienhard V JH. In: Design and optimization of hybrid ED-RO systems for the treatment of highly saline brines. International Desalination Association World Conference 2013, Tianjin, China; 2013.
[19] Minhas MB, Jande YAC, Kim WS. Combined reverse osmosis and constant-current operated capacitive deionization system for seawater desalination. Desalination 2014;344:299–305. https://doi.org/10.1016/j.desal.2014.03.043.
[20] Li W, Krantz WB, Cornelissen ER, Post JW, Verliefde ARD, Tang CY. A novel hybrid process of reverse electrodialysis and reverse osmosis for low energy seawater desalination and brine management. Appl Energy 2013;104:592–602. https://doi.org/10.1016/j.apenergy.2012.11.064.
[21] Tufa RA, Curcio E, Brauns E, van Baak W, Fontananova E, Di Profio G. Membrane distillation and reverse Electrodialysis for near-zero liquid discharge and low energy seawater desalination. J Membr Sci 2015;496:325–33. https://doi.org/10.1016/j.memsci.2015.09.008.
[22] Zhang F, Brastad KS, He Z. Integrating forward osmosis into microbial fuel cells for wastewater treatment, water extraction and bioelectricity generation. Environ Sci Technol 2011;45:6690–6. https://doi.org/10.1021/es201505t.
[23] Zhang M, Hou D, She Q, Tang CY. Gypsum scaling in pressure retarded osmosis: experiments, mechanisms and implications. Water Res 2014;48:387–95. https://doi.org/10.1016/j.watres.2013.09.051.
[24] Shaffer DL, Werber JR, Jaramillo H, Lin S, Elimelech M. Forward osmosis: where are we now? Desalination 2015;356:271–84. https://doi.org/10.1016/j.desal.2014.10.031.
[25] Altaee A, Zaragoza G, Sharif A. Pressure retarded osmosis for power generation and seawater desalination: performance analysis. Desalination 2014;344:108–15. https://doi.org/10.1016/j.desal.2014.03.022.
[26] Zhang S, Wang KY, Chung TS, Chen H, Jean YC, Amy G. Well-constructed cellulose acetate membranes for forward osmosis: minimized internal concentration polarization with an ultra-thin selective layer. J Membr Sci 2010;360:522–35. https://doi.org/10.1016/j.memsci.2010.05.056.

[27] Yangali-Quintanilla V, Li Z, Valladares R, Li Q, Amy G. Indirect desalination of Red Sea water with forward osmosis and low pressure reverse osmosis for water reuse. Desalination 2011;280:160–6. https://doi.org/10.1016/j.desal.2011.06.066.
[28] Zhang Y, Pinoy L, Meesschaert B, Van Der Bruggen B. A natural driven membrane process for brackish and wastewater treatment: photovoltaic powered ED and FO hybrid system. Environ Sci Technol 2013;47:10548–55. https://doi.org/10.1021/es402534m.
[29] Wang P, Cui Y, Ge Q, Fern Tew T, Chung TS. Evaluation of hydroacid complex in the forward osmosis-membrane distillation (FO-MD) system for desalination. J Membr Sci 2015;494:1–7. https://doi.org/10.1016/j.memsci.2015.07.022.
[30] Zhao S, Zou L, Mulcahy D. Brackish water desalination by a hybrid forward osmosis-nanofiltration system using divalent draw solute. Desalination 2012;284: 175–81. https://doi.org/10.1016/j.desal.2011.08.053.
[31] Mabrouk AN, Fath HES. Technoeconomic study of a novel integrated thermal MSF–MED desalination technology. Desalination 2015;371:115–25. https://doi.org/10.1016/j.desal.2015.05.025.
[32] Shahzad MW, Ng KC, Thu K, Saha BB, Chun WG. Multi effect desalination and adsorption desalination (MEDAD): a hybrid desalination method. Appl Therm Eng 2014;72:289–97. https://doi.org/10.1016/j.applthermaleng.2014.03.064.
[33] Sadri S, Ameri M, Haghighi Khoshkhoo R. Multi-objective optimization of MED-TVC-RO hybrid desalination system based on the irreversibility concept. Desalination 2017;402:97–108. https://doi.org/10.1016/j.desal.2016.09.029.
[34] Tri-hybrid NF/RO/MED desalination system, ACWA Power SASAKURA; 2015. http://198.58.93.40/~acwapowr/?page_id=2281.
[35] Morillo J, Usero J, Rosado D, El Bakouri H, Riaza A, Bernaola FJ. Comparative study of brine management technologies for desalination plants. Desalination 2014;336:32–49. https://doi.org/10.1016/j.desal.2013.12.038.
[36] D. Zhou, L. Zhu, Y. Fu, M. Zhu, L. Xue, Development of lower cost seawater desalination processes using nano filtration technologies—a review, DES 376 (2015) 109–116. doi:https://doi.org/10.1016/j.desal.2015.08.020.
[37] Marcovecchio MG, Mussati SF, Scenna NJ, Aguirre PA. Optimization of hybrid desalination processes including multi stage flash and reverse osmosis systems, desalination. Trends Technol 2011;182:311–34.
[38] El-Sayed E, Ebrahim S, Al-Saffar A, Abdel-Jawad M. Pilot study of hybrid systems. Desalination 1998;120:121–8. https://doi.org/10.1016/S0011-9164(98)00209-4.
[39] El-Sayed E, Abdel-Jawad M, Ebrahim S, Al-Saffar A. Performance evaluation of two RO membrane configurations in a MSF/RO hybrid system. Desalination 2000;128: 231–45. https://doi.org/10.1016/S0011-9164(00)00038-2.
[40] Cardona E, Piacentino A. Optimal design of cogeneration plants for seawater desalination. Desalination 2004;166:411–26. https://doi.org/10.1016/j.desal.2004.06.096.
[41] Cardona E, Culotta S, Piacentino A. Energy saving with MSF-RO series desalination plants. Desalination 2003;153:167–71. https://doi.org/10.1016/S0011-9164(02)01121-9.
[42] Helal AM, El-Nashar AM, A1-Katheeri ES, Al-Maler SA. Optimal design of hybrid RO/MSF desalination plants part III: sensitivity analysis. Desalination 2004;169:43–60.
[43] Environmental Agency—Abu Dhabi. Environmental snapshot Abu Dhabi 2015, http://www.ead.ae/resources/publications/; 2015.
[44] Energy and water statistics. Abu Dhabi, UAE: Statistics Centre; 2015.
[45] H. Fath, Z. Abbas, A. Khaled, Techno-economic assessment and environmental impacts of desalination technologies, 266 (2011) 263–273. doi:https://doi.org/10.1016/j.desal.2010.08.035.

[46] Codemo G, Awadalla AAR, Parker MJ, Banham J, Rybar S. Hamriyah SWRO Desalination Plant—largest sea water IMS plant. The Palm—Dubai, UAE: IDA World Congress—Atlantis; 2009. November 7–12.

[47] Al-Karaghouli A, Kazmerski LL. Energy consumption and water production cost of conventional and renewable-energy-powered desalination processes. Renew Sust Energ Rev 2013;24:343–56. https://doi.org/10.1016/j.rser.2012.12.064.

[48] Andrianne J, Alardin F. Thermal and membrane process economics: optimized selection for seawater desalination. Desalination 2003;153:305–11. https://doi.org/10.1016/S0011-9164(02)01128-1.

[49] Wade N, Willis J, Mcsorley J. The Taweelah A2 independent water and power project. 125:1999;191–202.

[50] El-Dessouky HT, Ettouney HM. Fundamentals of salt water desalination. 1st ed. Elsevier; 2002.

[51] Kader BA. Abu Dhabi revises water and electricity tariff, http://gulfnews.com/news/uae/government/abu-dhabi-revises-water-and-electricity-tariff-1.1411802; 2014.

[52] ADWE. Statement of future capacity 2016–2022 revision. 2015.

[53] Environment Agency. Greenhouse gas inventory for Abu Dhabi Emirate. 2012. 27. http://www.ead.ae/wp-content/uploads/2014/03/AD-Greenhouse-gas-inventory-Eng.pdf.

[54] Al-Khudhiri AI. Optimal design of hybrid MSF/RO desalination plant. KSA: MSc Thesis, King Saud University; 2006.

[55] Lachish U. Osmosis and thermodynamics. Am J Phys 2007;75:https://doi.org/10.1119/1.2752822.

[56] Lattemann S, Höpner T. Environmental impact and impact assessment of seawater desalination. Desalination 2008;220:1–15. https://doi.org/10.1016/j.desal.2007.03.009.

[57] Shahzad MW, Burhan M, Ang L, Choon Ng K. Energy-water-environment nexus underpinning future desalination sustainability. Desalination 2017;413:52–64.

[58] Kamal I. Myth and reality of the hybrid desalination process. Desalination 2008;230:269–80.

CHAPTER 4

Trigeneration and Polygeneration Configurations for Desalination and Other Beneficial Processes

Francesco Calise*, Massimo Dentice d'Accadia*, Edoardo Quiriti*, Maria Vicidomini*, Antonio Piacentino†
*University of Naples Federico II, Naples, Italy
†University of Palermo, Palermo, Italy

NOMENCLATURE

A	area (m^2)
ACH	absorption chiller
AH	auxiliary heater
AHU	air handling units
BISTs	building integrated solar systems
BOP	balance of the plant
BP	biomass polygeneration
BPE	boiling point elevation (°C)
c	specific heat (J/kg/K)
C	cost (€)
$\dot{C}$	thermal capacity (J/s/K)
CDI	capacitance deionization
CRF	capital recovery factor
CHW	chilled water
COP	coefficient of performance
C_{op}	operating cost (€)
cp	heat capacity at constant pressure (kJ/kg°C)
CPVT	concentrating photovoltaic/thermal collectors
C_{PVT}	concentration ratio
CSP	concentrated solar power plant
CW	cooling water
D	diverter
DEC	Discrete Ericsson Cycle
DHW	domestic hot water
DW	desalinized water
e	specific energy consumption indicator
E	energy (kWh)
EES	engineering equation solver
EHP	electrical heat pump
EU	European Union

Sustainable Desalination Handbook
https://doi.org/10.1016/B978-0-12-809240-8.00004-6

Ex	exergy (kW_{ex})
$\dot{Ex}_{d,i}$	exergy flow rate
F_i	exergy fuel
FFM	falling film evaporator
FO	forward osmosis
G	incidence radiation (W)
GH	gas hydrates
GP	geothermal polygeneration
G_{PVT}	incident radiative flow (W)
GW	geothermal well
H	height (m)
HD	hydrodistillation
HDH	humidification-dehumidification
HE	heat exchanger
HF	hot fluid
HW	hot water
h_c	convective heat transfer coefficient ($W/m^2/K$)
h_f	fluid-specific enthalpy (J/kg)
i	interest rate
I_b	beam radiation (W/m^2)
I_{tot}	total radiation (W/m^2)
IAM	incidence angle modifier
I_0	system capital cost (€)
J	component capital cost (€)
LHV	lower heating value (kWh/kg)
m	monthly mass flow rate (kg/month)
M	mixer
$\dot{M}$	mass flow rate (kg/s)
$\dot{m}_f$	fluid mass flow rate (kg/s)
MD	membrane distillation
MED	multieffect distillation
MM	molar weight (kg/kmol)
MSF	multistage flash
MVC	mechanical vapor compression
$\dot{N}_{w,fresh}$	molar flow rate (kmol/s)
NTU	number of transfer units
ORC	Organic Rankine Cycle
p	pressure
P	power (kW)
P_{aux}	electrical energy required by auxiliary devices (kJ/year)
P_{PVT}	electrical energy provided by PVT (kJ/year)
P_{CPVT}	electrical energy provided by CPVT (kJ/year)
P_i	pump
PE	primary energy (kWh)
PES	primary energy saving (kWh/year)
PTC	parabolic trough collector
PV	photovoltaic
PVT	photovoltaic/thermal collectors

q	flow rate (kg/s)
Q	thermal power (kW)
r	area-specific thermal resistance (m^2K/W)
$R_{universal}$	universal constant of gases (kJ/kmol K)
RED	renewable energy desalination
RES	renewable energy source
R_i	exergy residual
RO	reverse osmosis
RPS	renewable polygeneration system
RS	reference system
SCF	solar collector fluid
SHC	solar heating and cooling
SW	sea water
SPB	simple pay-back period (years)
T	temperature (°C)
TK	tank
TVC	thermal vapor compression
U	overall heat transfer coefficient ($kW/m^2°C$)
v	specific volume
V	volume (m^3)
VMD	vacuum membrane distillation
X	water salinity (parts per million, ppm)
$\dot{Z}_{unit}$	cost for unit of time (€/s)
$\tilde{Z}$	cumulative cost related to the capital cost
α	absorptance
ε	emittance
Ψ	dissociation factor
ρ	density (kg/m^3)
ρ_{PVT}	PVT reflectance
σ	Stephan-Boltzmann constant
λ	latent heat of vaporization/condensation (kJ/kg)
η	efficiency
η_{inv}	inverter efficiency
η_{mod}	module efficiency
η_{PV}	PV efficiency
η_{opt}	optical efficiency
η_{th}	thermal efficiency
η_{el}	electrical efficiency
ω	minimum work of separation in case of null recovery ratio (kJ/kmol)
$\tau_{operation}$	length of the basis of time adopted in the analysis (h)

SUBSCRIPTS

a	ambient
act	activation
ap	aperture
aux	auxiliary

b	beam
B	brine
bot, bott	bottom side
c, cool	cooling
ch	chilled
conc	concentrator
cw	cooling water
d	destruction
D, i	distillate by evaporation at *i*th effect
D*, flash *i	distillate produced by flash at brine inlet at *i*th effect
el	electrical
ext	external
f	fluid
ft	feed-in tariffs
h	heating
in	inlet
inv	inverter
mod	module
motive	related to the hot water stream supplying energy to the MED plant
o, out	outlet
op	operating
opt	optimal
ph	physical
s	summer
set	set point
sub	substrate
t, th	thermal
top	top side
tot	total
ue	unit cost of electricity (€/kWh)
ut	unit cost of thermal energy (€/kWh)
uwat	unit cost of desalinated water (€/m^3 or €/kg)
w	winter
wat	water

4.1 INTRODUCTION

In the last few years, several innovative and sustainable technologies were developed in order to face energy and environmental issues related to the use of fossil fuels, aiming at reducing CO_2 emissions through the replacement of fossil fuels and the increase of energy efficiency. A further challenge, also linked to climate changes, is represented by the scarcity of freshwater resources, affecting every continent and more than 2 billion people around the world; presently, according to the United Nations, more than 1 billion people lack access to clean drinking water. So, new paradigms about the management of energy and water resources should be considered for a more sustainable future. In this context, polygeneration systems integrating water desalination, based on renewable energies (biomass, geothermal, wind, solar), can represent a good solution.

In the following, a literature review on polygeneration systems based on solar, geothermal, and biomass is presented, with special focus on cases in which such systems are used for space heating and cooling (Solar Heating and Cooling, SHC) and desalination. In addition, special sections are dedicated to the last research articles concerning: (i) the simultaneous production of power and thermal energy by Concentrating Photovoltaic/Thermal Collectors (CPVT), and (ii) the multiple-effect desalination (MED) process, a desalination technique widely adopted for freshwater production.

4.1.1 Polygeneration With Solar-Driven Multiple-Effect Desalination (MED)

Solar polygeneration system integrating water desalination results to be a promising technology for energy providing and drinkable water supply. In particular, multiple-effect desalination (MED) system is a technology, which is often used in case of freshwater scarcity by desalting seawater. In fact, desalination systems may increase the availability of potable water in places where water scarcity is or will be a crucial issue, especially if one considers that more than 70% of the word population lives within 70 km from the sea [1].

Thermal-based MED systems [2] have been adopted for large quantities of drinkable water supply, reaching average freshwater flow rates even higher than 20.000 m^3/day; Unfortunately, one of the main problems of thermal desalination is the high consumption of thermal energy, equal to 50–60 kWh_{th}/m^3 of freshwater, causing low energy efficiency and high providing cost.

Due to this aspect, systems involving request of mechanical or electrical energy have been widely introduced thanks to their lower energy consumption, equal to 4–6 kWh_{el}/m^3, necessary to pump water at high pressures, providing seawater desalination, such as in reverse osmosis (RO) technologies [3]. Sometimes a polygeneration scheme, based on gas turbine exhaust heat conversion into mechanical power, was adopted [4]: this power drives a seawater reverse osmosis in an Arabian Gulf off-shore oil field, with a specific energy consumption equal to 4.1 kWh/m^3 and a exergy efficiency equal to 29%. But it is not always possible to use or produce itself mechanical or electrical energy needed to supply a desalination system. For this reason, in many cases, thermally based technologies could be economically advantageous when MED systems are fed by heat wastes recovered from power plants through cogeneration [5] or by renewable energy sources [6] such as solar and geothermal energies. A prototype of a compression ignition engine, with a shaft power equal to 9,9 kW, has been tested to provide distillated water thanks to a multiple-effect distillation, thermally fed by the wastes of heat produced by the engine [7].

Better results can be obtained through the use of renewable sources for thermal energy production. A prefeasibility technoeconomic study of geothermal MED process using hot sedimentary aquifers in Western Australia was developed in [8]; the preheated boosted MED system, applied to a well field, resulted to be the best economic configuration. In particular, flat plate, evacuated tube, or Parabolic Trough collectors are mature technologies and can be used to provide thermal energy for the freshwater production in arid places, where the solar irradiation is high and drinkable water availability is low. Obviously, some special arrangements must be found in order to provide water 24 h a day regardless of solar irradiation availability, as shown in [9].

A hybrid renewable polygeneration system combining Parabolic Trough Collector (PTC) field, geothermal wells, Organic Rankine Cycle (ORC) unit, and a Multieffect Desalination (MED) has been studied in Refs. [10] and [11], through dynamic simulations. The plant provides electrical, thermal, and cooling energy; domestic hot water (DHW); and a certain amount of desalinated water in order to cover the whole need of the island of Pantelleria.

A similar plant was investigated by Sharaf et al. [12], who analyzed a combined solar organic cycle with MED process. The authors found that, to increase the gain ratio, the number of MED effects should be up to 16–20 stages and the temperature of the operating condition should range between 70°C and 75°C. With a case study requiring 100 m^3/day of distillate water, the only water production scenario resulted to be more economically convenient compared to the desalination and power production scenario, due to the higher gain ratio and lower solar field need, consisting of Parabolic Trough Collectors (PTC).

Another study concerning the dynamic simulation of a novel hybrid solar and geothermal polygeneration system was presented and discussed in [13], where MED unit was developed in Engineering Equation Solver. After an accurate energy, economic, exergy, and exergoeconomic analysis of the system, it was found out that the global exergy efficiency varies from 40% to 50% during the thermal mode and from 16% to 20% during the cooling one [14]. However, the exergoeconomic costs of electricity, cooling water, chilled water, and desalinated water resulted very interesting, in the ranges 0.1475–0.1722 €/kW h, 0.01612–0.01702 €/kW hex, 0.1863–0.1888 €/kW hex, and 0.5695–0.6023 €/kW hex, respectively.

A detailed study regarding the methodology to be adopted in the desalination systems exergy analysis is provided by Sharqawy et al. [15].

A solar thermal polygeneration system integrating membrane distillation was analyzed in [16] for United Arab Emirates weather conditions; annual dynamic simulations, performed in TRNSYS environment, have shown that the daily freshwater productivity goes from 70 to 94 kg/day. Three different solar technologies were compared: flat plate, evacuated tube, and compound parabolic collectors. The best pay-back value, equal to 6.75 years, was achieved with evacuated tube collectors having gross area of 216 m^2. The novel solar thermal polygeneration pilot plant was analyzed in four different strategies [17], according to the typical seasonal requirements: (i) solar cooling strategy, (ii) potable water and domestic hot water cogeneration mode, (iii) cooling energy and potable water providing cogeneration mode, and (iv) trigeneration system. Another interesting study [18], still conducted in United Arab Emirates, highlighted that direct solar with thermal store integration was the best economic integration strategy, among the three analyzed, of a solar thermal cogeneration system producing potable water from brackish water, through a membrane distillation; the system provide 43 L/day of freshwater and, simultaneously, 277 kWh/day of hot water, thanks to a total gross solar collector area equal to 96 m^2. An interesting study was conducted in order to analyze the economic feasibility of a solar MED polygeneration both in Venezuela and in northern Chile [19]. The plant was dynamically simulated in TRNSYS environment and it was found that it can provide in both cases electricity and water to more than 85,000 inhabitants at favorable prices, increasing the total plant annual cost of only 6%–12%.

To further reduce the thermal energy supply for water desalination, it is possible to work below the atmospheric pressure, so that the saline water evaporation pressure decreases [20], reaching a maximum desalination unit efficiency equal to 40%.

In isolated places or small islands, where it is difficult to provide electricity and drinkable water, conventional desalination technologies, such as reverse osmosis ones, are not suitable to treat raw water; when biomass (algae, energy crops, or residual biomass) is available in these places, a polygeneration system activated by biomass thermal oil boilers can be an interesting solution [21]. Hybrid solar-biomass system can produce steam at 60 bar of pressure and at a temperature equal to 500°C, able to supply a turbine with a flow rate equal to 5 kg/s [22]. This plant provides electricity, cooling energy, and water desalination, achieving very good results in term of primary energy saving (PES), overcoming 50%, and an energy output increase up to 78.1% compared to a simple turbine power plant. Another hybrid solar-biomass power plant in India [23], with cooling and multieffect dehumidification, realized water desalination by using limited thermal energy amount taken from the refrigerant of a vapor absorption refrigerant cooling system, through a heat exchanger.

In another study, optimization works were also developed. For example, in [24], a parametric analysis was presented in order to maximize the energy performance and the economic feasibility of the solar MED polygeneration system: the lowest pay-back value was achieved with 5.9 m^2/kW of optimal ratio between solar field area and ACH nominal power, 151 per unit solar field area for storage tank volume, and an outlet solar field water temperature equal to 85°C in summer and 55°C in winter, through a set of implemented controllers. In [25], instead, a mathematical optimization process was implemented in order to find the optimal size of a polygeneration plant, which can provide freshwater from seawater, fueled by natural gas, solar energy, and gasified biomass, realized in a tourist resort in Spain. It was found out that natural-gas-fueled engines were the best solution from an economic point of view, but, by using renewable resources, best value of CO_2 emission reduction was obtained. By a two-step optimization procedure of a polygeneration unit [26], an 18% primary energy saving and more than 850 tonCO_2/year avoided emissions were achieved.

An innovative layout including double-effect absorption heat pump was investigated in [27], finding an economic and sustainable solution to the thermally based MED system during cloudy days. Accurate energy and exergy analyses of a small-scale solar MED plant in Almeria were conducted in [28]; a complex 14-effects design was realized in order to achieve high conversion efficiency. In [29] an existing large-scale Multiple-Effect Distillation along with Thermal Vapor Compression (MED-TV) with a 36,000 m^3/day capacity is analyzed in order to retrofit the plant with a top steam cycle with condensation and steam extraction used to drive the steam ejector.

4.1.2 Concentrating Photovoltaic/Thermal Collectors (CPVT)

The technology of Photovoltaic/Thermal solar collectors (PVT) consists of devices producing at the same time electrical and thermal energy from the sun.

In fact, a PVT solar collector is obtained by a traditional solar thermal collector whose absorber is covered by a apposite PV layer [30]. The absorbed thermal energy is distributed to a cooling fluid (typically water or air), while the PV produces electricity [31,32]. Therefore, this solar collector allows one to obtain the simultaneous production of electricity and heat and, as a consequence, the improvement of electrical efficiency, by means the cooling of PV cells. In addition, the combination of PV panels and solar thermal collectors allows one to reduce the required surface for the installation of the collectors, to improve the system power density and, therefore, to potentially reduce the technology capital cost. The electrical efficiency of the PVT collectors may be even higher than the one of conventional PV panels, based on the same PV material [31–33] because the cooling fluid allows one to reduce the PV operating temperature, thereby increasing the energy efficiency of the PV system. For this reason, PVT collectors are mainly used for applications in which the thermal energy is demanded at low temperature (floor heating, domestic hot water, desiccant cooling). Thus, although PVTs for medium-temperature applications (60–80°C), with high electrical efficiency, were developed about 40 years ago [34], several novel PVT arrangements were recently investigated [35–39]. The key question, for a future considerable commercialization of PVT collectors, is the decrease of system capital costs, obtained, for the same peak power, by reducing the PV area per kW_p. This result is reached by increasing the radiative flux on PV receiver and, in this framework, the concentrating systems are defined as one of the most encouraging PVT technologies. Concentrating PVT collectors (CPVT) consist of PVT collectors located in the focus of some reflectors (Fresnel, dish, parabolic, etc.) [31,32,34,40–42]. The CPVT-specific cost is dramatically lower than a flat-plate PVT one, due to the lower amount of PV needed per unit area. Additionally, the higher radiative flux also allows one to increase the fluid average temperatures of these devices. Anyway, CPVT collectors convert only the beam radiation and cannot exploit the diffuse radiation; therefore, in case of low beam-to-total radiation ratio, CPVT collectors become dramatically inconvenient with respect to the flat-plate PVT. Then, the CPVT operating temperature is usually higher than the one of flat-plate PVT and, as a consequence, this leads to a decrease of their electrical efficiency. Typically, this reduction is not excessively significant (0.45%/K for silicon cells), if the increase of temperature is fair (a silicon cell showing at 25°C, an efficiency of 15 %, will have at 75°C, an efficiency of 11.6%). Conversely, for high concentration ratios, the PVT operating temperature and the corresponding inefficiencies may considerably increase (for silicon cells, the voltage drops to zero at 270°C) [33]. Therefore, the use of CPVT collectors is particularly interesting if coupled to novel PV solar cells, such as multijunction ones. In fact, by adopting these novel materials, ultra-high electrical efficiencies up to 100°C can be achieved [40,43] by approaching a nominal efficiency of 40% [41,44]. The use of multijunction cells in CPVT collectors could lead to operating temperatures up to 240°C with acceptable electrical efficiency (slightly lower than 20%) [41]. Operating temperatures up to 200°C in CPVT solar field are technically and academically possible. In such circumstance, the possibility of adopting high-temperature CPVT is very significant, since it increases the number of feasible applications. In fact, the high-temperature thermal energy produced by CPVT collectors could be also adopted to supply Solar Heating and Cooling system (SHC) [41], heat engines [43], or an Organic Rankine Cycle (ORC) [44]. Unfortunately, precommercially or commercially available CPVT collectors are usually a small amount of the PVT under development [45] and the related experimental and theoretical research investigating CPVT collectors performance is limited. The performance of parabolic trough CPVT collectors with a geometric concentration ratio of 37× was experimentally investigated by Coventry [46]. In this study, the effect of the radiation flux distribution on the CPVT performance is investigated in detail. In particular, the impact of nonuniform illumination on the PV cells was analyzed by adopting purpose-built equipment moving a calibrated solar cell along the line of the receiver and measuring short-circuit current. The impact on efficiency due to open-circuit voltage reduction was discussed. Measured results under typical operating conditions show thermal and electrical efficiency around 58% and 11%, respectively. Finally, it is also seen that the thermal efficiency of flat-plate solar collectors is higher than the CPVT one only when radiation is not scarce and/or ambient temperature is high. Rosell et al. [47] developed the thermal model, validated by an experimental prototype, of a low solar CPVT collector based on a linear Fresnel concentrator coupled with a channel PVT collector. Measured thermal performance of the solar system gives values above 60%. Mittelman et al. [48] presented a novel miniature CPVT based on a dish concentrator ($0.95\,m^2$) with the PV module based on triple-junction cells (nominal conversion efficiency of 32%). A lumped-parameter thermal model of the CPVT collector was developed in order to calculate its performance. The system provides 140–180W of electricity and 400–500W of heat. The cost of the system was evaluated at 2.5 $ per peak electric Watt. The simultaneous production of electricity and thermal energy at high temperature (about 120°C) by CPVT collectors is investigated in [41]. In such work, the performance and the cost of a CPVT system with single-effect LiBr-H_2O absorption cooling are investigated in detail. In particular, triple-junction cells are integrated into CPVT system by achieving an electrical efficiency ranging about between 23% and 19% (depending on the PVT operating temperature, varying between 120°C and 65°C). The obtained thermal efficiency is about 60%. The possibility to use the CPVT thermal energy for driving a multieffect evaporation desalination plant is proposed in [51]. The combined system produces solar electricity and simultaneously exploits the waste heat of the photovoltaic cells to desalinate water. A detailed simulation was performed to compute the annual production of electricity and water. The cost of desalinated water was estimated and compared to that of alternative conventional and solar desalination plants. The results indicate that the proposed coupled plant can have a significant advantage relative to other solar desalination approaches. In some cases, CPVT desalination is even more cost effective than conventional desalination. Parabolic Trough CPVTs were also studied by Li et al. [49–51]. They compared the system thermal and electrical efficiencies by analyzing several PV technologies and concentration ratios. The results obtained show that GeAs cells have higher electrical efficiency and lower thermal efficiency with respect to silicon cells. The authors also concluded that the cost of unit area of the silicon cell is 131.34 $\$/m^2$ versus 3067.16 $\$/m^2$ of the GeAs. Bernardo et al. [52] developed and validated the thermal model of a low-concentrating parabolic trough CPVT collector, by investigating the Swedish weather conditions. The system, distributed by the Swedish company Absolicon, consists of one axis tracking parabolic concentrator and silicon cells. Al-Alilli et al. [53] developed the model of a CPVT collector producing electricity and thermal energy, employed to drive a vapor compression and a desiccant wheel cycle, respectively, for air conditioning application. A detailed finite-volume model of a parabolic trough CPVT collector with linear triangular receiver equipped with triple-junction cells is developed by Calise et al. [54]. The model evaluates the thermodynamic and electrical parameters along the axis of CPVT collector and, by the exergy balances, also the corresponding exergy destruction rate and exergy efficiency. The effect of the variation of the main design/environmental parameters on the CPVT energy and exergy performance is also investigated. Buonomano et al. [55] simulated the performance of evacuated tube and CPVT collectors for producing heat and electricity in an SHC system. Single-stage LiBr–H_2O absorption chillers were adopted in the model. In addition, the simulation code for the CPVT collectors is also validated by experimental data. The obtained primary energy savings evaluated by the model are about 74% and over 100% for evacuated tube collectors and CPVT, respectively. Calise et al. [56,57] modeled and analyzed a parabolic dish CPVT collector implemented in an SHC system. The novelty of this collector is that it is able to produce high-temperature thermal energy (up to 180°C) with respect to the conventional PVT collector operating typically below 45°C. Therefore, the cooling energy is produced by a two-effect absorption chiller. Results showed that both electrical and thermal efficiencies are very good in a wide range of operating conditions. A desiccant-based Air Handling Unit (AHU) was coupled with a novel parabolic trough CPVT collector with linear triangular receiver in [58]. Electricity produced by CPVT collector is used to power the auxiliaries of the AHU, the chiller and to balance the users load, while thermal energy to heat the regeneration air flow during the summer and the process air in the winter. The heat provided by the CPVT covers about 60% of thermal energy required by regeneration air and 30% of process air in winter operating mode. The yearly Primary Energy Saving ranges between 81% and 89%. The experimental and numerical study of a parabolic trough linear CPVT system was carried out by Del Col et al. [59]. The system is featured by a two-axis tracking system and exhibits a geometrical concentration ratio around 130. The receiving module placed on the focus line displays a secondary optics made of two flat mirrors to gather some reflected radiation and to contribute to the concentrated flux on two lines of triple junction PV cells soldered on a ceramic substrate in thermal contact with a heat exchanger cooling the PV cells. The design and the experimental procedure developed to evaluate the performance of a novel parabolic-trough CPVT system are discussed by Karathanassis et al. [60]. Three variations of the system receiver incorporating different PV-module and heat-sink designs were evaluated. The prototype CPVT system was found to achieve a thermal and electrical efficiency approximately equal to 44% and 6%, respectively.

Other CPVT systems were developed and investigated by employing the technology of the Building Integrated Solar Systems (BISTs). In fact, in the last decade, besides standard stand-alone CPVT applications (parabolic dish [57], Fresnel [61], Fresnel and concentrating parabolic trough system [62]), novel building integrated BI-CPVT ones were analyzed. As an example, a façade and roof integrated air-gap-lens-walled compound parabolic concentrator incorporated with a static solar CPVT system without tracking was analyzed in [63]. Chemisana et al. [64] designed and constructed

Fresnel PVT collectors for façade integration. The system, compared with a nonconcentration reference unit, consists of double-side reflective strips, which concentrate the incident beam toward a static PVT receiver. The reflectors are placed vertically at the façade and track the sun by rotating axially. By this analysis, the thermal output of the BI-CPVT almost doubles the reference one, whereas the BI-CPVT electrical power registered is more than 4.5 times. The experimental system electrical and thermal efficiencies were 6.0% and 35.0%, respectively.

4.1.3 Desalination

In the last years, in order to overcome the scarcity of drinkable water particularly in semiarid and costal/island areas, several seawater desalination processes [65–68] were implemented.

The desalination is basically defined as the process of removing dissolved salts and minerals from saline water to produce potable water [69]. Major desalination technology options are based on:

(i) thermal processes, using both thermal energy or electricity (Multistage Flash (MSF), Multieffect Distillation (MED), Thermal or Mechanical Vapor Compression, (TVC), (MVC);

(ii) membrane technologies, using electricity only (Reverse Osmosis, (RO), Electrodialysis, (ED)).

In addition, several novel technologies, which are growing rapidly, are Forward Osmosis (FO), Membrane Distillation (MD), Capacitance Deionization (CDI), Gas Hydrates (GH), freezing, and Humidification Dehumidification (HDH).

The MSF technique [70,71] involves the evaporation and condensation of water in several stages with a pressure lower than the previous stage, by recovering the heat developed during the evaporation for preheating incoming water. The system requires higher energy component because of the necessity to boil the sea water.

The MED method [2] desalinates sea water by heating and allowing it to evaporate in different stages until fresh water is produced. The energy requirement is very less when compared to MSF method [72].

In VC, the heat for water evaporation comes from compression rather than from direct heating. This process is generally used in combination with other processes (MED) to improve overall efficiency.

Thermal systems processes are involved mainly in large-scale plants, with an average production typically higher than 10–20 thousands m^3/day. Anyway, the main disadvantages of thermal desalination processes are the high consumption (about 50–60 $kWh_{thermal}$ per m^3) due to their low energy efficiency and, as a consequence, the high cost of desalinated water.

Instead, in membrane technologies, semipermeable selective membranes, by acting as physical barriers, separate dissolved salts and other impurities from salted water by producing desalinated water.

In RO method, fresh water penetrates through semipermeable membrane under high pressure and leaves behind highly concentrated brine solution [73].

In ED process, the electrically charged salt ions are separated using selective ion-exchange membranes, which leaves low-salinity water. RO can be used for both brackish as well as sea water distillation, whereas ED is used only for brackish water distillation [74].

More than 50% of the installed desalination plants worldwide use RO technology because of its simplicity and relatively low energy cost and energy consumption (in the order of 4–6 kWh_e per m^3) in comparison with thermal desalination processes [75]. The spread of such systems is in middle-small scale desalination plant, and it is basically due to their modular design.

Anyway, the energy requirement (electricity and thermal energy) for desalination processes for the separation of dissolved salts from brackish, saline and brine solution, is very high [72]. Electricity is used to drive RO, ED, and VC desalination processes whereas MSF, MED, and TVC systems are supplied by thermal energy and also electricity (used to pump water through the system). On average, in order to desalinate salt water with a salinity equal to 34,500 ppm, the minimum amount of theoretical energy required is 0.86 kWh/m^3, but the plants use 5–26 times more than this minimum required energy based on the type of process used for desalination [76]. Therefore, it is crucial for their spread to supply the technologies implemented in desalination processes, by the heat recovery of power plants operating in cogeneration model [5] or by eco-friendly energy sources [6]. In desalination processes, the renewable energy as the geothermal [77], solar [78], wind [79] is often used to meet their need of energy and reduce the energy demand from conventional fossil fuels. In fact, the renewable energy desalination (RED) systems are receiving an increasing interest worldwide and more than 130 RED plants have opened within the last few years [80].

In this framework, the solar source is more and more used, mainly in arid regions featured by a high solar radiation, seawater availability, and those that are hardly connected to the conventional production systems of drinking water (islands).

By focusing on the desalination processes driven by thermal energy, several different direct and indirect thermal systems were developed, as in detail reviewed in [81]. The indirect systems consist of two subsystems: the solar collector (such as evacuated tube, flat plate, heat pipe, etc.) and desalination unit (such as MD, MSF, VC, MED). Instead, in the direct systems, both evaporation and condensation take place in the same device (solar still). They are more efficiently suitable for household and small communities and are suitable for meeting freshwater demand up to 200 m^3/day [82].

In the last years, several studies [83,84] reported in literature have been developed on the topic of the solar desalination. This is a technique to desalinate water using solar energy, converted in electricity by PV panels or heat by solar thermal collectors, as main fuel to drive the desalination process. Among all the available studies, in the following, some of those focusing on the solar MED processes are reported. In fact, the MED method is featured by a flexible temperature level and a wide range of drinkable water capacity. This makes the technology suitable to be coupled to the large solar fields for producing significant amounts of freshwater, for small-medium scale communities in remote zones such as islands. Different configurations of a MED/TVC system with a water production rate of 9,000 m^3/day driven by Linear Fresnel solar field were economically and technically analyzed in Ref. [85], for the weather condition of Kish Island (South Iran). The analysis results show that the system water production cost with 67.77% contribution of solar energy and without thermal storage is higher (3.32 \$/m^3) than the cost of the conventional fossil-fuel-powered MED plants (1.26 \$/m^3). Talebbeydokhti et al. [86] developed a novel parabolic trough Concentrated Solar Power plant with a Discrete Ericsson Cycle, which uses air as heat transfer fluid (CSP-DEC) for producing electricity and heat for lower-temperature (LT) MED processes. By their analysis, it resulted that 0.062 kW/daym^3 source of freshwater is the value obtained by the novel technology versus 0.195 kW/daym^3 of a conventional LT-MED/CSP plant. The tracking mode effect on the freshwater production of a parabolic trough collector-MED plant is investigated in Ref. [87]. From obtained results, it is shown, for instance, that at the winter solstice the plant with a full tracking system produces 341% more freshwater than plant with fixed collectors. An experimental analysis of the MED system coupled to large-aperture flat-plate collectors is presented in [67]. It was found that the thermal storage allows an autonomy of up to 6 h when the MED works at 65°C, by allowing a continuous operation even in presence of clouds. In addition, from a theoretical study, an annual distillate production between 6701 m^3 and 11,314 m^3, respectively, resulted.

Geothermal source, featured on average by a constant temperature and availability, and solar source are excellent candidates for desalination units supplied by thermal energy. These features make the geothermal energy reliable for a safe operation of the desalination unit. For this reason, the arrangement of geothermal energy coupled to the desalination systems is diffusely investigated in the relevant literature presenting both theoretical and experimental studies. Examples are the prototypes of geothermal desalination plant based on the Hydrodistillation (HD) method [88], Falling Film Evaporator (FFM) [89], and Vacuum Membrane Distillation (VMD) [90]. Goosen et al. [91] carried out an exhaustive review of desalination processes coupled to the geothermal source. They also presented several applications of these systems in Greece, Algeria, and Mexico and concluded that such technology may be of interest in many areas of the world.

For the usual temperature range of geothermal source at low enthalpy (70–90°C), the geothermal heat is suitable to drive low-temperature MED units, as examined in Ref. [92]. By focusing on MED process, Manenti et al. [93,94] designed and simulated a MED system fueled by the geothermal heat available in Sicily Island (South Italy). In particular, Karystsas [95] described a system consisting of a MED unit driven by geothermal groundwater with temperatures ranging from 75°C to 90°C on the Milos island in Greece, able to produce about 600–800 m^3/day of drinkable water. A technoeconomic analysis of MED and RO desalination powered by low-enthalpy geothermal energy is carried out in [96]. Authors concluded that these processes still show potential to be a cost-effective method of RED in the coastal countries of Arabian Gulf.

4.2 LAYOUT OF THE SYSTEMS INVESTIGATED

4.2.1 Geothermal Polygeneration Plant (GP)

The first system investigated is a Geothermal Polygeneration (GP) plant, which combines Solar Heating and Cooling (SHC), Concentrating PhotoVoltaic-Thermal collectors (CPVT), Geothermal Well (GW), and Multiple-Effect Distillation (MED) for desalination purpose. It can provide simultaneously electricity, heating and cooling, and freshwater thanks to brackish water treatment. The main components and operating fluids, with their corresponding loops, are shown in the system layout in Fig. 4.1, with set point and/or limit temperatures in both summer (s) and winter (w) seasons, in order to clarify the operation of the system.

Nine different operating fluids are involved: Solar Collector Fluid (SCF), Hot Fluid (HF), Cooling water (CW), Chilled Water (CHW), Domestic

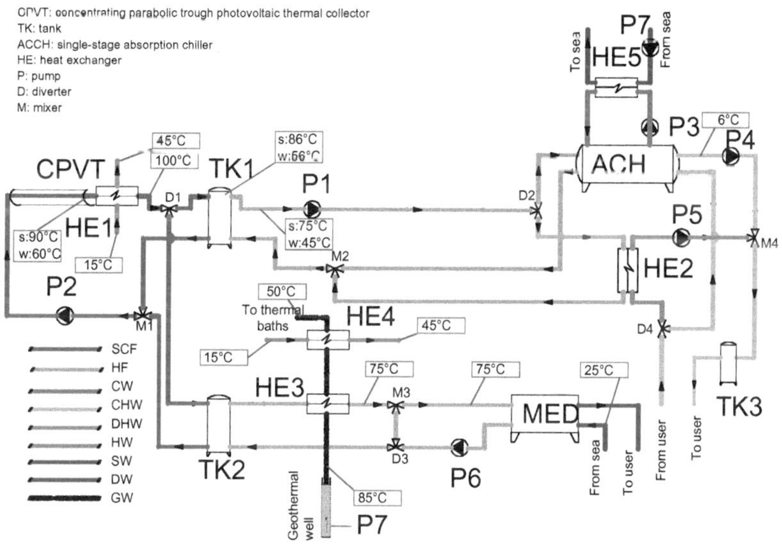

Fig. 4.1 SHC, CPVT, Geothermal Well, and MED system layout [11].

Hot Water (DHW), Hot Water (HW), Seawater (SW), Desalinate Water (DW), and Geothermal Well Fluid (GW).

The solar field consists in concentrating parabolic trough solar collectors (CPVT); it can operate up to 100°C and it is equipped with a one-axis tracking system. A triple-junction PV layer is placed on its absorber and a triangular receiver, located on the focus of the parabola that is covered with a multijunction PV panel, catches the beam radiation concentrated on it; inside the receiver, there is a pipe conveying a cooling fluid.

SCF is pressurized water that flows from the source sides of tanks (TK) 1 and 2 to the solar field; TK1 and TK2 load sides, instead, are connected to all devices requiring solar thermal energy through pressurized hot water HF. In particular, TK1 supplies heat for heating purpose or it thermally feeds the Absorption Chiller (ACH). The tank consists in a set of vertical stratified hot storage tanks; the source water varies the entering position and it is equipped with inlet stratification devices, in order to keep inlet water and tank temperatures equal. TK2, instead, stores heat provided by the solar field and needed to desalinate water through MED system, reporting the same characteristics of TK1.

CW is cooling water needed to remove heat from the fluids involved in the ACH, which provides chilled energy, required for cooling purpose, through water flowing in its evaporator, namely, CHW.

In particular, the ACH adopted is a LiBr-H_2O single-effect absorption chiller, thermally fed by the solar field through HF and cooled by seawater. TK3, an inertial chilled water storage tank, is required in order to reduce the number of ACH start-ups and shutdowns.

Heat recovered by Heat Exchangers (HE) 1 and 4 is used to produce DHW. In particular, HE1 is a plate-fin heat exchanger (as well as all the other heat exchangers) and it provides DHW only when ACH/HE2/MED thermal request is lower than the solar irradiation; HE4, instead, provides DHW through geothermal well fluid coming out from HE3, in which hot fluid is heated by geothermal source to thermally supply MED system.

HW, produced thanks to HE2, in which HF transfers TK1 stored heat, satisfies heating purpose.

SW is salt water that can be desalinated by the MED system, in order to provide freshwater, DW, otherwise is used in HE5 in the ACH cooling loop through seawater.

Also, GW is hot geothermal water that supplies MED through HE3, otherwise HE4 for DHW production, as mentioned before. In particular, one of the main components is the geothermal well, composed of 10 wells at a depth of 80 m and equipped with submerged pumps that can take hot water from the geothermal source.

The plant requires five single-speed pumps and two variable-speed pumps: single-speed pumps P1, P3, P4, P5, and P6 for, respectively, HF, HW, SW, CHW, and CW loops and variable-speed pumps P2 and P7 for SCF loop and Geothermal Wells. For sake of brevity, other minor components, necessary for carrying out system operations, such as valves, mixers (M), diverters (D), pipes, controllers, etc. called Balance of the Plant (BOP), are not shown in Fig. 4.1. At the same time, a simulation model requires additional components such as integrators, printers, weather databases, daily and seasonal schedulers, and some controllers (proportional, feedback and on/off).

The photovoltaic component of CPVT converts solar irradiation in electricity, while the thermal component in heat, necessary to reach the outlet temperature set point value $T_{set,CPVT}$ ($T_{set,CPVT,s}$ in summer and $T_{set,CPVT,w}$ in winter, respectively). The provided electricity satisfies the electrical plant request, due to all the electrical devices, and the electricity surplus is delivered to the national grid.

A feedback controller, which works according to the secant method, regulates the variable speed pump P2 in order to reach set point temperatures and turns P2 off when the solar irradiation is low, since it could determine a cooling effect of the CPVT operating fluid. For this reason, when the solar irradiation is not sufficient or when the user demand is very high, CPVT outlet temperature can be lower than the set point value. When the solar irradiation is high, instead, or when the user demand is low, CPVT outlet temperature is equal to or higher than the set point value. In particular, when it is higher than 100°C, not reaching the boiling temperature value (equal to 120°C), HE1 turns on, exploiting heat surplus to provide DHW and, so, cooling SCF, in order to equal the set point temperature. HE1 is bypassed, instead, when CPVT outlet temperature is lower than 100°C.

The hot solar collector fluid that exits from CPVT, thanks to the Diverter D1 and the Mixer M1, can supply to tank 1 or 2. In particular, a sensor checks the temperature at the hottest zone of TK1, i.e., the top side of TK1, and it compares it with a fixed set point temperature, $T_{TK,set}$, equal to $T_{set,CPVT} - \Delta T_{TK}$. When the temperature measured by the sensor is lower than $T_{TK,set}$, and/or it is lower than 75°C, that is the MED system design temperature, SCF is conveyed only to TK1 by D1. Vice versa, when such temperature reaches the maximum value between MED system design temperature and $T_{TK,set}$, SCF supplies TK2, storing heat for the seawater desalination purpose. Actually, the MED system design temperature, equal to 75°C, is not always achieved only by the heat stored in TK2, but mostly thanks to the heat exchanged in HE3 by the hot geothermal fluid. The variable speed pump P7 is managed by a feedback controller, which varies the hot flow geothermal rate in order to reach, and not exceeding, 75°C. GW, exiting from HE3, is still hot enough to provide DHW at 45°C through HE4; finally, after HE4, GW is set equal to 50°C and may be sent to thermal baths. Obviously, when TK2 outlet temperature is higher than 75°C, P7 turns off and no domestic hot water is provided.

MED system is supplied by both solar and geothermal heat to produce freshwater from sea water. In particular, in cloudy days characterized by scarce solar irradiation or when TK1 thermal energy is mostly used for heating and cooling purposes, MED is predominantly activated by hot geothermal fluid.

Pump P1 conveys HF from TK1 load side to final users. Diverter D2 and Mixer M2 manage the different summer and winter operating modes. In fact, HF is sent from the top of the tank TK1 to the ACH in summer, to satisfy space-cooling demand and to the heat exchanger HE2 in winter, to satisfy space-heating purpose. In particular, a controller manages pump P1 monitoring the temperature at the top of TK1. Two limit values for the monitored temperature are set for summer and winter operating modes, respectively, equal to 75°C and 45°C. Thanks to the controller, pump P1 turns off when such temperature is lower than the seasonal limit temperature value. In this way, CPVT can bring TK1 top temperature up to the minimum required values. Therefore, ACH is always supplied by hot water at least 75°C and, in winter, HE2 is always supplied by hot water at least 45°C.

Seawater cools ACH absorber and condenser indirectly, in order to avoid corrosion phenomenon; in this way, a cooling tower is not necessary. Thanks to the Mixer M4 and the Diverter D4, chilled water, provided by ACH, is pumped by P4 to the user in summertime; P5, instead, regulates hot water flow rate in wintertime. Tank TK3 is necessary to represent the volume of the water flowing inside the piping system.

All the main components, including MED system, turn off when the incident solar irradiation is zero.

Different types of controllers manage such control systems; in particular, on/off with hysteresis, feedback and proportional controllers have been implemented, using around 2–4°C of appropriate dead bands in order to reduce the number of start-ups and shut-downs events.

4.2.2 Biomass Polygeneration Plant (BP)

The second system investigated is a Biomass Polygeneration (BP) plant, which combines Solar Heating and Cooling (SHC), Concentrating PhotoVoltaic-Thermal collectors (CPVT), Auxiliary biomass-fired Heater (AH), and Multiple-Effect Distillation (MED) for desalination purpose.

As well as for the first case, it can provide simultaneously electricity, heating and cooling, and freshwater thanks to seawater treatment. The main components and operating fluids, with their corresponding loops, are shown in the system layout in Fig. 4.2.

Eight different operating fluids are involved: Solar Collector Fluid (SCF), Hot Fluid (HF), Cooling water (CW), Chilled Water (CHW), Domestic Hot Water (DHW), Hot Water (HW), Seawater (SW), and Desalinate Water (DW).

It is possible to see that plant layout, components, operating fluids loops and operation strategies are identical to the ones described, except for thermal energy integration to be supplied to the MED system. In fact, instead of geothermal wells, an auxiliary wood-chip-fired heater provides the required amount of heat.

In particular, when TK2 top side temperature is lower than 75°C, AH heats the fluid up to 75°C; when the TK2 outlet water temperature is higher

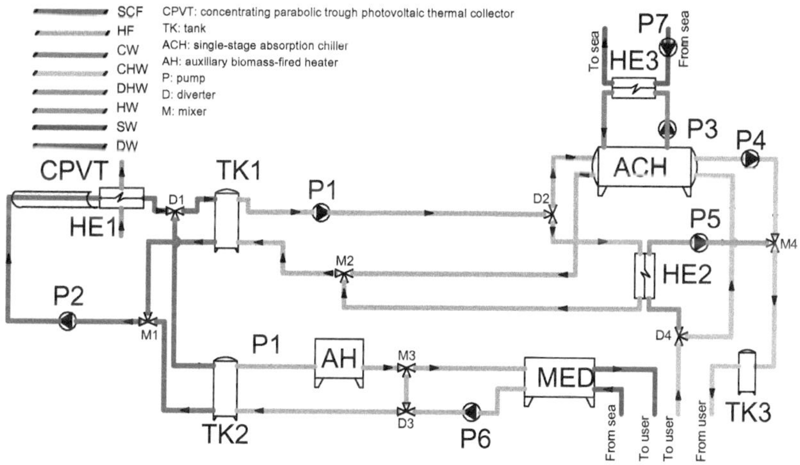

Fig. 4.2 SHC, CPVT, Biomass Heater, and MED system layout [24].

than 75°C, instead, AH is bypassed. In this case, to dissipate the heat surplus, Mixer M3 and Diverter D3 mix the MED outlet source water flow rate with the MED inlet one in order to reach permanently 75°C and let MED unit work as much stable as possible.

4.3 SIMULATION MODEL

The solar MED polygeneration systems analyzed before, were simulated dynamically in TRNSYS environment, a widely adopted software for academic and commercial purposes. It includes a wide library of preset components incorporating experimental data [97]. Most of components models have been taken from TRNSYS library, such as diverters, mixers, valves, pumps, pipes, controllers, and absorption chiller. Other new models, named "types" in TRNSYS, have been developed in Fortran by the authors and inserted into TRNSYS environment, such as energy and economic analyses calculators, MED, CPVT, and others. In particular, MED and CPVT have been properly developed by the authors and will be described in detail, while for the others new TRNSYS types created by authors, further details are listed in Refs. [98–102]. CPVT model was validated against experimental data in [103]; ACH was validated intrinsically because it is based on manufacturer's data [99]; MED unit was completely developed by authors and it was validated by comparisons with available data in literature.

4.3.1 Energy Model (TRNSYS)

4.3.1.1 CPVT

The CPVT model utilized in this work refers to the study of Bernardo et al. [52,104], considering a recently commercialized Absolicon prototype [45,52,104].

The CPVT, which operates according to the same principles of Solar Trough Collectors (PTC) [100,102,105], consists in concentrating parabolic trough solar collectors equipped with a one-axis tracking system; a triangular receiver, placed on the focus of the parabola, catches the beam radiation concentrated on it. A triple-junction PV layer is placed on the two bottom sides facing the reflector while a thermal absorber is located on the top side of the receiver; inside the receiver, a cooling fluid is conveyed by a circular inner channel. The parabolic receiver reflects only the beam radiation, and not the total radiation, while the top surface reflects both beam and diffuse radiations (nonconcentrating solar radiation). In the solar system, no covering glass is installed (since it reduces thermal losses and the overall heat transfer coefficients) and InGaP/InGaAs/Ge triple-junction solar cells are adopted [44]: thanks to these two characteristics, CPVT system differs from the one analyzed in [52,104], increasing the electrical efficiency significantly. Actually, covering glasses imply a reduction of the incident solar radiation on the PV cells; for this reason, the covering glass removal may let the thermal efficiency decrease, especially with high operating temperatures. In conclusion, a decrease of few percentage points of thermal efficiency allows one to obtain a higher electrical efficiency, thanks also to the triple-junction PV cells, much more efficient than silicon cells.

CPVT-simplified zero-dimensional models are also available in literature [106]; but a 1-Dimensional simulation model of the CPVT is the goal of this study. The CPVT model has to provide the values of thermodynamic parameters, such as pressure and temperature, electrical and thermal energies, and exergy parameters. It is based on a computational domain homogeneous discretization along the longitudinal axis. The system is split into *n* parts and energy, mass, and exergy balances for each part are developed. Engineering Equation Solver (EES), a mathematical software that contains detailed routines for thermophysical calculations of the air and the water (working fluids involved), was used in order to implement the model. Many assumptions have been adopted in the model: steady state, thermodynamic equilibrium, in the energy balances kinetic and gravitational terms have been considered negligible, uniform radiation concentrated on photovoltaic surface.

Even though theoretically CPVT can work up to 240°C [41], the system reaches a maximum temperature of 100°C, thanks to the cooling fluid, in order to prevent any damage of PV cells. Liquid water was assumed as cooling fluid among several choices and its thermophysical properties were provided through temperature-dependent equations.

Firstly, it is important to define the concentration ratio, C_{PVT}, which is equal to the ratio between the aperture area A_{ap} and the area of the receiver, i.e., the triangular receivers, A_{PVT}:

$$C_{PVT} = \frac{A_{ap}}{A_{PVT}} \tag{4.1}$$

The incident radiation on the PV area, G_{PVT}, is equal to:

$$G_{PVT} = A_{PVT} I_b C_{PVT} \eta_{opt} \mathrm{IAM}_{th} \tag{4.2}$$

The optical efficiency, η_{opt}, was assumed to be constant [41] and only the beam radiation, I_b, was considered. The Incidence Angle Modifier, IAM [107], related to the thermal energy production, was evaluated on experimental data [52,104] and, together with the receiver optical efficiency, allows to correct the radiation, considering that the radiation decreases when the incidence angle increases.

Top thermal absorber takes additional thermal energy, Q_{top}:

$$Q_{top} = A_{top} I_{tot} \alpha_{top} \tag{4.3}$$

where A_{top} is the top surface and can convert the total radiation, I_{tot}, namely, both diffuse and beam radiation, when it is not concentrated. The radiative heat transfer between absorber on the top and the sky, $Q_{top\text{-}sky}$, with a much higher area, is equal to [107]:

$$Q_{top\text{-}sky} = A_{top} \varepsilon_{top} \sigma \left(T_{top}^4 - T_{sky}^4\right) \tag{4.4}$$

where T_{sky}, i.e., the sky equivalent temperature, is provided by TRNSYS routine, while T_{top} is the top surface temperature.

In the same way, the surface of the concentrator was assumed much bigger than the receiver area and, being PVT and concentrator gray surfaces, the radiative heat transfer between the PVT and the concentrator [107] is:

$$Q_{PVT\text{-}conc} = A_{PVT} \sigma \varepsilon_{PVT} \left(T_{PVT}^4 - T_{conc}^4\right) \tag{4.5}$$

where T_{PVT} is the PVT surface temperature and T_{conc} is the concentrator surface temperature.

The following equations provide: (i) the convective heat transfer between the air and the PVT [108], (ii) the forced convection heat between the air and the top absorber [108], and (iii) the PV gross electrical power:

$$Q_{\text{conv,PVT}} = A_{\text{PVT}} h_{c,\text{PVT}} \left(T_{\text{PVT}} - T_a\right) \tag{4.6}$$

$$Q_{\text{conv,top}} = A_{\text{top}} h_{c,\text{top}} \left(T_{\text{top}} - T_a\right) \tag{4.7}$$

$$P_{\text{PVT,gross}} = C_{\text{PVT}} A_{\text{PVT}} I_b \eta_{\text{opt}} \eta_{\text{PV}} \text{IAM}_{\text{el}} \tag{4.8}$$

The electrical power equation takes into account the incident concentrated beam radiation on PV cells, considering the optical efficiency of the concentrator and the coefficient IAM, and it is corrected by, i.e., the PV electrical efficiency, equal to [41]:

$$\eta_{\text{PV}} = 0.298 + 0.0142 \ln\left(C_{\text{PVT}}\right) + \left[-0.000715 + 0.0000697 \ln\left(C_{\text{PVT}}\right)\right] \left(T_{\text{PVT}} - 298\right) \tag{4.9}$$

Triple-junction PV electrical efficiency is related experimentally to the concentration ratio, C_{PVT}, and the temperature T_{PVT}. As usual for multijunction PV cells, the equation returns very high electrical efficiency values, up to 40%.

It is important to note that part of the provided electricity is lost for the module connections, η_{mod}, and inverter, η_{inv}, efficiencies [41]; for this reason, it is possible to write:

$$P_{\text{PVT,net}} = P_{\text{PVT,gross}} \eta_{\text{mod}} \eta_{\text{inv}} \tag{4.10}$$

Also, the cooling fluid absorbed heat is:

$$Q_f = \dot{m}_f \left(h_{f,\text{out}} - h_{f,\text{in}}\right) \tag{4.11}$$

The inlet and outlet cooling fluid enthalpies, h_{in} and h_{out}, have been found through the previously mentioned thermophysical property subroutine.

Considering a control volume, which includes the whole triangular receiver, the overall energy balance is equal to:

$$\begin{aligned} A_{\text{PVT}} I_b C_{\text{PVT}} \eta_{\text{opt}} \text{IAM}_{\text{th}} + A_{\text{top}} I_{\text{tot}} \alpha_{\text{top}} &= \dot{m}_f \left(h_{\text{out}} - h_{\text{in}}\right) \\ &+ C_{\text{PVT}} A_{\text{PVT}} I_b \eta_{\text{opt}} \eta_{\text{PV}} \text{IAM}_{\text{el}} \\ &+ A_{\text{PVT}} I_b C_{\text{PVT}} \eta_{\text{opt}} \text{IAM}_{\text{th}} \rho_{\text{PVT}} + A_{\text{top}} \varepsilon_{\text{top}} \sigma \left(T_{\text{top}}^4 - T_{\text{sky}}^4\right) \\ &+ A_{\text{PVT}} \sigma \varepsilon_{\text{PVT}} \left(T_{\text{PVT}}^4 T_{\text{conc}}^4\right) \\ &+ A_{\text{PVT}} h_{c,\text{PVT}} \left(T_{\text{PVT}} - T_a\right) + A_{\text{top}} h_{c,\text{top}} \left(T_{\text{top}} - T_a\right) \end{aligned} \tag{4.12}$$

With another control volume, which considers the fluid channel and the metallic substrate, seen as a heat exchanger, the second energy balance is:

$$\dot{m}_f\left(h_{\text{out}} - h_{\text{in}}\right) = \varepsilon \dot{m}_f c_f \left(T_{\text{sub}} - T_{\text{in}}\right) \tag{4.13}$$

where T_{sub} is the metallic substrate temperature.

The boundary conditions are known and these are beam and total radiations and relative angle of incidence, inlet mass flow rate and temperature, ambient pressure, wind velocity, environment and sky temperature. There are five unknowns: substrate temperature, PVT temperature, concentrator temperature, outlet fluid temperature, and top receiver surface temperature. So other three equations are required, besides [12] and [13]. The third one comes from another energy balance, considering a volume control with the PVT layer and the metallic substrate:

$$A_{\text{PVT}} \frac{T_{\text{PVT}} - T_{\text{sub}}}{r_{\text{PVT-sub}}} = \dot{m}_f\left(h_{\text{out}} - h_{\text{in}}\right) + A_{\text{top}} \frac{T_{\text{sub}} - T_{\text{top}}}{r_{\text{top}}} \tag{4.14}$$

Considering, instead, a control volume including the top surface of the triangular receiver and the top side of the substrate, it is possible to write a fourth energy balance:

$$\begin{aligned} A_{\text{top}} \frac{T_{\text{sub}} - T_{\text{top}}}{r_{\text{top}}} + A_{\text{top}} I_{\text{top}} &= A_{\text{top}} I_{\text{top}} \rho_{\text{top}} + A_{\text{top}} \varepsilon_{\text{top}} \sigma \left(T_{\text{top}}^4 - T_{\text{sky}}^4\right) \\ &\quad + A_{\text{top}} h_{c,\text{top}} \left(T_{\text{top}} - T_a\right) \end{aligned} \tag{4.15}$$

With a control volume, lastly, including only the parabolic concentrator, the fifth energy balance is:

$$\begin{aligned} A_{\text{PVT}} \sigma \varepsilon_{\text{PVT}} \left(T_{\text{PVT}}^4 - T_{\text{conc}}^4\right) + I_{\text{tot}} A_{\text{conc}} \alpha_{\text{conc}} &= A_{\text{conc}} \sigma \varepsilon_{\text{conc,back}} \left(T_{\text{conc}}^4 - T_{sky}^4\right) \\ &\quad + A_{\text{conc}} h_{c,\text{conc,front}} \left(T_{\text{conc}} - T_a\right) \\ &\quad + A_{\text{conc}} h_{c,\text{conc,back}} \left(T_{\text{conc}} - T_a\right) \end{aligned} \tag{4.16}$$

4.3.1.2 MED

Small-scale MED systems, feeding by cogeneration systems heat cascades [109] or solar thermal energy [110], are not so common in literature. Different solar fields types and desired power-to-ratio values strongly affect the operation and the design of MED systems. Hot storage tanks volume influences the duration of small-scale solar desalination systems; in fact, they can provide freshwater up to 8–12 h/day depending on storage capacities, unlike large-scale desalination systems that work 24 h/day [111]. However, it was chosen to assume a stationary model for the MED unit, avoiding increasing the model complexity with its transient simulation, since shut-downs and start-ups don't affect significantly daily freshwater flow rate production values.

Different equations are required by the model: energy balances, water and dissolved salt mass balances, heat transfer equations. One of the aims of the study is to reach very high Performance Ratios values (PRs), since that renewable energy sources are utilized for the potable water production; besides, the total number of effects was assumed equal to 8, the Mediterranean seawater salinity equal to 38,000 ppm, last effect temperature $T_{B,8}$, equal to 40°C at nominal conditions and rejected brine salinity, $X_{B,8}$, equal to 72,000 ppm. The model description is analyzed in three parts: (i) first effect, where the solar heat stored inside the tank feed the desalination process, (ii) generic effects from 2 to 8, and (iii) condenser.

First Effect

Fig. 4.3 shows a simplified scheme of the first effect:

Hot water, coming from the storage tank, enters the left side of the unit with a mass flow rate equal to $\dot{M}_{motive}$ and an inlet temperature equal to $T_{motive,in}$. Its heat is utilized to preheat feed water entering on the top of the unit from $T_{f,2}$ to $T_{B,1}$, that is the effect temperature level, and to realize the distillation of a fraction of the brackish water, with a mass flow rate equal to $\dot{M}_{D,1}$. For these two reasons, hot water cools down up to $T_{motive,out}$. Since the distillate salinity $X_{D,1}$ is only between 5 and 15 ppm, it can be assumed equal to zero, that is completely salt-free water.

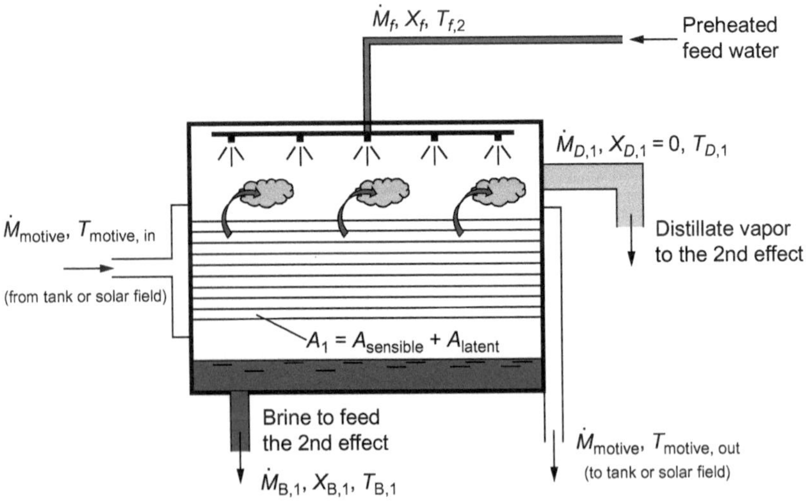

Fig. 4.3 Hot water fired first-effect simplified scheme.

The following two equations express the water and salt mass balances, referring to the first-effect level:

$$\dot{M}_{D,1} + \dot{M}_{B,1} = \dot{M}_f \quad (4.17)$$

$$\dot{M}_{D,1} \cdot X_{D,1} + \dot{M}_{B,1} \cdot X_{B,1} = \dot{M}_f \cdot X_f \quad (4.18)$$

Two different equations (based on the effectiveness-NTU method) sets are reported for the sensible heat transfer area and the latent heat transfer area. The sensible heat refers to the top section of the bundle of tubes, where feed water passes from $T_{f,2}$ to $T_{B,1}$, while the latent heat refers to the feed water evaporation on the bottom of the bundle of tubes.

Referring to the sensible heat, the equations are:

$$\dot{C}_{\min} = \min\left\{\dot{M}_{\text{motive}} c_{p,\text{motive}}, \dot{M}_f \cdot c_{p,f}\right\} \quad (4.19)$$

$$\text{NTU}_{\text{sensible}} = \frac{U_{\text{sensible}} \cdot A_{\text{sensible}}}{\dot{C}_{\min}} \quad (4.20)$$

To correlate the crossflow heat exchangers and tubes side unmixed flow, the relation is:

$$\varepsilon_{\text{sensible}} = \varepsilon_{\text{sensible}}\left(\frac{\dot{M}_f \cdot c_{p,f}}{\dot{M}_{\text{motive}} c_{p,\text{motive}}}, \text{NTU}_{\text{sensible}}\right) \quad (4.21)$$

The final sensible heat is equal to:

$$\dot{Q}_{\text{sensible}} = \varepsilon_{\text{sensible}} \cdot \dot{Q}_{\text{sensible,max}} = \varepsilon_{\text{sensible}} \cdot \dot{C}_{\min} \cdot \left(T_{\text{motive,in}} - T_{f,2}\right) \quad (4.22)$$

Motive hot water and feed water energy balance, instead, is equal to:

$$\begin{aligned}\dot{Q}_{\text{sensible}} &= \dot{M}_{\text{motive}} c_{p,\text{motive}} \cdot \left(T_{\text{motive,in}} - T_{\text{motive,intermediate}}\right) \\ &= \dot{M}_f \cdot c_{p,f} \cdot \left(T_{B,1} - T_{f,2}\right)\end{aligned} \quad (4.23)$$

Referring to the latent heat, the shell side fluid follows an evaporation process that is almost isothermal; so, ε-NTU conventional evaporators expression is used:

$$\varepsilon_{\text{latent}} = 1 - e^{-\text{NTU}_{\text{latent}}} \quad (4.24)$$

And the final latent heat is equal to:

$$\dot{Q}_{\text{latent}} = \varepsilon_{\text{latent}} \cdot \dot{Q}_{\text{latent,max}} = \varepsilon_{\text{latent}} \cdot \dot{M}_{\text{motive}} c_{p,\text{motive}} \left(T_{\text{motive,intermediate}} - T_{B,1}\right) \quad (4.25)$$

The energy balance, considering the motive hot water outlet temperature and distillate mass flow rate, is calculated as follows:

$$\dot{Q}_{\text{latent}} = \dot{M}_{\text{motive}} c_{p,\text{motive}} \left(T_{\text{motive,intermediate}} - T_{\text{motive,out}} \right) = \dot{M}_{D,1} \cdot \lambda_{D,1} \quad (4.26)$$

where $\lambda_{D,1}$ is the latent heat of vaporization at the distillate temperature $T_{D,1}$. Such temperature is equal to the difference between $T_{B,1}$ and Boiling Point Elevation, BPE, due to the dissolved salt absence in distillate water, and precisely:

$$T_{D,1} = T_{B,1} - \text{BPE}_1 \quad (4.27)$$

In particular, BPE can be obtained through a polynomial equation of temperature and water salinity:

$$\text{BPE}_1 = a_1 T_{B,1} X_{B,1} + a_2 X_{B,1} + a_3 T_{B,1} + a_4 \quad (4.28)$$

that is a quadratic equation sufficiently reliable in the range of the temperature of interest, i.e., from 40°C to 80°C.

Temperature values are expressed in Celsius degrees, while concentration values in part per million; pressure drops are negligible. The coefficient of the polynomial equations is equal to: $a_1 = 7 \times 10^{-8}$, $a_2 = 8.67 \times 10^{-6}$, $a_1 = 7.74 \times 10^{-4}$, $a_4 = 8.14 \times 10^{-2}$.

Generic Effects from 2 to 8

A simplified generic effect scheme is illustrated in Fig. 4.4:

All the effects between the second and the eighth one are very similar and, so, they can be modeled in the same way; the generic effect will be indicated with the subscript "i" that goes from 2 to 8.

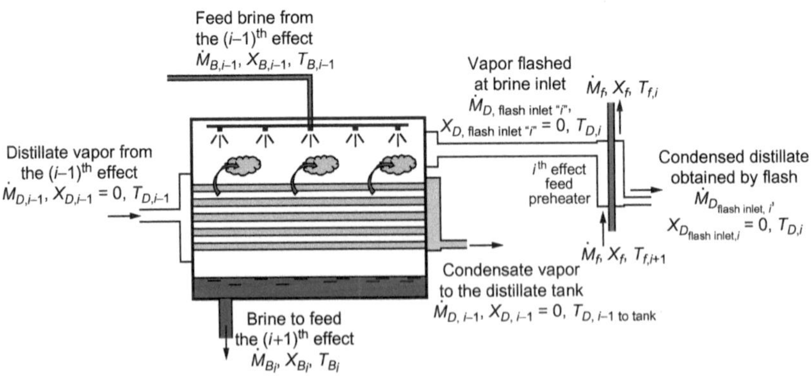

Fig. 4.4 Simplified generic effect scheme from 2 to 8.

The *i*th distillate mass flow rate, $\dot{M}_{D,i-1}$, provided at the previous effect, $(i-1)$th, releases its condensation latent heat and condensates inside the tubes; besides, $(i-1)$th produced brine is sprayed around the bundle of tubes, providing $\dot{M}_{D,i}$, i.e., a new flow rate of distillate water.

For each *i*th effect, the pressure is equal to the saturation pressure referred to $T_{B,i}$ temperature.

The temperature difference between $T_{D,i-1}$ and $T_{B,i}$ is the driving force for the transfer of the heat; from the $(i-1)$th effect, it is possible to determine $T_{D,i-1}$, while through evaporator heat transfer equation, $T_{B,i}$.

Since the inlet motive hot water temperature is lower than 100°C and it decreases from the first effect to the eighth one, each effect is performed with vacuum conditions during morning start-ups, thus needing vacuum pumps for small-scale MED systems or vacuum ejectors in large-scale ones; however, as mentioned before, MED unit is analyzed under stationary condition, excluding the necessity of such vacuum components.

$\dot{M}_{B,i-1}$ flow rate is at a temperature equal to $T_{B,i-1}$, that is a little bit higher than the *i*th effect saturation temperature, $T_{B,i}$; subsequently, such water flow rate is subjected to an initial flash process, where flash separates an amount of distillate equal to:

$$\dot{M}_{D,\text{flash inlet``}i\text{''}} \cong \frac{c_{p,B,i-1}\left(T_{B,i-1}-T_{B,i}\right)}{\lambda_{D,i}} \tag{4.29}$$

On the bundle of tubes outer side, the remaining brine falls down and it absorbs the condensation latent heat provided by the previous effect distillate water, separating, therefore, some distillate. For this reason, the same heat load, $\dot{Q}_{\text{latent}}$, is shared for each effect from the second to eighth. In fact, sensible heat is not transferred in these effects, differently from first one, and fluids, which exchange heat, follow a process almost isothermal. The distillate water produced by the evaporation process is equal to:

$$\dot{Q}_{\text{latent}} = \dot{M}_{D,i-1}\cdot\lambda_{D,i-1} = \dot{M}_{D,i}\cdot\lambda_{D,i} \tag{4.30}$$

Since the latent heat of vaporization decreases with the temperature, distillate flow rate produced in each effect is lower than the previous one and so $\lambda_{D,i-1} < \lambda_{D,i}$.

The heat transfer relation is:

$$\dot{Q}_{\text{latent}} = U_i\cdot A_i\cdot\left(T_{D,i-1}-T_{B,i}\right) \tag{4.31}$$

It is quite hard to evaluate analytically the overall heat transfer coefficient, U_i, because, in "falling film evaporators," to reflect properly all the variables, which influence the heat transfer coefficient on tubes bundle outer side, very complex correlations are necessary; besides, it is quite impossible to prevent the scaling formation on the surface of the tubes, although feed water pretreatment is performed. Such correlations vary with $CaSO_4$ layer thickness and chemical treatments frequency.

Values obtained thanks to empirical observations are utilized, more and more frequently, for MED models. Empirical correlations are indicated in [112], based on a high number of data, referring to both "clean" or "fouled" evaporators. Considering plants with adequate maintenance and cleaning frequency, the overall heat transfer coefficient is equal to:

$$U_i = U^* \left[1 - \alpha \frac{i-2}{6} \right] \quad \text{for } i = 2-8 \tag{4.32}$$

It is possible to note that U_i decreases gradually from second to eighth effect; U^* is assumed equal to 3.8 kW/m^2°C and α equal to 0.3. Therefore, U_i goes from 3.8 kW/m^2°C, for $i=2$, and 2.66 kW/m^2°C, for $i=8$.

Water and salt mass balances, for each effect, are:

$$\dot{M}_{D,i} + \dot{M}_{D,\text{flash inlet``}i\text{''}} + \dot{M}_{B,i} = \dot{M}_{B,i-1} \tag{4.33}$$

$$\left(\dot{M}_{D,i} + \dot{M}_{D,\text{flash inlet``}i\text{''}} \right) X_{D,i} + \dot{M}_{B,i} \cdot X_{B,i} = \dot{M}_{B,i-1} \cdot X_{B,i-1} \tag{4.34}$$

Brine salinity increases heavily from second to eighth effect, being $X_{D,i} \cong 0$. Because of environmental issues, $X_{B,8}$ cannot be higher than 70,000–75,000 ppm, since last effect brine in conveyed back to the sea.

At last, on the right side of Fig. 4.4, there is the feed water preheating section. Here, condensed distillate obtained by flash is considered, where an amount of heat is released for the feed preheating, which passes from $T_{f,i+1}$ to $T_{f,i}$. It is important to note that increasing the first-effect inlet feed water temperature, $T_{f,2}$, its sensible heat reduces and thus the distillate production for each kWh provided by the motive hot water increases. Feed preheaters energy balance can be written as follows:

$$\dot{Q}_{\text{preheater},i} = \dot{M}_{D,\text{flash inlet``}i\text{''}} \cdot \lambda_{D,i} = \dot{M}_f \cdot c_{p,f} \left(T_{f,i} - T_{f,i+1} \right) \tag{4.35}$$

Condenser

A simplified scheme of the condenser is shown in Fig. 4.5:

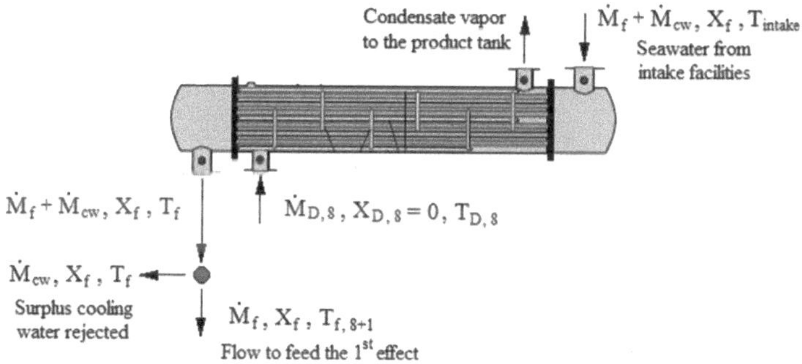

Fig. 4.5 Condenser simplified scheme.

Part of the latent heat release for the distillate condensation in the eighth effect, $\dot{M}_{D,8}$, is recovered by an economizer serving the MED condenser. Such heat is utilized to preheat water, which passes from T_{intake} to $T_{f,8+1}$, increasing feed final temperature admitted to first effect, $T_{f,2}$.

A supplementary cooling water flow rate, $\dot{M}_{cw}$, is usually supplied to remove the heat surplus. Generally, seawater, which is the sum of cooling and feed water, circulates in tube side, while condensing fluid in the shell side, as illustrated in Fig. 4.5.

No separation process are involved in condenser; thus, mass balances are not reported.

Appropriate heat transfer coefficients expressions on the shell and tube side, instead, are necessary for the energy balance and the heat transfer equations.

The energy balance and the heat transfer equations are reported:

$$\dot{Q}_{\text{condenser}} = \dot{M}_{D,8} \cdot \lambda_{D,8} = \left(\dot{M}_f + \dot{M}_{cw}\right) c_{p,f} \left(T_{f,8+1} - T_{\text{intake}}\right) \tag{4.36}$$

$$\dot{Q}_{\text{condenser}} = U_{\text{condenser}} \cdot A_{\text{condenser}} \cdot \Delta T_{\text{mL,condenser}} \tag{4.37}$$

$U_{condenser}$ has been evaluated by considering empirical correlations shown in [112] and it is assumed equal to 2.6 kW/m^2°C, considering $T_{D,i}$ approximately equal to 40 °C.

4.3.1.3 System Level

All the operating results have been determined thanks to the presented analytical model, once adopted all settings, indicated in Tables 4.1 and 4.2:

However, other system-level variables must be introduced. Firstly, the total freshwater flow rate provided, $\dot{M}_{D,total}$, is introduced:

Table 4.1 Main design parameters for the BP

Parameter	Description	Value	Unit
c_f	Water-specific heat	4.187	kJ/kgK
N_{CPVT}	Number of CPVT collector	400	/
A_{CPVT}	CPVT Aperture Area	12	m^2
v_{TK1}	TK1 Volume/CPVT Area	15	L/m^2
f_{TK2}	TK2 Volume/TK1 Volume	1	/
$T_{set,CPVT,summ}$	CPVT outlet set-point temperature, summer	90	°C
$T_{set,CPVT,wint}$	CPVT outlet set-point temperature, winter	60	°C
ΔT_{TK}	TK2 activation dead band	4	°C
V_{TK3}	TK3 Volume	20	m^3
$Q_{P4}=Q_{P5}$	P4, P5 flow	1.20×10^5	kg/h
q_{P4}	P4 flow/Number of CPVT collectors	300	kg/h
ΔT_{rated}	ACH-rated CHW temperature difference	5	°C
P_{ACH}	ACH-rated cooling capacity, $P_{ACH}=1.3Q_{P4}c_f\Delta T_{rated}$	700	kW
q_{P2}	P2 flow/CPVT area	100	$kg/h\,m^2$
n_{TK}	Number of Tank nodes	5	/
H_{TK}	Tank height	2	m
Q_{P1}	P1 flow	0.624×10^5	kg/h
$Q_{P3}=Q_{P7}$	P4, P5 flow	2.80×10^5	kg/h

$T_{act,summ}$	TK1 temperature minimum temperature for system activation, summer	75	°C
$T_{act,wint}$	TK1 temperature minimum temperature for system activation, winter	45	°C
$T_{TK,set,\ summer}$	TK1 temperature for switching to TK2, summer	86	°C
$T_{TK,set,winter}$	TK1 temperature for switching to TK2, winter	56	°C
$T_{set,MED}$	MED nominal inlet temperature	75	°C
Q_{P6}	P6 flow	1.80×10^5	kg/h
P_{AH}	AH-rated heating capacity	1200	kW
$COP_{ACH,rated}$	Rated ACH Coefficient of Performance	0.80	/
$\dot{M}_{motive}$	Nominal mass flow rate of motive hot water	50	kg/s
$A_{sensible}$	Sensible heat transfer area at first effect	10	m^2
A_{latent}	Latent heat transfer area at first effect	50	m^2
A_i	Heat transfer area in effects from 2 to 8	80	m^2
X_f	Salinity of feed water	38,000	ppm
$X_{B,8}$	Maximum salinity of disposed brine	72,000	ppm
$A_{condenser}$	Heat transfer area at the condenser	62	m^2
$\dot{M}_{cw}$	Mass flow rate of cooling water	60	kg/s
T_{intake}	Temperature of seawater at intake facilities	25	°C

Table 4.2 Main design parameters for the GP

Parameter	Description	Value	Unit
N_{CPVT}	Number of CPVT collector	400	/
A_{CPVT}	CPVT Aperture Area	12	m^2
v_{TK1}	TK1 Volume/CPVT Area	15	L/m^2
f_{TK2}	TK2 Volume/TK1 Volume	1	/
$T_{set,CPVT,summ}$	CPVT outlet set-point temperature, summer	90	°C
$T_{set,CPVT,wint}$	CPVT outlet set-point temperature, winter	60	°C
ΔT_{TK}	Solar heat is switched from TK1 to TK2 when TK1 top temperature is higher than $T_{set,CPVT}-\Delta T_{TK}$	4	°C
q_{P4}	P4 flow/Number of CPVT collectors	300	kg/h
q_{P2}	P2 flow/CPVT area	100	kg/h m^2
n_{TK}	Number of Tank nodes	5	/
$Q_{P3}=Q_{P8}$	P4, P5 flow	2.80×10^5	kg/h
Q_{P7}	P7	1.20×10^5	kg/h
T_{GW}	Geothermal Well fluid temperature	85	°C
$T_{GW,HE4}$	GW set point temperature exiting from HE4	50	°C
$T_{act,summ}$	TK1 temperature minimum temperature for system activation, summer	75	°C
$T_{act,wint}$	TK1 temperature minimum temperature for system activation, winter	45	°C
$T_{TK,set,summer}$	TK1 temperature for switching to TK2, summer	86	°C
$T_{TK,set,winter}$	TK1 temperature for switching to TK2, winter	56	°C
Q_{P6}	P6 flow	1.80×10^5	kg/h
$COP_{ACH,rated}$	Rated ACH Coefficient of Performance	0.80	/
$\dot{M}_{motive}$	Nominal mass flow rate of motive hot water	50	kg/s

$A_{sensible}$	Sensible heat transfer area at first effect	10	m^2
A_{latent}	Latent heat transfer area at first effect	50	m^2
A_i	Heat transfer area in effects from 2 to 8	80	m^2
X_f	Salinity of feed water	38,000	ppm
$X_{B,8}$	Maximum salinity of disposed brine	72,000	ppm
$A_{condenser}$	Heat transfer area at the condenser	62	m^2
M_{cw}	Mass flow rate of cooling water	60	kg/s
T_{intake}	Temperature of seawater at intake facilities	25	°C

$$\dot{M}_{D,\text{total}} = \beta\left(\sum_{i=1}^{8}\dot{M}_{D,i} + \sum_{j=2}^{8}\dot{M}_{D,\text{flash inlet "}j\text{"}}\right) \tag{4.38}$$

The constant β is necessary to take into account the process of remineralization; in fact, the distillate water is assumed completely salt free and, so, it cannot be provided as potable water.

For this reason, a preliminary mixing is necessary, through an appropriate quantity of chemically pretreated seawater. In this way, the total freshwater amount produced is slightly increased.The coefficient β can be obtained through:

$$\beta \cong 1 + \frac{X_{\text{drinkable}} - X_D}{X_f - X_{\text{drinkable}}} \tag{4.39}$$

The World Health Organization recommends an $X_{\text{drinkable}}$ value equal to 420 ppm. X_f and X_D have been assumed equal to 38,000 ppm and 0 ppm, respectively; with these values, β is 1.011.

Considering the same heat transfer surface for each effect, the total heat transfer area can be obtained as follows:

$$A_{\text{total}} = A_{\text{sensible}} + A_{\text{latent}} + \sum_{i=2}^{8} A_i + A_{\text{condenser}} \tag{4.40}$$

A specific energy consumption indicator, e_{MED}, is evaluated:

$$e_{\text{MED}} = \frac{\dot{M}_{\text{motive}} \cdot c_{p,\text{motive}}\left(T_{\text{motive,in}} - T_{\text{motive,out}}\right)}{3.6 \cdot \dot{M}_{D,\text{total}}} \left[\frac{\text{kWh}}{\text{m}^3}\right] \tag{4.41}$$

The investment cost of the plant is obtained as a function of the total installed heat transfer area, according to [113] and through a prudential approach, given by a lack of data in the literature:

$$J^1_{\text{MED}} = c \cdot \left(A_{\text{total}}\right)^{\gamma} \tag{4.42}$$

where the constant γ is assumed equal to 0.95 and c equal to 300 €/$m^{1.9}$.

By fixing a literature-derived and average unit freshwater cost, an alternative estimation of the plant investment cost can be obtained. A typical value is 800 € per m^3/day, compatible to the results shown in [114]. Although an intermittent operation is adopted due to the solar source availability, a 24-h/day operation must be considered to evaluate such investment cost value.

Therefore, it is possible to evaluate:

$$J_{\mathrm{MED}}^{2} = 800\left(\dot{M}_{D,\mathrm{total}} \cdot 3.6 \cdot 24\right) \tag{4.43}$$

The two analyzed approaches are very similar; so, the best solution is represented by the average value:

$$J_{\mathrm{MED}} = \frac{J_{\mathrm{MED}}^{1} + J_{\mathrm{MED}}^{2}}{2} \tag{4.44}$$

Primary Energy Savings

From an energy point of view, the last step is to calculate the Primary Energy Savings, PES or ΔPE, between the both polygeneration systems and a Reference System, RS, considering all the energy flows. In the RS, electricity is produced by a conventional thermoelectric power plant with an electric efficiency, $\eta_{\mathrm{el,t}}$, equal to 46.1%, or rather by a conventional diesel engine with $\eta_{\mathrm{el,RS}}$ equal to 0.35, thermal energy is supplied by a conventional natural gas-fired heater characterized by an average conversion efficiency equal to 80%, while cooling energy is provided by an electric heat pump characterized by an average Coefficient of Performance, COP, equal to 3.

The ΔPE value can be obtained through the following equation, summing the terms over all the time steps:

$$\Delta\mathrm{PE} = \sum_{i}\left(\frac{Q_{c,i}}{\eta_{\mathrm{el,RS}}\mathrm{COP}_{\mathrm{EHP,RS},i}} + \frac{Q_{h,i}}{\eta_{t,\mathrm{RS}}} + \frac{Q_{\mathrm{DHW},i}}{\eta_{t,\mathrm{RS}}} + \frac{E_{\mathrm{el,CPVT}} - E_{\mathrm{el,aux}}}{\eta_{\mathrm{el,RS}}}\right) \tag{4.45}$$

The first term of the equation represents the primary energy required to provide the same amount of cooling energy produced by the GP and BP systems through an Electrical Heat Pump, EHP; the primary energy required for space heating and Domestic Hot Water production is represented by the second and the third terms of the equation, considering a conventional boiler with an average efficiency equal to $\eta_{t,RS}$; the fourth term, lastly, represents the primary energy needed to provide the electricity difference between CPVT production and the amount utilized for auxiliary electrical components.

4.3.2 Economic Model

In this section, the economic model developed in order to carry out the system analysis from the operating and capital costs point of view is presented.

Eqs. (4.46)–(4.50) represent the economic relations implemented to estimate the capital cost (J_i) of the several components (CPVT collectors, absorption chiller, tanks, pumps, heat exchangers, and geothermal well) included in both plants:

$$J_{\mathrm{CPVT}} = 600 A_{\mathrm{CPVT}} \tag{4.46}$$

$$J_{\mathrm{ACH}} = 10^{-5} P_{\mathrm{ACH}}^{3} - 0.0393 P_{\mathrm{ACH}}^{2} + 244.53 P_{\mathrm{ACH}}^{3} + 95,494 \tag{4.47}$$

$$J_{\mathrm{TK}} = 494.9 + 0.808 V_{\mathrm{TK}} \tag{4.48}$$

$$J_{\mathrm{pump}} = 1.08\left(-0.00000002 Q_{\mathrm{pump}}^{2} + 0.0285 Q_{\mathrm{pump}} + 388.14\right) \tag{4.49}$$

$$J_{\mathrm{HE}} = 150\left(\frac{A_{\mathrm{HE}}}{0.093}\right)^{0.78} \tag{4.50}$$

The MED desalination unit capital cost was previously provided in the Section 4.3.1.

In case of GP system, the geothermal well (80-m depth) consists of 10 wells, each one equipped with a submerged pump. The total rated capacity of pumped geothermal water by each well is $12\,\mathrm{m}^3/\mathrm{h}$. The cost of whole geothermal system also takes into account the cost of insulation, pipes, heat exchangers, and is estimated on the basis of manufacturer data and is equal to 416 k€. In particular, it is derived from the unit cost of wells (200 €/m) and from the cost of each submerged pump (16 k€).

The yearly economic saving, ΔC_{op}, due to operating costs, achievable by the analyzed polygeneration plants (GP and BP) is particularly correlated to the obtained energy savings. In particular, such economic saving, evaluated by Eq. (4.51), calculates the yearly economic saving of both analyzed polygeneration plant (GP and BP) with respect to the reference system RS as follows:

$$\Delta C_{\mathrm{op}} = \sum_i \left(\frac{Q_{c,i}}{\mathrm{COP}_{\mathrm{EHP,RS},i}} + \frac{E_{\mathrm{el,CPVT}} - E_{\mathrm{el,aux}}}{\eta_{el,t}} \right) c_{ue} + c_{ut}\left(Q_{h,i} + Q_{\mathrm{DHW},i}\right) + m_{\mathrm{wat},i} c_{u\mathrm{wat}} - \frac{PE}{\eta_{t,\mathrm{AH},i} \mathrm{LHV}_{\mathrm{biomass}}} c_{u,\mathrm{biomass}} \tag{4.51}$$

In Eq. (4.51), the first term represents the electrical economic saving, the second one the thermal economic saving (space heating and DHW production), the third one considers the economic saving due to production of desalinated water and, finally, the last term is related to operating cost in case of the polygeneration plant with biomass auxiliary heater (BP).

In the presented economic analysis in this chapter, the unit costs of electrical energy (c_{ue}) and thermal energy (c_{ut}) were assumed equal to 0.15 €/kWh and 0.075 €/kWh, respectively, whereas the unit cost of desalinated water (c_{uwat}) and of biomass (wood-chip) was set to 1.2 €/m^3 and 0.06 €/kg (LHV = 3.7 kWh/kg), respectively. The last term is obviously zero in case of polygeneration plant without biomass auxiliary heater.

It is significant noting that in GP, a potential supplementary economic revenue could be obtained by the hot water produced for domestic hot water purposes at HE4 outlet by the geothermal hot water. Anyway, the data concerning the total demand of the water required to the user are not available. Therefore, such revenue is, in this analysis, conservatively neglected.

The economic feasibility of the system was assessed in terms of Simple Pay-Back period (SPB). This evaluation was carried out by taking into account two different scenarios: with and without public founding, as defined in Ref. [115].

In the scenario with public founding, the following feed-in tariffs were supposed:

- 0.45 €/kWh for electric energy;
- 0.20 €/kWh for cooling energy;
- 0.20 €/kWh for thermal energy;
- 1.0 €/m^3 for desalinated water.

The considered feed-in tariffs proposed in this chapter are on average equal to ones assumed in EU for electrical, cooling, and thermal production by solar energy [115]. Instead, the economic incentive for the desalinated water production is assumed only to take into account a potential future spread of desalination technology, since generally no EU Country provides public financial to support it.

4.3.3 Exergy Model

The set of equations shown in the paragraph "energy model" allows one to define the energy flows and the thermodynamic parameters of the components included in the investigated polygeneration plants. Once these data are obtained, an additional submodule of whole developed simulation model evaluates the local exergy performance for each component of plant. Such exergy model consists of exergy balances considering the physical exergy correlated to the energy and material streams incoming and leaving each system component, and chemical exergy due to significant chemical reactions, occurring in components as the biomass auxiliary heater and the MED unit.

Eq. (4.52) calculates the physical exergy for all the fluids involved in the analyzed polygeneration systems:

$$\begin{aligned} \mathrm{ex}_{\mathrm{ph},f}\left(t_f, p_f\right) &= \left[h_f\left(t_f, p_f\right) - h_f\left(t_a, p_a\right)\right] - T_a\left[s_f\left(t_f, p_f\right) - s_f\left(t_a, p_a\right)\right] \\ &= c_f\left[\left(T - T_a\right) + v_f\left(p - p_a\right)\right] - c_f T_a \ln\left(\frac{T}{T_a}\right) \end{aligned} \quad (4.52)$$

Note that all the considered fluids are in liquid phase and, therefore, the term $v(p-p_a)$ is on average negligible with respect to the other terms. This result is also confirmed by the results obtained in Ref. [54] Here it is shown that physical exergy due to the pressure drops is completely negligible with respect to the exergy associated to differences of temperature.

Consequently, Eq. (4.52) is simplified in Eq. (4.53) and is used to determine exergy per unit mass for all the streams of system.

$$\begin{aligned} \mathrm{ex}_{\mathrm{ph},f}\left(t_f, p_f\right) &= \left[h_f\left(t_f, p_f\right) - h_f\left(t_a, p_a\right)\right] - T_a\left[s_f\left(t_f, p_f\right) - s_f\left(t_a, p_a\right)\right] \\ &= c_f\left(T - T_a\right) - c_f T_a \ln\left(\frac{T}{T_a}\right) \end{aligned} \quad (4.53)$$

where

T and T_a are the temperatures of the considered state point and the temperature of the dead state, respectively;

p and p_a are the pressure of the considered state point and the pressure of the dead state, respectively;

c_f is the fluid-specific heat.

In the carried out exergy analysis, T_a was assumed equal to the fix value of 25°C, a value reasonably representative of the yearly average outdoor ambient temperature, whereas a constant dead state pressure, $p_a = 1.013$ bar, was assumed. This assumption is supported also by results achieved in Ref. [116], where it is declared that "the results of energy and exergy analysis are usually not significantly sensitive to reasonable variations in these properties."

Concerning the evaluation of chemical exergy, the chemical exergy of reaction is needed to take into account. As example, the exergy flow due to the inlet biomass at the biomass auxiliary heater is computed as proposed by Song et al. [117], i.e.:

$$\dot{E}x_{\mathrm{biomass}} = \dot{m}_{\mathrm{biomass}} \varepsilon_{\mathrm{biomass}} \quad (4.54)$$

The biomass used in the investigated case is the wood chip and its specific chemical exergy, $\varepsilon_{biomass}$, is assumed equal to 19.37×10^3 kJ/kg [121].

Each flow of exergy correlated to the heat transfer is calculated by correcting the heat flow through the corresponding Carnot factor [118].

$$\dot{E}x_{\dot{Q}} = \dot{Q}\left(1 - \frac{T_a}{T}\right) \tag{4.55}$$

Note that in Eq. (4.55), by taking into account that heat transfers are associated to losses toward the outdoor ambient, the state point temperature T is coincident to the temperature of the dead state T_a and, then, the associated flow of exergy correlated to the heat transfer is null.

In this paragraph, the equations of the exergy balances related to the CPVT collectors and to the MED unit are shown, whereas for sake of brevity, the equations of exergy balances for other components are omitted.

4.3.3.1 CPVT Collectors

In case of CPVT collectors, the following exergy flows are identified:

(i) the exergy flow due to the solar radiation

This exergy flow can be calculated by adopting several approaches (Petela, Jeter, and others) [119] belonging to the category of the "technical boundary approaches." In particular, the technical boundary approaches are different on the basis of the selected sun temperature.

A totally different approach is the "physical boundary approach," which neglects the conversion of solar radiation into thermal energy in the solar collector. Agreeing to the common practice presented in works of literature for the "technical boundary approach" [119], in this analysis, the calculation of exergy flow due to the solar radiation is carried out by the Petela theorem, in which the sun temperature is fixed at 4077 °C (3/4 of the corresponding black body temperature) [120]:

$$\dot{E}x_{\text{sun}} = A \cdot I \cdot \left(1 - \frac{T_a}{T_{\text{sun}}}\right) \tag{4.56}$$

(ii) The exergy flow due to produced electrical energy. This exergy flow is numerically coincident with the electricity produced [118].

(iii) The physical exergy due to the fluid stream, entering and exiting from the ith slice of the computational domain, by adopting Eq. (4.53)

(iv) The physical exergy due to the thermal losses (convective and radiative) to the environment

By considering a control volume containing the triangular receiver of the ith slice of the computational domain, the exergy balance can be written as:

$$\left[A_{\text{PVT},i} I_b C_{\text{PVT}} \eta_{\text{opt}} \text{IAM}_{\text{th}} \left(1-\rho_{\text{PVT},i}\right) + A_{\text{top},i} I_{\text{tot}} \alpha_{\text{top},i}\right]\left(1-\frac{T_a}{T_{\text{sun}}}\right)$$
$$= \dot{m}_f \left[ex_{ph,f}\left(t_{i+1},p_{i+1}\right) - ex_{ph,f}\left(t_i,p_i\right)\right]$$
$$+\dot{P}_{\text{PVT,gross},i} + \left(\dot{Q}_{\text{PVT-conc},i} + \dot{Q}_{\text{conv,PVT},i}\right)\left(1-\frac{T_a}{T_{\text{PVT}}}\right) \tag{4.57}$$
$$+\left(\dot{Q}_{\text{conv,top},i} + \dot{Q}_{\text{top-sky},i}\right)\left(1-\frac{T_a}{T_{\text{top}}}\right) + \dot{E}x_{d,i}$$

where $\dot{E}x_{d,i}$ is the destroyed exergy at the ith slice of the computational domain.

As concerns, the conventional calculation of the exergy balance [118], the exergy fuel F_i for the ith slice is coincident to the incident solar radiation on the CPVT:

$$F_i = \left[A_{\text{PVT},i} I_b C_{\text{PVT}} + A_{\text{top},i} I_{\text{tot}}\right]\left(1-\frac{T_a}{T_{\text{sun}}}\right) \tag{4.58}$$

Instead, the *exergy product* P_i is due to both produced electricity and to the increase of fluid physical exergy, as shown in the following equation:

$$P_i = \dot{m}_f \left[ex_{ph,f}\left(t_{i+1},p_{i+1}\right) - ex_{ph,f}\left(t_i,p_i\right)\right] + \dot{P}_{\text{PVT,gross},i} \tag{4.59}$$

Finally, the *exergy residual*, R_i which corresponds to the exergy flow dissipated to the outdoor ambient, is obtained as:

$$R_i = \left(\dot{Q}_{\text{PVT-conc},i} + \dot{Q}_{\text{conv,PVT},i}\right)\left(1-\frac{T_a}{T_{\text{PVT}}}\right) + \left(\dot{Q}_{\text{conv,top},i} + \dot{Q}_{\text{top-sky},i}\right)\left(1-\frac{T_a}{T_{\text{top}}}\right)$$
$$+\left[A_{\text{PVT},i} I_b C_{\text{PVT}}\left(1-\eta_{\text{opt}} \text{IAM}_{\text{th}}\left(1-\rho_{\text{PVT},i}\right)\right) + A_{\text{top},i} I_{\text{tot}} \rho_{\text{top},i}\right]\left(1-\frac{T_a}{T_{\text{sun}}}\right) \tag{4.60}$$

It is important to note that R_i is an exergy loss because it is not possible to convert such exergy flow in additional work. Consequently, the local exergy performance can be estimated by using the local exergy efficiency, calculated as:

$$\eta_{ex,i} = \frac{P_i}{F_i} \tag{4.61}$$

In the same way, the exergy destruction for each slice of the computational domain is computed by the sum of destroyed exergy and the exergy residual.

4.3.3.2 MED Unit

In the following, the exergy balances of the MED unit are presented.

The chemical exergy related to the material flow $\dot{m}_{\text{fresh}}$ of desalinated water could be evaluated as in Eq. (4.62), [121]:

$$\dot{E}x^{C}_{\text{fresh}} = \dot{N}_{w,\text{fresh}} \cdot \omega_{\text{th},0} \tag{4.62}$$

where

- $\dot{N}_{w,\text{fresh}}$ is the molar flow rate (kmol/s) evaluated as:

$$\dot{N}_{w,\text{fresh}} = \frac{\dot{m}_{\text{fresh}}}{MM_{\text{water}}}$$

with MM_{water}, molar weight of water in kg/kmol, equal to 18 kg/kmol.

- $\omega_{\text{th},0}$ is the minimum work of separation in case of null recovery ratio (in kJ per kmol of freshwater), expressed as follows:

$$\omega_{\text{th},0} = \varphi \cdot R_{\text{universal}} \cdot T_0 \cdot X_{s,\text{feed}}$$

where

φ is the dissociation factor (assumed equal to 2 for NaCl)

$R_{\text{universal}}$ is the universal constant of gases (kJ/kmol K)

$X_{s,feed}$ is the molar concentration of salts in the seawater. Supposing that the salinity of the feed is 38,000 ppm (reasonable value of the Mediterranean area), a molar concentration in the order of 0.01158 is achieved.

4.3.4 Exergoeconomic Model

The exergoeconomic model here presented is developed only for the biomass polygeneration plant (BP) and is based on the notation implemented in Fig. 4.6.

Here, all the representative material streams related to the corresponding exergy flows are sequentially numbered in order to better comprehend the equations in the following presented. In the model, the average exergoeconomic costs were calculated:

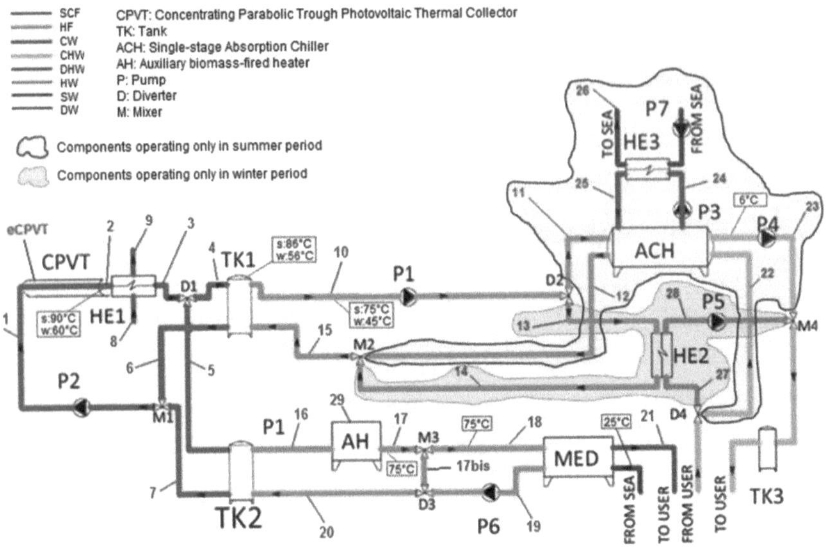

Fig. 4.6 SHC, CPVT, Geothermal Well, and MED system layout [126].

(i) on weekly basis, in order to identify the variations of the unit exergoeconomic costs due to the variability of the total weekly solar energy;

(ii) on seasonal basis, by taking into account the "winter" season, in which the produced hot water by HE2 is used for space heating, and a "summer" season, in which the produced cold water by ACH is used for space cooling;

(iii) on annual basis.

The aim to calculate the seasonal and weekly exergoeconomic costs is well known in literature. In fact, when a system supplied by a Renewable Energy Source (RES), featured by an irregular availability, is analyzed from the exergoeconomic point of view, the unit cost of the produced outputs experiments [123] significant variations, which may be better interpreted, if an adequate temporal base is selected.

The method implemented in the developed exergoeconomic model is based on the definition of a unit cost for each of the exergy flows connected to the material streams (as numbered in Fig. 4.6). This methodology allows to reduce the uncertainty of the results, since it is not required to specify "Products" and "Fuels" for each component. In literature, the exergoeconomic analysis is commonly implemented for systems with quasisteady or steady operational conditions, and therefore, cost balances, as ones after presented, were formulated in this form:

$$\sum_{i\in\{\text{inlets}\}} c_i \cdot \dot{E}x_i + \dot{Z}_{\text{unit}} = \sum_{j\in\{\text{outlets}\}} c_j \cdot \dot{E}x_j \tag{4.63}$$

where $\dot{Z}_{unit}$ and $\dot{E}x$ are expressed in €/s and kW_{ex} (or in €/h and kJ_{ex}/h), respectively.

Such methodology, confirmed reliable in large power plants applications, integrally adopts that average or instantaneous hourly exergy flows continue to be constant during the operating period. Anyway, because of operating conditions extremely dynamic of the examined polygeneration plants, in this chapter a different technique is implemented, by substituting the exergy flows $\dot{E}x$ in Eq. (4.63) with the integrated exergy flows over a well-established basis of time:

$$Ex_i = \int_{\text{period}} \dot{E}x_i \cdot dt \tag{4.64}$$

where the period can be fixed alternatively equal to the whole year, the season, or the week. As regards the cumulative cost related to the capital cost to install and purchase each component or unit, it will be expressed as follows:

$$\tilde{Z}_{\text{unit}} = Z_{\text{unit}} \cdot \text{CRF}\frac{\tau_{\text{operation}}}{8760} = Z_{\text{unit}} \frac{i\cdot(1+i)^n}{(1+i)^n - 1}\frac{\tau_{\text{operation}}}{8760} \tag{4.65}$$

where $\tau_{\text{operation}}$ (h) is the length of the basis of time adopted in the analysis, i is the interest rate, and CRF is the Capital Recovery Factor.

Concerning the exergoeconomic model, it must be principally noted that the system works 19 weeks per year during the "winter season," 32 weeks per year in "summer season," and 1 week per year in "switching season" by which we can expect within the same week both cold water production for space cooling and hot water production for space heating. Therefore, since a single model could allocate incorrectly the costs when comprehending some null exergy flows, it was definite to separate the model into three submodels:

- Submodel 1 in the "summer season" for seasonal and weekly analyses;
- Submodel 2 in the "winter season" for seasonal and weekly analyses;
- Submodel 3 in the "switching season" for the annual analysis.

This last submodel is less important, since it contains both the flows occurring in the summer and ones in winter season; therefore, the details of this model will be not shown.

4.3.4.1 Submodel 1—"Summer Season"

The basic system layout assumed for the summer season could be simply achieved in Fig. 4.6, by not including the shaded area with thin dotted red line contour counting the "components operating only in winter period." For each component included in the exergoeconomic submodel, cost balances and "$n_{\text{exits}} - 1$" auxiliary equations for each component associated with multiple (i.e., $n_{\text{exits}} > 1$) exiting exergy flows are written. The development of such auxiliary equations is a key point of the model [124,125]; therefore, it will be briefly described for each case.

In order to obtain the highest level of precision and descriptiveness of the exergoeconomic results, each component should be hypothetically modeled as a distinct unit. Anyway, a slightly simplified method was assumed, based on the following hypotheses:

- The pumps are not considered as "units," but their investment cost as well as the cost related to their consumption of electricity are assigned to the components "consuming" (by using part of their exergy) the exergy of the circulated fluid. This work hypothesis is acceptable by taking into account the high incidence of pumps thermal exergy with respect to their exergy consumption and the mechanical exergy of fluids;
- A single unit containing the tank TK3 (i.e., the volume of the hydraulic circuit), the ACH, and the open cooling loop with seawater is modeled. This unit also includes the pumps P1, P3, P4, and P7 with the related pipes. Therefore, the exergy flow nos 24 and 25, which are neither exiting nor entering this "macrounit," are not included in cost balances. The advantage of this methodology is evident, when it is considered that the streams 24, 25, and 26 and the component HE3 are only connected to the necessity to reject the heat produced at the absorber and condenser of the ACH, toward the environment. Therefore, the heat exchanger HE3 could be considered as a "dissipative component" [126] and the low level of heat removed could be seen as "useful to the ACH operation." Anyway, taking into account in detail the streams 24, 25, and 26 are not particularly useful; therefore, the assumed aggregation grade appears a good compromise in order to achieve results easily to understand.

CPVT

The capital resource $\tilde{Z}_{CPVT}$ of CPVT collectors is due to the "consumption" of exergy for the increase of the sensible thermal exergy of stream 1 and contemporaneously for the production of the electricity flow Ex_{eCPVT} (supposed as pure exergy).

$$c_1 \cdot Ex_1 + c_{eCPVT} \cdot Ex_{P2} + \tilde{Z}_{eCPVT} + \tilde{Z}_{P2} = c_2 \cdot Ex_2 + c_{eCPVT} \cdot Ex_{eCPVT} \; (\text{cost balance}) \tag{4.66}$$

As concerns the auxiliary equation for CPVT collectors, since both the thermal exergy increase of the fluid stream and the electricity flow are considered as "useful products," it was assigned the same unit exergy cost [127] evaluated by Eq. (4.67):

$$c_{eCPVT} = \frac{c_2 \cdot Ex_2 - c_1 \cdot Ex_1}{Ex_2 - Ex_1} \; (\text{auxiliary equation}) \tag{4.67}$$

HE1

For the heat exchanger HE1, the cost balance is obtained by the following equation:

$$c_2 \cdot Ex_2 + c_8 \cdot Ex_8 + \tilde{Z}_{HE1} = c_3 \cdot Ex_3 + c_9 \cdot Ex_9 \; (\text{cost balance}) \tag{4.68}$$

For this component, two auxiliary equations ((4.69) and (4.70)) are required.

The unit cost of streams 2 and 3 is supposed equivalent, agreeing to the well-known principle of the "exergy extraction." In fact, the exergy Ex_3 is evaluated as "thermal exergy not removed from the material stream 2"and is considered as a residual exergy, which shares with Ex_2 the same *cost formation process* (called also "productive history"), and therefore, the unit cost of these streams must coincide:

$$c_3 = c_2 \; (\text{auxiliary equation } a) \tag{4.69}$$

Since the inlet water at HE1 shows the same conditions of the local water grid, the Ex_8 is featured by unit cost, which can be assumed equal to zero:

$$c_8 = 0 \; (\text{auxiliary equation } b) \tag{4.70}$$

Diverter 1, D1

The cost balance of diverter 1 is completely banal, in fact as shown in Eq. (4.71), its capital cost is neglected. Anyway, this equation is here presented but it was presented at the aim to correctly allocate the cost of stream 3 on the streams 4 and 5 in proportional mode to the diverted mass flows. Furthermore, the auxiliary equation of diverter D1 states that exiting streams 4 and 5 are featured by the same unit cost. This is due since the diverter D1 does not imply any exergy destruction or conversion process of energy.

$$c_3 \cdot Ex_3 = c_4 \cdot Ex_4 + c_5 \cdot Ex_5 \quad \text{(cost balance)} \tag{4.71}$$

$$c_4 = c_5 \quad \text{(auxiliary equation)} \tag{4.72}$$

Storage Tank TK1

The Storage Tank 1 cost balance is reported in Eq. (4.73):

$$c_4 \cdot Ex_4 + c_{12} \cdot Ex_{12} + \tilde{Z}_{\mathrm{TK1}} = c_6 \cdot Ex_6 + c_{10} \cdot Ex_{10} \quad \text{(cost balance)} \tag{4.73}$$

The auxiliary equation is achieved by applying the principle of the "exergy extraction," by considering that the Ex_6 is estimated as "thermal exergy not extracted from the material stream 4":

$$c_6 = c_4 \quad \text{(auxiliary equation)} \tag{4.74}$$

"ACH + HE3 + auxiliaries"

These components were modeled as a single components and in their cost balance a cumulative capital cost containing the costs of the pumps P3, P4, and P7, the HE3 and ACH is included. Such values were obtained by supposing "as $\tau_{\mathrm{operation}}$" in Eq. (4.65), the interval of time coincident with the summer period (i.e., the interval of time in which the produced cold water is used for space cooling purposes). Conversely, the cumulative capital costs related to the P1 and TK3 components were estimated by supposing as "$\tau_{\mathrm{operation}}$" in Eq. (4.65), the interval of time coincident with the whole year, since such units work both during the winter and summer periods:

$$\begin{aligned} &c_{10} \cdot Ex_{10} + c_{22} \cdot Ex_{22} + c_{e\mathrm{CPVT}} \left(Ex_{P1} + Ex_{P3} + Ex_{P4} + Ex_{P7} \right) \\ &\quad + \tilde{Z}_{\mathrm{ACH}} + \tilde{Z}_{\mathrm{HE3}} + \tilde{Z}_{\mathrm{TK3}} + \tilde{Z}_{P1} + \tilde{Z}_{P3} + \tilde{Z}_{P4} + \tilde{Z}_{P7} = c_{12} \cdot Ex_{12} \\ &\quad + c_{23} \cdot Ex_{23} + c_{26} \cdot Ex_{26} \quad \text{(cost balance)} \end{aligned} \tag{4.75}$$

Two auxiliary equations are here necessary:

$$c_{22} = c_{23} \quad \text{(auxiliary equation } a\text{)} \tag{4.76}$$

$$c_{26} = 0 \quad \text{(auxiliary equation } b\text{)} \tag{4.77}$$

Equation (*a*) is based on, as previously explained, the "exergy extraction" principle, and adopts a same unit cost for the streams 22 and 23; in fact, when the cooling exergy is delivered to the users, the exergy of stream 22 is equivalent to the "cooling exergy not extracted from stream 23."

The auxiliary equation (*b*) arises from fixing an exergoeconomic cost to the stream 26 equivalent to the zero, whose thermal exergy is wasted to the outdoor ambient, therefore representative of a "residue" (agreeing to the common nomenclature adopted in thermoeconomics) [128,129]:

Storage Tank TK2

The Storage Tank 2 auxiliary equation and cost balance are the same of those described for the Storage Tank 1:

$$c_5 \cdot Ex_5 + c_{20} \cdot Ex_{20} + \tilde{Z}_{\mathrm{TK2}} = c_7 \cdot Ex_7 + c_{16} \cdot Ex_{16} \quad \text{(cost balance)} \tag{4.78}$$

$$c_7 = c_5 \quad \text{(auxiliary equation)} \tag{4.79}$$

M1

The cost balance of the mixer M1 is also banal, by not involving such component significant conversion processes of energy:

$$c_6 \cdot Ex_6 + c_7 \cdot Ex_7 = c_1 \cdot Ex_1 \quad \text{(cost balance)} \tag{4.80}$$

Auxiliary Heater, AH

The cost balance of the biomass Auxiliary Heater is totally intuitive. In fact, this component is supplied by the sole input of energy (the biomass consumed) featured by a non-null cost and coincident with the stream 29; this cost was calculated by means of an *external assessment* (i.e., at its current price of market), which for the AH component represents an auxiliary equation:

$$c_{16} \cdot Ex_{16} + c_{29} \cdot Ex_{29} + \tilde{Z}_{\mathrm{AH}} = c_{17} \cdot Ex_{17} \quad \text{(cost balance)} \tag{4.81}$$

$$c_{29} = 0.01095€ / \mathrm{kWh_{ex}} \quad \text{(auxiliary equation)} \tag{4.82}$$

Mixer M3

The mixer M3 cost balance is obtained in parallel to that written for the mixer M1:

$$c_{17} \times Ex_{17} + c_{17bis} \times Ex_{17bis} = c_{18} \times Ex_{18} \quad \text{(cost balance)} \tag{4.83}$$

MED

The cost balance of the MED component is written as represented in Eq. (4.84):

$$c_{18} \cdot Ex_{18} + c_{e\mathrm{CPVT}} \cdot Ex_{P6} + \tilde{Z}_{\mathrm{MED}} + \tilde{Z}_{P6} = c_{19} \cdot Ex_{19} + c_{21} \cdot Ex_{21} \quad \text{(cost balance)} \tag{4.84}$$

It is significant to note that in order to write a complete balance, an additional exergy flow should be involved in the balance. Such exergy flow, neither numbered nor depicted in Fig. 4.6, is due to the high concentration brine rejected to the environment.

Anyway, since the concentration exergy from brine is not simply obtainable, this exergy flow should be considered in the same way of the thermal exergy of stream 26 (this means that a null exergoeconomic cost should be assigned). Therefore, the exergy of the discarded brine can be neglected without affecting the results stability.

Regarding the auxiliary equation, the principle "exergy extraction" is clearly again adopted:

$$c_{19} = c_{18} \quad (\text{auxiliary equation}) \tag{4.85}$$

Diverter D3

The cost balance for the diverter D3 and its auxiliary equation is easily written in analogy to those written for D1:

$$c_{19} \cdot Ex_{19} + c_{17bis} \cdot Ex_{17bis} + c_{20} \cdot Ex_{20} \quad (\text{cost balance}) \tag{4.86}$$

$$c_{17bis} = c_{20} \quad (\text{auxiliary equation}) \tag{4.87}$$

4.3.4.2 Submodel 2—"Winter Season"

The basic layout adopted for the winter season operation is simply obtained from Fig. 4.6, by eliminating the "Components operating only in the summer period," contained into shaded area contoured by bold continuous blue line.

By quickly analyzing Fig. 4.6 and the consequent layouts for summer and winter operation seasons, it is noted that the most of the components of the winter season can be modeled with the same equations described in the previous section. During the winter season, anyway, the model valid for the components HE3, ACH, and auxiliaries, which are included into a single unit, should not be here considered, whereas a new model must be presented for the heat exchanger HE2 (employed to produce the hot water for space-heating purposes).

Heat Exchanger HE2

The heat exchanger HE2 cost balance is written as follows:

$$\begin{aligned} c_{10} \cdot Ex_{10} + c_{27} \cdot Ex_{27} + c_{eCPVT} \cdot \left(Ex_{P1} + Ex_{P5}\right) \\ + \tilde{Z}_{HE2} + \tilde{Z}_{TK3} + \tilde{Z}_{P1} + \tilde{Z}_{P5} \quad (\text{cost balance}) \\ = c_{14} \cdot Ex_{14} + c_{28} \cdot Ex_{28} \end{aligned} \tag{4.88}$$

In order to define the auxiliary equation, the same method adopted for the exergy flows related to the exergy flows 22 and 23 is followed, by supposing, then, for streams 27 and 28 an equal unit exergoeconomic cost. This is due because the stream 27 has an exergy representing "cooling exergy not extracted from stream 28" at moment that the hot fluid is delivered to the users by producing space heating:

$$c_{27} = c_{28} \quad \text{(auxiliary equation)} \tag{4.89}$$

4.4 RESULTS

In this paragraph, the results of the thermoeconomic, exergy and exergoeconomic analyses are presented. In particular, the results of the polygeneration system including the solar collectors and the geothermal well (GP) are presented only from the thermoeconomic point of view, whereas for the polygeneration system including the solar collectors and the biomass auxiliary heater (BP), the results achieved by all the models described in the previous paragraph are shown in detail.

The weather zones selected as case study are Naples and Pantelleria (South of Italy) for the BP and GP systems, respectively. In case of GP system, Pantelleria was selected since it is a small volcanic island featured by scarcity of potable water and fossil fuels, but a significant availability of solar and geothermal energy, therefore ideal zone for the installation of this kind of polygeneration plant.

For both cases, (BP and GP systems), the key operational/design parameters are listed in Tables 4.1 and 4.2, respectively. Further details regarding some specific design parameters of the CPVT collectors are listed in Ref. [103] and for other components in Refs. [98–102].

Note that the systems are designed so that a variation of the solar collector number leads to a proportional variation of other design parameters (HE capacities, ACH, TK volumes, etc.). In fact, the included components were parametrically designed, so that in case a variation of the solar field size occurs, all components can be automatically resized as a function of the capacity of solar field. The same is for the flowrate q_{P4}: its variation determines a proportional variation in the capacity of HE, pumps, and ACH, without to affect the CPVT field capacity.

The GP and BP systems are designed in order to match the MED unit thermal energy demand equal to about 1.0 MW. The absorption chiller, solar collector, and the storage tanks sizes are obtained by considering this data, as shown in Tables 4.1 and 4.2.

As mentioned earlier, the dynamic simulation tool adopted in the developed cases study is able to provide the results for whatever time basis (year, months, weeks, days, and seconds); therefore, in order to better understand the variation of the main energy/exergy parameters, the temperature trends, or the system performance indexes, the results will be shown by considering the years, weekly, and daily integration base.

4.4.1 Thermoeconomic Analysis Results

4.4.1.1 BP System

In Fig. 4.7, the representative temperature trends for a summer day are displayed.

As depicted, Fig. 4.7 shows the trends of temperature from 4.00 am to 8.00 pm (5980–5996 h). In such figure, the operation strategy implemented to correctly manage the operation of the pump P2 and the diverter and mixer (D1 and M1) located upstream to the TK1 and TK2 is shown. In fact, in absence of radiation incident on the CPVT collectors, during the hours of the early morning (from 4.00 am to 6.00 am), the temperatures of both TK tanks (1 and 2) decrease since they are not heated by solar thermal energy (their thermal losses must be also taken into account). During the same operation hours, the MED unit is switched off, whereas at about 6.00 am, when the CPVT outlet temperature fast increases up to the fix set-point (equal to 90°C), both MED and P2 subsystems are activated by the

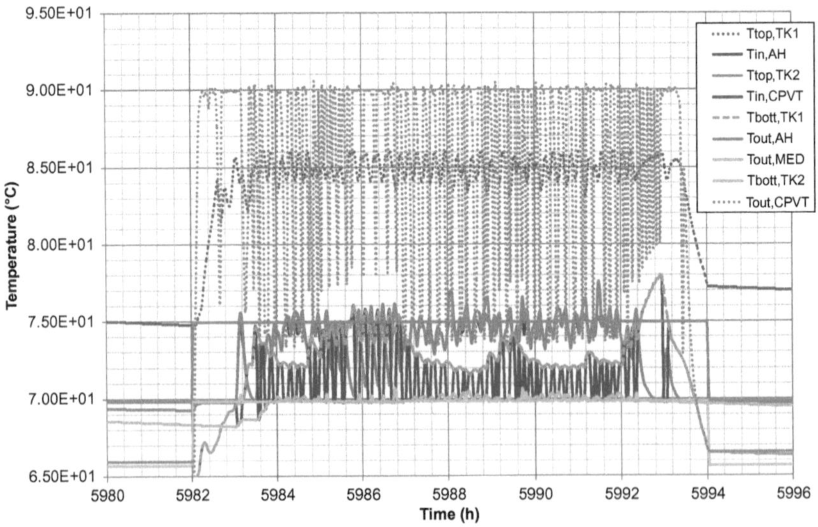

Fig. 4.7 Temperatures, summer day [24].

radiation sensors. In fact, during these hours, the pump P2 starts to supply the CPVT collectors by the water stored at bottom of tank TK1 at about 65°C and the hot water produced by CPVT field is stored into TK1 at about 75°C (see hour 6.00 am), as shown by trend of the temperatures of top nodes of TK1. In this case, for the adopted operation strategy, the CPVT supplies heat only to TK1 since the TK1 top temperature is lower than 86°C. As it is easily observed, the TK1 top temperature increases, whereas the bottom one decreases since no hot water flowrate is provided to the hot side of this tank.

Around 7.00 am, TK1 top temperature reaches 86°C, M1 and D1 switch from TK1 to TK2, and, in this circumstance, an important increase of TK2 top and bottom temperatures occurs. When TK1 top temperature is lower 84°C (this occur after some minutes), the controller supplies again TK1. This nonstop switching operation, observed for all day, leads to also a variation of the CPVT inlet water flow rate. Therefore, the CPVT outlet temperature depends also on the tank to be supplied. It is also worth noting that for the shown day, the HE1 is never activated, since CPVT outlet temperature is always lower than 100°C.

Instead, the biomass auxiliary heater AH is frequently switched on, and this means that top the temperature of TK2 is frequently lower than 75°C.

This is shown in Fig. 4.8, by observing that the thermal flow rate necessary to activate the MED unit (Q_{MED}) is in part due to the biomass auxiliary

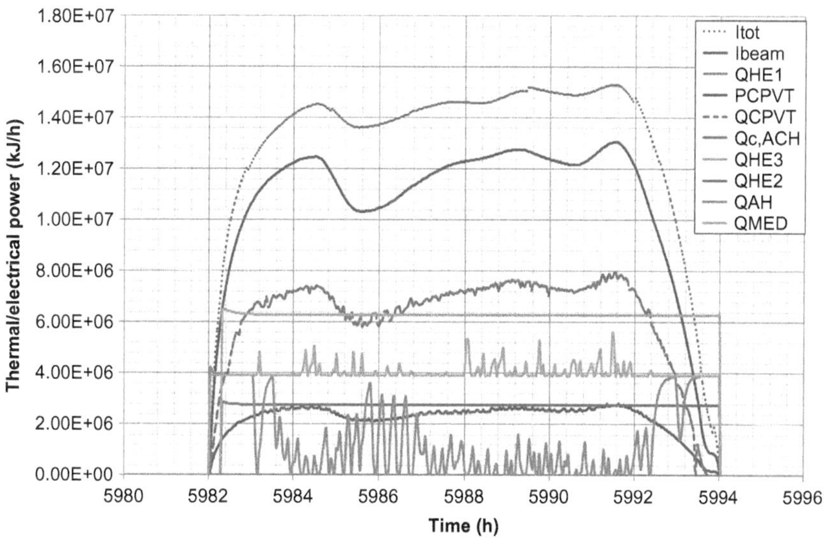

Fig. 4.8 Thermal and electrical power, summer day [24].

heater (Q_{AH}). The small fluctuations noted for the thermal input of MED are basically motivated by the oscillations of the inlet fluid temperature. In Fig. 4.8, the trends of both total (I_{tot}) and beam radiation (I_{beam}), and as a consequence, the trends of CPVT thermal power (Q_{CPVT}) and electrical power (P_{CPVT}) are also depicted. Here it can be noted that the ACH cooling power (Q_{ACH}) is almost constant for all operation hours and this is basically due to TK1 top temperature, which is constantly close to 85°C. This fact leads to the no deactivation of ACH (lower than 75°C) and as a consequence, it is able to provide an almost constant cooling power. Therefore, the thermal power necessary to the heat exchanger HE3 (Q_{HE3}) is also constant as well as the ACH coefficient of performance.

Fig. 4.8 also shows the excellent summer system performance in terms of large produced desalinated water and cooling energy, although a low quantity of biomass is used.

Conversely, during the winter season the performance of the system is unfortunately different, as shown in Fig. 4.9. Here, the temperature trend for a casual day of winter season (from 6.00 am to 6.00 pm) is reported.

It is evident as the CPVT outlet temperature is close to the set point (60°C) only during the first hours of the day, due to the reduction of the incident beam radiation, although the flowrate of P2 pump is reduced to the minimum possible value. Also TK1 top temperatures decrease with respect

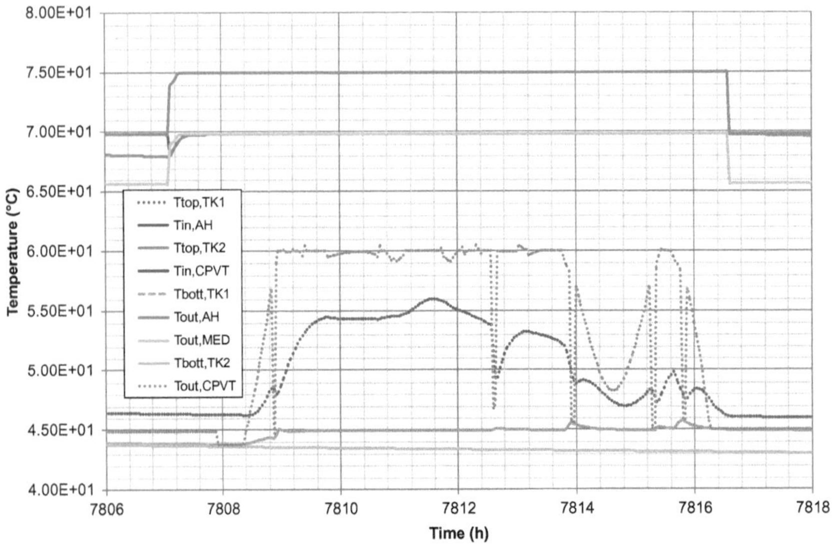

Fig. 4.9 Temperatures, winter day [24].

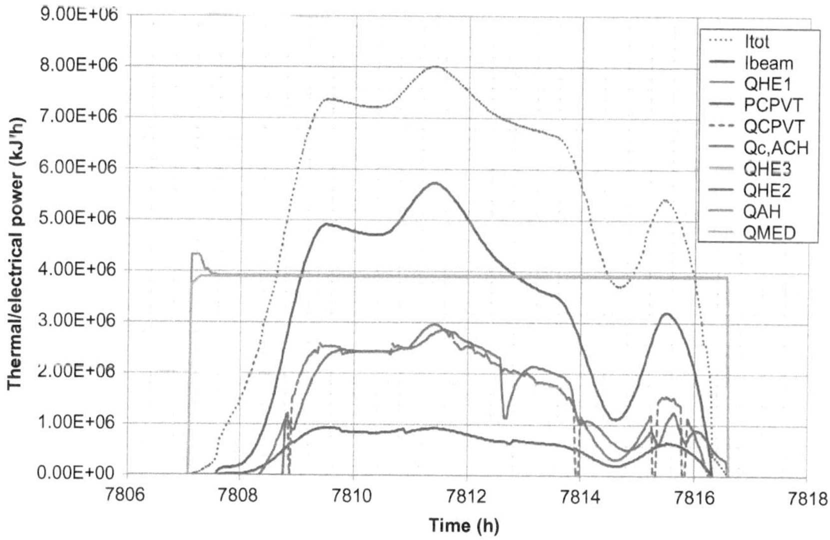

Fig. 4.10 Thermal and electrical power, winter day [24].

the summer ones, ranging between 55°C and 45°C. Therefore, the CPVT produced hot water supplies only in TK1 and as a consequence all the thermal power necessary to desalinate the water is obtained through the biomass auxiliary heater. Fig. 4.10 also displays that the solar thermal power (Q_{CPVT}) is essentially useful only for space heating by the heat exchanger HE2 by means of the produced heat (Q_{HE2}), which basically follows the availability of solar thermal energy (Q_{CPVT}). Conversely, no excess of the solar thermal energy is available for DHW production or to supply the MED unit.

The significant difference of operation between the summer and winter season is better motivated by the weekly analysis (Figs. 4.11–4.15). In Fig. 4.11, the weekly electrical and thermal energies associated to the solar loop are depicted.

With respect to the summer beam radiation, the winter one is dramatically lower and, therefore, the CPVT winter thermal and electrical production totally reduces. Fig. 4.11 also shows that the CPVT electricity is totally higher than the electricity used by auxiliary devices (e.g., pumps). The CPVT weekly thermal energy also affects the trends of the weekly thermal and cooling production (Fig. 4.12).

Fig. 4.12 shows that the production of thermal energy is normally lower than the one of cooling energy, in spite of the fact the ACH COP ranges around 0.80. In this figure, the plot of the ACH chilled energy ($Q_{c,ACH}$) appears comparable - although scaled - to the trend of the ACH

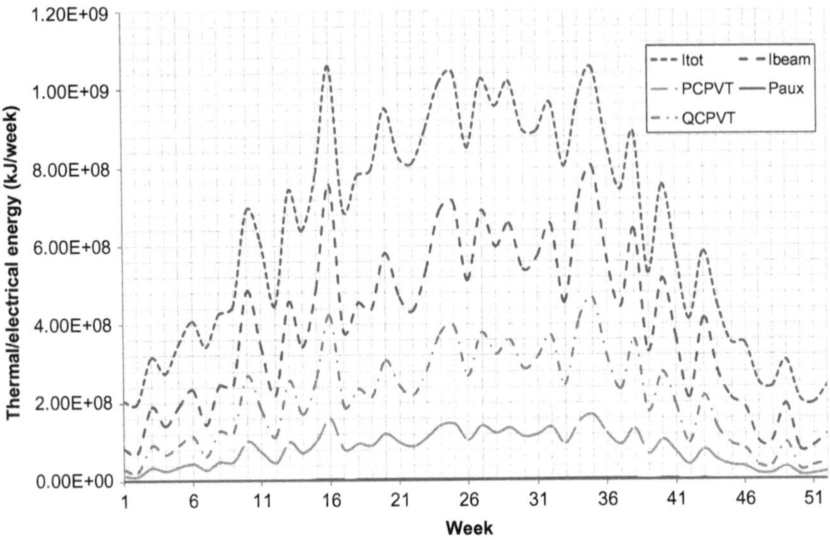

Fig. 4.11 Thermal and electrical energies, weekly analysis [24].

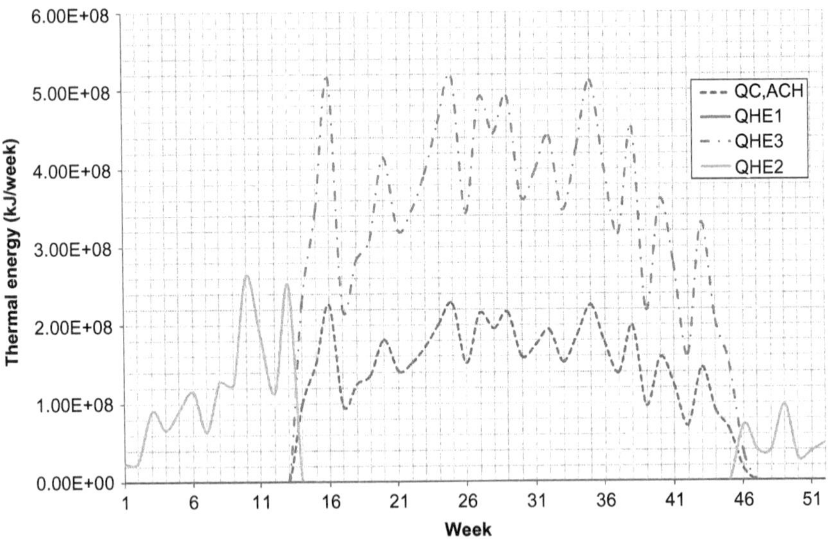

Fig. 4.12 Thermal energies, weekly analysis [24].

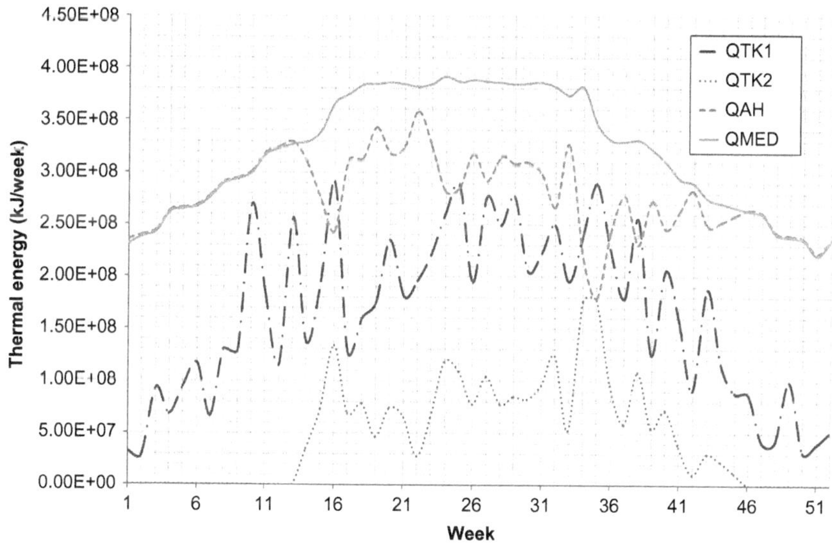

Fig. 4.13 Thermal energies, weekly analysis [24].

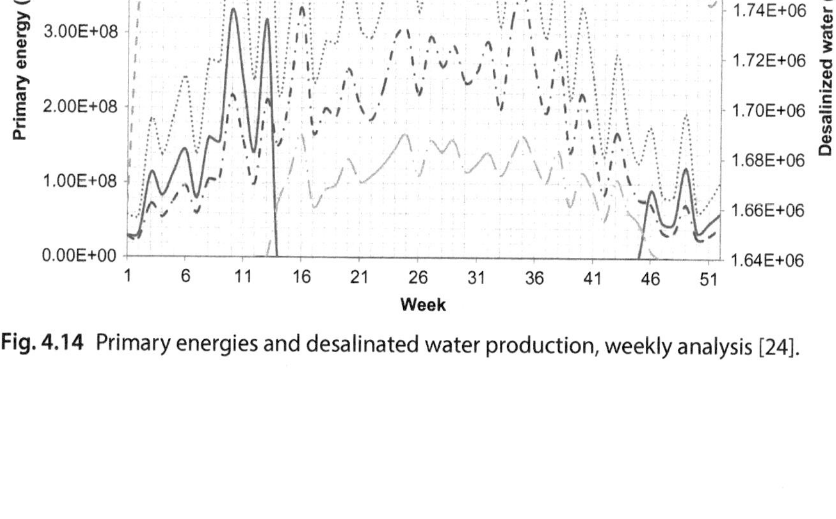

Fig. 4.14 Primary energies and desalinated water production, weekly analysis [24].

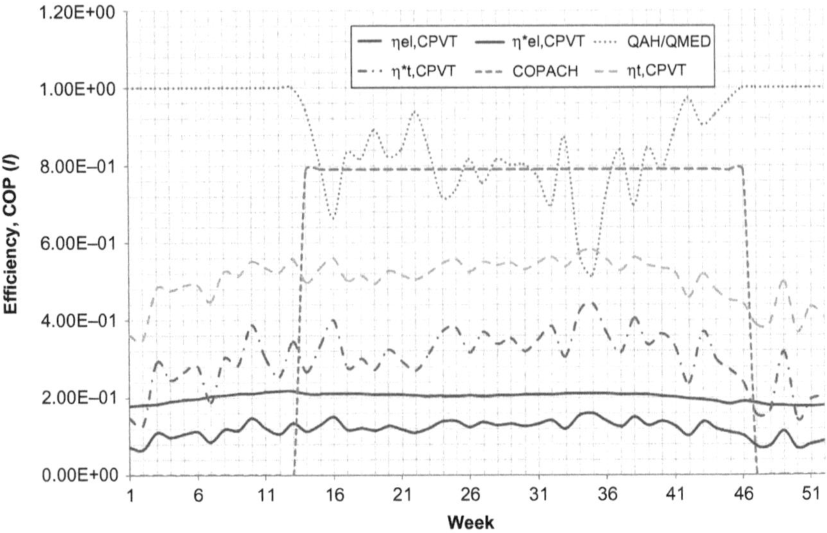

Fig. 4.15 Efficiencies and performance parameters, weekly analysis [24].

cooling energy (Q_{HE3}). The solar radiation availability is also a key factor to evaluate the amount of solar thermal energy provided to the MED unit. In fact, as previously discussed, the CPVT thermal energy is first supplied to TK1, for the space-heating and cooling purposes, and subsequently only when this tank reaches the fixed set point temperature at top node, the solar thermal energy is supplied to TK2 (necessary to drive the MED unit). Therefore, in case of low availability of solar radiation, the thermal energy for MED activation (Q_{MED}) is totally produced by the auxiliary heater (Q_{AH}), and during the winter season, this circumstance always occurs (Fig. 4.13).

Conversely, in summer weeks, an increase of the thermal energy produced by CPVT field supplied to the tank TK2 occurs, and therefore, the AH thermal energy demand reduces.

Fig. 4.13 shows that in summer season, the MED thermal energy demand (Q_{MED}) increases. This is mainly due to the greater number of operating hours per day occurring in summer, as a consequence of a longer interval of daylight, the thermal energy demand of MED unit being almost constant (see Fig. 4.10). This result also affects the desalinated water production (Fig. 4.14), which increases during summer season.

In this figure, the trend of Primary Energy Savings (ΔPE), obtained as the difference between the primary energy required in the reference system and the one used in the investigated system, is also reported. The highest contribution to ΔPE is principally correlated to the thermal (PE_{heat}) and electrical production (PE_{el}), whereas the DHW one is negligible.

The performance of the investigated polygeneration plant is synthesized in Fig. 4.15.

In such figure, the CPVT thermal and electrical efficiencies ($\eta^*_{t,CPVT}$ and $\eta^*_{el,CPVT}$), evaluated with respect to the beam radiation, are about 50% and 20%, respectively. The trend of the CPVT thermal and electrical efficiencies, calculated with respect to the total radiation ($\eta_{t,CPVT}$ and $\eta_{el,CPVT}$), are also reported. The obtained efficiencies of CPVT efficiencies are in accordance with the ones reported in literature for analogous CPVT systems featured by III–V cells. Example are:

- The work of Mittelman et al. [40] reporting for dish CPVT an electrical efficiency ranging between 19% and 23%, and a thermal efficiency equal to 60%;
- The study of Coventry [46] that for a linear parabolic CPVT equipped with silicon cells measures an electrical efficiency around 10%–12% and a thermal efficiency ranging between 55% and 65%;
- Li et al. [49–51] evaluating an electrical and thermal efficiency, by a linear parabolic CPVT (equipped with silicon cells) collector equal to 7.5% and 42%, respectively;
- Bernardo et al. [52] by studying a linear parabolic CPVT (equipped with silicon cells) featured by an electrical efficiency around 6% and a thermal efficiency of 40%

The coefficient of performance COP of the ACH is basically comparable with the same obtained in previous works [98,99]. It is worth noting that the amount of thermal energy to drive MED unit with respect to the one provided by the biomass AH definitely reduces during the warmest weeks of summer season. Anyway, the weekly support of solar energy to the activation of MED unit is on average lower than 50%.

In Table 4.3, the energy and economic results of the yearly simulations are shown.

Due to the use of III-V PV cells, the CPVT electrical efficiency as well as the thermal efficiency, calculated with respect to the beam solar radiation, is very high. As discussed, the produced cooling energy is much higher than the thermal energy production: this is basically due to the lower performance of the CPVT collectors in the winter operation period.

The thermal energy related to the DHW production (QHE1) is completely negligible, since the outlet temperature of CPVT is few times greater than 100 °C. This is due since the excess of CPVT thermal energy is sent by the tank TK2 to the MED unit.

Table 4.3 Yearly results [24]

Param.	Value	Unit	Param.	Value	Unit	Param.	Value	Unit
I_{tot}	3.32E+10	kJ/year	Q_{TK2}	2.44E+09	kJ/year	C_{fund}	9.33E+05	€/year
I_{beam}	2.04E+10	kJ/year	Q_{AH}	1.45E+10	kJ/year	J_{tot}	3.61E+06	€
P_{CPVT}	4.19E+09	kJ/year	Q_{MED}	1.66E+10	kJ/year	SPB	1.3.6E+01	years
P_{aux}	1.76E+08	kJ/year	PE_{DHW}	5.07E+04	kJ/year	SPB_{fund}	3.01 E+00	years
Q_{CPVT}	1.08E+10	kJ/year	PE_{el}	8.73E+09	kJ/year	$\eta_{el,CPVT}$	2.05E−01	/
$Q_{C,ACH}$	5.07E+09	kJ/year	PE_{heat}	2.39E+09	kJ/year	$\eta^*_{el,CPVT}$	1.26E−01	/
Q_{HE1}	4.05E+04	kJ/year	PE_{cool}	3.67E+09	kJ/year	$\eta_{t,CPVT}$	5.29E−01	/
Q_{HE3}	1.15E+10	kJ/year	ΔEP	1.48E+10	kJ/year	$\eta^*_{t,CPVT}$	3.26E−01	/
Q_{HE2}	1.91E+09	kJ/year	m_{wat}	4.15E+07	kg/year	COP_{ACH}	7.90E−01	/
Q_{TK1}	8.47E+09	kJ/year	ΔC	2.64E+05	€/year	Q_{AH}/Q_{MED}	8.68E−01	/

The system capital cost results about 3.6 M€. Without any type of incentive, the obtained economic yearly saving ΔC is around 0.26 M€/year and as a consequence the SPB is equal to 13.6 years. However, the economic feasibility definitely improves when funding policy scenarios were assumed: in this case the SPB reduces about 3 years.

It is worth noting that the yearly economic saving ΔC is mainly based on the electricity production, although noteworthy savings are also due to the cooling and thermal energy production. In this investigated polygeneration plant without the assumption of public funding, the desalinated water production is not profitable from the economic point of view.

This principally occurs because the biomass cost necessary to the activation of the AH is higher than the value assumed for the desalinated water production.

A detailed sensitivity analysis was carried out at the aim to analyze the system performance by varying the main design parameters and/or operation conditions. In Fig. 4.16, the variation of key energy ratios and of SPB period as a function of the specific tank volume (v_{TK}), i.e., the ratio between the tank volume (both TK1 and TK2) and CPVT solar collectors area, is depicted.

It is observed that higher TK volumes are in general negative, as consequence of lower temperatures into tanks. Anyway, this leads to a very small growth of CPVT thermal efficiency, while no difference is noted regarding the trend of CPVT electrical efficiency, which is unaffected by the tank volumes variation. Anyway, the lower the TK1 temperature, the rarer the case to switch the hot water in excess to TK2. This is happens since TK1 top

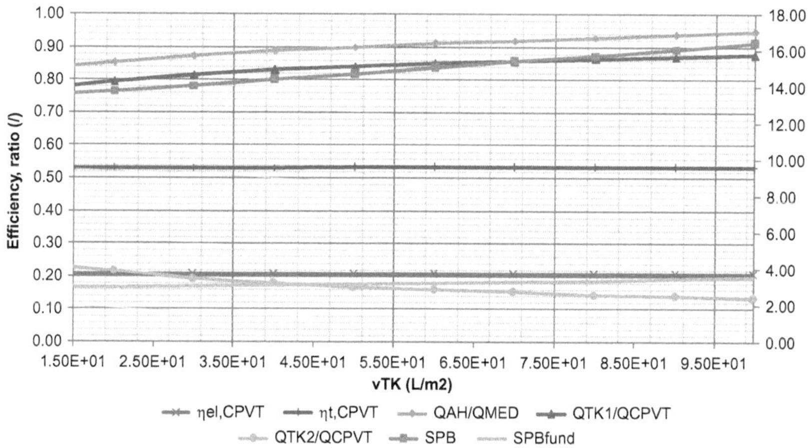

Fig. 4.16 Sensitivity analysis: TK-specific volume [24].

temperature with difficulty will be higher than the set point temperature imposed for the switching.

This is very evident by observing Fig. 4.16 showing the increasing of the CPVT thermal energy supplied to TK1 and a parallel increasing of the thermal energy required to biomass AH to produce desalinated water. This result affects the economic profitability, which decreases for higher tank volumes. This operation condition must be excluded if the selling price assumed for the desalinated water is lower than the biomass purchase cost.

In Fig. 4.17, it is possible to note the effect of the flowrate q_{P4} on the ratio between the cooling/thermal energy demand of user and the capacity of solar collectors. In fact, the higher q_{P4}, the higher the ACH and HE2 capacities, without affecting the size of the other components located in the solar loop, i.e., both tanks (TK1 and TK2) and the CPVT collectors. As reported in Fig. 4.17, the increase of q_{P4} leads to a reduction of the quantity of thermal energy exceeding to be supplied to tank TK2. This occurs since for lower space heating/cooling energy demands, the investigated plant is able to deliver to the MED unit a higher amount of heat. For a solar field consisting of 400 collectors, the best SPB period is achieved with a nominal ACH cooling capacity equal to 1.2 MW. The SPB is featured by a non-monotonic trend, since if the ACH capacity is further augmented, the capital cost increases, without obtaining any improvement from the economic point of view. In fact, at constant solar collector capacity, a higher ACH capacity cannot be supplied by an adequate thermal energy.

As previously discussed, the increase of the CPVT collector number N_{SC} leads to a resizing of all the components of the plant (thermal/cooling

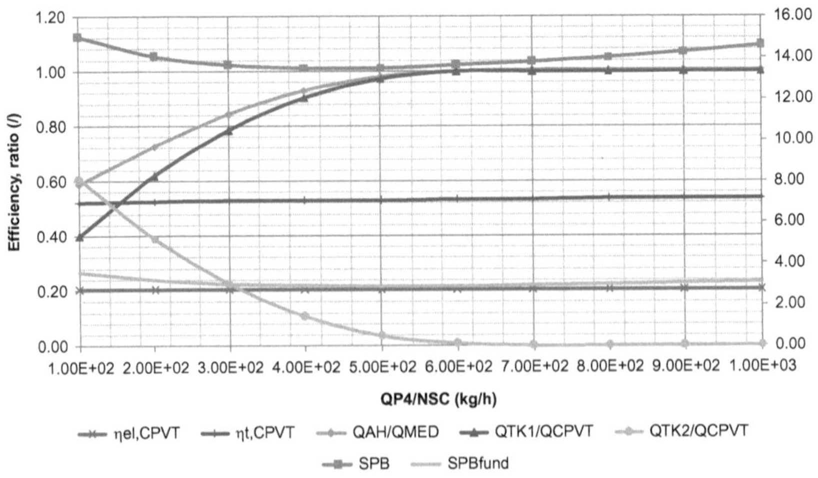

Fig. 4.17 Sensitivity analysis: P4 specific flow rate [24].

energy demand, pumps, TKs, ACH, etc.), except the MED unit, featured by a fix nominal thermal capacity equal to 1.0 MW. Therefore, the increase of N_{SC} does not significantly affect the ratio between the thermal energy provided to TK1 and TK2 (Fig. 4.18) Anyway, the increasing of the solar field capacity affects the amount of thermal energy sent to the TK2 and, therefore, increases the heat driving the MED. As a result, the AH contribution also decreases, and this leads to a better system economic performance.

Other key parameter varied in sensitivity analysis, in order to optimize the plant operation, was the CPVT summer set point temperature ($T_{set,CPVT,summer}$), as depicted in Fig. 4.19.

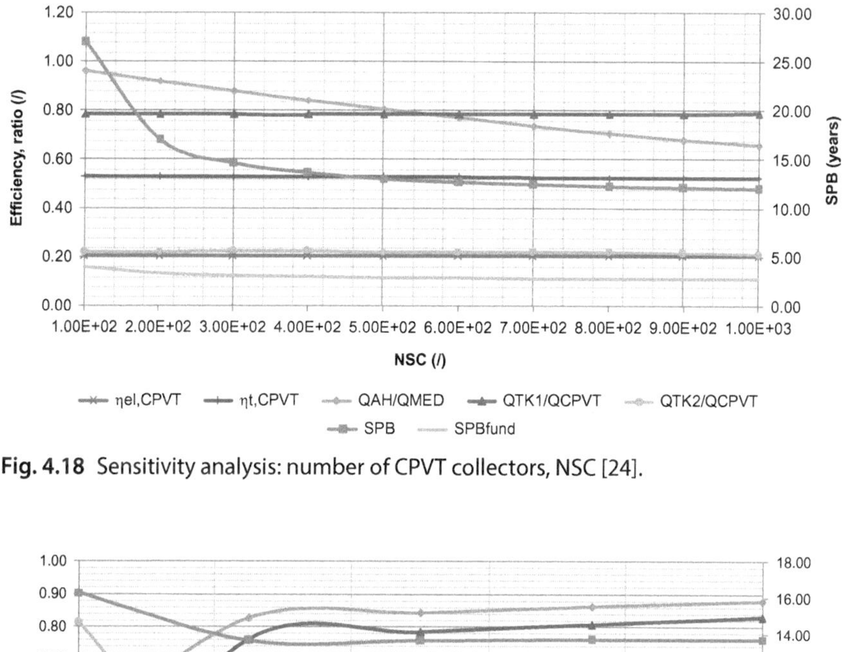

Fig. 4.18 Sensitivity analysis: number of CPVT collectors, NSC [24].

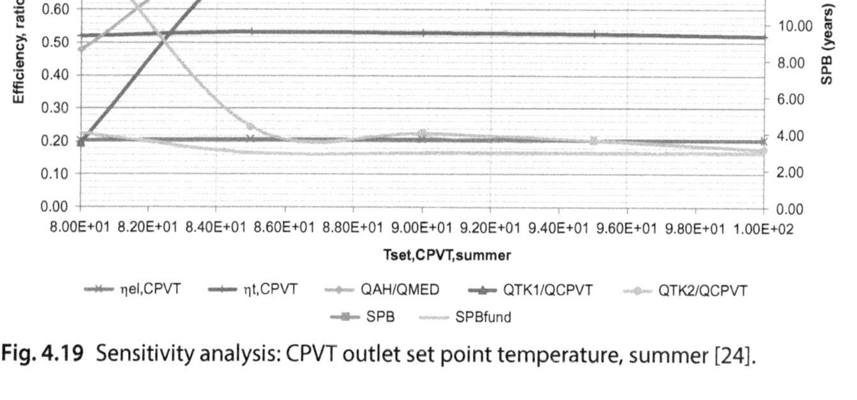

Fig. 4.19 Sensitivity analysis: CPVT outlet set point temperature, summer [24].

The results of variation of this parameter are shown in Fig. 4.19. Here, it is evidently reported that lower $T_{set,CPVT,summer}$ values are economically positive for the MED unit, since the amount of thermal energy required to the biomass AH is reduced. This occurs since lower values of $T_{set,CPVT,summer,}$ lead to lower temperatures required to switch the produced thermal energy to TK2 ($T_{set,CPVT,summer}$ - 4.0°C). Anyway, this circumstance avoids the activation of ACH, which is off for TK1 temperatures lower than 75°C, by reducing the yearly cooling energy production. In fact, at equal to 80°C, the RPS does not produce any cooling energy.

The optimum value of $T_{,CPVT,summer,}$ which results in a compromise between the both effects from the thermoeconomic point of view, is about 86 °C. With this value, it is possible to obtain at the same time a large cooling energy production by ACH and a decrease of heat required by AH.

To conclude, the effect of variation of the number of MED unit effects (using 3, 5, 8, 11, and 14 effects) on the system thermoeconomic performance was also analyzed.

Some graphs related to this parameter are shown in Figs. 4.20 and 4.21.

In Fig. 4.20, it is evidently reported that the number of MED effects does not considerably affect the other thermodynamic parameters of the plant. This is basically due to the issue discussed, related to the fact that only the MED first effect is coupled, from the thermal point of view, to the solar collectors loop. Therefore, all the considered thermal energies (solar energy, the energy stored into tanks TK1, TK2, etc.) are not affected by number of MED effects. Instead, such parameter certainly affects the total production

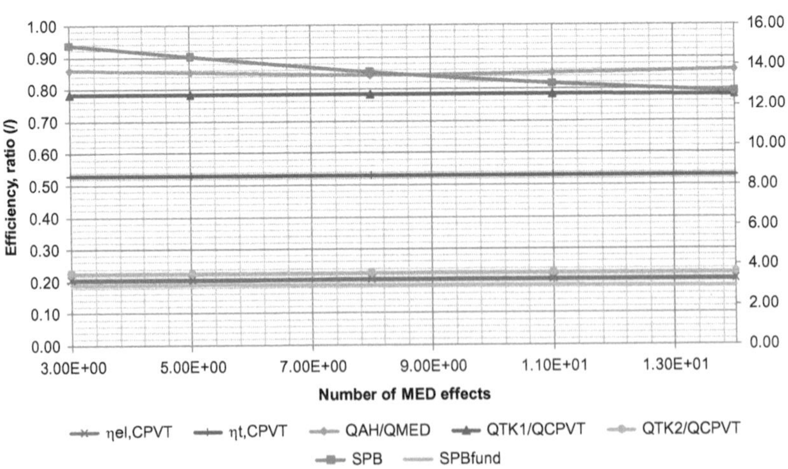

Fig. 4.20 Sensitivity analysis: number of MED effects [24].

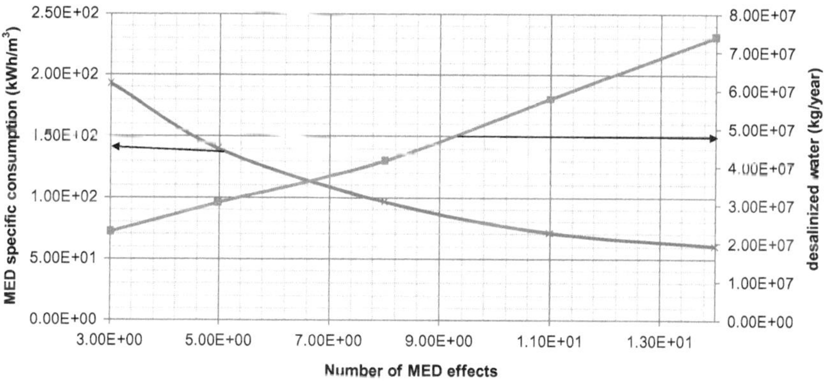

Fig. 4.21 MED-specific consumption and desalinated water produced vs number of effects [24].

of desalinated water (Fig. 4.21), which considerably grows for a higher number of MED effects.

Anyway, by considering the operating criterion of the MED unit, the MED input thermal energy does not significantly depend by the number of effects. Therefore, MED units, which are featured by higher number of effects, will produce higher volumes of desalinated water, but with the same amounts of input thermal energy. This condition also leads to a reduction of MED energy-specific consumption (Fig. 4.21), defined as the heat required to produce 1 m^3 of desalinated water.

By concluding, it is seen that a higher number of MED effects lead to an increase of desalinated water volumes but, unavoidably, to an increase of plant capital cost. The trade-off between these contrasting effects is reported in Fig. 4.20. Here, it is visible that, although the number of MED effects increases, the SPB period decreases. This result is simply explained since the increase of capital cost, when higher number of effects are taken into account, is balanced by the economic revenues due to the growth of desalinated water volumes.

4.4.1.2 GP System

In this section, the dynamic results for two random days of the winter and summer season are presented.

As shown in Fig. 4.22, during the winter, the control system is not constantly able to reach the CPVT outlet set point temperature. This occurs since the CPVT solar loop performance is very affected by the solar radiation availability, scarce during such season.

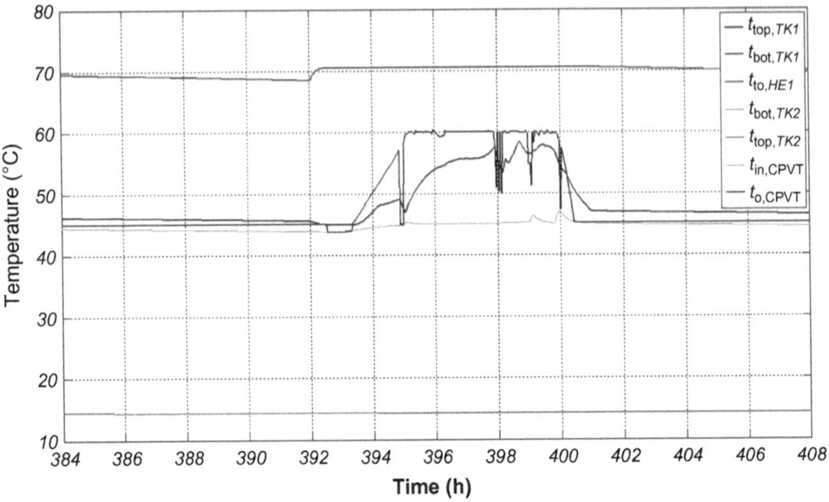

Fig. 4.22 Temperatures (1), winter day [11].

In fact, the CPVT outlet temperature approaches the set point value, fixed at 60°C, only few hours during the day and this occurs by means of the control system lowering the mass flow rate inside the CPVT loop, i.e., reducing the P2 pump speed. As consequence of a lower flowrate, during the winter, the CPVT thermal energy production dramatically decreases.

In addition, for the same reason, the solar thermal energy sent to TK1 also decreases, as evidently shown in Fig. 4.22. In fact, in this figure, it is reported that the top temperature of TK1 ($t_{top,TK1}$) is much lower than the CPVT outlet temperature ($t_{o,CPVT}$) and that the bottom temperature of TK1 ($t_{bot,TK1}$) is about 45°C. Regarding the tank TK2, the top temperature ($t_{top,TK2}$) and bottom one ($t_{bot,TK2}$) are equal to about 70°C and 15°C, respectively. Because the top temperature of TK1 does not reach the fixed value of 75°C, the solar heat is never sent to TK2. Therefore, TK2 is never heated by the CPVT solar collectors, and its bottom temperature remains at minimum value. This result is evident in Fig. 4.22, since the CPVT inlet temperature ($t_{in,CPVT}$) is constantly equal to TK1 bottom one ($t_{bot,TK1}$). This means that for the P2 pump the fluid always flows from the TK1 bottom. Instead, TK2 maximum temperature is very near to temperature coming back from MED unit. The high difference of TK2 temperatures is due to its thermal stratification: the flow from MED unit is supplied to the node of TK2 whose temperature is equal to the temperature of inlet flow. The final result of such strategy is that only the TK2 top is hot, while the bottom

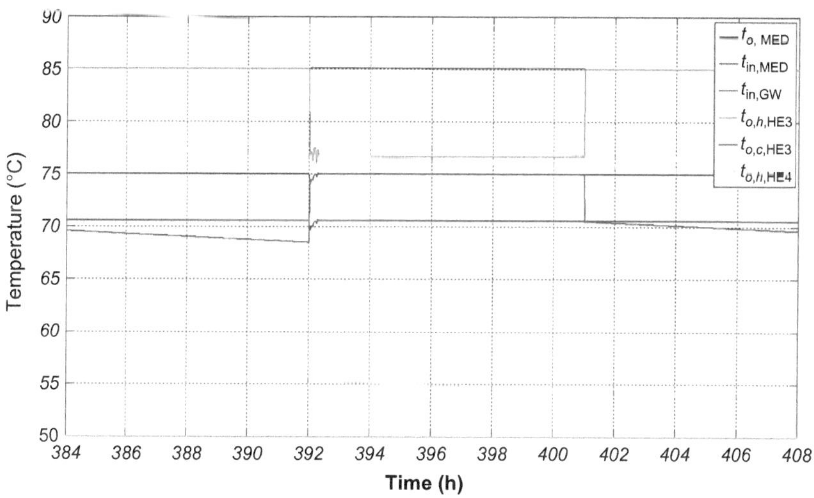

Fig. 4.23 Temperatures (2), winter day [11].

one is cold. This prevents the whole tank volume from heating, thereby also reducing the thermal losses toward the environment.

The MED loop temperatures are shown in Fig. 4.23.

As expected for all the days of operation, the MED unit starts to work only when the controller managing the availability of solar radiation is able to activate the P2 pump so that the outlet fluid from CPVT collectors is heated. Therefore, MED unit is activated after sunshine and deactivated before sunset. During these hours, the sole source able to drive the MED unit is the geothermal fluid, continuously pumped from geothermal well at temperature of 85°C, by heating the water supplying the MED at 75°C ($T_{in,MED} = T_{o,c,HE3}$). This occurs since the hot water produced by CPVT is able to supply only the TK1 and never the TK2. The inlet geothermal fluid to heat exchanger HE4 heats the water flowing on the cold side, by producing DHW at 50 °C (an appropriate temperature value for thermal baths).

Fig. 4.24 evidently reports the control strategy regarding the HE2 producing the space heating in this representative winter day: HE2 produces hot water only if the TK1 top temperature is higher than the fixed set point temperature of 45°C, in order to avoid to supply the user with water at low temperature.

All the figures showing trends of system temperatures (Figs. 4.22, Fig. 4.23, Fig. 4.24) are better understood by observing the graphs reporting the thermal and electrical powers, as shown in Figs. 4.25 and 4.26.

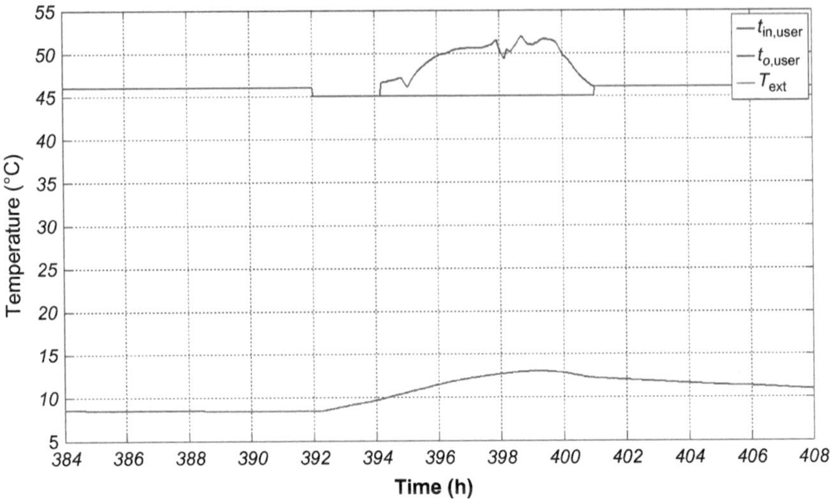

Fig. 4.24 Temperature (3), winter day [11].

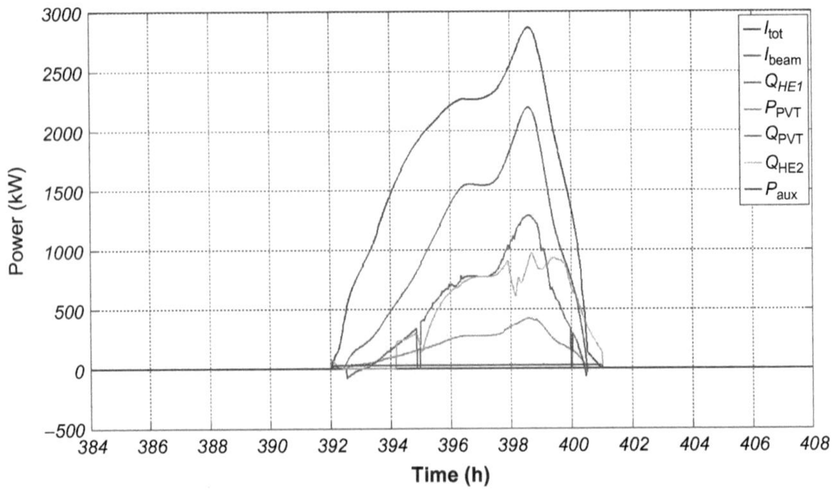

Fig. 4.25 Powers (1), winter day [11].

In Fig. 4.25, the beam radiation in the selected winter day appears very inconstant, being completely very low during the hours of the morning and afternoon.

Here, the thermal storage effect of TK1 is also shown. In fact, the CPVT heat (Q_{PVT}), mostly obtained during the central hours, is stored in TK1 but will be available for the user (Q_{HE2}) also in the afternoon hours with a value

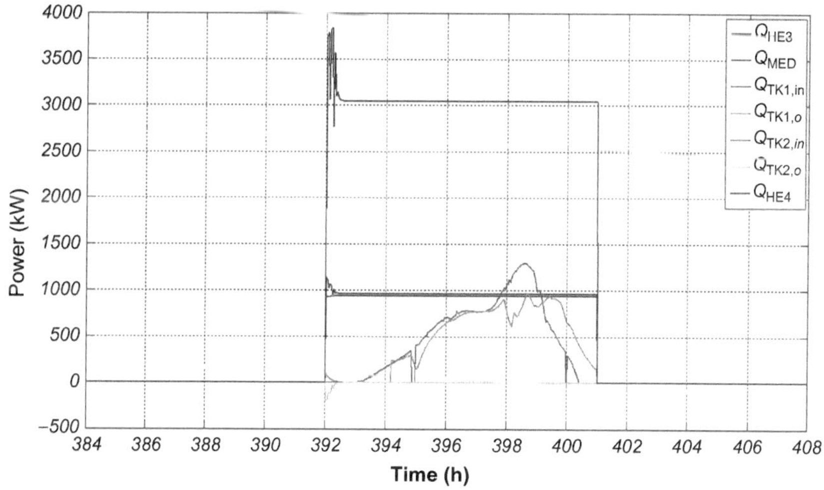

Fig. 4.26 Powers (2), winter day [11].

higher than the heat produced by CPVT (Q_{PVT}). This effect is clearly visible in Fig. 4.26, by comparing the inlet and the outlet thermal flow $Q_{TK1,in}$ and $Q_{TK1,o}$ of TK1, respectively.

As discussed, Fig. 4.26 proves also that, during this winter day, TK2 is never charged by the heat of CPVT (s$Q_{TK2,in}=0$) and, as a consequence, the MED unit is supplied only by geothermal power ($Q_{HE3}=Q_{MED}$) and that the thermal flow recovered by heat exchanger HE4 (Q_{HE4}) is significantly higher than all the others.

Conversely, during the summer season, the obtained temperatures and powers are totally different as evidently shown in the following figures. Fig. 4.27 reports the summer temperatures of the solar loop, visibly oscillating around the fixed set point temperatures: TK1 and TK2 top temperatures ($t_{top,TK1}$ and $t_{top,TK2}$) oscillate about 85°C and 75°C, respectively, while CPVT outlet temperature ($t_{o,CPVT}$) is stably around 90°C.

The oscillations are produced since the control strategy expected for the solar loop uninterruptedly switches the thermal flow produced by CPVT to TK1 or to TK2, as a function of the TK1 top temperature. Such oscillating trend is more evident for CPVT outlet temperature ($t_{o,CPVT}$). This occurs since the controller managing P2 pump flowrate reads the inevitable oscillations of CPVT inlet temperature, which abruptly changes if the system switches from TK1 to TK2 (or vice versa).

The less evident fluctuation of the MED inlet temperature ($t_{in,MED}$) is due to the oscillating trend of the hot water supplying TK2 (Fig. 4.28).

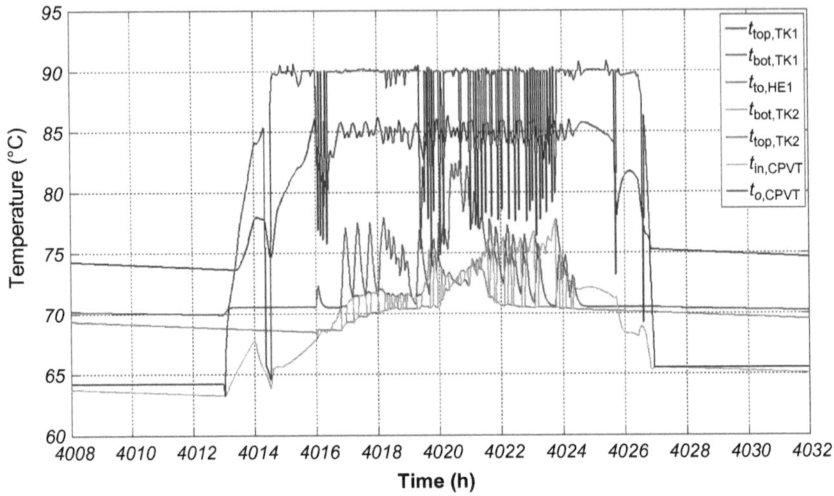

Fig. 4.27 Temperature (1), summer day [11].

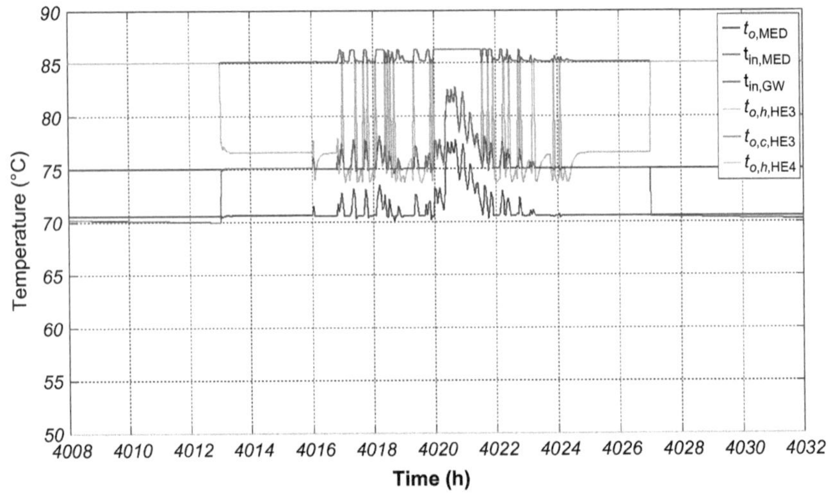

Fig. 4.28 Temperatures (2), summer day [11].

Anyway, these small variations in the temperature values do not produce significant variations in MED unit performance and in the pressure of each effect, but are responsible of the HW return temperature trend ($T_{o,HW,ACH}$) of the ACH, as display in Fig. 4.29.

On the contrary, the outlet cooling water temperature ($T_{o,CW,ACH}$) and chilled water return ($T_{o,user}$) and supply ($T_{in,user}$) temperatures are definitely constant.

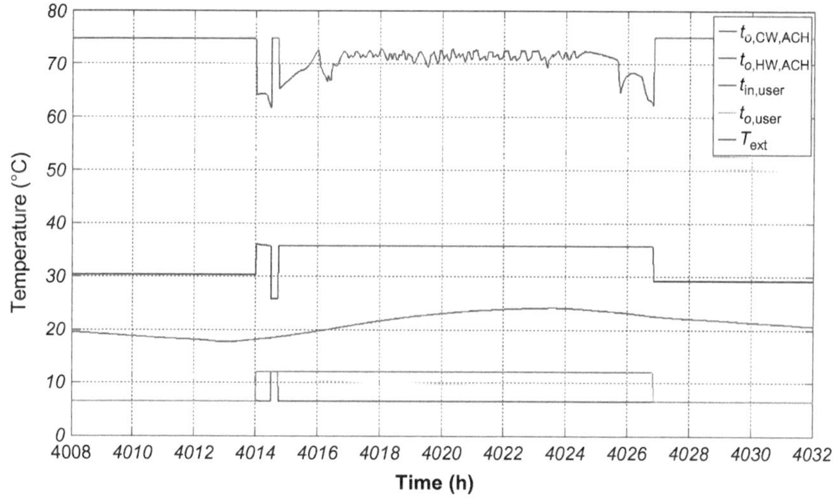

Fig. 4.29 Temperatures (3), summer day [11].

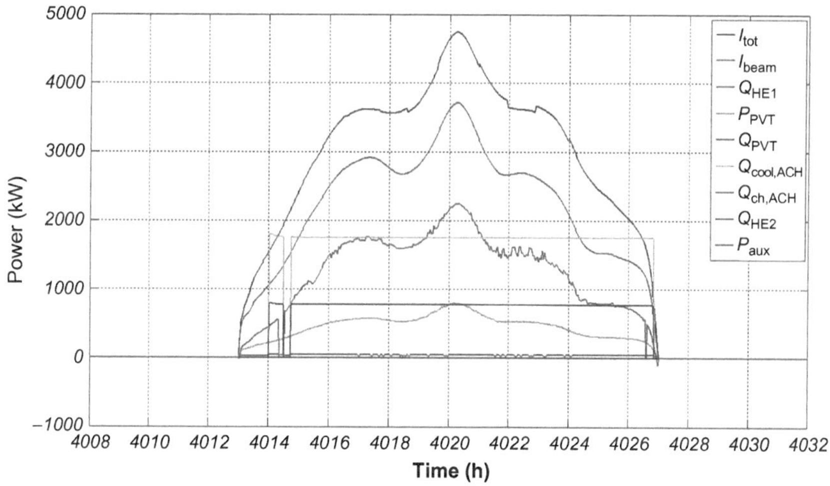

Fig. 4.30 Powers (1), summer day [11].

As shown in Fig. 4.30, in the reported summer day, the beam radiation (I_{beam}) is highly available for a lot of hours of day, and this determines a very stable CPVT production both electrical power and thermal one (P_{CPVT} and Q_{CPVT}, respectively).

In such graph, the electricity consumption of the auxiliary devices (P_{aux}) is also reported, by resulting negligible with respect to the CPVT electricity production. In addition, it is seen that the CPVT electricity production is always consumed by the users, and it is considerably lower than the electricity demand of the simulated weather zone, Pantelleria (the minimum and maximum electricity demand are around 3.0 MW and 7.0 MW, respectively).

Finally, in Fig. 4.31, the thermal flows trends in MED loop are reported. Here, it is evident that the heat need of MED unit is totally provided by heat exchanger HE3 only during the hours of the early morning and late afternoon. Instead, during the central hours of the day, the heat required to supply the MED unit is provided by the solar loop (via the tank TK2).

In order to better analyze the previous dynamic results, the graphs in Fig. 4.32, Fig. 4.33, Fig. 4.34, Fig. 4.35, and Fig. 4.36 report the monthly analysis of the main parameters of the investigated plant, by integrating the energy flows on a monthly basis.

Fig. 4.32 highlights the significant difference between the beam and total radiation occurring during the year. This obviously leads to the winter reduction of winter electrical energy (P_{PVT}) and thermal energy (Q_{PVT} in Fig. 4.33) of CPVT collectors.

According to the dynamic results, the monthly demand of Pantelleria is much higher than the electrical energy production. As shown, the Pantelleria monthly electricity demand is about 3.5 GWh/month for all year months, except for the month of July, August, and September, in which an increase of demands occurs until to 4.5, 5.0, and 4.0 GWh/month, respectively.

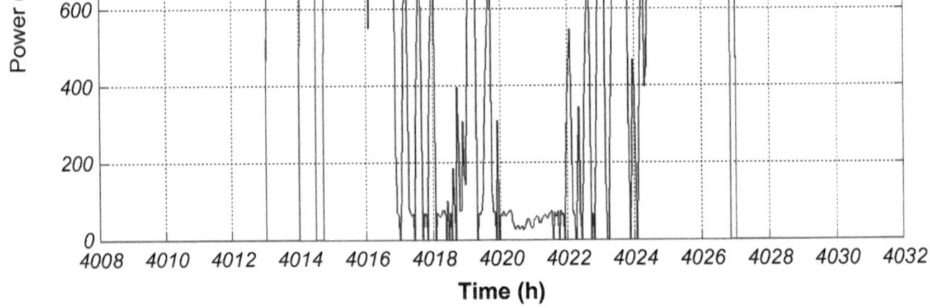

Fig. 4.31 Powers (2), summer day [11].

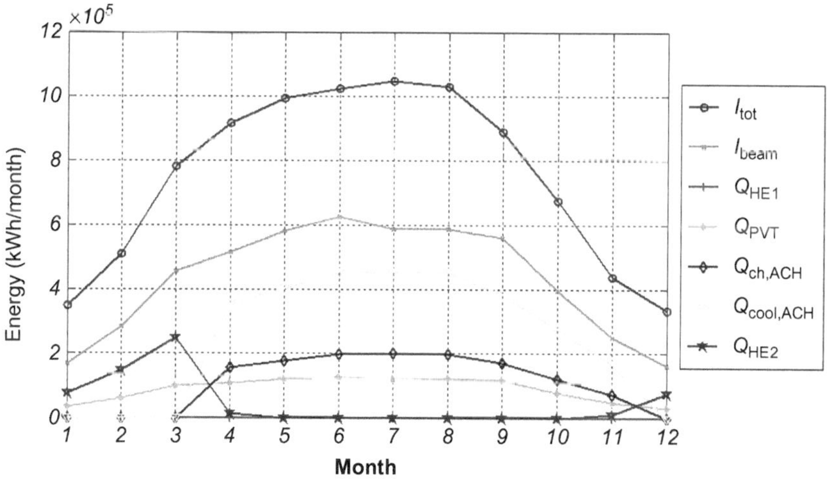

Fig. 4.32 Monthly thermal energy: CPVT, ACH, and HE 2 loops [11].

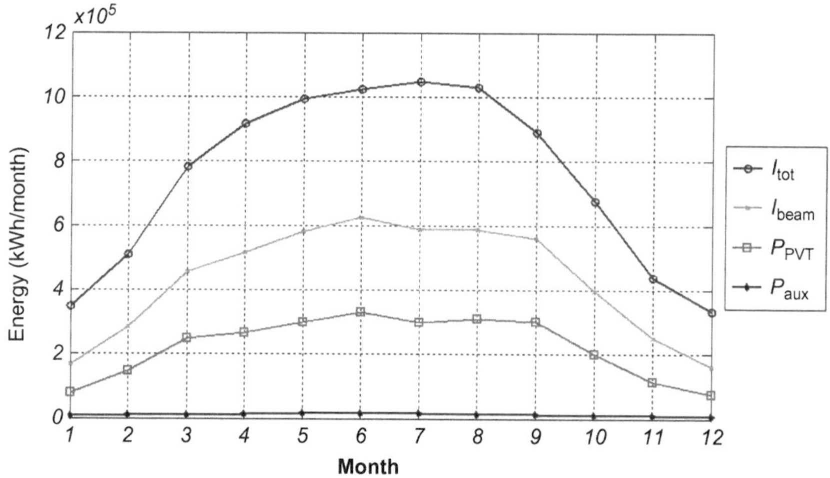

Fig. 4.33 Monthly electrical energy [11].

It is worth noting that the investigated polygeneration plant is able to produce more cooling energy ($Q_{ch,ACH}$) than the thermal energy used for the space heating, Q_{HE2}.

Again it is seen that the thermal energy produced by heat exchanger HE1 is null or insignificant for all the months of year.

In Fig. 4.34, the main thermal energies of the MED and TK MED loops are represented.

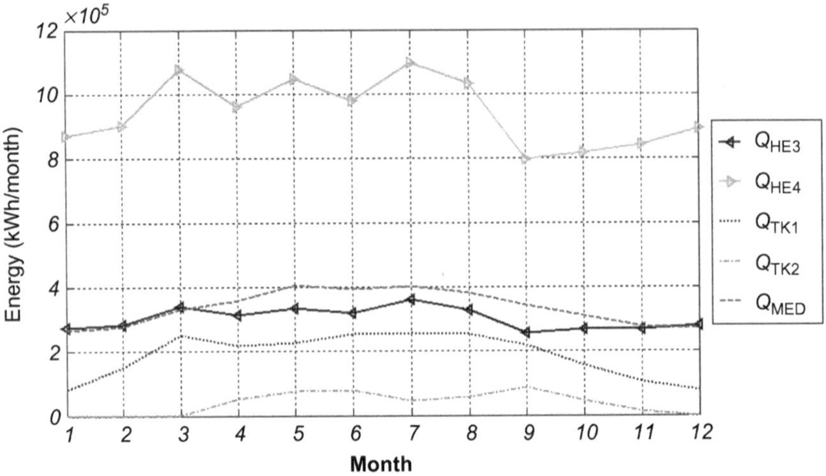

Fig. 4.34 Monthly thermal energy: TK and MED loops [11].

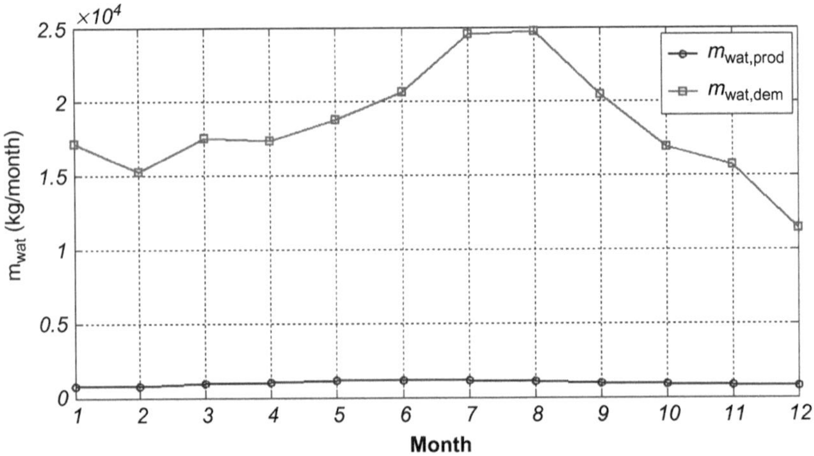

Fig. 4.35 Monthly water production and demand [11].

Here, it is important to note that the thermal energy demand of the MED unit (Q_{MED}) is higher during the summer months, since in these months a higher number of operating hours are obtained. Anyway, such energy increase does not perforce lead to a consequent growing of the thermal energy supplied by the geothermal well through the heat exchanger HE3 (Q_{HE3}), since in the summer months the solar support of MED production by thermal energy stored into TK2 (Q_{TK2}) is significantly higher.

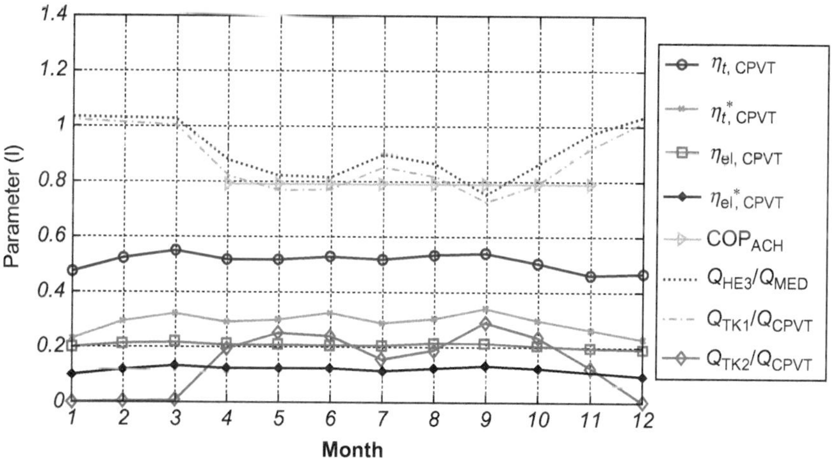

Fig. 4.36 Monthly efficiency parameters [11].

The higher number of operation hours leads to a corresponding increase of the desalinated water production (Fig. 4.35). Anyway, the desalinated water production covers only about 5%–6% of the global water user demand. This result highlights that the investigated system is clearly undersized from the desalinated water production point of view, since it is always necessary to satisfy the island needs by additional amounts of freshwater (by ships).

The maximum water production by MED unit occurs in summer months, but its maximum contribution is achieved in May and in December when the summer demand increases because of the high number of tourists.

Finally, some system energy performance parameters are reported in Fig. 4.36. It is shown that the amount of solar energy provided to the MED unit (Q_{TK2}/Q_{CPVT}) is negligible or null in coldest winter months (January, February, March, and December), reaching about 0.30 in the middle season. During the summer months, this ratio lowers to 0.20. This reduction is basically due to the higher thermal energy required by the ACH in order to produce space cooling. As a consequence, the ratios Q_{TK1}/Q_{CPVT} and Q_{HE3}/Q_{MED} are specular with respect to ratio Q_{TK2}/Q_{CPVT}.

Fig. 4.36 shows that the monthly CPVT electrical ($\eta_{el,CPVT}$) and thermal ($\eta_{t,CPVT}$) efficiencies are greater than 20% and 50%, respectively. This is a very positive result by taking into account that the CPVT solar collectors operating temperatures are too high, by ranging from 60°C to 90°C, depending on the winter and summer season, respectively. It is worth noting that such efficiencies are evaluated by considering the beam solar radiation, while if the total radiation is considered, the thermal ($\eta^*_{t,CPVT}$) and electrical ($\eta^*_{el,CPVT}$) efficiencies reduce around 30% and 12%, respectively.

The carried out yearly thermoeconomic analysis is summarized in Table 4.2, where the main economic profitability and energy indexes of the investigated polygeneration plant are reported.

As shown, the further exposed results, with respect the ones widely discussed, regard principally the economic aspect. In fact, in this table, the total system capital cost (J_{tot}) equal to about 3.95 M€ is indicated. It is worth noting that the highest capital cost is due to the solar field, around 2.9 M€. The yearly economic saving ΔC, basically due to the electricity production, space cooling and heating, and desalinated water, is 1.14 M€.

The evaluated SPB period is about 3.45 years. In case of a potential politic scenarios based on the public funding (feed in tariffs) for the combined production of desalinated water and energy by renewable sources is supposed, the SPB period would reduce to value lower than 2.0 years. This highlights that the investigated polygeneration plant is totally profitable and this occurs also in case of no public funding.

Anyway, the obtained economic indexes are significantly affected by the assumption of a full utilization of the cooling, electrical, thermal energy, and water produced.

The discussed results regard the analysis carried out for the island of Pantelleria.

However, further additional volcanic Mediterranean islands were investigated. These were selected since they are featured by the same availability of geothermal energy and solar radiation, and also because present the same problem regarding the lack of freshwater and energy, particularly in the summer months.

Then, on the basis of the same design parameters presented in Table 4.2, the simulations were carried out by taking into account two further small Italian volcanic islands, namely, the Aeolian Islands (38° 32' N, 14° 54' E, North of Sicily), and Ischia (40° 43' N, 13° 54' E, close to Naples).

The obtained results are summarized in Table 4.4.

Unexpectedly, the beam solar radiation in Pantelleria is lower than the ones available in Aeolian Islands and in Ischia. Therefore, in these two additional islands, a higher electrical and thermal yearly productivity of CPVT solar collectors is achieved (from 15% to 10%). The highest space-heating production Q_{HE2} is achieved curs in Pantelleria, while the highest production of cooling energy $Q_{C,ACH}$ occurs in the Aeolian islands.

Table 4.4 Annual results [11]

Param.	PANTEL.	AEOLIAN	ISCHIA	Unit
I_{tot}	8.98E+06	9.79E+06	9.23E+06	kWh/y
I_{beam}	5.18E+06	5.97E+06	5.69E+06	
P_{CPVT}	1.06E+06	1.22E+06	1.17E+06	
P_{aux}	1.45E+05	1.50E+05	1.46E+05	
Q_{CPVT}	2.67E+06	3.18E+06	3.01E+06	
$Q_{C,ACH}$	1.29E+06	1.62E+06	1.41E+06	
Q_{HE1}	2.65E+02	1.33E+02	5.27E+02	
Q_{HE5}	2.94E+06	3.70E+06	3.20E+06	
Q_{HE2}	5.70E+05	5.67E+05	5.37E+05	
Q_{TK1}	2.25E+06	2.66E+06	2.36E+06	
Q_{TK2}	4.55E+05	5.33E+05	6.77E+05	
Q_{HE3}	3.63E+06	3.56E+06	3.40E+06	
Q_{MED}	4.02E+06	4.02E+06	4.00E+06	
Q_{HE4}	1.13E+07	1.09E+07	1.09E+07	
ΔPE	1.87E+07	1.90E+07	1.85E+07	
m_{wat}	4.18E+07	4.18E+07	4.16E+07	kg/y
ΔC	1.14E+06	1.15E+06	1.13E+06	€/y
J_{tot}	3.95E+06	3.95E+06	3.95E+06	€
SPB	3.45E+00	3.42E+00	3.50E+00	y
SPB_{ft}	2.00E+00	1.87E+00	1.96E+00	
$\eta_{el,CPVT}$	2.05E−01	2.05E−01	2.05E−01	/
$\eta^{*}_{el,CPVT}$	1.18E−01	1.25E−01	1.26E−01	
$\eta_{t,CPVT}$	5.16E−01	5.32E−01	5.29E−01	
$\eta^{*}_{t,CPVT}$	2.98E−01	3.24E−01	3.26E−01	
COP_{ACH}	7.90E−01	7.90E−01	7.90E−01	
Q_{HE3}/Q_{MED}	9.05E−01	8.85E−01	8.50E−01	

The produced volumes of desalinated water are similar in the three investigated zones, as well as in the economic parameters. The lowest SPB period, equal to 3.4 years, is reached in Aeolian islands. By this analysis, it can be established that the investigated plant is economically feasible for a large number of localities. Anyway, it is worth noting that the presented economic analysis is performed by taking into account a CPVT collectors-specific cost valid only by supposing a large commercial production. Conversely, by evaluating the current costs of the several CPVT prototypes, totally higher, the economic results previously discussed inevitably change.

4.4.2 Exergy Analysis Results

All the components of the exergy balances (products, fuel and exergy destroyed), with all the parameters of exergy performance, such as exergy efficiency and efficiency defect, can be calculated through dynamic simulations, once chosen a time basis, for each component or the whole system, referring to the exergy model seen before. Exergy flows and efficiencies are dynamically reported in the following plots for a representative summer and winter day, as adopted in [24], in which energy flows and temperature plots are also shown and, for sake of brevity, are not here reported. The exergy flows of products, fuel, and exergy destroyed for a representative summer day are shown in Fig. 4.37:

It is clear that the solar exergy, $\dot{E}x_{sun}$, reaches the highest values of exergy; in fact, AH biomass exergy $\dot{E}x_{biomass}$, is much lower than $\dot{E}x_{sun}$. Consequently, $\dot{E}x_{d,total}$, which is the overall exergy destruction rate is predominantly given by the exergy of CPVT and AH, respectively, $\dot{E}x_{d,CPVT}$ and $\dot{E}x_{d,\mathrm{AH}}$, while other components exergy destruction rates, such as MED one, $\dot{E}x_{d,\mathrm{MED}}$, are marginal. $\dot{E}x_{d,\mathrm{AH}}$ depends strongly on the amount of heat required by MED system. In fact, it is worth nothing that CPVT and AH exergy destruction rates are highly affected by the electricity and/or thermal energy provided. At the same time, the electrical energy provided by CPVT, $\dot{P}_{el,\mathrm{CPVT}}$, is the highest exergy product, much higher than the Domestic Hot Water one, $\dot{E}x_{DHW}$, desalinated water one, $\dot{P}_{MED}$, and the one related to thermal and cooling energies provided to the user, $\Delta\dot{E}x_{user}$, not shown in Fig. 4.37 since they are orders of magnitude lower than $\dot{P}_{el,\mathrm{CPVT}}$.

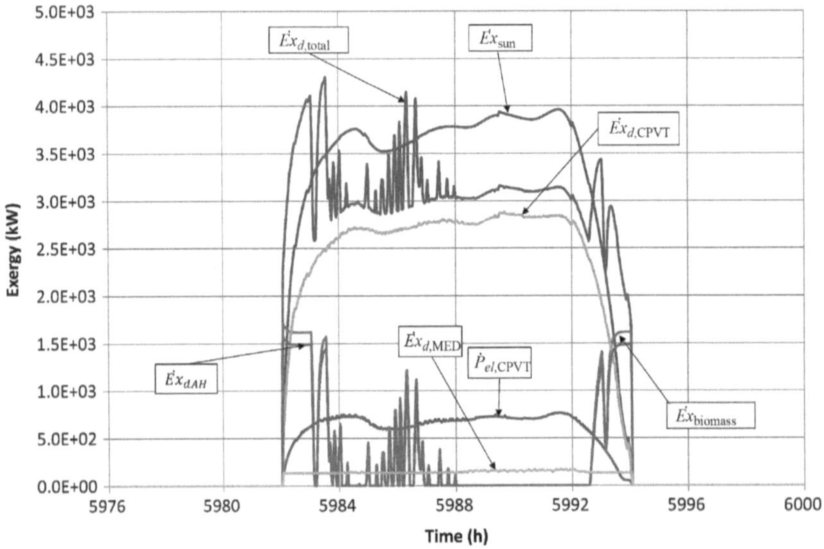

Fig. 4.37 Exergy flows for a representative summer day [126].

In particular, water chemical exergy, whose variation affects MED exergy product, takes about only 1 kWh of exergy for each m^3 of desalted water; then, Domestic Hot Water heat request is negligible when it is not null; also, since the fluid average temperature is quite similar to the ambient one, "Carnot factor" values are negligible too, and so $\Delta \dot{E}x_{user}$ is very small, although thermal and cooling energies provided to the user are higher than the CPVT electricity production.

For the same representative summer day, in Fig. 4.38, the exergy efficiency, which is the ratio between exergy product and fuel, of the whole system and of all the components is shown.

Considering the daylight of such day, the ambient temperature varies from 18°C to 27°C.

System exergy efficiency, $\eta_{ex,total}$, which is inversely proportional to the component irreversibilities, is roughly equal to 20%. Thus, about 80% of the inlet fuel exergy is destroyed in such system. Small exergy efficiencies values are reached by the components of the system; in fact, for example, MED exergy efficiency, $\eta_{ex,MED}$, is slightly above 5%, while AH exergy efficiency, $\eta_{ex,AH}$, reaches maximum 8%, a typical value for the heaters. The highest value is detected by the ACH exergy efficiency, $\eta_{ex,AH}$, approximately equal to 30%. The exergy efficiency of CPVT, $\eta_{ex,CPVT}$, varies from 25 to 30%; in fact, CPVT collectors can provide simultaneously two exergy products,

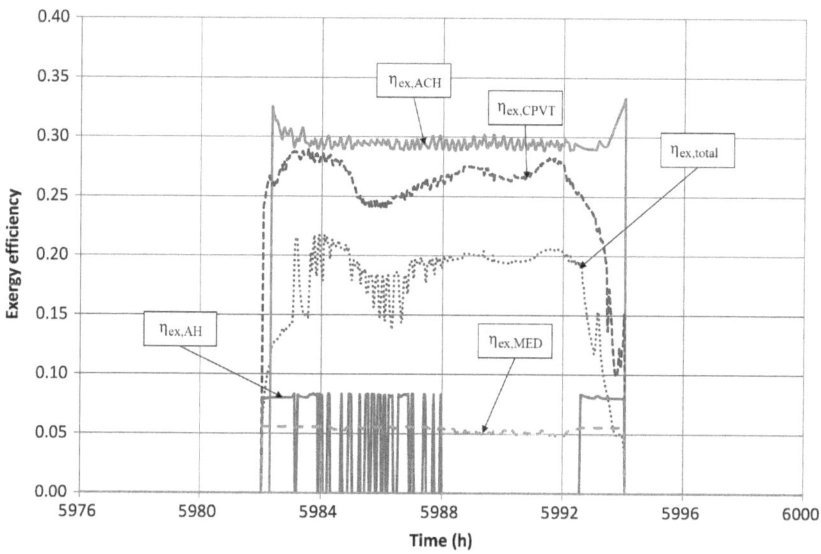

Fig. 4.38 Exergy efficiencies for a representative summer day [126].

i.e., electricity and heat, used to increase the solar fluid physical exergy. For this reason, $\eta_{ex,CPVT}$ is strictly dependent on its energy source, that is the solar energy, and on the electricity and heat provided trends.

In particular, the more the solar irradiation increases, the more electrical and thermal productions increase; obviously, the ambient temperature is higher too, causing thermal losses decrease. These results determine a significant $\eta_{ex,CPVT}$ increase during sunny days.

The exergy flows and efficiencies for a representative winter day, instead, are illustrated in Figs. 4.39 and 4.40.

Comparing the winter plots with the summer case, it is evident that the solar irradiation is much lower and so biomass is consumed continuously by the system to supply thermal energy. For this reason, in the winter case, $\dot{E}x_{d,AH}$ is absolutely similar to $\dot{E}x_{d,CPVT}$, unlike summer case where solar exergy is prevailing over the other exergy fuels and biomass heater rarely is activated and consume biomass. For such representative winter day, the solar exergy reaches maximum 2 MW, against 4 MW in summer days, and the biomass exergy request constantly approaches 1.6 MW. Consequently, the Auxiliary Heater exergy destruction rate becomes higher, very similar to the one of CPVT. Analyzing the efficiencies plots in Fig. 4.40, it is clear that all the exergy efficiencies reduce dramatically, especially for $\eta_{ex,CPVT}$ that is strongly influenced by the scarce solar irradiation availability and the ambient temperature decrease.

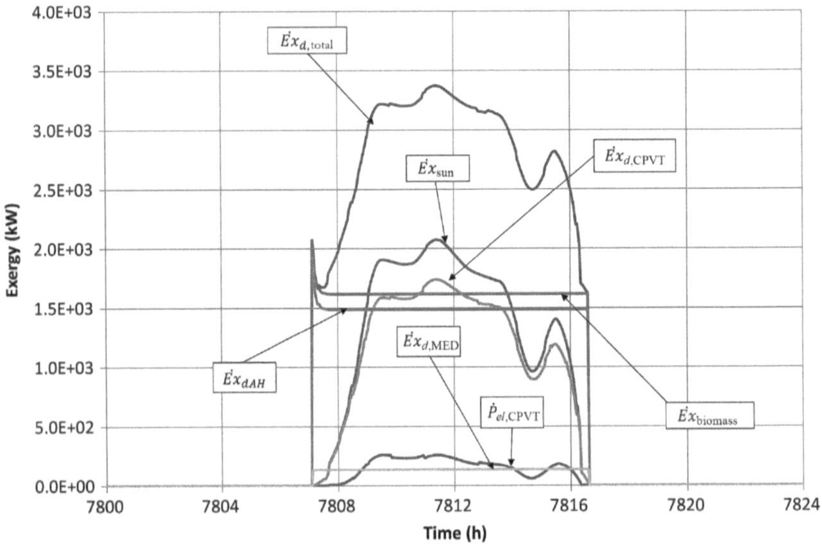

Fig. 4.39 Exergy flows for a representative winter day [126].

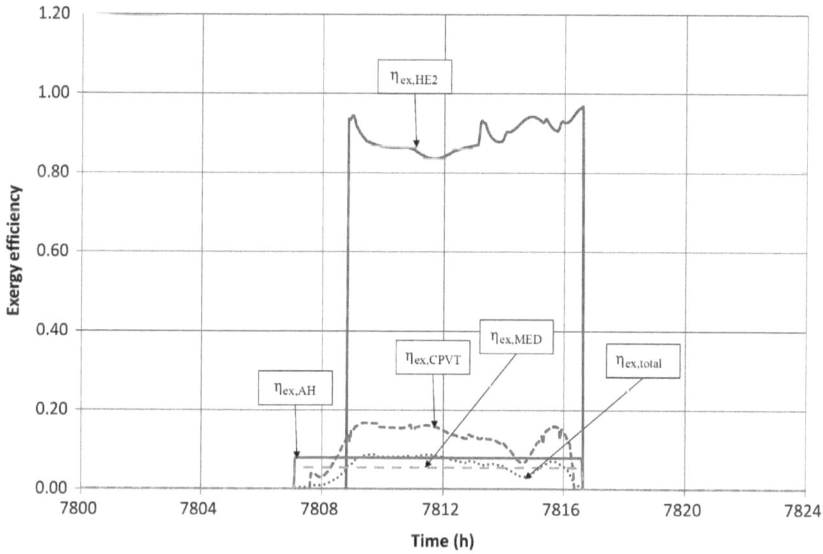

Fig. 4.40 Exergy efficiencies for a representative winter day [126].

$\eta_{ex,MED}$ and $\eta_{ex,AH}$, instead, are similar in winter and summer plots; in fact, these devices are not influenced significantly by weather conditions. In conclusion, $\eta_{ex,total}$ reaches 10% approximately, the system exergy performance being particularly sensible to the solar irradiation availability.

Results, based on weekly integration time step, are reported in Fig. 4.41, Fig. 4.42, Fig. 4.43, and Fig. 4.44, highlighting clearly the overall system exergy performance.

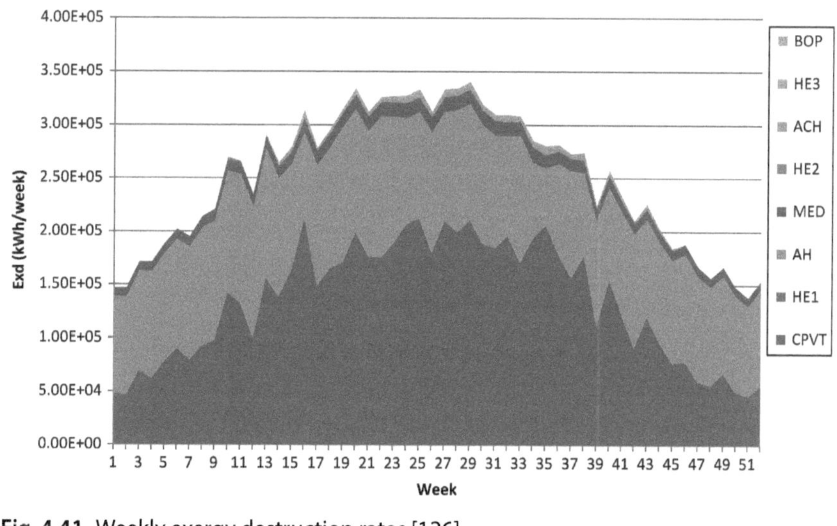

Fig. 4.41 Weekly exergy destruction rates [126].

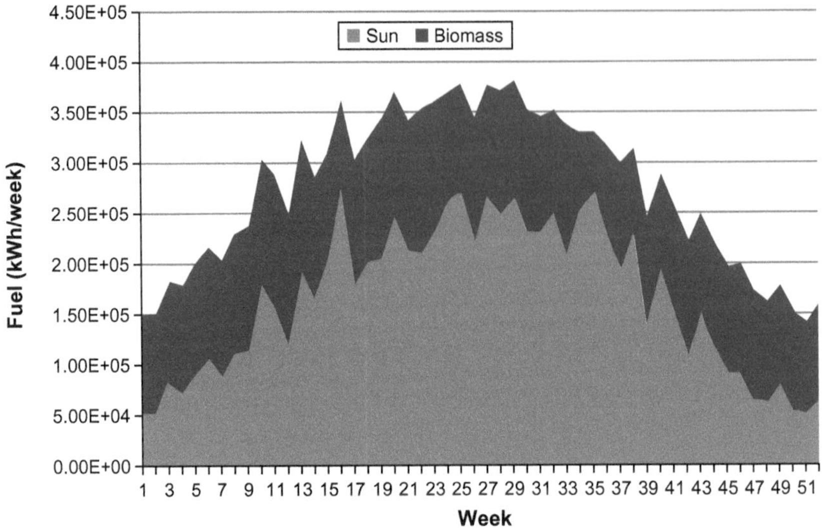

Fig. 4.42 Weekly exergy fuels [126].

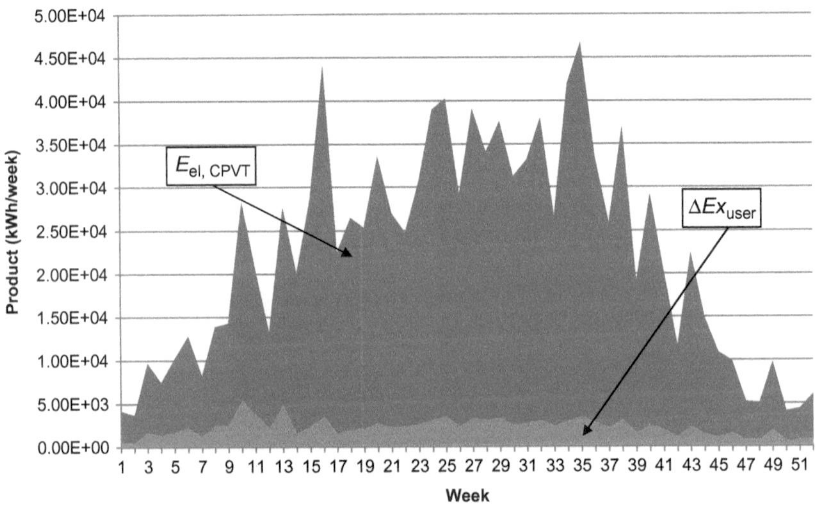

Fig. 4.43 Weekly exergy products [126].

In particular, in Fig. 4.41, it is clear that, during the summer, the total exergy destruction rate increases considerably due to CPVT and AH contributions. However, such increase is due to the conversion rates of thermal, electrical and cooling energy, and desalinated water, which are higher in the summer season.

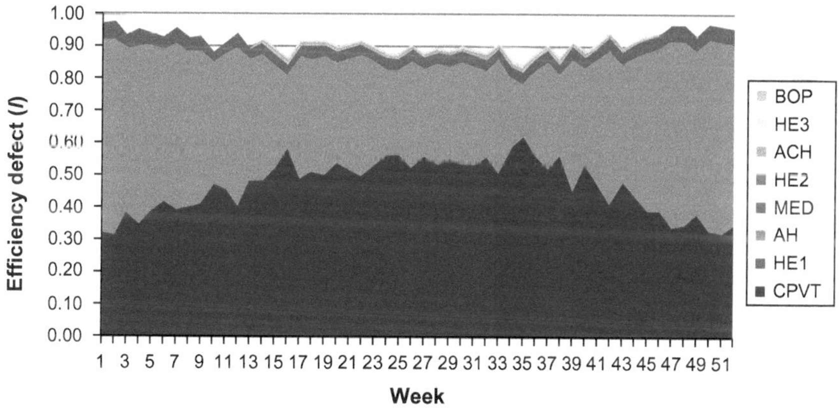

Fig. 4.44 Weekly efficiency defects [126].

Exergy products and fuels considerably increase in summer too, as highlighted in Fig. 4.42 and 4.43. Referring to the exergy fuel, during the summer, it consists substantially in solar energy, while the auxiliary fuel related to the electricity not covered demand is about two orders of magnitude lower than the other ones and, for this reason, it is not reported in the plot. During the winter, instead, solar and biomass exergy are comparable. Electricity provided by CPVT gives the most significant contribution for the products, since the exergy flow related to the thermal and cooling energies is low due to the low difference between its temperature and the ambient one.

Besides, water chemical exergy variation is the only factor influencing MED exergy product, much lower than electrical, thermal and cooling energies exergy products. In conclusion, DHW exergy, Ex_{DHW}, and MED exergy product, P_{MED}, are much inferior than the user exergy, ΔEx_{user}, and electricity provided by CPVT, $E_{el,CPVT}$, and so are not reported in Fig. 4.43.

In Fig. 4.44, the exergy results are reported through the efficiency defects of the components, such as the ratio between the destroyed exergy in the components and the system overall exergy fuel. CPVT reaches the highest value of efficiency defect, especially in summer, since it produces the biggest irreversibility amount, since most of the energy conversions happen in the solar field of CPVT. In winter season, instead, AH causes high efficiency defects due to the solar irradiance scarcity. MED efficiency defect, instead, is quite constant and low during the whole year. The average exergy efficiency of the system, which is the complement to 1 of the sum between efficiency defects of the components) oscillates from 3% (in the coldest week of winter) to 16% (in the hottest week of summer).

4.4.3 Exergoeconomic Analysis Results

In Fig. 4.45, the exergoeconomic results of the system are shown, considering the exergy flows integrated on annual basis:

The cumulated exergy flows, the capital recovery rate due to all the components of the system, and the numbered streams unit exergoeconomic costs are reported. It is possible to note that the produced electrical and thermal exergy unit cost for the CPVT is equal to $c_{eCPVT}=c_2=0.1748$ €/$kWh_{ex.}$ Such cost is mostly due to the high solar field capital and installation cost, equal to 2.4 M€ (figures with detailed cost information are reported in [24]), and to the its scarce utilization in many months characterized by low solar radiation.

The average unit cost of thermal exergy utilized for winter space heating, $c_{28,}$ is equal to 0.4394 €/kWh_{ex} and it is lower than the one related to cooling exergy utilized for summer space cooling, c_{23}, equal to 1.188 €/$kWh_{ex.}$ Such values are too high because of the low content of exergy in the thermal and cooling energy. It is important to evaluate the total cost flows connected to all the streams in €/kWh_{en}, instead of €/kWh_{ex}, in order to know the equivalent average costs per unit of energy.

This is possible through the ratio between energy and exergy of the heating or cooling energy that can be considered equal to the "Carnot factors."

Since the ACH inlet and outlet set point temperatures are equal to11.5°C and 6.5°C, the average value is assumed equal to 9°C; in the same way, the inlet and outlet temperatures for the space heating are equal to 40°C and 45°C, and thus the average value is assumed equal to 45°C. Thanks to these values of temperature, and once fixed the ambient temperature, T_a, equal to 25°C, it is possible to calculate easily average "Carnot factors" values, equal to 0.055 for winter space heating and 0.057 for summer space cooling. Also, the equivalent average costs per unit of energy are 0.024 €/kWh_{en} and 0.068 €/kWh_{en}, respectively.

Freshwater average unit cost, c_{21}, provided by MED unit, is equal to 3.457 €/kWh_{ex}. Considering the produced freshwater salt-free and a specific exergy desalted water content equal to 0.795 kWh_{ex}/m^3, it was found an average production cost equal to 2.748 €/m^3, which is an interesting unit cost for the freshwater supply especially in insulated and arid places.

The MED capital cost, being $\tilde{Z}_{MED}$ =15534 €/winter, is relatively low with respect to the desalted total water cost, as already handled in Ref. [130].

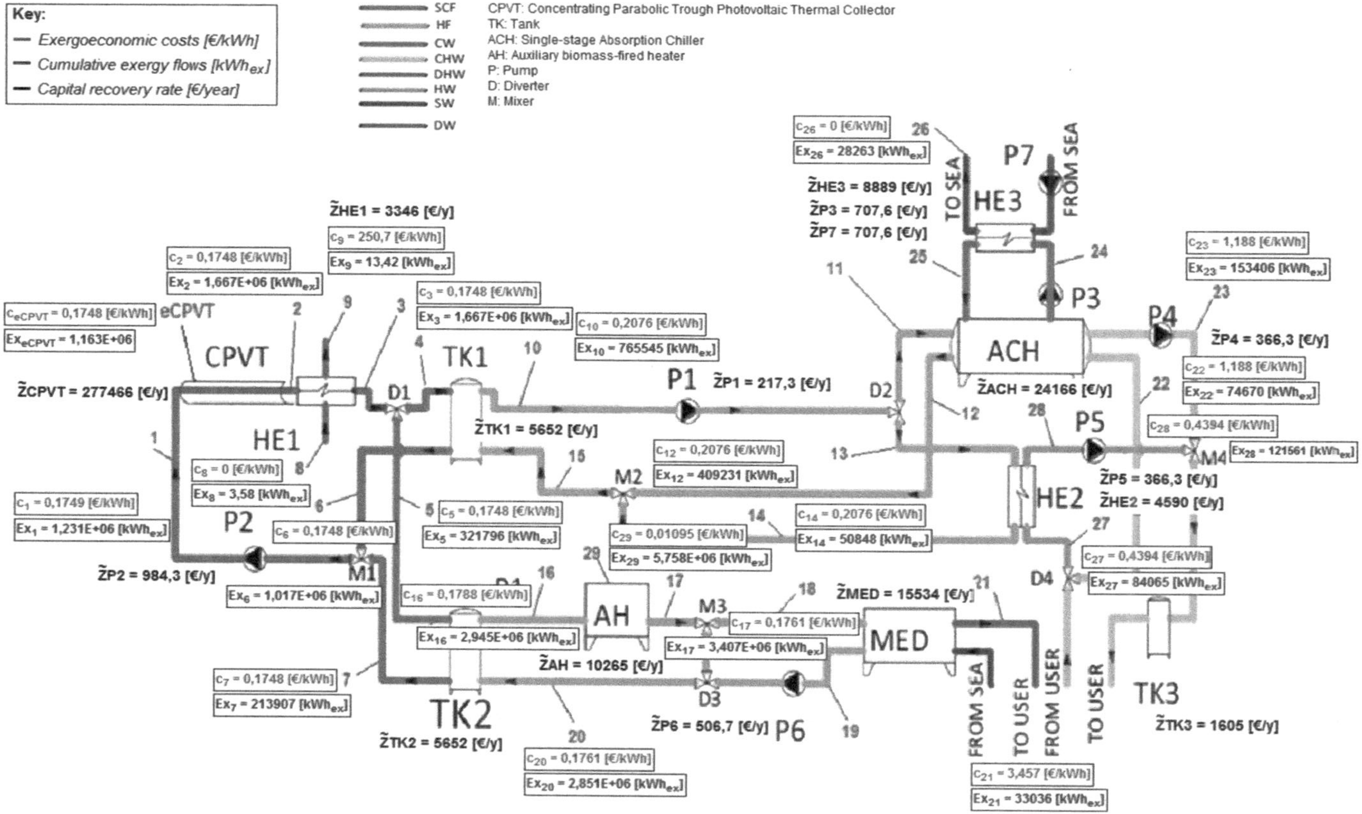

Fig. 4.45 Exergoeconomic cost obtained for annual cumulative exergy flows [126].

It is important to note that space cooling and heating are not simultaneous and so the capital recovery costs of the system components, which work for 12 months/year, can be assigned erroneously on the streams 22 and 23 for the cooling and 27 and 28 for the heating. To better understand the exergoeconomic results obtained, two different submodels with seasonal time basis of integration of exergy flows are presented in Fig. 4.46 and Fig. 4.47, for summer and winter season, respectively.

It is important to note that, considering the model during only one season, the thermal and cooling exergy unit costs for heating and cooling purposes, are equal to 0.8057 €/kWh$_{ex}$ and 0.9858 €/kWh$_{ex}$, respectively, sensibly different from the one calculated during the whole year; for this reason submodels were considered in order to better analyze the final products capital recovery costs. Space heating, referring only to HE2 and P5 components, requires a much lower capital cost with respect to the cooling one, referring to ACH, P3, P4, P7, and HE3; conversely, the heating unit exergoeconomic cost, c_{28}, is slightly lower than cooling one, c_{23}, and precisely:

$$\sum_{\text{winter}} Z = Z_{HE2} + Z_{P2} = 4596.3\ \text{€/season} \tag{4.90}$$

$$\sum_{\text{summer}} Z = Z_{ACH} + Z_{HE3} + Z_{P3} + Z_{P4} + Z_{P7} = 34836.5\ \text{€/season} \tag{4.91}$$

The main reason of such aspect is unavoidably related to the much higher solar irradiance and to the production of power rate during the summer season.

Other exergoeconomic analyses have been realized on weekly time basis in order to follow more clearly the yearly trend of the unit exergoeconomic costs, considering submodels referring to winter or summer weeks; a unique and detailed model has been considered for just one week in which both space heating and space cooling occur, where both Ex_{23} and Ex_{28} exergy flows are not null on a weekly basis, and the same goes for P4 and P5 energy consumptions.

The weekly-based exergoeconomic cost of the electrical energy provided is shown in Fig. 4.48:

It is worth nothing that very high fluctuations of the electricity unit cost occur, varying from 0.075 €/kWh$_{ex}$ in summer weeks up to 1.26 €/kWh$_{ex}$ in winter ones, not considering any type of incentive. In such case, some support mechanisms would make the analyzed polygeneration system more and more competitive compared to other alternative technologies.

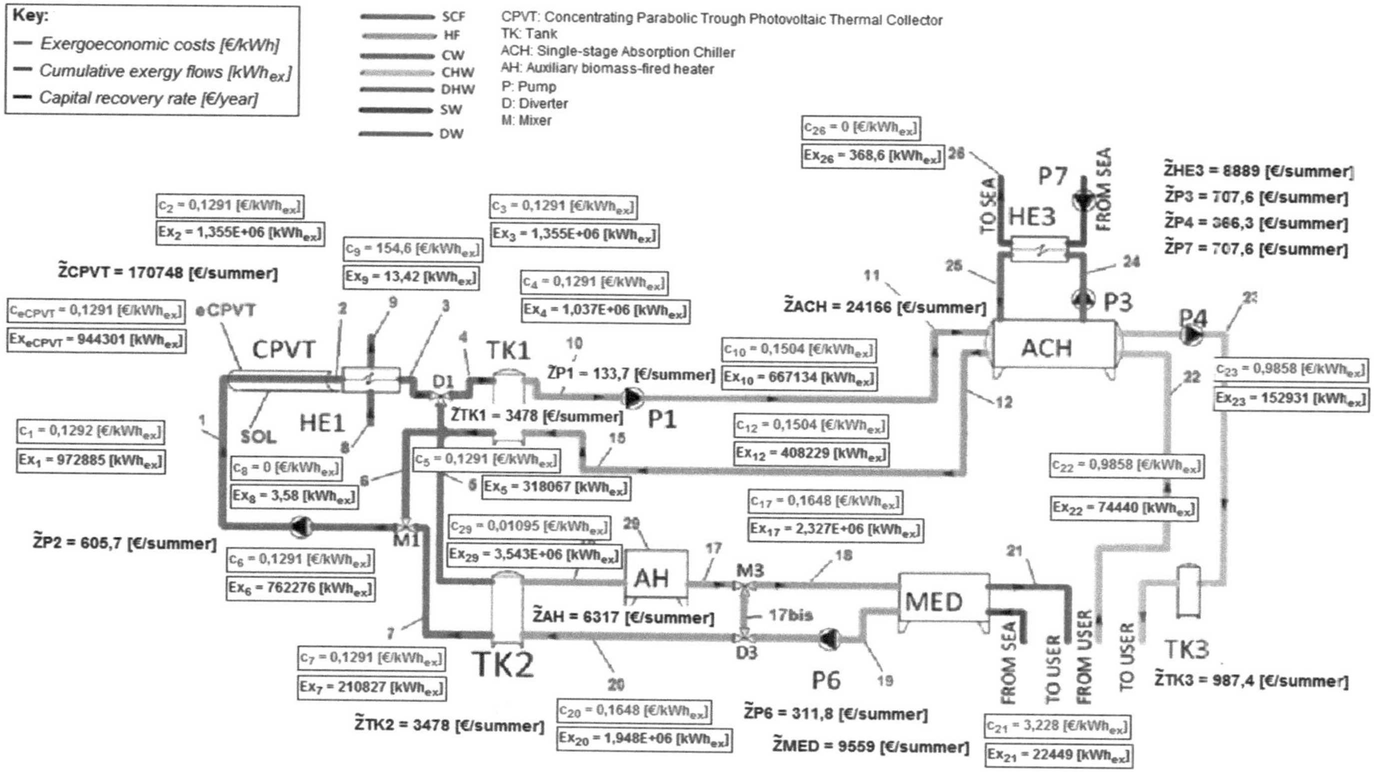

Fig. 4.46 Exergoeconomic costs obtained for cumulative exergy flows calculated over the whole summer season (space cooling period) [126].

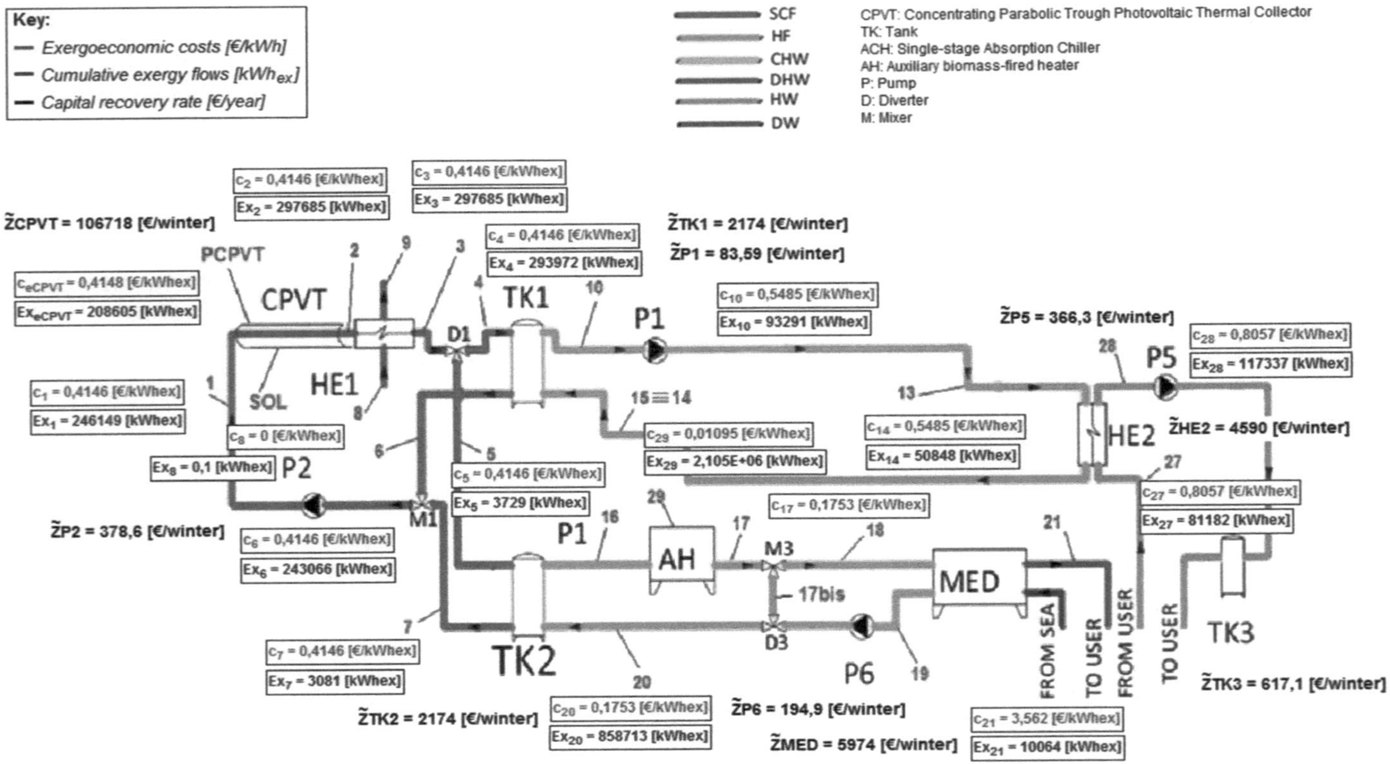

Fig. 4.47 Exergoeconomic costs obtained for cumulative exergy flows calculated over the whole winter season (space-heating period) [126].

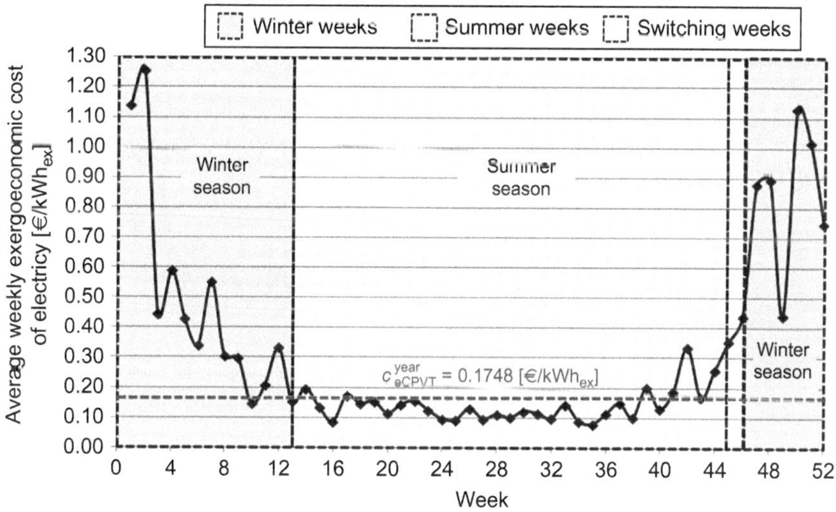

Fig. 4.48 Annual trend of weekly exergoeconomic costs of electricity [126].

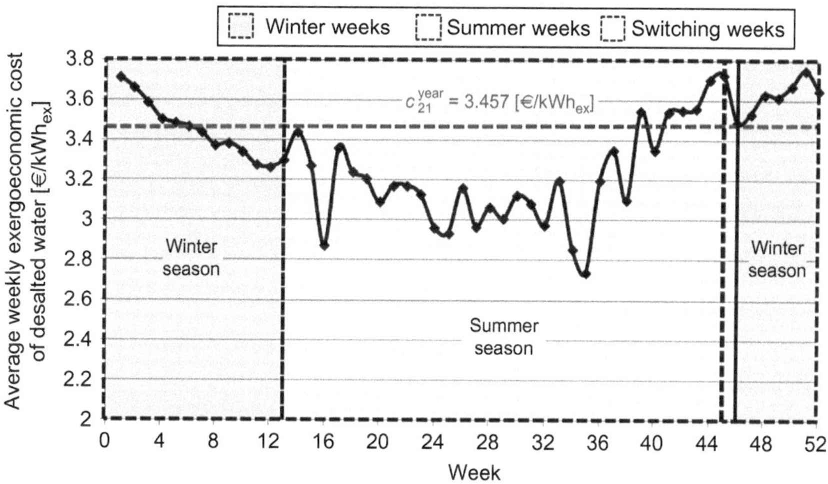

Fig. 4.49 Annual trend of weekly exergoeconomic costs of desalted water [126].

The weekly exergoeconomic cost of the desalinated water provided by MED unit is illustrated, through an annual trend, in Fig. 4.49. The freshwater production cost presents a smaller number of fluctuations along the year compared to the electricity one. It varies from 2.8 to 3.0 €/kWh$_{ex}$ for the summer season, up to 3.6–3.8 €/kWh$_{ex}$ for the winter season; such unit cost increase is due to the lower solar energy contribution toward TK2, compensated by biomass utilization.

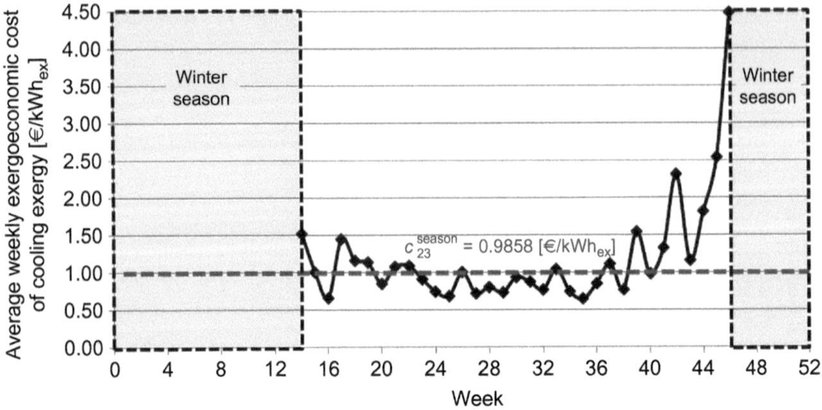

Fig. 4.50 Seasonal trend of weekly exergoeconomic costs of cooling exergy [126].

The weekly exergoeconomic cost trend, related to the cooling energy production for space cooling purpose, and named c23, is shown in Fig. 4.50. It is worth nothing that the unit cost is almost constant and mostly lower than the average seasonal unit cost, increasing only in the last cooling, requiring weeks up to 4.481 €/kWhex in the 46th week where solar irradiation is much lower and cooling to heating switch occurs. During the winter season, P5 is off and D2 conveys only hot water to the Absorption Chiller; thus, no values are reported.

Lastly, in Fig. 4.51, the weekly exergoeconomic cost trend of the thermal energy required for heating purpose, c28, is shown. It is clear that it is

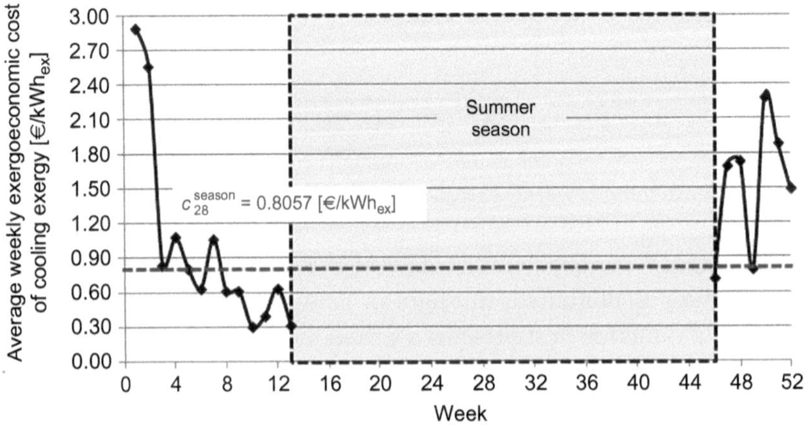

Fig. 4.51 Seasonal trend of weekly exergoeconomic costs of thermal exergy for space heating [126].

strongly dependent to the solar irradiation trend and, thus, the cost increases in the coldest weeks, such as first and second, in which the values are up to 10 times higher than in the other weeks.

In conclusion, flexible fees should be imposed by the plant operator/owner on annual basis, because of the high fluctuations experienced by the different products in the unit costs of production along the whole year. Besides, the unit price to be respected, in order to provide the profits expected, should be: 0.1748 €/kWh for the electricity, 0.9858 €/kWh_{ex} for the space cooling and, considering the average ratio kWh_{ex}/kWh_{en} deriving from the cold streams 22 and 23 level of temperature, 0.056 €/kWh as average cooling energy production cost, 0.8057 €/kWh_{ex} for the space heating and, considering the average ratio kWh_{ex}/kWh_{en} deriving from the hot streams 27 and 28 level of temperature, 0.044 €/kWh as average thermal energy production cost, 2.748 €/m^3 for the freshwater production.

4.4.3.1 Economic Feasibility of the Renewable Polygeneration System

In order to find out the real economic feasibility of such renewable polygeneration system (RPS), it was compared with four different reference scenarios based on different technologies, efficiencies, and productions costs. The first scenario is named "Extremely high efficiency," findable only in highly developed countries, where the electricity is produced thanks to natural-gas fuelled combined-cycle power plants for 70% and characterized by an overall efficiency, η_{el}, equal to 0.52, and renewable energy sources for 30%. Geothermal reversible heat pumps are considered, instead, for the heating and cooling with a seasonal average COP equal to 5.8 and 5.2, respectively. For the freshwater production, natural sources, easy to access, are considered, with a unit cost equal to 1.4 €/m^3 for the final user.

The second scenario is named "High efficiency," and it is based on the electricity production thanks to

natural-gas-fueled power plants for 80% and characterized by an electrical efficiency, η_{el}, equal to 0.41, and renewable energy sources for 20%. Cooling is provided by an air cooled chiller characterized by a COP equal to 3.1, heating by natural-gas-fueled condensation boiler characterized by a nominal efficiency equal to 98% and, lastly, freshwater from easy-to-access natural source, as seen in the previous scenario. The third scenario is named "Moderate efficiency," and it is based on the electricity production thanks to natural-gas-fuelled power plants for 30% and characterized by an electrical efficiency, η_{el}, equal to 0.52, oil-fuelled power plants for 30% and coal-fuelled power plants for 40%, through steam production and characterized by an electrical efficiency equal to 0.41. Reversible heat pumps are considered for heating and cooling, with a seasonal average COP equal to 3.5 and 3.1, respectively. Freshwater is considered supplied by easy-to-access natural sources again.

The last scenario is named "Disadvantaged" and it is common in remote or insulated places, such as small and arid islands. The electrical energy is provided thanks to diesel-oil-fuelled power plants, with a nominal electrical efficiency equal to 0.36. Heating and cooling purposes are provided by the same technology of the third scenario. Freshwater is supplied with shipping tankers with a unit cost equal to 8 €/m^3 for the final user, as indicated in Refs. [131] and [132] by considering European and Italian levels.

The following fuels unit costs have been assumed for the comparisons, referring to historical data registered in Refs. [133] and [134]: 10.81 €/GJ for natural gas, 20.95 €/GJ for fuel oil, 8.82 €/GJ for coal, and 26.83 €/GJ for diesel oil.

Average unit exergoeconomic costs for electricity, heating, cooling and freshwater production, are reported in Table 4.5, in comparison with the four reference scenarios:

By observing the last row of the Table 4.5, it is possible to know synthetically when the RPS is economically viable, that is in the fourth scenario, because of the high electricity and water costs. For this reason, the results seem to be very interesting especially is island applications, as noticed in

Table 4.5 Unit cost comparison [126]

	"Extremely high efficiency" scenario	"High efficiency" scenario	"Moderate efficiency" scenario	"Disadvantaged" scenario	Renewable polygeneration system
Electricity [€/kWh]	0.042	0.061	0.087	0.268	0.175
Heating [€/kWh]	0.007	0.032	0.025	0.077	0.044
Cooling [€/kWh]	0.008	0.020	0.028	0.086	0.056
Desalted water [€/m^3]	1.4	1.4	1.4	8.0	2.748
Economic viability of RPS	NO	NO	NO	YES	-

[139–141], and, by considering the Italian island Pantelleria, in [11], unless capital costs of CPVT and other main components go down; in such case, it may be economically viable also in other scenarios.

4.5 CONCLUSION

In this chapter, two polygeneration plants based on the integration of several technologies (CPVT, SHC, GW, AH, and MED) were presented and dynamically simulated in TRNSYS. In particular, the two systems, the GP and BP, produce solar space heating and cooling, electricity, domestic hot water, and desalinated water, by using the solar energy, the geothermal energy, and the biomass energy, respectively.

For both plants GP and BP, a numerical case study located in volcanic island of Pantelleria and in Naples (two South Italy localities) was developed and widely discussed, by highlighting the potential thermoeconomic saving realizable by these systems. In addition, in case of BP, an exergy and an exergoeconomic model was also presented in order to show the exergy efficiency and exergy destructions of main system components and to identify the exergoeconomic costs of the different energy/material products.

The achieved results can be summarized as follows:

- The winter performance of CPVT collectors is significantly lower than the summer performance, due to the reduction of beam radiation observed during this period. This determines a corresponding reduction of electrical and thermal energy production, and as a consequence, a reduction of the space heating energy delivered to the user.
- In winter season, the majority of energy used to supply the MED unit is provided by the biomass auxiliary heater and by the geothermal well, respectively, in case of BP and GP system.
- The energy performances of both plants are excellent during the summer, since the systems are able to produce large amounts of space cooling energy, DHW, electricity and desalinated water by integrating to the solar energy the biomass/low-enthalpy geothermal energy, respectively, for BP and GP.
- The carried-out sensitivity analysis showed that the GP can be particularly promising for several Mediterranean Volcanic islands, featured by higher summer energy and water demands, since the obtained SPB periods are resulted excellent.
- In order to maximize the system thermoeconomic performance (i) the optimal ratio between solar field area and ACH capacity was found around

$5.9\,m^2/kW$; (ii) small storage tank volumes should be preferred (around 15 L per unit area of solar field); (iii) the outlet temperatures from the solar field should be around 85°C in the summer and 55°C in the winter; (iv) a high number of effects [14] should be used for the MED unit.

- From the economic point of view, the obtained SPB periods are similar to those reported for other systems based on renewable energy sources.
- The plants are particularly interesting in case of feed-in tariffs similar to those currently assumed for PV technologies in many EU areas. Such feed-in tariffs are currently obligatory for most technologies based on renewable energy, in order to encourage their diffusion and to reduce their capital cost.
- From the exergy point of view, the highest exergy destructions occur at the CPVT collectors and at the biomass auxiliary heater.
- Significant exergy destructions resulted at the MED unit, where only a very low exergy input fraction through hot water is converted into chemical exergy of the desalinated water product, as is typical in thermal desalination units.

REFERENCES

[1] Serra LM, Lozano M-A, Ramos J, Ensinas AV, Nebra SA. Polygeneration and efficient use of natural resources. Energy 2009;34(5):575–86.

[2] Tian L, Tang Y, Wang Y. Economic evaluation of seawater desalination for a nuclear heating reactor with multi-effect distillation. Desalination 2005;180:53–61.

[3] Wang LK, Chen JP, Hung YT, Shammas NK. Membrane and desalination technologies. New York: Humana Press; 2010.

[4] Eveloy V, Rodgers P, Qiu L. Performance investigation of a power, heating and seawater desalination poly-generation scheme in an off-shore oil field. Energy 2016;98:26–39.

[5] Kronenberg G. Cogeneration with the LT-MED desalination process. Desalination 1997;208:287–94.

[6] Fernández-López C, Viedma A, Herrero R, Kaiser AS. Seawater integrated desalination plant without brine discharge and powered by renewable energy systems. Desalination 2009;235:179–98.

[7] Hossain AK, Thorpe R, Vasudevan P, Sen PK, Critoph RE, Davies PA. Omnigen: providing electricity, food preparation, cold storage and pure water using a variety of local fuels. Renew Energy 2013;49:197–202.

[8] Christ A, Rahimi B, Regenauer-Lieb K, Chua HT. Techno-economic analysis of geothermal desalination using Hot Sedimentary Aquifers: a pre-feasibility study for Western Australia. Desalination 2017;404:167–81.

[9] Szacsvay T, Hofer-Noser P, Posnansky M. Technical and economic aspects of small-scale solar pond powered seawater desalination systems. Desalination 1999;122:185–93.

[10] Calise F, Macaluso A, Piacentino A, Vanoli L. A novel hybrid polygeneration system supplying energy and desalinated water by renewable sources in Pantelleria Island. Energy 2017;37:1086–1106.

[11] Calise F, Cipollina A, Dentice d'Accadia M, Piacentino A. A novel renewable polygeneration system for a small Mediterranean volcanic island for the combined production of energy and water: dynamic simulation and economic assessment. Appl Energy 2014;135:675–93.
[12] Sharaf MA, Nafey AS, García-Rodríguez L. Exergy and thermo-economic analyses of a combined solar organic cycle with multi effect distillation (MED) desalination process. Desalination 2011;272(1–3):135–47.
[13] Calise F, Dentice d'Accadia M, Macaluso A, Vanoli L, Piacentino A. A novel solar-geothermal trigeneration system integrating water desalination: design, dynamic simulation and economic assessment. Energy 2016;115(Part 3):1533–47.
[14] Calise F, d'Accadia MD, Macaluso A, Piacentino A, Vanoli L. Exergetic and exergoeconomic analysis of a novel hybrid solar—geothermal polygeneration system producing energy and water. Energy Convers Manag 2016;115:200–20.
[15] Sharqawy MH, Lienhard VJH, Zubair SM. On exergy calculations of seawater with applications in desalination systems. Int J Therm Sci 2011;50(2):187–96.
[16] Mohan G, Kumar U, Pokhrel MK, Martin A. A novel solar thermal polygeneration system for sustainable production of cooling, clean water and domestic hot water in United Arab Emirates: dynamic simulation and economic evaluation. Appl Energy 2016;167:173–88.
[17] Mohan G, Uday Kumar NT, Pokhrel MK, Martin A. Experimental investigation of a novel solar thermal polygeneration plant in United Arab Emirates. Renew Energy 2016;91:361–73.
[18] Uday Kumar NT, Mohan G, Martin A. Performance analysis of solar cogeneration system with different integration strategies for potable water and domestic hot water production. Appl Energy 2016;170:466–75.
[19] Mata-Torres C, Escobar RA, Cardemil JM, Simsek Y, Matute JA. Solar polygeneration for electricity production and desalination: case studies in Venezuela and northern Chile. Renew Energy 2017;101:387–98.
[20] Ibrahim AGM, Dincer I. Experimental performance evaluation of a combined solar system to produce cooling and potable water. Sol Energy 2015;122: 1066–79.
[21] Maraver D, Uche J, Royo J. Assessment of high temperature organic Rankine cycle engine for polygeneration with MED desalination: a preliminary approach. Energy Convers Manag 2012;53(1):108–17.
[22] Sahoo U, Kumar R, Pant PC, Chaudhary R. Development of an innovative polygeneration process in hybrid solar-biomass system for combined power, cooling and desalination. Appl Therm Eng 2017;120:560–7.
[23] Sahoo U, Kumar R, Pant PC, Chaudhury R. Scope and sustainability of hybrid solar–biomass power plant with cooling, desalination in polygeneration process in India. Renew Sust Energ Rev 2015;51:304–16.
[24] Calise F, Dentice d'Accadia M, Piacentino A. A novel solar trigeneration system integrating PVT (photovoltaic/thermal collectors) and SW (seawater) desalination: dynamic simulation and economic assessment. Energy 2014;67:129–48.
[25] Rubio-Maya C, Uche-Marcuello J, Martínez-Gracia A, Bayod-Rújula AA. Design optimization of a polygeneration plant fuelled by natural gas and renewable energy sources. Appl Energy 2011;88(2):449–57.
[26] Rubio-Maya C, Uche J, Martínez A. Sequential optimization of a polygeneration plant. Energy Convers Manag 2011;52(8–9):2861–9.
[27] Alarcón-Padilla DC, García-Rodríguez L, Blanco-Gálvez J. Design recommendations for a multi-effect distillation plant connected to a double-effect absorption heat pump: a solar desalination case study. Desalination 2010;262:11–4.
[28] García-Rodríguez L, Gómez-Camacho C. Exergy analysis of the SOL-14 plant (Plataforma Solar de Almería, Spain). Desalination 2001;137:251–8.

[29] Tamburini A, Cipollina A, Micale G, Piacentino A. CHP (combined heat and power) retrofit for a large MED-TVC (multiple effect distillation along with thermal vapour compression) desalination plant: high efficiency assessment for different design options under the current legislative EU framework. Energy 2016;115(Part 3):1548–59.
[30] Calise F, Palombo A, Vanoli L. In: Design and dynamic simulation of a novel solar trigeneration system based on photovoltaic/thermal collectors. ECOS 2011; June 2011; Novi Sad, Serbia; 2011.
[31] Zondag HA. Flat-plate PV-thermal collectors and systems: a review. Renew Sust Energ Rev 2008;12(4):891–5.
[32] Chow TT. A review on photovoltaic/thermal hybrid solar technology. Appl Energy 2010;87(2):365–79.
[33] Skoplaki E, Palyvos JA. On the temperature dependence of photovoltaic module electrical performance: a review of efficiency/power correlations. Sol Energy 2009;83:614–24.
[34] Ibrahim A, Othman M, Ruslan MH, Mat S, Sopian K. Recent advances in flat plate photovoltaic/thermal (PV/T) solar collectors. Renew Sust Energ Rev 2011;15: 352–65.
[35] Tripanagnostopoulos Y. Aspects and improvements of hybrid photovoltaic/thermal solar energy systems. Sol Energy 2007;81(9):1117–31.
[36] Zhang X, Zhao X, Smith S, Xu J, Yu X. Review of R&D progress and practical application of the solar photovoltaic/thermal (PV/T) technologies. Renew Sust Energ Rev 2011;16(1):599–617.
[37] Daghigh R, Ruslan MH, Sopian K. Advances in liquid based photovoltaic/thermal (PV/T) collectors. Renew Sust Energ Rev 2011;15(8):4156–70.
[38] Kumar R, Rosen MA. A critical review of photovoltaic–thermal solar collectors for air heating. Appl Energy 2011;88(11):3603–14.
[39] Zhao J, Song Y, Lam WH, Liu W, Liu Y, Zhang Y, et al. Solar radiation transfer and performance analysis of an optimum photovoltaic/thermal system. Energy Convers Manag 2011;52:1343–53.
[40] Mittelman G, Kribus A, Mouchtar O, Dayan A. Water desalination with concentrating photovoltaic/thermal (CPVT) systems. Sol Energy 2007;83(8):1322–34.
[41] Mittelman G, Kribus A, Dayan A. Solar cooling with concentrating photovoltaic/thermal (CPVT) systems. Energy Convers Manag 2007;48:2481–90.
[42] Nishioka K, Takamoto T, Agui T, Kaneiwa M, Uraoka Y, Fuyuki T. Annual output estimation of concentrator photovoltaic systems using high-efficiency InGaP/InGaAs/Ge triple-junction solar cells based on experimental solar cell's characteristics and field-test meteorological data. Sol Energy Mater Sol Cells 2006;90:57–67.
[43] Vorobiev Y, González-Hernández J, Vorobiev P, Bulat L. Thermal-photovoltaic solar hybrid system for efficient solar energy conversion. Sol Energy 2006;80: 170–6.
[44] Kosmadakis G, Manolakos D, Papadakis G. Simulation and economic analysis of a CPV/thermal system coupled with an organic Rankine cycle for increased power generation. Solar Energy 2011;85:308–24.
[45] Zhang X, Zhao X, Smith S, Xu J, Yu X. Review of R&D progress and practical application of the solar photovoltaic/thermal (PV/T) technologies. Renew Sust Energ Rev 2011;16:599–617.
[46] Coventry JS. Performance of a concentrating photovoltaic/thermal solar collector. Sol Energy 2005;78(2):211–22.
[47] Rosell JI, Vallverdú X, Lechón MA, Ibáñez M. Design and simulation of a low concentrating photovoltaic/thermal system. Energy Convers Manag 2005;46(18, 19): 3034–46.

[48] Kribus A, Daniel K, Mittelman G, Hirshfeld A, Flitsanov Y, Dayan A. A miniature concentrating photovoltaic and thermal system. Energy Convers Manag 2006;47(20):3582–90.
[49] Li M, Li GL, Ji X, Yin F, Xu L. The performance analysis of the Trough Concentrating Solar Photovoltaic/Thermal system. Energy Convers Manag 2011;52(6):2378–83.
[50] Li M, Ji X, Li G, Wei S, Li YF, Shi F. Performance study of solar cell arrays based on a Trough Concentrating Photovoltaic/Thermal system. Appl Energy 2011;88(9): 3218–27.
[51] Li M, Ji X, Li GL, Yang ZM, Wei SX, Wang LL. Performance investigation and optimization of the Trough Concentrating Photovoltaic/Thermal system. Sol Energy 2011;85(5):1028–34.
[52] Bernardo LR, Perers B, Håkansson H, Karlsson B. Performance evaluation of low concentrating photovoltaic/thermal systems: a case study from Sweden. Sol Energy 2011;85(7):1499–510.
[53] Al-Alili A, Hwang, Y., Radermacher, R., Kubo, I. A high efficiency solar air conditioner using concentrating photovoltaic/thermal collectors, Appl Energy. 2011; In Press, Corrected Proof, May 2011.
[54] Calise F, Palombo A, Vanoli L. A finite-volume model of a parabolic trough photovoltaic/thermal collector: energy and exergetic analyses. Energy 2012;46(1):283–94.
[55] Buonomano A, Calise F, Palombo A. Solar heating and cooling systems by CPVT and ET solar collectors: a novel transient simulation model. Appl Energy 2013;103: 588–606.
[56] Buonomano A, Calise F, Dentice d'Accadia M, Vanoli L. A novel solar trigeneration system based on concentrating photovoltaic/thermal collectors. Part 1: design and simulation model. Energy 2013;61:59–71.
[57] Calise F, Dentice d'Accadia M, Palombo A, Vanoli L. Dynamic simulation of a novel high-temperature solar trigeneration system based on concentrating photovoltaic/ thermal collectors. Energy 2013;61:72–86.
[58] Calise F, Dentice d'Accadia M, Roselli C, Sasso M, Tariello F. Desiccant-based AHU interacting with a CPVT collector: simulation of energy and environmental performance. Sol Energy 2014;103:574–94.
[59] Del Col D, Bortolato M, Padovan A, Quaggia M. Experimental and numerical study of a parabolic trough linear CPVT system. Energy Procedia 2014;57:255–64.
[60] Karathanassis IK, Papanicolaou E, Belessiotis V, Bergeles GC. Design and experimental evaluation of a parabolic-trough concentrating photovoltaic/thermal (CPVT) system with high-efficiency cooling. Renew Energy 2017;101:467–83.
[61] Crisostomo F, Taylor RA, Surjadi D, Mojiri A, Rosengarten G, Hawkes ER. Spectral splitting strategy and optical model for the development of a concentrating hybrid PV/T collector. Appl Energy 2015;141:238–46.
[62] Otanicar TP, Theisen S, Norman T, Tyagi H, Taylor RA. Envisioning advanced solar electricity generation: parametric studies of CPV/T systems with spectral filtering and high temperature PV. Appl Energy 2015;140:224–33.
[63] Li G, Pei G, Ji J, Su Y. Outdoor overall performance of a novel air-gap-lens-walled compound parabolic concentrator (ALCPC) incorporated with photovoltaic/thermal system. Appl Energy 2015;144:214–23.
[64] Chemisana D, Rosell JI, Riverola A, Lamnatou C. Experimental performance of a Fresnel-transmission PVT concentrator for building-façade integration. Renew Energy 2016;85:564–72.
[65] van der Bruggen B. Desalination by distillation and by reverse osmosis—trends towards the future. Membr Technol 2003;2003(2):6–9.
[66] Alsadaie SM, Mujtaba IM. Dynamic modelling of Heat Exchanger fouling in multistage flash (MSF) desalination. Desalination 2017;409:47–65.

[67] Chorak A, Palenzuela P, Alarcón-Padilla D-C, Ben AA. Experimental characterization of a multi-effect distillation system coupled to a flat plate solar collector field: empirical correlations. Appl Therm Eng 2017;120:298–313.

[68] Chen L, Xu Q, Gossage JL, Lou HH. Simulation and economic evaluation of a coupled thermal vapor compression desalination process for produced water management. J Nat Gas Sci Eng 2016;36(Part A):442–53.

[69] Alkaisi A, Mossad R, Sharifian-Barforoush A. A review of the water desalination systems integrated with renewable energy. Energy Procedia 2017;110:268–74.

[70] Abduljawad M, Ezzeghni U. Optimization of Tajoura MSF desalination plant. Desalination 2010;254:23–8.

[71] Cardona E, Piacentino A. Optimal design of cogeneration plants for seawater desalination. Desalination 2004;166:411–26.

[72] Manju S, Sagar N. Renewable energy integrated desalination: a sustainable solution to overcome future fresh-water scarcity in India. Renew Sust Energ Rev 2017;73: 594–609.

[73] Mehta D, Gupta L, Dhingra R. Forward osmosis in India: status and comparison with other desalination technologies. Int Scholarly Res Notices 2014;.

[74] Abdelmoez W, Mahmoud MS, Farrag TE. Water desalination using humidification/dehumidification (HDH) technique powered by solar energy: a detailed review. Desalin Water Treat 2014;52:4622–40.

[75] Shenvi SS, Isloor AM, Ismail AF. A review on RO membrane technology: developments and challenges. Desalination 2015;368:10–26.

[76] Energy Requirements of Desalination Processes. Available from: http://www.desware.net/Energy-Requirements-Desalination-Processes.aspx; n.d. [Accessed 16 September 2015].

[77] Noorollahi Y, Taghipoor S, Sajadi B. Geothermal sea water desalination system (GSWDS) using abandoned oil/gas wells. Geothermics 2017;67:66–75.

[78] Zhang L-Z, Li G-P. Energy and economic analysis of a hollow fiber membrane-based desalination system driven by solar energy. Desalination 2017;404:200–14.

[79] Loutatidou S, Liosis N, Pohl R, Ouarda TBMJ, Arafat HA. Wind-powered desalination for strategic water storage: techno-economic assessment of concept. Desalination 2017;408:36–51.

[80] Hasan E. Desalination integration with renewable energy for climate change abatement in the MENA Region. In: Recent progress in desalination, environmental and marine outfall systems. vol. 1. Cham, Switzerland: Springer; 2015. p. 159–73.

[81] Chandrashekara M, Yadav A. Water desalination system using solar heat: a review. Renew Sust Energ Rev 2017;67:1308–30.

[82] Saleh A, Qudeiri JA, Al-Nimr MA. Performance investigation of a salt gradient solar pond coupled with desalination facility near the Dead Sea. Energy 2011;36(2): 922–31.

[83] Malik M, Tiwari G, Kumar A, Sodha M. Solar distillation: a practical study of a wide range of stills and their optimum design construction and performance. New York: Pergamon Press; 1996.

[84] Fath H, Ghazy A. Solar desalination using humidification—dehumidification technology. Desalination 2002;142:119–33.

[85] Askari IB, Ameri M. Techno economic feasibility analysis of Linear Fresnel solar field as thermal source of the MED/TVC desalination system. Desalination 2016;394:1–17.

[86] Talebbeydokhti P, Cinocca A, Cipollone R, Morico B. Analysis and optimization of LT-MED system powered by an innovative CSP plant. Desalination 2017;413: 223–33.

[87] Gholinejad M, Bakhtiari A, Bidi M. Effects of tracking modes on the performance of a solar MED plant. Desalination 2016;380:29–42.

[88] Mohamed AMI, El-Minshawy NAS. Humidification—dehumidification desalination system driven by geothermal energy. Desalination 2009;249(2):602–8.
[89] Bourouni K, Deronzier JC, Tadrist L. Experimentation and modelling of an innovative geothermal desalination unit. Desalination 1999;125(1):147–53.
[90] Sarbatly R, Chiam C-K. Evaluation of geothermal energy in desalination by vacuum membrane distillation. Appl Energy 2013;112:737–46.
[91] Goosen M, Mahmoudi H, Ghaffour N. Water desalination using geothermal energy. Energies 2010;3:1423–42.
[92] Gude VG, Nirmalakhandan N, Deng S. Desalination using solar energy: towards sustainability. Energy 2011;36(1):78–85.
[93] Manenti F, Masi M, Santucci G. Start-up operations of MED desalination plants. Desalination 2013;329:57–61.
[94] Manenti F, Masi M, Santucci G, Manenti G. Parametric simulation and economic assessment of a heat integrated geothermal desalination plant. Desalination 2013;317:193–205.
[95] Karytsas C. In: Mediterranean conference on renewable energy sources for water production. European commission, EURORED Network. Santorini, Greece CRES; 1996. p. 128–31.
[96] Loutatidou S, Arafat HA. Techno-economic analysis of MED and RO desalination powered by low-enthalpy geothermal energy. Desalination 2015;365:277–92.
[97] Klein SA. Solar Energy Laboratory. In: TRNSYS: a transient system simulation program. Madison: University of Wisconsin; 2006.
[98] Calise F, Dentice d'Accadia M, Palombo A. Transient analysis and energy optimization of solar heating and cooling systems in various configurations. Sol Energy 2010;84(3):432–49.
[99] Calise F. Thermoeconomic analysis and optimization of high efficiency solar heating and cooling systems for different Italian school buildings and climates. Energy Build 2010;42:992–1003.
[100] Calise F, Dentice d'Accadia M, Palombo A, Vanoli L. In: Dynamic simulation of high temperature solar heating and cooling systems. Eurosun 2010; September 2010; Graz (AU); 2010.
[101] Calise F, Ferruzzi G, Vanoli L. In: Transient simulation of polygeneration systems based on fuel cells and solar cooling technologies. ECOS 2010; Lousanne; 2010.
[102] Calise F. Design of a hybrid polygeneration system with solar collectors and a Solid Oxide Fuel Cell: dynamic simulation and economic assessment. Int J Hydrog Energy 2011;36(10):6128–50.
[103] Calise F, Dentice d'Accadia M, Vanoli L. Design and dynamic simulation of a novel solar trigeneration system based on hybrid photovoltaic/thermal collectors (PVT). Energy Convers Manag 2012;60:214–25.
[104] Bernardo R. Retrofitted solar thermal system for domestic hot water for single family electrically heated houses. Development and testing: Lund University; 2010.
[105] Calise F. High temperature solar heating and cooling systems for different Mediterranean climates: dynamic simulation and economic assessment. Appl Therm Eng 2011;32:108–24.
[106] Florschuetz LW. Extension of the Hottel-Whillier model to the analysis of combined photovoltaic/thermal flat plate collectors. Sol Energy 1979;22:361–6.
[107] Duffie JA, Backman WA. Solar engineering of thermal processes. John Wiley and Sons; 1980.
[108] Incropera FP, DeWitt DP. Fundamentals of heat and mass transfer. 5th ed. John Wiley & Sons; 2001.
[109] Cardona E, Piacentino A, Marchese F. Performance evaluation of CHP hybrid seawater desalination plants. Desalination 2007;205:1–14.

[110] El-Nashar AM, Samad M. The solar desalination plant in Abu Dhabi: 13 years of performance and operation history. Renew Energy 1998;14:263–74.
[111] Abu-Jabal MS, Kamiya I, Narasaki Y. Proving test for a solar-powered desalination system in Gaza-Palestine. Desalination 2001;137:1–6.
[112] El-Dessouky HT, Ettouney HM. Fundamentals of salt water desalination. Amsterdam, The Netherlands: Elsevier; 2002.
[113] Kemp IC. Pinch analysis and process integration. Oxford, UK: Butterworth-Imprint of Elsevier; 2007.
[114] El-Nashar AM. The economic feasibility of small solar MED seawater desalination plants for remote arid areas. Desalination 2001;134:173–86.
[115] Calise F, Dentice d'Accadia M, Vanoli L. Thermoeconomic optimization of solar heating and cooling systems. Energy Convers Manag 2010;52(2):1562–73.
[116] Rosen MA, Dincer I. Effect of varying dead-state properties on energy and exergy analyses of thermal systems. Int J Therm Sci 2004;43(2):121–33.
[117] Song G, Shen L, Xiao J. Estimating specific chemical exergy of biomass from basic analysis data. Ind Eng Chem Res 2011;50(16):9758–66.
[118] Kotas TJ. The exergy method of thermal plant analysis. London: Butterworths; 1995.
[119] Torío H, Angelotti A, Schmidt D. Exergy analysis of renewable energy-based climatisation systems for buildings: a critical view. Energy Build 2009;41(3):248–71.
[120] Ajam H, Farahat S, Saehaddi F. Exergetic optimization of solar air heaters and comparison with energy analysis. Int J Thermodyn 2005;8(4):183–90.
[121] Spiegler K, El-Sayed Y. The energetics of desalination processes. Desalination 2001;134(1–3):109–28.
[122] Calise F, Dentice d'Accadia M, Piacentino A. Exergetic and exergoeconomic analysis of a renewable polygeneration system and viability study for small isolated communities. Energy 2015;92(Part 3):290–307.
[123] Baskut O, Ozgener L. Exergoeconomic assessment of a wind turbine power plant (WTTP): Cesme, Izmir, example. Energy 2012;47(1):577–81.
[124] Lazzaretto A, Tsatsaronis G. SPECO: a systematic and general methodology for calculating efficiencies and costs in thermal systems. Energy 2006;31(8–9):1257–89.
[125] Lozano M, Valero A. Theory of the exergetic cost. Energy 1993;18(9):939–60.
[126] Agudelo A, Valero A, Torres C. Allocation of waste cost in thermoeconomic analysis. Energy 2012;45(1):634–43.
[127] Bejan A, Tsatsaronis G, Moran M. Thermal design and optimization. New York: Wiley and Sons; 1996.
[128] Torres C, Valero A, Rangel V, Zaleta A. On the cost formation process of the residues. Energy 2008;33(2):144–52.
[129] Seyyedi S, Ajam H, Farahat S. A new criterion for the allocation of residues cost in exergoeconomic analysis of energy systems. Energy 2010;35(8).
[130] Janghorban Esfahani I, Yoo C. A highly efficient combined multi-effect evaporation-absorption heat pump and vapor-compression refrigeration part 2: thermoeconomic and flexibility analysis. Energy 2014;75:327–37.
[131] B. M. Opening the oceans as a supply source. Water 21, 2001; 22–24.
[132] Scientifiche" FFdATe. Cogeneration: implementation in the Mediterranean islands and coastal areas. With support of European Commission. Handbook Water and Power—Work Pagckage No 7. 2014.
[133] EIA, Energy Information Administration, Analysis and projections: price summary (historical and forecast), Annual Energy Outlook, 2014, EIA Department of Washington.
[134] DECC. Fossil fuel price projections 2013. UK: Department of Energy and Climate Change; 2014.

[135] Karellas S, Terzis K, Manolakos D. Investigation of an autonomous hybrid solar thermal ORC–PV RO desalination system. The Chalki Island case. Renew Energy 2011;36(2):583–90.
[136] Kalogirou SA. Chapter 8—Solar desalination systems. Solar energy engineering, (2nd ed.) Boston: Academic Press; 2014. p. 431–79.
[137] Schwantes R, Cipollina A, Gross F, Koschikowski J, Pfeifle D, Rolletschek M, et al. Membrane distillation: solar and waste heat driven demonstration plants for desalination. Desalination 2013;323:93–106.

CHAPTER 5

Design and Construction of Open Intakes

Nikolay Voutchkov
Water Globe Consultants, LLC, Winter Springs, FL, United States

5.1 INTRODUCTION

The saline source water intake is one of the key components of every desalination plant. The primary purpose of the intake is to ensure adequate and consistent flow and quality of source water over the entire useful life of the desalination plant. The type and configuration of intake selected for a given desalination project has a significant impact on the nature and quantity of foulants contained in the source water and, therefore on the complexity of the pretreatment system needed to control RO membrane fouling [1].

Currently, there are two widely used types of desalination plant source water collection facilities—surface (open) and subsurface (groundwater, well) intakes. Open intakes collect source water directly from a surface water body (e.g., ocean, sea, brackish river, or lake) via onshore or offshore intake structure and pipeline connecting this structure to the desalination plant. The source water collected by this type of intakes usually contains debris, coarse and fine solids and silt, and aquatic organisms (e.g., fish, algae, bacteria, etc.), which make the pretreatment of this water more complex.

Subsurface intakes tap into brackish groundwater aquifer, or saline coastal aquifer to collect source water for the desalination plant. Source water obtained using subsurface intakes is conveyed through the soils of the aquifer, which naturally filter out debris, marine organisms, and some organics, and thereby the RO membrane fouling potential of the source water is reduced. As a result, the use of subsurface intakes for source water collection usually requires simpler pretreatment than source water collected by open intakes.

Sustainable Desalination Handbook
https://doi.org/10.1016/B978-0-12-809240-8.00005-8

5.2 OPEN INTAKES

5.2.1 Types and Configurations

Based on the location of their inlet structure, open intakes are typically classified as onshore and offshore. The inlet structure of onshore intakes is constructed on the banks of the source water body, while the inlet structure of offshore intakes is typically located several hundred to several thousand meters away from the shore.

5.2.1.1 Onshore Open Intakes

To date, onshore intakes have found application mainly for very large thermal or hybrid seawater desalination plants. Such intakes typically consist of large and deep intake canal ending into a concrete forebay structure equipped with coarse bar screens followed by fine screens and intake pump station.

The largest seawater reverse osmosis desalination plant with onshore intake is the 155,000 m^3/day (40 MGD) Point Lisas desalination plant in Trinidad. In this case, the intake consists of concrete forebay structure located on the shore of industrial port's ship turning basin (the Gulf of Paria) and is equipped with coarse and fine screens, and vertical turbine pumps.

Another desalination plant, which uses source water collected by an onshore open intake, is the Tampa Bay desalination plant. This plant collects most of its source seawater from the discharge of cooling water of the power plant with which it is colocated. A portion of the source seawater, however, is collected directly from onshore intake located on one side of an intake canal (see Fig. 5.1) and mixed with cooling water from the power plant to maintain desalination plant feed water temperature below 38°C at all times.

5.2.1.2 Offshore Open Intakes

Offshore open intakes typically consist of velocity-cap type inlet structure (Fig. 5.2), one or more intake water conduits (pipelines or intake tunnel), onshore intake chamber, trash racks, fine screens, and source water intake pump station.

The inlet structure of offshore open intakes is usually a vertical concrete, copper-nickel or steel well (vault), or wedgewire screen located 4–10 m (13–33 ft) above the floor of the water body and submerged between 4 and 20 m (13 and 66 ft) below the water surface. Open intakes are the most commonly used type of source water collection system for medium and large desalination plants. Baudish et al. [2] presents examples of open offshore intakes of large seawater desalination plants [2].

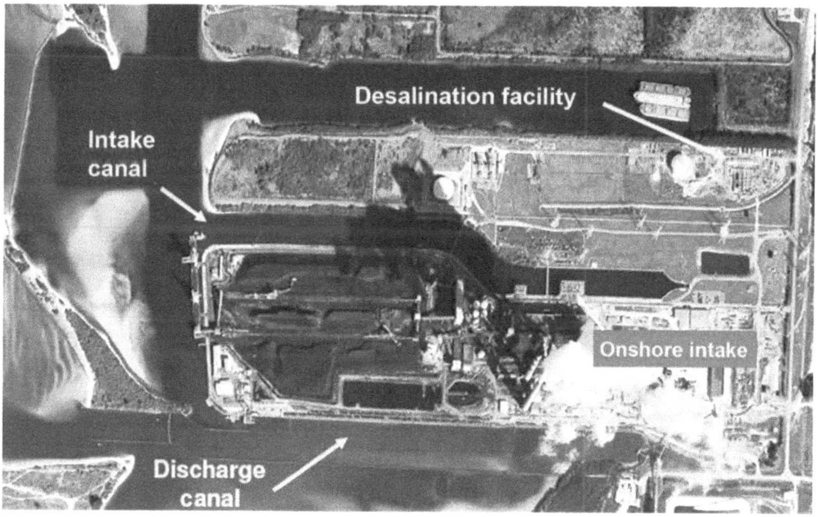

Fig. 5.1 Intake of Tampa Bay desalination plant.

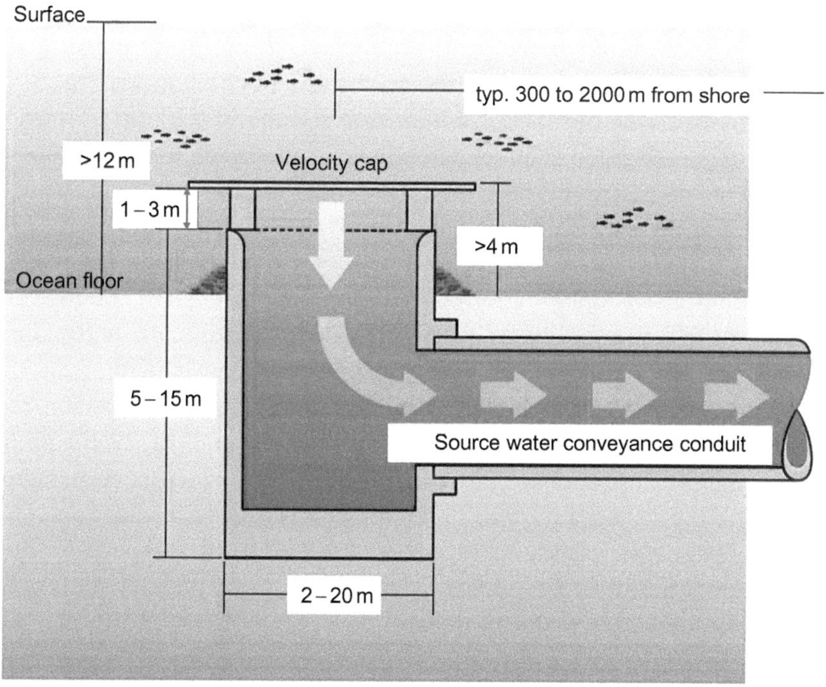

Fig. 5.2 Velocity cap intake inlet structure.

In Table 5.1, the maximum entrance velocity is the through screen velocity through the intake bars. The depth below the water surface is the distance between the mean water level and the top of the velocity cap of the intake. Distance from the bottom is the distance between the bottom of the ocean and the top of the velocity cap.

Offshore Intakes, which usually extend several 100 m away from the shoreline and 8–20 m below the water surface, are typically not influenced by the fresher coastal aquifers. Therefore, such intakes usually yield saline water of the same TDS (total dissolved solids) content as that of ambient ocean water.

One exception is offshore intakes located near the entrance of large rivers or other fresh water bodies into the ocean. Since old river beds and associated alluvial aquifers could often extend far beyond the tidal zone, the water quality of such offshore intakes may be influenced by the groundwater quality in the alluvial aquifers, which often is inferior than open ocean seawater in terms of content of iron, manganese, and other undesirable contaminants.

5.2.1.3 Colocated Intakes

Configuration

Colocated intakes are connected to the discharge of adjacent power generation station using saline water for once-through cooling (see Fig. 5.3). At present, this type of intake configuration has found application mainly for seawater desalination facilities cosited with once-through cooling coastal power generation stations. However, such configuration could also be applied for inland brackish water desalination plants where the host power plant uses brackish intake and discharge wells for once-through cooling.

The colocation configuration allows using the power plant cooling water as both saline source water for the desalination plant and as blending water for salinity reduction of the desalination plant concentrate.

The desalination plant intake is connected at a location between the point of exit of the warm cooling water from the power plant and the point of discharge of this water to the receiving water body (ocean, river, saline aquifer, etc.) to collect a portion of the cooling water for desalination.

The desalination plant concentrate is conveyed to the power plant discharge downstream of the point of desalination plant intake connection. As a result, colocated desalination plants do not have their own intake and discharge.

Table 5.1 Examples of large open intakes for seawater desalination plants

Desalination plant/ production capacity	Max entrance velocity (m/s/fps)	Depth below water surface (m/ft)	Distance from the bottom (m/ft)	Number of inlet structures and conduits	Inlet structure diameter, (m/ft) and screen size (mm/in.)	Conduit diameter (m/ft), material and distance from shore (m/ft)
Adelaide, Australia 300,000 m^3/day	0.15/0.50	18/59	5.0/16.4	1/1	9.5/31.2 100/4	2.8/9.2 tunnel 1000/3300
Sydney, Australia 500,000 m^3/day	0.15/0.50	24/79	6.0/20.0	4 inlet structures on common tunnel	8.5/27.9 340/13	3.4/11.2 tunnel 300/980
Gold Coast, Australia 136,000 m^3/day	0.05/0.16	22/72	4.4/14.4	1/1	5.8/19.0 140/5.5	2.8/9.2 tunnel 1400/4600
Perth I, Australia 130,000 m^3/day	0.10/0.33	8/26	4.0/13.0	1/1	100/4	2.8/9.2 GRP pipe 300/1000
Perth II, Australia, 300,000 m^3/day	0.15/0.50	10/33	4.0/13.0	2	7.0/23 100/4	2.4/9.1 HDPE pipes 500/2600
Fujairah I, UAE 170,000 m^3/day	0.10/0.33	10/33	6.0/19.7	3/3	3.0/9.8 80/3	2.0/6.6 GRP pipes 380/1250
Al Dur, Bahrain 240,000 m^3/day	0.10/0.33	4/13	2.3/7.5	4/4	7.2/23.6	2.4/7.9 GRP pipes 1500/4920

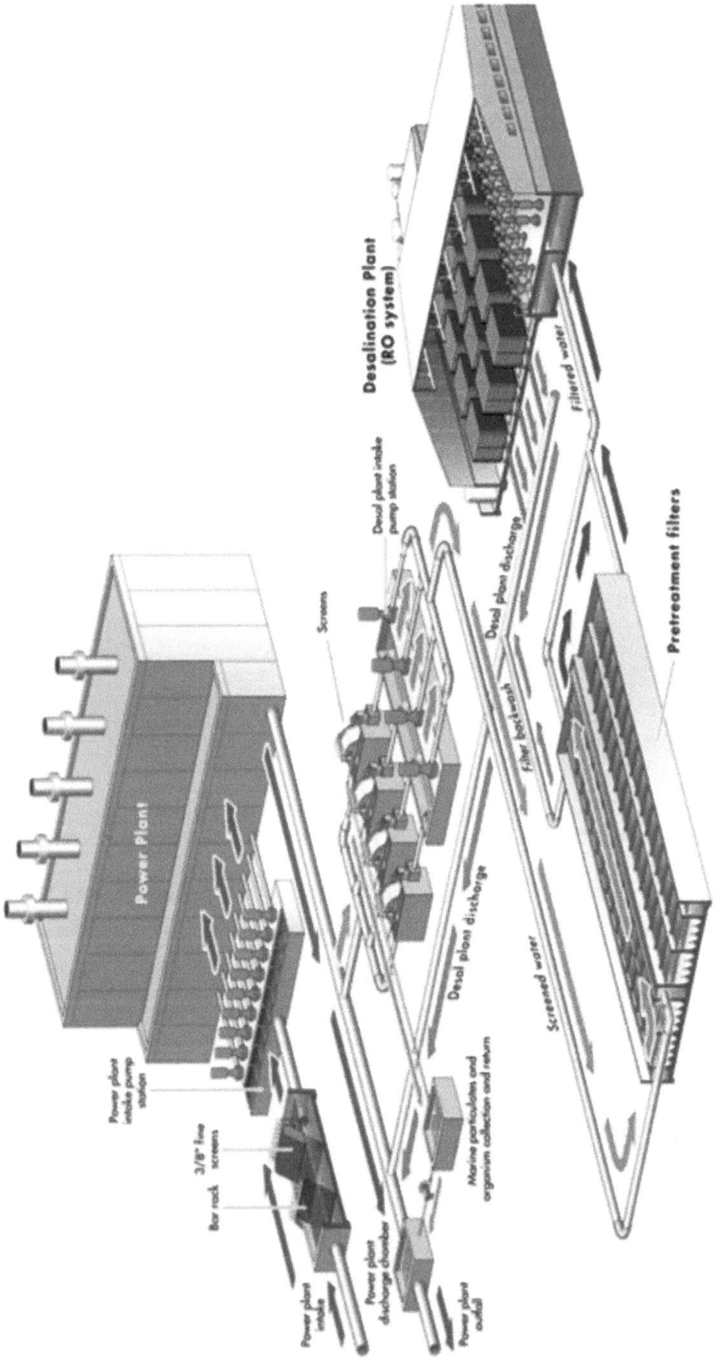

Fig. 5.3 Colocation of desalination plant and power generation facility.

Potential Colocation Benefits

Sharing intake infrastructure has environmental benefits because it avoids the need for new construction in the open water body area near the desalination plant. The construction of a separate new open intake structure and pipeline for the desalination plant could cause significant disturbance of the benthic marine organisms inhabiting the water body.

Usually, coastal power plants with once-through cooling systems use large volumes of seawater. Because the power plant intake seawater has to pass through the small-diameter tubes (typically 10-mm/(3/8-in.) or less) of the plant condensers to cool them, the plant discharge cooling water is already screened through bar racks and fine screens similar to those used at surface water intake desalination plants (see Fig. 5.3). Therefore, a desalination plant, whose intake is connected to the discharge outfall of a power plant, usually does not require the construction of a separate intake structure, intake pipeline, and screening facilities (bar racks and fine screens). Since the construction cost of a new surface water intake structure for a desalination plant is typically 5%–30% of the total plant construction expenditure, power plant colocation could yield significant construction cost savings.

The need for installation of additional fine screening facilities for the desalination plant intake is driven by the screenings disposal practice adopted by the power plant and the type of desalination plant pretreatment system. Typically, power plants remove the screenings retained at their intake bar racks and fine screens, and dispose these waste debris to a landfill or return them back to the ocean. However, in some power plants, the screenings collected at the power plant's mechanical screens are discharged into the cooling water downstream from the plant's condensers. In this case, the power plant discharge would contain screenings that need to be removed at the desalination plant intake.

Another clear environmental benefit of the colocation of power and desalination plants is the reduced overall impingement and entrapment of marine organisms as compared to construction of two separate open intake structures—one for the power plant and one for the desalination plant. This benefit stems from the fact that total biomass of the impacted marine organisms is typically proportional to the volume of the intake saline source water. By using the same intake saline water twice (once for cooling and second time for desalination) the net intake inflow of saline water and aquatic organisms is minimized.

Potential Colocation Challenges

In order for the collocation concept to be cost effective and possible to implement, the power plant cooling water discharge flow has to be larger than the capacity of the desalination plant with which it is colocated, and the power plant outfall configuration has to be adequate to avoid entrainment and recirculation of concentrate into the desalination plant intake. It is preferable that the length of the power plant outfall downstream of the point of connection of the desalination plant discharge is adequate to achieve complete mixing prior to the point of entrance into the ocean.

A special consideration has to be given to the effect of the power plant operations on the cooling water quality, since this discharge is used as source water for the desalination plant. For example, if the power plant discharge contains levels of copper, nickel, or iron significantly higher than these of the ambient seawater, this power plant discharge may be not be suitable for colocation because these metals may cause irreversible fouling of the membrane elements. In such cases, the pretreatment costs for colocated desalination plant are likely to be higher than those for plant with separate open ocean intake.

Another potential problem could be the location of where the power plant intake screenings are disposed. In most power plants, the screening debris are removed from the intake cooling water and disposed off site to a landfill. However, this disposal practice may change during the course of the power plant and desalination plant operations.

For example, in the case of the Tampa Bay seawater desalination plant, during the final phase of the desalination plant construction, the power plant decided to change their screenings disposal practices and to discharge their intake screenings just upstream of the desalination plant intake rather than to continue disposing them off site. This change in power plant operations had a dramatic effect on the Tampa Bay Water desalination plant start up and operations, and especially on the pretreatment system performance.

Since the desalination plant was pilot tested and designed around the original method of power plant operations under which all screenings were removed from the cooling water, the desalination plant was not built with its own separate intake screening facilities. The presence of power plant waste screenings in the desalination plant intake water had a detrimental effect of the pretreatment filter operations because the screening debris frequently clogged the filter distribution piping, airlifts, and sand media.

Although this problem has a significant effect on the desalination plant operations, it also has relatively straightforward solutions—either installing separate fine screening facilities for the desalination plant or moving the point of the power plant screening debris discharge downstream of the desalination plant intake. A summary of key advantages and disadvantages of the colocation approach is presented in Table 5.2.

Table 5.2 Key advantages and disadvantages of desalination plant collocation

Advantages	Disadvantages
• Capital cost savings by avoiding the construction of separate intake pipeline and structure, and new discharge outfall • Decrease of the required RO system feed pressure and power cost savings as a result of using warmer water • Reduction of unit power cost by connecting directly to power plant generation facilities and avoiding power transmission charges • Accelerated permitting process as a result of avoidance of construction of new intake and discharge outfalls in the ocean • Reduction of marine organism impingement and entrainment because the desalination plant does not take additional seawater from the ocean • Reduction of the impact on marine environment as a result of faster dissipation of thermal plume and concentrate • Reduction of the power plant thermal discharge to the ocean because a portion of this discharge is converted to potable water • Use of already disturbed land at the power plant minimizes environmental impact	• Use of warmer seawater may accelerate membrane biofouling, especially if the source water is rich in organics • RO membranes may be exposed to iron, copper, or nickel fouling if the power plant condensers and piping are built of low-quality materials • Source seawater has to be cooled if its temperature increases above 40°C in order to protect RO membrane integrity • Permeate water quality diminishes slightly with the increase of source water temperature • Use of warmer water may result in lower boron rejection and require feed water pH adjustment to meet stringent boron water quality targets • RO plant source water screening may be required if the power plant disposes off its screenings through their outfall and the point of disposal is upstream of the desalination plant intake • Desalination plant operations may need to be discontinued during periods of heat treatment of the power plant facilities

5.2.2 Selection of Open Intake Type

5.2.2.1 Onshore versus Offshore Intake

Onshore intakes have one key advantage—they are usually the lowest cost type of intake, especially for large desalination plants. However, such intakes typically produce the worst source water quality because in most cases they are designed to collect water from the entire depth of the water column, including the surface, and because they are located in the surf zone where breaking waves continuously lift particles from the bottom into suspension thereby significantly increasing water turbidity as compared to deeper waters offshore outside of the zone of tidal influence.

The first 8–10 m (26–33 ft) of water from the surface in the surf zone typically contain several times higher levels of turbidity, algae, hydrocarbon contamination, silt and organics than deeper waters. The quality of water collected by this type of intakes could vary significantly because in many cases it is influenced by wind, tides, currents, ship traffic, storms, and fresh water runoff. Onshore intakes could also be exposed to beach erosion and direct wave action with irreversibly damaging consequences.

Because of the poor quality of the source water they collect, onshore open intakes have found very limited application at present and unless the specific site location or costs dictate the need to use this type of intake, they are less desirable for reverse osmosis desalination plants. However, onshore intakes are often the prime choice for thermal desalination plants, where source water quality in terms of suspended solids, organic, and algal content is of secondary importance and has minimal influence on the desalination process.

5.2.2.2 Wedgewire Screens vs. Conventional Inlet Structure

Wedgewire screens are preferred over conventional offshore velocity-cap type of intakes when suitable site conditions exist. According to one source [3], they can be used for desalination plants with intake capacity of up to 20,000 m^3/h (127 MGD).

Three key criteria for suitability of the site are: (1) the sweeping velocity of currents naturally occurring in the vicinity of the intake should be at least 0.3 m/s (1.0 fps); (2) the minimum depth of the water above the screens at all times (including low-tide conditions) should be at least 50% of the screen diameter; (3) the available distance between the bottom of the water body and the screens should be a minimum of 50% of the screen diameter or 1.0 m (whichever is higher).

The key benefits of wedgewire screens as compared to conventional intakes with offshore towers are: (1) lower construction and O&M costs; (2) reduced entrainment of aquatic species; and (3) faster and more simplified installation.

5.2.3 Selection of Open Intake Location

Key factors that determine the selection of the location of open intakes are potential for beach erosion in the intake area, location and direction of underwater currents, presence and location of active seismic faults, topography and geology of the floor of the water body, location of environmentally sensitive habitats in the along the intake pipe route and near the intake inlet, location and size of municipal and industrial wastewater discharges within 1 km (0.6 miles) radius from the intake, size of waves and depth of wave impacts, ship and boat traffic, tide and wind characteristics in the intake area.

Open intakes should be located away from coastal areas of active beach erosion and seismic faults; high waves; strong underwater currents carrying debris, silt, plankton, sea grass, weeds, and other stringy materials; and locations with heavy ship and boat traffic. If sensitive marine habitats are encountered, the intake route should either be modified or the intake conduit/s should be installed via directional drilling or tunneling under the sensitive area instead of using open trench construction or laying the conduit on the bottom of the water body.

The proper design of open ocean intakes requires the collection of detailed source water quality data from the proposed site of the intake, characterization of aquatic life in the vicinity of the intake, and completion of detailed sanitary survey assessing the potential sources of SWRO desalination plant source water quality contamination in the vicinity of the intake location (such as waste discharges of industrial and municipal wastewater plants, storm water discharges or large industrial port or municipal marina activities, which may result in oil and gasoline spills and in other ocean water contamination).

The exact location and depth of the offshore intake structure must be determined based on a hydrological study to ensure that the intake will be adequately submerged at low tide; protected from the damaging orbital storm wave motion; and will be far enough offshore to avoid the near-shore sediment transport zone where storms can cause suspension of large quantities of silt and sediment, and can ultimately damage the intake structure and the interconnecting piping.

Diurnal and seasonal source water quality fluctuations should also be considered when determining the location of the intake structure. At minimum low-tide conditions, inlet mouth should be submerged at least 3 m (10 ft) below the water surface. In addition, the distance between the inlet mouth and the ocean floor should be no <3 m (10 ft) to prevent excessive sand carryover into the downstream intake facilities. The intake water supply can be protected against large aquatic organisms and large floating debris by installation of wire net across the intake mouth.

The proper design of open ocean intakes requires the collection of detailed source water quality data from the proposed site of the intake, characterization of aquatic life in the vicinity of the intake, and completion of detailed sanitary survey assessing the potential sources of RO plant source water quality contamination in the vicinity of the intake location (such as waste discharges of industrial and municipal wastewater plants, storm water discharges, or large industrial port or municipal marina activities, which may result in oil and gasoline spills and in other ocean water contamination).

5.2.4 Minimization of Impingement and Entrainment Impacts

Open ocean intakes would result in some entrainment of aquatic organisms as compared to subsurface intakes because they collect source water directly from the saline water body (e.g., ocean, sea, or bay) rather than collecting water prefiltered through the coastal aquifer formations. In addition, some marine organisms may be impinged on the screening facilities and ultimately could be entrained in the plant source water flow.

Impingement and entrainment of marine organisms is not an exception for saline water intakes. The same phenomena and extent of environmental impact may be observed at the intakes of conventional water treatment plants collecting surface water from rivers, lakes, and estuaries as well. Impingement and entrainment of open intakes and their most common mitigation methods are discussed elsewhere [4].

5.2.5 Design Considerations

5.2.5.1 Onshore Intakes

As indicated previously, onshore intakes have found limited application for collection of source water for reverse osmosis desalination plants. However, depending on the site-specific source water body conditions, they may offer a low-cost construction alternative to offshore intakes. It should be noted that because of the inferior water quality they collect, the use of onshore instead of offshore intakes is likely to result in increased pretreatment system complexity and costs and may require the construction of multiple pretreatment facilities in series (e.g., DAF clarifiers followed by two-stage granular media filers). Therefore, the most suitable combination of intake and pretreatment system should be determined based on a detailed source water quality evaluation and lifecycle cost-benefit analysis.

Depending on the coastal conditions, onshore intakes could be installed on a sandy coast with low gradient, on a rocky coast, or in natural or artificial enclosure (i.e., ship turning basin, marina, industrial port, or lagoon). Of the three coastal environments, rocky bottom conditions are the most favorable for construction of onshore intakes. Key factors associated with the design of such intakes are wind and swell regimes, water level variations, tidal regime, bathymetry, and coastal currents.

Onshore intakes on sandy coast with low gradient are usually constructed with a long entrance canal, which is designed to protect the intake from littoral sediment transport by prevailing near shore currents, winds, swells, and tides. The intake canal is constructed with protective jetties, which are oriented such that they create a protective shield of the intake against the prevailing current and prevent the littoral drift form carrying sediments into the intake area. Without the jetty, the canal would fill up with sand and would need to be dredged frequently. The jetties are constructed of stone rock blocks and usually extend to elevation of 2–3 m (6.6–10 ft) above the mean water level. In order to avoid fish entrapment, the canal is designed for average velocity at mean water level of 0.3 m/s (1 fps) and at 0.15 m/s (0.5 fps) at high water level. The canal ends into an onshore inlet structure with trash racks.

Rocky-coast onshore intakes typically are concrete or metal structures, which are open directly to the water body. Depending on the site-specific conditions, such intakes may be designed with jetty protection. If possible, it is preferable to design the water entrance at a depth or at least 2 m (6.6 ft) below the low-tide level and to protect the entrance with wave-breaking jetties. The main difference between rocky bottom and sandy bottom shoreline conditions is that wave action typically does not cause a significant stir up of sediments and elevated source water turbidity.

Onshore intakes in natural or artificial water body enclosures are typically well protected against wave and wind action and therefore have more consistent water quality. However, one such intake location is that the embayment accumulates silt and sand and unless the area in front of the intake is dredged periodically, the intake capacity decreases over time. Dredging operations have both cost and source water quality implications.

5.2.5.2 Offshore Intakes

Selection of the location, depth, and configuration of the offshore intakes is of critical importance for the source water quality of the desalination plant and its performance. These intake design parameters are established based on series of onsite investigations of the intake area conditions, including bathymetric profile, geotechnical survey, source water quality profile, and beach erosion and wave analysis.

Bathymetric Profile

The bathymetric profile provides a topographic representation of the floor of the water body and indicates bottom slope and distance to the water surface and the shore. Usually, several initial topographic profiles are completed within the vicinity of the target intake location, and one of these profiles is selected as a preferred alternative based on the floor topography. As a second step of the engineering investigation of the intake area, geotechnical survey and water quality characterization are completed along the route of the preferred water conduit route.

Geotechnical Survey

This survey aims to determine geological formations, presence and location of seismic faults, and seabed conditions along the preferred intake conduit route. This information is used to decide whether the intake pipeline could be installed directly on the bottom of the water body, in open-cut trench, and/or it would need to be directionally drilled in the bottom strata.

Flat sandy bottom conditions usually allow installation of the intake pipeline directly on the bottom or in a trench. Pipeline installed directly on the bottom is typically anchored by concrete saddles or engineered backfill to protect it against destructive current and wave action.

For rocky bottom conditions, the intake pipeline in trenches is usually costly and laying it on the surface of the ocean bottom is not suitable because of the inadequate support and risks for pipe abrasion and damage. In such conditions, the intake conduit is directionally drilled or tunneled under the ocean bottom.

Wave and Tide Survey

The purpose of this survey is to determine the magnitude of wave heights that could occur in the area of the intake and to identify the wave height with 1000-year average recurrence interval (ARI), which is typically used for design purposes, if long-term record of wave heights is available. Such projection is important to assess the impact the horizontal currents, which high waves could create in shallow water and the force such currents could exert on the intake structure, as well as the depth of sediment resuspension the horizontal current could trigger. In addition, the survey would need to also include review of tidal fluctuation information in order to estimate the submergence of the intake inlet structure under minimum, maximum, and average tide elevations.

Underwater Current Survey

Underwater currents could have a measurable impact on source water quality and the ability of wedgewire screens to function properly. The direction and velocity of prevailing underwater currents are also of critical importance for determining the location of intake and outfall facilities and the minimum distance them. Underwater current information is typically available from navigation (nautical) maps.

Biological (Ecological) Survey

The purpose of this survey is to identify the presence of environmentally sensitive habitats that may be located in the intake inlet area or along the pipeline conduit route. Habitats of high biological productivity such as seagrass beds (meadows), kelp forests, algal mats, seaweed bays, coral reefs, salt marshes, and mangrove flats should be avoided because they are used as spawning grounds for many marine species and often are very fragile habitats.

Source Water Quality Profile

Source water quality profile is typically completed as a minimum at 3–5 locations along the preferred route of the intake conduit—typically at distances of 300 m (1000 ft); 500 m (1640 ft); 1000 m (3280 ft); 1500 m (4920 ft); and 2000 m (3785) from the shore. The exact location of each water quality sampling point is representative of potential intake location and is selected such that it avoids strong underwater currents (unless wedgewire screens are used), bottom irregularities, environmentally sensitive habitats, or active beach erosion zones.

Once the sampling locations are determined, a sampling protocol is developed identifying the sampling depth and frequency and the water quality parameters for which each sample would be tested. Some of the sampling could be automated by installing buoys equipped with water quality monitoring instruments for data collection of parameters such as conductivity, temperature, dissolved oxygen content, turbidity, and pH of the source water.

Most water sampling, however, is completed manually at intervals of once to twice per month to once every quarter for a period of 1 year to gain a thorough understanding of the seasonal water quality variations in the potential intake locations. Additional sampling is usually carried out to capture extreme water quality events such as heavy storms, algal blooms, intake area dredging, periods of elevated shift traffic, occurrence of seasonal currents, and winds.

Once the topographic, water quality, and geologic data are collected and analyzed, the actual location of the desalination plant offshore inlet structure is selected such that it yields the best source water quality (the lowest turbidity, silt density index, algal content, and least seasonal source water quality variations) and at the same time is the closest to the shore.

The selected location and depth of the intake inlet structure should be such that the intake is adequately submerged at low tides, protected from the damaging storm wave motion, and is far enough offshore to avoid the near-shore surf zone where storms can cause suspension of large quantities of silt and sediment, and can ultimately damage the intake structure and the interconnecting piping.

For the selected intake location, the optimal intake depth and distance from the bottom are determined by preparation of depth profiles of turbidity, temperature, algal content, salinity and temperature, and selection of depth from the surface and distance from the bottom, which yields the best source water quality (lowest SDI, turbidity, algal content, and highest temperature). Such profiles should be prepared for worst-case water quality conditions such as heavy storm/high wave conditions, high-intensity winds, lowest water temperatures, algal blooms with algal counts over 2000 cells/mL, seasonal underwater currents, intake area dredging, and heaviest ship traffic conditions. The optimum intake depth is determined as the shortest distance below the water surface, where the various extreme environmental conditions that may occur in the intake area have minimal or no impact on the source water quality and plant operations.

5.2.5.3 Offshore Intake Inlet Structure—Design and Construction Considerations

General guidelines for the depth of inlet structures for offshore intakes are presented in Fig. 5.2 and Table 5.3.

Inlet Depth

Typically, the optimum depth from mean surface water level to the top of the intake velocity cap is 12–20 m (39–66 ft). The top of the inlet velocity cap is submerged at a minimum of 8 m (26 ft) and 4 m (13 ft) below the

Table 5.3 Key design criteria for inlet structures of offshore intakes

Feature	Recommended size	Notes
Number of inlet structures	2 (minimum)	Number of intake structures is dictated by plant design availability factor and size of intake conduit
Diameter (size)	2–20 m (13–130 ft)	Most inlet structures have circular shape If inlet is located in current, the shape may be rectangular in with shorter side against current
Distance from mean surface water level to top of velocity cap	8 m/26 ft (minimum at mean water level) 4 m/13 ft (minimum at low water level) 20 m/66 ft (maximum) 12–20 m/40–66 ft (optimum)	Shallower intake structures may be needed if desirable depth cannot be achieved within 2000 m (6600 ft) from the shore Structures deeper than 20 m (66 ft) may not be cost effective
Distance from bottom to top of velocity cap	One seabed pipe diameter or 4 m/13 ft (minimum)	This distance is determined by the depth of sweeping current at the bottom
Distance from the shore	300–2000 m 1000–6600 ft	Distance determined by the length of the tidal or active beach erosion zones
Distance between coarse screen bars	50–300 mm (2–12 in.)	Wider distance is preferred for tropical environments with anticipated heavy shellfish/coral growth
Through screen velocity	0.10–0.15 m/s (0.3–0.5 fps)	Screen velocity of 0.10 m/s is recommended for source waters with jellyfish content higher than 1 organism/m^3 Through screen velocity should be determined for 50% of the area between the bar screens

water surface at mean and lowest tide water levels, respectively. The minimum water depth should be confirmed with the authorities responsible for regulating naval vessel navigation in the vicinity of the intake.

Intakes with depths lower than 8 m (26 ft) from the mean water surface level usually produce source water of inferior quality due to accelerated wave- and wind-driven mixing. Since water quality does not improve significantly beyond depth of 20 m (65 ft), and the water temperature decreases with depth, the construction of significantly deeper intakes typically is not cost effective because construction and maintenance costs of intake inlets increase with depth. In addition, deeper water is colder, and because of its higher viscosity membrane salt separation by reverse osmosis needs more energy. Since the relationship between water temperature and viscosity becomes curvilinear for temperatures below 8–12°C (46–54°F), the use of colder source water would result in almost exponential increase in energy consumption for desalination of source water with temperature below 10°C (50°F).

One benefit of colder waters is their lower biofouling potential. However, this benefit is most pronounced for waters warmer than 20°C (68°F) and rich in organics (TOC > 1.5 mg/L), and is practically negligible for waters colder than 12°C (54°F).

The distance between the top of the inlet mouth (the velocity cap) and the ocean floor should be at least 50% of the length of the bar screens but no <4 m (13 ft) in order to prevent excessive carryover of bottom sediments into the downstream intake facilities. If the intake area is affected by frequent naval traffic, then the minimum depth from the bottom should account for the depth of bottom sediment resuspension by large ships/boats.

Inlet Screens

Inlet screens are typically coarse bar screens with distance between the bars of 50–300 mm (2–12 in.). The bar length is usually between 1 and 3 m (3–10 ft) and is determined based on the selected maximum distance between the screen velocity cap and the water surface and the bottom.

Design maximum through screen velocity varies depending on the content of jellyfish in the source water and is usually in a range of 0.10–0.15 m/s (0.3–0.5 fps). It is important to note that the design through-screen velocity should be calculated for 50% of the total area between the screen bars to allow for growth of shellfish and accumulation of debris over time. It is anticipated that the inlet screening area will be reduced by 30%–50% every 18–24 months and therefore the screens would need to be cleaned periodically by divers.

A maximum through-screen velocity of 0.10 m/s or less is typically selected if the source water contains jellyfish load of 0.5 kg/m^3 of source water or more. This lower flow through-screen velocity allows to minimize the intake of jellyfish into the plant and thereby to minimize their negative impact on downstream screening facilities. Jellyfish outbreaks are typical for warmer, shallower, and polluted waters, especially if the intake is located in the middle of an underwater current through which jellyfish travel most frequently.

Typically, small jellyfish specimen that can enter the intake weigh between 200 and 400 g per individual and for large plants with few fine screens jellyfish could add significant load to the screens, which in extreme conditions could cause equipment damage. In addition, jellyfish are very difficult to clean from the screens and often standard debris jet sprays with which fine screens are equipped cannot remove them. From this perspective, drum screens are usually more difficult to clean once the jellyfish attach on the screens and therefore reducing their intake by reducing the inlet through-screen velocity is of critical importance, especially for plants with fine drum screens. If large quantity of jellyfish enters the intake, they could render the fine screens inoperable or reduce significantly plant source water capacity.

Inlet Materials

Inlet shafts are typically built from corrosion-resistant materials such as copper-nickel alloy, concrete, or stainless steel. The screen bars are either made of 90/10 Cu/Ni or super duplex stainless steel for seawater applications and duplex stainless steel for brackish water intakes.

Inlet Configuration

Small plants are usually designed with a single inlet structure and conveyance pipeline. However, intake configuration with two or more inlets connected to individual conduits (pipelines or tunnels) is preferable in order to avoid shutting down the entire plant when the intake system is taken out of service for cleaning. For plants with two or more inlets, it is important to note that the inlets should be located in an adequate distance from each other to avoid mutual impacts on their flow patterns.

As a rule of thumb, the individual inlets are recommended to be located at a distance of two to three inlet diameters from each other. In addition, intake inlets should be located at an adequate distance (typically 400–1000 m/1300–3300 ft) away and always upstream of the discharge so prevailing currents can dissipate the discharge concentrate away from the intake and short circuiting could be avoided. The minimum distance between the intake inlet and discharge diffuser field locations should be determined based on hydrodynamic modeling.

If the plant intake is designed with a single conduit, usually the conduit diameter is sized for 120%–130% of the annual average intake flow in order to accommodate biogrowth on the conduit walls and silt and debris accumulation on the bottom. If two or more conduits are installed, they are commonly designed for 60% of the plant intake flow.

Typically, intake inlets and conduits have to be cleaned once every 1–2 years in order to maintain their conveyance capacity. Intake tower and conduit cleaning typically takes 2–5 days. Therefore, desalination plants with product water capacity larger than 150,000 m^3/day (40 MGD) are commonly designed with three to five inlets, which can be serviced individually. Exception are inlets located in very deep waters (15 m/49 ft or more), where the process of aquatic growth on the inlet bar screens and conduits is significantly slower because of the limited abundance of marine life and nutrients, and colder temperatures.

Intake Water Conduit Configuration

Depending on the plant capacity, intake water conduits are either pipes or tunnels. Most reverse osmosis desalination plants with fresh water production capacity smaller than 300,000 m^3/day are designed with pipeline conduits. Intake pipeline could be constructed using one of three methods: (1) it could be directly installed on the bottom of the water body and anchored with concrete blocks; (2) it could be buried in trenches; or (3) it could be directionally drilled or microtunneled in the bottom sediments.

Pipelines are laid directly on the bottom of the water body if this bottom is sandy and flat, and the intake is located in low-energy shoreline (i.e., bays, semienclosed lagoons, and deep quiescent waters) with minimal exposure to beach erosion, heavy storms, typhoons, tsunamis, or strong underwater currents. For example, the intakes of the desalination plant in Ashkelon, Israel, are constructed in this manner. Typically, HDPE pipes are used in such configuration. The intake pipe is delivered to the site in long sections, welded on shore, and fitted with concrete blocks at regular intervals. A trench is dredged on the bottom to flatten the pipeline route before the pipe is installed. The pipe with the attached concrete blocks is closed on one end, towed and adjusted to fit the design route, and then sunk by filling it up with water. Finally, the inlet is installed and connected to the pipe.

Pipelines are installed in trenches (typically at depth of 1–5 m/3–16 ft) when the water body's bottom is uneven, or it is exposed to strong currents and wind or tidal action. Usually open trench excavation is mainly suitable for relatively shallow sandy or soft rock bottom. Examples of seawater desalination plants with such intakes are Perth, Australia; Al Dur, Bahrain; Fujairah I, UAE; Barka III in Oman; and others. Directional drilling or microtunneling is the most viable method of intake pipe installation in rocky coastal environment.

Dredge and cover pipeline construction is an alternative to pipeline installation in trenches and is typically applied for deeper waters. In this case, a relatively shallow trench is constructed by dredging of the bottom, the pipe is installed in the trench, and then covered with several layers of suitable size, gravel and rock and then covered with the dredged material. Depending on the site-specific bottom conditions, rock armor cover may also be added on the top of the pipe. Similar to the installation of HDPE pipe directly on the bottom, in this case the pipe position is secured along its desired trench route by using concrete weigh blocks.

Intake Pipeline Materials

Materials that have found wide application for desalination plant intake pipelines to date are glass-reinforced plastic (GRP), high-density polyethylene (HDPE), polyvinyl chloride (PVC), concrete, cement-lined ductile iron, and cement- or epoxy-lined mild steel. GRP and HDPE are the two most commonly used pipeline materials.

HDPE has found widest application for intake pipelines installed directly on the bottom of the water body because this material is very durable and resistant to corrosion, it can handle moderate wave and wind action without the need to be buried, it is more flexible than GRP, and it can be easily transported, installed, and fused in long sections. The two key disadvantages of HDPE as compared to GRP are that HDPE pipe is only available in limited size diameters (up to 2000mm/78-in.) and it is more costly than GRP.

GRP piping is practically not restricted in terms of size (largest diameter is 4000mm/157in.). However, it has lower flexibility than HDPE, it is lighter and tends to float, and cannot carry high dynamic loads. Therefore, GRP pipe has to be buried in a trench and covered with armor rock (crushed stone) or other material to protect it against floating, currents, and other natural destructive forces.

While concrete and steel pipe are also used for desalination applications they are significantly more costly, have lower durability, and are more difficult to install. Therefore, they are not commonly used in new desalination projects.

Intake conduits for large membrane desalination plants are typically concrete tunnels directionally drilled at depth between 5 and 20 m (16–66 ft) below the bottom of the source water body, and connected to the inlet screen structure via vertical riser shaft. The tunneled intakes could be of two types—a tunnel with a single intake inlet structure at the end (see Fig. 5.4) or a tunnel with multiple inlet structures located along its length (see Fig. 5.5). Examples of tunnel intakes with single inlet are the Gold

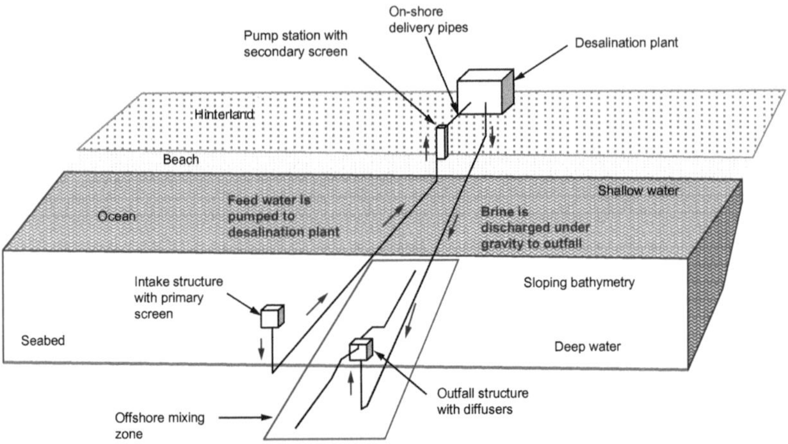

Fig. 5.4 Offshore intake with single inlet structure.

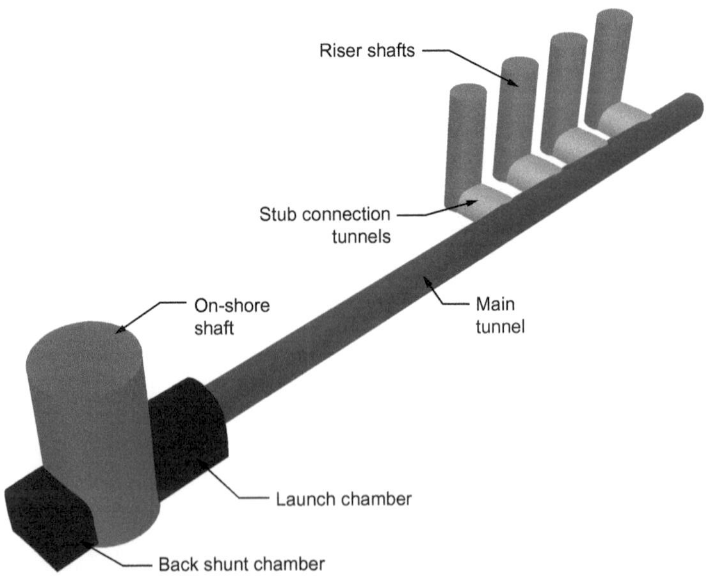

Fig. 5.5 Intake with multiple inlet structures.

Coast and Adelaide plants in Australia, while Sydney and Melbourne desalination plants have intake tunnels with four inlet structures connected to them via vertical shafts.

Tunnel construction usually begins with the installation of onshore shaft and launch chamber for the tunnel-boring machine. This machine allows constructing the tunnel in concrete segments (rings). In one segment installation cycle, the machine drills space adequate to install one segment, then its rams are retracted and the new tunnel ring segment is installed. The tunnel rings are waterproofed and grouted before the drilling machine moves forward to drill space for installation of the next segment.

Once the horizontal tunnel is built, one or more riser shafts are constructed from an offshore platform by first installing an external caisson into the seabed that acts like a protective shell and then completing the actual shaft drilling and excavation. Once the shaft reaches the tunnel level, it is sealed at the top and stub connection tunnels are constructed by drilling multiple horizontal holes to form the tunnel connector and the tunnel connector walls are grouted and water proofed. As a final step, the inlet structure is installed on the top of the riser shaft.

5.2.6 Costs of Open Intakes

5.2.6.1 Construction Costs of Onshore Intakes

Fig. 5.6 provides graphic representation of the construction costs of onshore intakes as a function of the desalination plant intake flow expressed in year 2017 US dollars. Because of the significant impact of the site-specific condition of the actual intake costs, such costs may vary within ±30% of the values indicated on this figure.

5.2.6.2 Construction Costs of Offshore Intakes

The construction costs for intake systems with offshore inlet structures and HDPE pipelines, and with concrete tunnels are depicted in Fig. 5.7. These costs are presented as a function of the plant intake flow and are expressed in year 2017 US$ per meter of intake conduit.

Similar to the onshore intake construction costs, the costs for this type of intakes will vary from one location to another. Based on site-specific conditions such as water depth, geology, and currents, these costs could be within a 30% envelope of the values indicated in Fig. 5.7. Analysis of this figure indicates that the construction of desalination plant intakes with deep tunnels is typically several times more costly than the installation of HDPE pipeline on the bottom of the source water body.

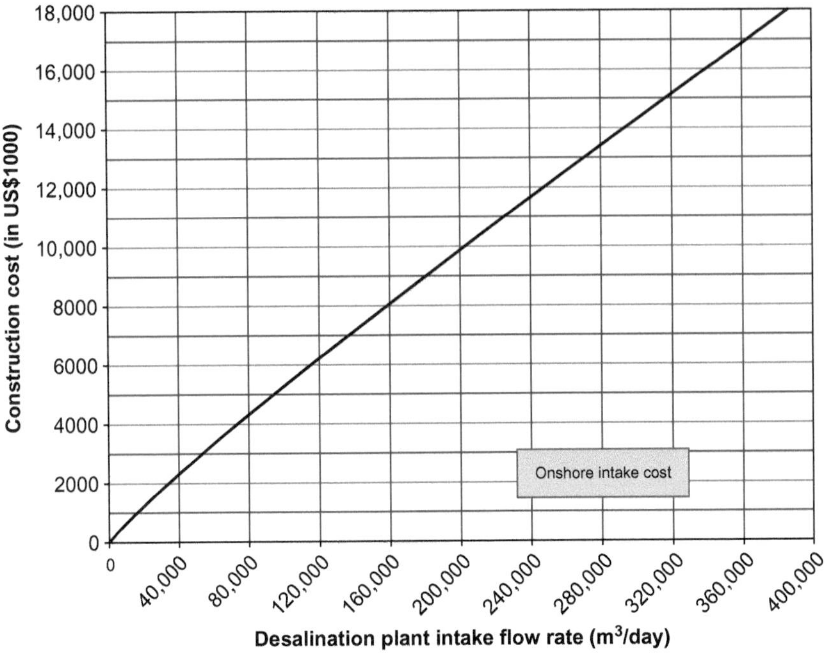

Fig. 5.6 Onshore intakes—construction costs.

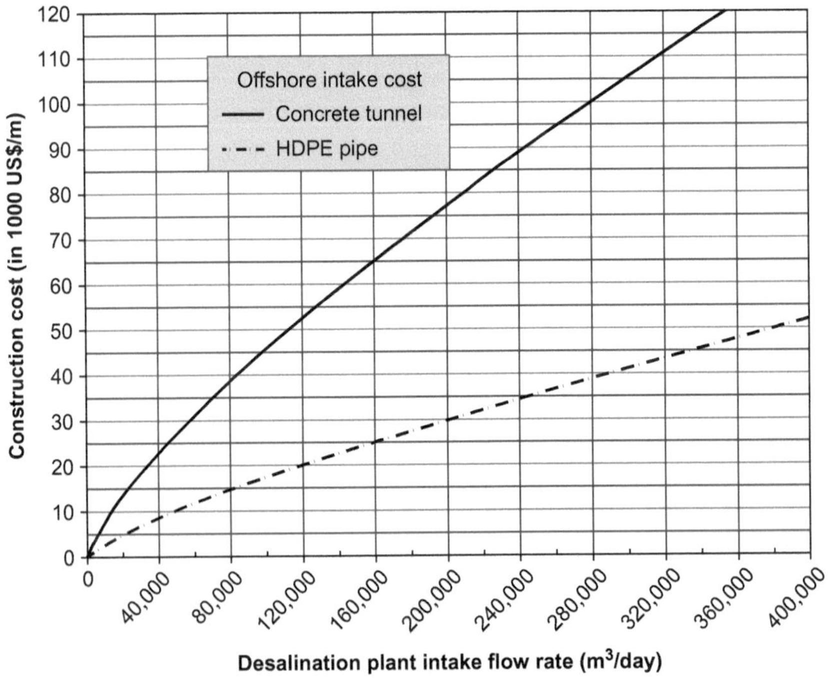

Fig. 5.7 Offshore intakes—construction costs.

REFERENCES

[1] Henthorne L, Boysen B. State-of-the-art of reverse osmosis desalination pretreatment, Desalination, 356, Elsevier, 129-139, 2015.

[2] Baudish PA, Lavery HA, Burch RN, Pain DD, Franklin DG, Banks PJ. In: Design considerations and interactions for tunneled seawater intake and brine outfall systems. IDA World Congress, Perth, Australia, September 4–9, 2011, 2011. REF: IDAWC/PER11–242.

[3] Gille D. Seawater intakes for desalination plants. Desalination 2003;156:249–56.

[4] WRA. Desalination plant intakes—impingement and entrainment impacts and solutions. White Paper Alexandria, VA: WateReuse Association (WRA); 2011.

CHAPTER 6

Design and Construction of Subsurface Intakes

Dennis E. Williams
GEOSCIENCE Support Services, Inc., La Verne, CA, United States

6.1 SUBSURFACE INTAKE SYSTEMS FOR DESALINATION FEEDWATER SUPPLY

Properly designed subsurface intake systems for desalination feedwater supply reduce or completely eliminate pretreatment requirements for particulate matter—thus diminishing impacts to the ocean environment from entrainment or impingement of marine life. The number of subsurface intakes throughout the world is relatively small compared to open ocean intakes. This is largely due to capacity differences between the two types. Subsurface intakes average approximately 12 mgd per facility, as compared to approximately 52 mgd per facility for open ocean intakes [1,2]. However, subsurface feedwater supply systems near and beneath the ocean floor utilize the natural properties of permeable subsea aquifer sediments to filter out organic matter and suspended sediment rather than relying on expensive pretreatment systems. This results in a significant cost savings to the seawater reverse osmosis (SWRO) plant in the form of reduction in capital expenditure for construction, operation, and maintenance of a pretreatment system. Research over the past 11 years has shown that total yield is a function of scale and field tests show that properly designed and constructed slant wells with engineered artificial filter packs allow production of marine groundwater with low turbidity and silt density indexes (SDIs) less than one. Most importantly, subsurface intakes are quickly becoming the favored feedwater supply method due to the reduced environmental impacts they offer. They are even recommended in the State of California's Water Quality Control Plan for Ocean Waters of California [3]. As a result, subsurface intakes can be much easier to get permitted for SWRO desalination plants as compared to open ocean intakes.

Sustainable Desalination Handbook
https://doi.org/10.1016/B978-0-12-809240-8.00006-X

The most common types of subsurface intake systems include:

1. Vertical Beach Wells
2. Ranney Collector Wells
3. Horizontal Directionally Drilled Wells (HDD Wells)
4. Slant Wells
5. Beach Infiltration Galleries
6. Offshore Infiltration Galleries

The following sections provide a brief description of these alternatives. A summary of desalination subsurface feedwater supply systems world-wide is provided in Table 6.1.

6.1.1 Vertical Beach Wells

Vertical water supply well construction and maintenance is a well-established technology, having been developed and refined for >70 years as a reliable source of municipal and agricultural supply. Modern well design and construction techniques including engineered artificial filter packs result in low turbidity water with SDIs acceptable to minimize or eliminate pretreatment. Vertical wells could be constructed on the beach in the project area to produce water from shallow alluvial formations (Fig. 6.1).

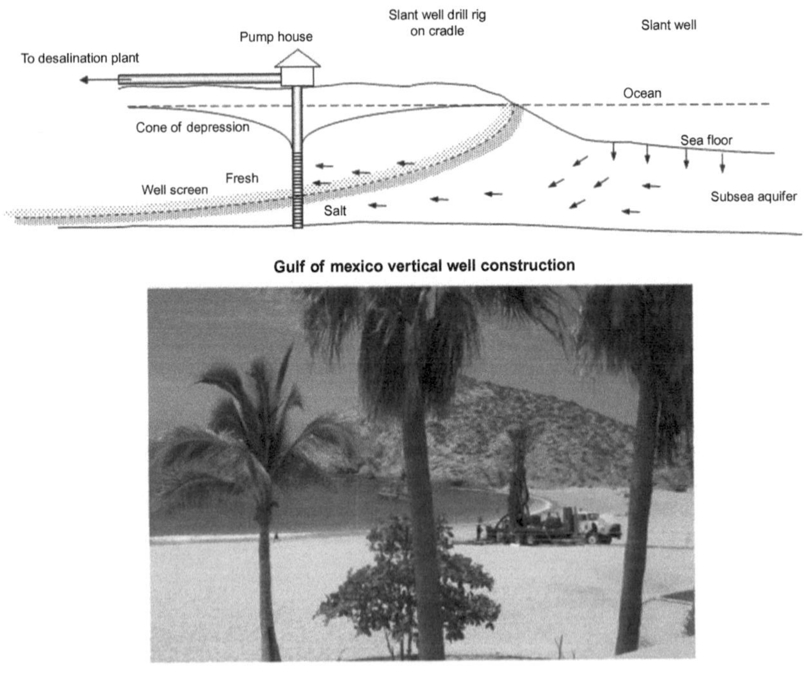

Fig. 6.1 Vertical beach well intake [4].

Table 6.1 Summary of desalination subsurface feedwater supply systems

Project name		Intake type	Construction date	Feedwater supply MGD	Relative cost	SDI feedwater	Subsea aquifers	Source
Adelaide desalination plant	Lonsdale, South Australia	Open Ocean	2008–2012	72	AU $1.83 billion			[2]
Binningup desalination plant	Binningup, Western Australia	Open Ocean	2011		AU $955 million			[2]
Carlsbad SWRO desalination plant	Carlsbad, California	Open Ocean						[2]
Perth desalination plant	Western Australia	Open Ocean	Completed 2006	38	AU $387 Million			[2]
Gold Coast desalination plant	Queensland, Australia	Open Ocean		35				[2]
Dubai SWRO desalination plant	Dubai, UAE	Open Ocean		75				[2]
Perth desalination plant	Perth, Western Australia	Open Ocean		38				[2]
Perth II desalination plant	Perth, Western Australia	Open Ocean		40	AU $900 Million			[2]
Victorian desalination plant	Victoria, Bass Coast of Australia	Open Ocean						[2]
Sydney desalination plant	Sydney, Eastern Australia	Open Ocean		66	AU $1.7 Billion			[2]
Olympic Dam Expansion	Southern Australia	Open Ocean			AU $1.2 Billion			[2]

Continued

Table 6.1 Summary of desalination subsurface feedwater supply systems—cont'd

Project name		Intake type	Construction date	Feedwater supply MGD	Relative cost	SDI feedwater	Subsea aquifers	Source
Long Beach	Long Beach, California	Beach Infiltration Gallery—Pilot Test						
Dana Point	Dana Point, California	Slant Well		11,400 m^3/d		0.35	Alluvial Sand	[14]
	Fukuoka, Japan	Off Shore Infiltration Gallery	Completed 2005	103,000 m^3/d	40.8 million yen	2–2.5	Ocean floor sand	[1,14]
Al Birk	Al Birk, Saudi Arabia	Vertical Wells				0–4	Carbonate silty sand with shell and coral	[14]
Arucas-Moya	Arucas-Moya, Gran Canaria	Vertical Well	Expansion Completed in 2009	8000 m^3/d		<1	Basalt	[14]
Pemex Salina Cruz Refinery	Salina Cruz, Mexico	Ranney Collector Wells (3)		15,000 m^3/d each		<2		[14]
	San Pedro del Pinatar, Spain	HDD Wells (10)		>150,000 m^3/d		<5		[14]
LBWD Pilot Study	Long Beach, California, USA	Beach Infiltration Gallery					Engineered Sand/ Beach Sand	

Sur	Oman	Beach wells (28), unspecified		160,000 m^3/d combined		0.8–1.2		[1]
	Alicante, Spain	Beach wells (30), unspecified		130,000 m^3/d combined				[1]
Tordera	Blanes, Spain	Beach wells (10), unspecified		128,000 m^3/d combined				[1]
Pembroke	Malta	Beach wells, number and type unspecified	Completed 1991	120,000 m^3/d combined				[1]
Bajo Almanzora	Almeria, Spain	Beach wells (14), unspecified		120,000 m^3/d combined				[1]
Bay of Palma	Mallorca, Spain	Beach wells (16), unspecified		89,600 m^3/d combined				[1]
WEB	Aruba	Beach wells (10), unspecified		80,000 m^3/d combined				[1]
Lanzarote IV	Canary Islands, Spain	Beach wells (11), unspecified		60,000 m^3/d combined				[1]
Sureste	Canary Islands, Spain	Beach wells, unspecified type and number		60,000 m^3/d combined				[1]
Blue Hills	New Providence Island, Bahamas	Beach wells (~12), unspecified		54,600 m^3/d combined				[1]

Continued

Table 6.1 Summary of desalination subsurface feedwater supply systems—cont'd

Project name		Intake type	Construction date	Feedwater supply MGD	Relative cost	SDI feedwater	Subsea aquifers	Source
Santa Cruz de Tenerife	Canary Islands, Spain	Beach wells (8), unspecified		50,000 m^3/d combined				[1]
Ghar Lapsi	Malta	Vertical Wells (18)	Completed in 1986	45,000 m^3/d combined				[1]
Cirkewwa	Malta	Vertical Wells, unspecified number		42,000 m^3/d combined				[1]
CR Aguilas	Murcia, Spain	Beach wells, unspecified type and number		41,600 m^3/d combined				[1]
SAWACO	Jeddah, Saudi Arabia	Beach wells, (10), unspecified	Completed 1994	31,250 m^3/d combined				[1]
Dahab	Red Sea, Egypt	Beach wells (15), unspecified		25,000 m^3/d combined				[1]
Turks & Caicos Water Company	Providenciales, Turks & Caicos Islands	Beach wells (6), unspecified		23,260 m^3/d combined				[1]
Windsor Field	Bahamas	Beach wells, unspecified type and number		20,000 m^3/d combined				[1]
North Side Water Works	Grand Cayman	Beach wells, unspecified type and number		18,000 m^3/d combined				[1]

Ibiza	Spain	Beach wells (8), unspecified		15,000 m^3/d combined				[1]
North Sound	Grand Cayman	Beach wells, unspecified type and number		12,000 m^3/d combined				[1]
Red Gate	Grand Cayman	Beach wells, unspecified type and number		10,000 m^3/d combined				[1]
Abel Castillo	Grand Cayman	Beach wells, unspecified type and number		9000 m^3/d combined				[1]
Al-Birk	Saudi Arabia	Vertical Wells (3)		5100–8700 m^3/d combined				[1]
Lower Valley	Grand Cayman	Beach wells (3), unspecified		8000 m^3/d combined				[1]
West Bay	Grand Cayman	Beach wells, unspecified type and number		7000 m^3/d combined				[1]
Britannia	Grand Cayman	Beach wells (4), unspecified		5400 m^3/d combined				[1]
Bar Bay	Tortola, BVI	Beach wells, unspecified type and number		5400 m^3/d combined				[1]
Morro Bay	California, USA	Beach wells (5), unspecified		4500 m^3/d combined				[1]
Ambergris Caye	Belize	Beach wells, unspecified type and number		3.600 m^3/d combined				[1]

However, because the wells would not extend beneath the ocean floor, vertical wells rely on both onshore and offshore lateral recharge. Water-level interference may limit the production capacity of a beach wellfield and there is a good chance the wells would induce a greater amount of groundwater from onshore than offshore. This may limit production capacity due to regulatory, permitting, and water rights issues. In addition, if the near-shore aquifer is fairly thin, then a large number of beach wells would be required to produce the design feedwater supply from the target aquifer.

6.1.2 Ranney Collector Wells

Horizontal radial collector wells were developed by Leo Ranney in the 1920s for the oil industry in Texas and Ohio as a method to drill horizontally for oil recovery. The technology was later modified in 1933 in London, where large amounts of water were centrally pumped from a single large-diameter vertical shaft or caisson. In the United States, the first radial collector well for water supply was constructed in 1936. Today, there are approximately 220 active collector wells in the country [5].

Prior to construction of a radial collector well, it is necessary to drill a number of vertical exploratory borings throughout the area to determine the subsurface geohydrologic characteristics. These characteristics include the location, lateral extent, type and thickness of permeable aquifers, the location of potential faults in the area, and groundwater quality and quantity. During construction of a radial collector well, the central caisson, which may range from 8 to 20 ft in diameter, must be located very close to a surface water source (Fig. 6.2). The caisson can be 120–260 ft in depth with six to eight (or more) horizontal laterals radiating outward from near the bottom of the vertical shaft [5].

The laterals contain sections of screen and are placed by either jacking outward from the vertical shaft under hydraulic pressure, or by jetting them into place. Compressive screen strength and type of formation material encountered are limiting factors that determine the maximum distance that the laterals can extend outward. Currently, the length of the laterals used for water supply is typically 127–240 ft [5]

The central caisson consists of sections of large-diameter steel-reinforced concrete pipe that are either prefabricated or are fabricated onsite. They are assembled with water-tight joints and are designed to withstand the forces under which they will be subjected. Soil is excavated or undermined from beneath the structure as it is being pulled downward with hydraulic rams, causing the caisson to settle downward to the designed depth. At desired

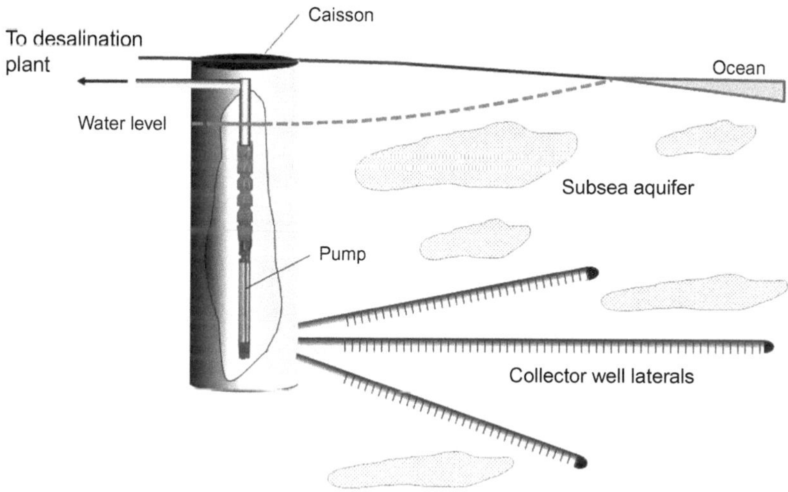

Fig. 6.2 Ranney collector well.

locations, specially designed "windows" or portholes are placed in the walls of the caisson through which the horizontal laterals are placed. It should be noted that laterals can be placed in more than one plane, if the aquifer thickness allows. As the laterals are placed, packers are used to control the inflow of water from each lateral until construction is complete. Additionally, the interior of the caisson must be continually dewatered so that work on the laterals can take place. Development is minimal as the laterals are naturally developed and are designed so that the velocity of fluid entering the screens remains very low. A gravel-packing procedure for collector wells has been developed by Preussag in Germany, and has been used in the United States via license agreement [6]. The use of filter-packed screens makes collector well technology more adaptable to all aquifer grain sizes—enabling sand-free production.

In permeable aquifers with a constant recharge source, collector wells are capable of very high production rates of 2–92 mgd and have been used for both fresh water and seawater intakes. Other uses include cooling water intake for power plants, industrial and municipal purposes, irrigation, and for climate control through the use of heat pumps.

Recent advances in gravel-packed screens placed in the laterals have allowed collector wells to be successful in a wide range of geologic environments. However, most radial collector wells still utilize slotted well screens and development of a natural filter pack by removal of fine-grained sediments from the formation materials, and do not have the benefit of a gravel envelope to filter out sand and sediment. Radial collector wells rely on very low entrance velocities to keep sand production and turbidity to a minimum.

There are several advantages to using collector wells over vertical wells. The cost of a single collector well can be less than the equivalent of six or eight vertical wells. Operation and maintenance (O&M) costs for pumps, pipelines, roadways, buildings, electrical switching, and control facilities are reduced as only one pumping facility is needed per collector well. The required lot size for the well facility is also relatively small and only slightly larger than the completed well (though the area needed during construction is much larger).

Limitations to the use of collector wells for desalination plant feedwater supply include restricted lateral lengths and construction-related issues in the beach environment. The length of the laterals is currently limited to approximately 127–240 ft for the traditional Ranney-type collector well and 350–375 ft for collector wells using the Sonoma method of construction. The construction of a large caisson on the beach and dewatering operations requires a large construction footprint. The timeframe for constructing a collector well is approximately 6–9 months—two to three times the time needed to construct a slant well.

6.1.2.1 Ranney Collector Well—Sonoma Type

The "Sonoma Method" refers to a traditional Ranney-type radial collector well that has been modified to include the addition of one or two larger diameter collector arms that are drilled using mud rotary technology rather than being hydraulically jacked into place. Sonoma County Water Authority's newest collector well (Well #6) is located at a greater distance from the Russian River than previous collector wells in order to minimize impacts on riparian habitat. In addition to the standard tier of laterals, the Sonoma collector well has an added tier that has two 18-in-diameter lateral arms that radiate outward (toward the river) from the bottom of a 110-ft-deep vertical caisson that is 30 ft in diameter. The laterals were constructed at distances of 350 and 375 ft. from the vertical caisson, and each produce 18 mgd. They consist of laser-slotted well screen with a natural filter pack.

The contractor's personnel worked under pressure and were submerged inside the caisson using SCUBA gear. Two months of time were spent in developing the laterals due to the use of drilling mud and the poorly sorted aquifer material. The current capacity of the Sonoma method radial collector well is 40–45 mgd; older collector wells in the area produce 90–92 mgd.

Friction caused by the flow of water within each lateral reduces its flow. To remedy this, future Sonoma-style Ranney wells may be designed with larger-diameter laterals [7].

6.1.3 Horizontal Directional Drilling Wells

Horizontal directional drilling (HDD) wells utilize "open hole" technology—meaning that drilling fluid is required to hold the borehole open during drilling and construction. Drill pipe and downhole tools are used to advance the borehole while drilling fluid is used to cool and lubricate the bit, stabilize the borehole, and carry cuttings (formation material) to the surface (Fig. 6.3). Drilling of the borehole is generally achieved in two stages: drilling the small-diameter pilot borehole, followed by enlarging the pilot borehole in one or more reaming passes to the diameter required to contain the casing, screen, and filter pack.

HDD wells are classified as being either blind or continuous. Blind wells are generally used in deep subsurface oilfield applications to increase recovery of oil and gas, or in relatively shallow environmental remediation applications where the target formation is located under a building or some other obstacle. Blind wells drilled for oilfield applications may start either vertically or on an incline, while those drilled for environmental applications, due to their shallow true vertical depth (TVD), generally start only on an incline. Once the targeted depth is reached, the drill string is rotated to

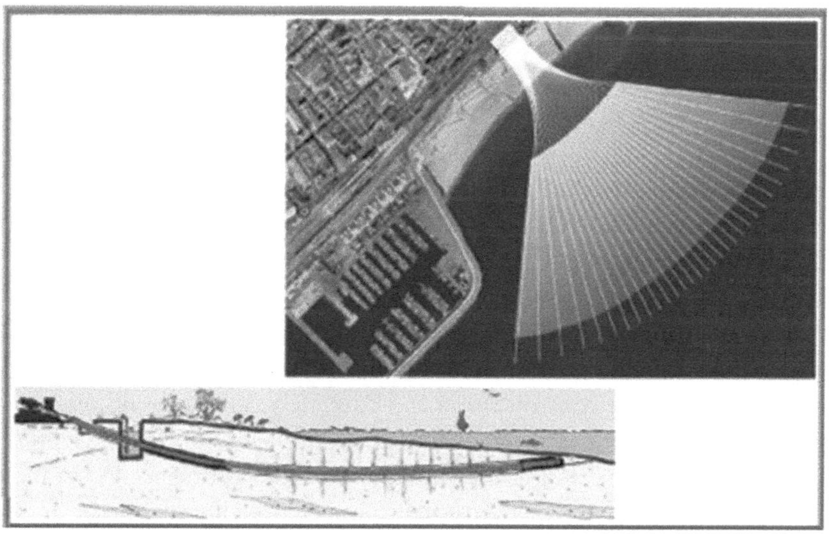

Fig. 6.3 HDD feedwater supply, Mataró, Spain [9].

horizontal for the required distance before being terminated in the subsurface at the targeted point (Fig. 6.3). Reaming is accomplished by pushing and rotating the drill bit as it follows the pilot bore to its completion depth. In most cases with blind wells, the borehole is stabilized by casing and cementing through the curvature prior to drilling the horizontal portion of the borehole.

Once the borehole has been reamed to the required diameter, well completion materials are pushed into place in the open borehole. The well intake structure typically consists of either standalone screen or prepacked screen. Artificial filter-packing technology is not yet perfected for HDD wells. Therefore, most successful HDD wells have been completed in competent formations such as limestone. Another completion option using HDD is the use of a porous pipe, such as Schumasoil filter screen constructed of sintered polyethylene resin beads. Schumasoil was developed to maximize the open area of the screen while maintaining filtration properties that meet or exceed prepack performance [8]. It achieves excellent filtration characteristics, but the porous construction of the screen limits its tensile strength and the lack of a filter pack makes the screen prone to clogging with fine-grained materials—reducing production over time.

The required work area for HDD wells is approximately 120×120 ft ($14{,}400\ ft^2$) plus another 100×100 ft for staging. An advantage of the HDD method is the ability to construct relatively long boreholes and relatively long wells (up to 3000 ft). A major benefit of wells with long screens is higher production potential. Additionally, the salinity of water from an HDD well screened offshore within the saltwater wedge would more closely approximate that of seawater. Also, by being able to locate a screen further offshore with the HDD method, the cone of depression would be less likely to impact onshore groundwater levels. Another major benefit of the long boreholes drillable with the HDD method is the ability to avoid obstacles or sensitive areas. The HDD method ensures accurate borehole trajectory using remote sensing and measurement while drilling (MWD) technology.

The relatively short drilling and construction timeframe for an HDD well is advantageous in terms of duration and public access. An HDD well could be drilled and constructed in approximately 3 months, although development could take several additional months. Additionally, the HDD well construction method creates minimal surface disturbance. As in most other well construction methods, the completed wellhead for the HDD well can be buried to minimize visual impact.

6.1.4 Slant Wells

Construction of slant wells can be accomplished using the dual rotary drilling method. This method requires the mast of the drilling rig to be detached from the power unit and mounted on an angled cradle or platform at the required angle (Fig. 6.4). A slant well can be constructed at any shallow angle dictated by the platform angle, including horizontal—as long as the rig is lowered into an excavated and shored pit, or is otherwise positioned to target a horizontal zone.

A dual rotary drilling rig (formerly known as a Barber drilling rig) has two drives, comprising a lower hydraulic drive that clamps onto the exterior of the drive casing and an upper top-head drive to rotate the inner drill string. In the dual rotary drilling process, an outer drill casing is advanced using the lower drive that can be telescoped from larger to smaller casing diameters (e.g., 24–20-in., etc.) while simultaneously removing formation materials from inside the casing during advancement using a conventional dual tube reverse circulation system that is driven by the top-head drive. As formation materials are removed, the borehole is advanced by simultaneously rotating the outer drill casing with the lower drive (Fig. 6.4).

Fig. 6.4 Slant well construction at Dana Point, CA [10].

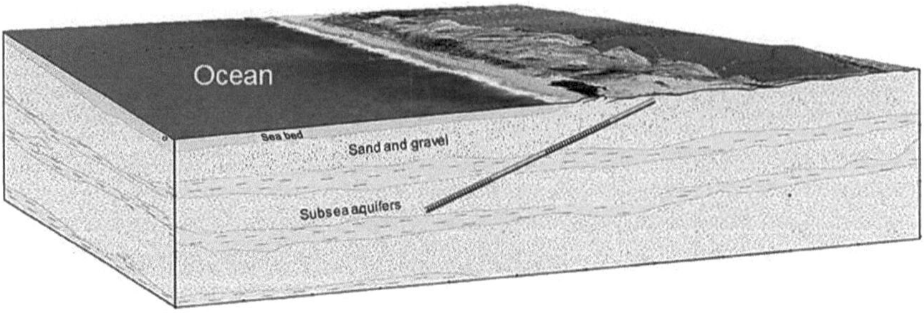

Fig. 6.5 Example of slant well desalination feedwater supply [4].

Slant wells can target specific aquifer zones, resulting in higher productivity than vertical wells (Fig. 6.5). Also, since slant wells can extend offshore into subsea aquifers, onshore impacts are minimized while well discharge is composed of a high percentage of seawater.

The dual-tube reverse-circulation rotary method uses flush-jointed double-walled drill pipe. Compressed air or air with a small amount of water is injected through an inlet on the side of the rotary head and forced downward under high pressure between the outer barrel (or wall) of the drill string and the inner barrel. At the leading end of the dual-tube drill string, a drill bit breaks up large-diameter formation materials (i.e., large gravel and cobbles) for removal during advancement of the borehole. After the outer drill casings are advanced to total depth, the dual tube drill string is removed and well casing and screen materials are then completed within the cased hole.

Slant well construction requires an area of approximately 7800 ft^2 (60 × 130 ft). The production potential of a dual rotary-drilled slant well is limited by the size of the submersible pump that can fit within the well's pump house chamber diameter. However, using the telescoping method, large pump house chambers can accommodate large capacity submersible pumps. A DR-40 dual rotary rig allows for drilling and placing temporary casing in boreholes up to 40 in. in diameter and lineal lengths of 1000 ft. A gravity-fed collector system manifolded to a common caisson-type pump chamber is also feasible to produce water from multiple slant wells. Slant wells minimize construction footprints by allowing construction of multiple wells at different azimuth angles from the same wellhead area. The length of a dual rotary-drilled well is limited by the ability of the rig's lower drive to pull back the outer drill casing from around the well at total depth. Currently, the 1000-ft long Cal-Am Test Slant Well in Monterey, California, is the longest dual rotary-drilled artificially filter-packed slant well. This well is anticipated to be used for desalination intake for the Monterey Peninsula Water Supply Project, but is still in a trial phase (as of 2017).

6.1.5 Subsurface Infiltration Galleries

An infiltration gallery consists of horizontal collector-type well screens placed in permeable aquifer materials, usually adjacent to a water body or beneath its bed. Infiltration galleries are used where aquifer thickness is insufficient to support groundwater extraction. Screens are placed in open trenches and backfilled with appropriate filter materials (Fig 6.6). Water entering the infiltration gallery is often collected in a sump constructed beneath the end of the screens; in galleries with high infiltration rates, a centrifugal pump is manifolded to the screens. Because they are constructed via open excavation, they are generally limited to depths of approximately 25 ft.

6.1.5.1 Engineered Beach Gallery

An engineered beach gallery consists of a single excavated pit. The base of the pit is equipped with horizontal collector pipes and covered with an engineered filter pack followed by backfilling with the excavated sand soil. The water is collected through the collector pipes placed at the bottom of the pit excavated on the beach. The excavation is dewatered to allow excavation of native sand deposits and placement of collector pipes, filter pack, and native sand backfill. A 2-year testing of an engineered beach gallery test site at Long Beach, California, showed that water can be produced over a wide range of tidal conditions at an infiltration rate of 0.10–0.15 gpm/ft^2. In previous work by Ocean Technologies, Inc. conducted in 1994, it was reported (using empirical data) that gallery infiltration rates may range from 0.33 to 3.47 gpm/ft^2 [12]. However, for general purposes infiltration rates of 0.10 gpm/ft^2, developed from the pilot testing at Long Beach, are considered reasonable.

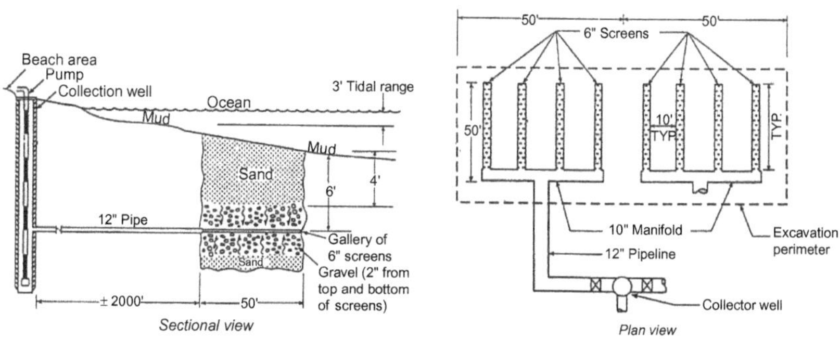

Fig. 6.6 Seabed infiltration gallery [11].

6.1.5.2 *Offshore Galleries*

An offshore engineered infiltration gallery also consists of a single excavated pit with horizontal collector pipes buried beneath an engineered filter pack. Water vertically infiltrates by gravity from the ocean to the collector pipes placed at the bottom of the pit. The offshore gallery typically uses coffer-dams to allow excavation of the offshore filtration pit without long-term interference or impact to the Beach. The excavation will need to be de-watered to allow removal of native sand deposits and placement of collector pipes, filter pack, and native sand backfill. Fukuoka District Waterworks Agency in Japan completed a seawater desalination plant in March 2005 that uses an infiltration-type intake system, in which a network of perforated pipes is buried approximately 8 ft under the seabed to collect clean seawater through the overlying sand layer at a slow infiltration speed. The seawater intake capacity of the system is 103,000 m^3/day (27 mgd), and the intake screens cover an area of approximately 20,000 m^3 (5 acres).

Infiltration galleries can be very difficult to maintain due to incrustation caused by iron bacteria that thrive in an aerated environment, and their tendency to clog if overpumped or if left inactive [13]. Seawater intake data from the offshore infiltration gallery at Fukuoka, Japan, showed infiltration rates that were in the range of those observed for the Long Beach gallery.

6.2 SELECTED CASE HISTORIES

6.2.1 Vertical Beach Wells

6.2.1.1 *Al-Birk Desalination Plant, Saudi Arabia*

The Al-Birk SWRO Desalination Plant is located in the southern region of the Red Sea coast of Saudi Arabia and was commissioned in 1983. The plant consists of two individual trains that were originally fed by seawater received from screened open intake systems positioned about 20–23 ft (6–7 m) below the surface of the ocean and roughly 1000 ft (300 m) from shore. The feedwater was then pretreated prior to desalination. This process included chlorination, the addition of a coagulant (polyelectrolyte mang-nifloc C-573), and filtration via dual media filters and fine carbon filter [15]. After filtration, the dual media filters had to be backwashed and de-chlorinated. Despite the pretreatment efforts, the plant was experiencing serious biofouling problems within 3 months of starting operation. This necessitated frequent cleaning of the RO membranes. The biofouling was attributed largely to the warm temperatures experienced in the area and which favor bacteriological activity [16].

In an effort to reduce the biofouling, an alternative intake system was investigated and implemented. This involved the construction of vertical beach wells to provide the feedwater from the ocean subsurface. Geologic and hydrogeologic studies were carried out by drilling test boreholes. The results showed that boreholes to the north contained mostly basaltic rock below 30–65 ft (10–20 m) and were inferior to locations to the west, which contained medium-to-dense carbonate silty sand with shell and coral fragment [15]. Therefore, three permanent wells were installed on the west side of the plant to depths of around 100–120 ft (30–36 m) below ground surface. After developing the wells and running them continuously for 3 months, the beach well water quality stabilized and showed a significant reduction in SDI and total dissolved solids (TDS) to <1 and 45,724 ppm (64,400 μs/cm), respectively. In April of 2004, the Al-Birk plant began operating on feedwater from the wells. The increased quality of the feedwater eliminated the need for pretreatment, eliminated membrane fouling, reduced membrane cleaning, increased production, and increased plant availability compared to open intake pretreated feedwater.

The total construction costs for the three wells and a standby fourth well was roughly $200,000 USD (SAR 0.7 million) while the estimated capital cost for pretreatment, open intake, and auxiliaries is about $1.2 million USD (SAR 4.5 million). Therefore, in the case of the Al-Birk plant, the use of beach wells to provide feedwater presented significant cost savings compared to the rigorous pretreatment needed to operate on feedwater from open ocean intakes. This was considered to be true even when taking into account the reduced life of vertical wells compared to open intake structures and the probable need for replacing the wells in an estimated 10 years.

6.2.2 Ranney Collector Wells

6.2.2.1 Salina Cruz, Mexico

The only known operating Ranney collector well system used for an SWRO desalination plant feedwater intake is located in Salina Cruz, Mexico [1]. Three wells are used to supply water for the cooling towers at the Pemex Salina Cruz oil refinery and subsequent desalination. Each well has a capacity of 15,000 m^3/d or 3.8 mgd [17]. The intake structures are located on the beach and represent an example of the visual impact this type of beach well can have. While subsurface intakes generally produce higher-quality water with a lower SDI value, there are still significant concerns with the feedwater quality from the Ranney wells at the Salina Cruz site. The feedwater at this site shows high concentrations of manganese and iron, which requires chemical conditioning prior to desalination [17]. The TDS concentration is also unpredictably variable. For example, TDS from one well at the Salina Cruz desalination plant ranges from 16.8 to 21. 8 ppt and between 17.8 and 19.8 in another collector well. This is explained by a freshwater influence. It presents constraints to the efficiency of the plant and often results in rises in the unit cost of water production over time.

6.2.3 Horizontal Directionally Drilled Wells

6.2.3.1 San Pedro del Pinatar, Spain

Due to permitting restrictions and local hydrogeological conditions, the Neodren HDD well technology was used to supply feedwater for the desalination plant at San Pedro del Pinatar. Also known as the "Nuevo Canal de Cartagena," San Pedro del Pinatar is located to the north of the Mar Menor (Minor Sea) in Spain. The location of the desalination plant is bounded on one side by the Regional Park of San Pedro del Pinatar and on the other by a brackish water desalination plant, which treats drainage from the "Campo de Cartagena" irrigation area. The entire surrounding area is designated as a protected zone due to the presence of sea grass [18].

Originally, the plant was designed to have subsurface seawater intake in the form of vertical wells. However, the Governing Body of the Environment Department required that the intake could not include source water from the aquifers of the "Campo de Cartagena" nor the wetlands of the San Pedro del Pinatar Regional Park. After some exploratory drilling and hydrogeological investigations, it was discovered that fresh water/seawater extended far offshore, making the use of vertical wells impossible. The investigation also indicated that the target aquifer was a 10-m-thick (33 ft) compact white marble unit underlying a thick layer of clay. Given that the bed was not very thick, the only possible way to extract the required amount of water and ensure the source is solely ocean water was to drill a horizontal well. Ten HDD production lines were installed in the area. Each of these wells can produce 8500 m^3/day (2.2 mgd) [18].

6.2.4 Slant Wells

6.2.4.1 Doheny Ocean Desalination Project-Dana Point California

The Doheny Ocean Desalination Project (formerly South Orange Coastal Ocean Desalination Project or SOCOD), established that slant wells are a viable subsurface intake alternative. Results from a 2-year extended pumping test demonstrated reliability of an ocean water supply (Fig. 6.7). Additional benefits include:

Fig. 6.7 Doheny Ocean Desalination Project showing seven 1000-ft slant wells with two standby wells to produce a 30-mgd feedwater supply.

- Avoidance of marine impacts;
- Viewed favorably by regulatory agencies and the environmental community;
- Easier to permit than other intake systems;
- The subsea aquifer is very effective in removing particulates (low SDI) reducing or eliminating the requirement for conventional pretreatment;
- Lower cost intake system (less than intakes requiring pretreatment);
- Typical high percentage ocean water yields (>95%); and
- Allows fully buried facilities to minimize visual and other environmental impacts.

During the 2-year pilot testing, the test slant well produced approximately 2100 gpm with relatively stable drawdowns. Geochemical tracer tests conducted during the extended pilot testing used groundwater age dating to derive relevant amounts of water sources to provide a direct indication of slant well connectivity to the ocean [19]. The multiple tracer approach quantified the groundwater source captured by the slant well intake during the extended pumping and supported the increased capture of shallow, young marine groundwater as demonstrated during the last 5 months of the pilot testing. Comparison of output from a variable density groundwater model to observed TDS produced realistic results with respect to young and older marine water capture. With time, the old marine ground water will be fully pumped out of the offshore formations and replaced by seawater—resulting in a feedwater supply with lower levels of dissolved iron and manganese. The full-scale project is designed to produce 30 mgd from seven 1000 ft slant wells (Fig. 6.7).

Silt Density Index

SDI is one of the major design parameters for a successful desalination feedwater supply well. SDI is a nonspecific measure of the colloidal fouling potential of water and is the most widely used method for determining feedwater quality for membrane desalting processes. As a general guideline, most membrane manufacturers recommend maintaining feedwater SDI less than values of 3–5, depending on the membrane type and manufacturer. Waters with SDI values consistently >5 generally require pretreatment before the membrane desalting process. Subsurface intakes constructed using properly engineered artificially filter-packed wells (vertical, Ranney Collector, HDD, or slant) typically meet the minimum SDI requirements for membrane fouling [4].

Dissolved Iron and Manganese in Old Marine Groundwater

Based on studies by Charette, an alluvial channel was incised offshore of San Juan Creek in the Dana Point area following the end of the last ice age [19]. Subsurface outflows from the Creek discharged beneath the ocean within the area shoreward of the seawater interface. On the ocean side of this interface, the ocean-filled alluvium groundwater has remained isolated since its inundation about 7500 years ago. This water was designated as "Old Marine Groundwater." During the 2-year pilot testing, the dissolved iron levels stayed relatively constant after reaching approximately 10 mg/L after 8 months of pumping and slightly increased to 11 mg/L by the end of the test. After 1 year of pumping, based on the natural isotope data, the pumped water was a mixture of brackish groundwater that steadily decreased, ocean water that had steadily increased, and old marine groundwater that initially had increased as the seawater interface was pulled to the well.

The increasing amount of "younger" ocean water and relatively consistent fraction of old marine groundwater kept the iron concentrations relatively flat over the last year of the test. The isotope data showed a slightly decreasing fraction of old marine groundwater being pumped over the test, as the "younger" ocean water recharged the marine aquifer area where brackish groundwater had discharged out under the ocean. It is estimated that under full-scale pumping (30 mgd), the old marine groundwater would be fully pumped out within approximately a year. Also, it is estimated that after the old marine groundwater is removed, the wells will produce approximately 95% "younger" ocean water, which has very low levels of dissolved iron/manganese (about 2 μg/L) and 5% brackish groundwater, which has about 2 mg/L of dissolved iron/manganese, resulting in a blended concentration of about 0.10 mg/L.

6.2.4.2 Monterey Peninsula Water Supply Project

In 2004, the California American Water Company (CalAm) filed an application for the Coastal Water Project (also referred to as the Moss Landing Project) to replace existing Carmel River water supplies for the CalAm Monterey District service area. The Coastal Water Project and its two alternatives, the North Marina Project and the Regional Project, involved producing desalinated water supplies, increasing the yield from the Seaside Groundwater Basin and building additional storage and conveyance systems to move the replacement supplies to the existing CalAm distribution system. The Coastal Water Project was sized to meet existing water demand and did not include supplemental supplies to accommodate growth. Subsequent to approval of the Regional Project, CalAm withdrew its support for the Regional Project and submitted an application for the Monterey Peninsula Water Supply Project (MPWSP). The MPWSP includes many of the same elements of the previous proposed projects; however, key components, including the seawater intake system and desalination plant, have been relocated and/or modified for the MPWSP.

The proposed MPWSP includes construction of a desalination plant located in unincorporated Monterey County northeast of the City of Marina and 10 subsurface slant wells at the CEMEX active mining area in the northern area of the City of Marina to produce approximately 10,750 afy. The proposed MPWSP Desalination Plant has a rated capacity of 9.6 mgd.

Alternative 5a is a variation of the proposed MPSWP with the capacity to produce 6.4 mgd and includes the construction and operation of a reduced-capacity desalination plant. Project components would be sited at the same locations as the proposed project and the only differences are the number of slant wells and the size of the desalination plant.

Development and implementation of any emerging technology project requires a cohesive blend of technical experts, environmental compliance, public perception, and regulatory awareness. As such, for a slant well feedwater supply in sensitive coastal areas of California, a collaborative effort was necessary, which included participation by recognized experts in geology, hydrogeology, and modeling. The collaborative group included stakeholders of groundwater usage and management in the MPWSP project area and led to development of the Hydrogeologic Working Group (HWG). The HWG first met in 2013 to discuss conceptual models and to form a collaborative plan of investigation to assess the hydrogeologic conditions in the project area. As with any collaborative group, individual opinions need to be evaluated against actual field data and testing to arrive at a conceptual model that reflects a common understanding at the areas of concern. A draft workplan was prepared, which provided a phased approach to progressively investigate the hydrogeology and the potential effects to aquifers from the use of subsurface slant wells for obtaining feedwater supply. The final workplan incorporated comments and recommendations by members of the HWG, and covered the investigative steps needed to evaluate the project impacts, which became the hydrogeology investigation roadmap.

The process adopted by the HWG for the workplan consisted of on-going steps of data collection and analysis. The data collected from this initial phase of investigation were used to construct the various groundwater models, which became the tools to evaluate the short- and long-term hydrogeologic impacts in the project area from operation of the MPWSP. Each step of data gathering was preceded by an update of the workplan as appropriate, describing the proposed work and desired outcomes. Results were documented by a technical memorandum, the Hydrogeologic Investigation Workplan (HWP), which became the main working document for all exploratory, testing, and modeling work, including: describing the methods of data collection, findings, and recommendations.

In order to maximize recharge from ocean sources, as well as minimize variations in salinity and impacts to inland resources, slant wells should be located as close as possible to the ocean where permeable materials extend offshore.

Locating subsurface intakes near and offshore maximizes both vertical leakage through the seabed and horizontal recharge from offshore aquifers. Other considerations include coastal erosion, 100-year flood zone, sea-level rise and proximity to sensitive habitat, well construction footprint, access to the drilling site and equipment staging area, and periodic access for monitoring and maintenance [20–22]. For example, in the Doheny project, the test well was located on a State Beach and subjected to strict compliance with recreational uses. For the Monterey project, siting considerations were within a strict footprint to comply with all of the factors listed.

Proximity to existing surface water bodies should also be considered during the siting process as infiltration from either seasonal or perennial surface water bodies may impact the water quality from the initial well testing phase used for long-term treatment planning. The location of wells should consider constraints (environmental and others) to site access as well as the appropriate footprint for well construction and staging.

Construction in sensitive coastal areas of California involves numerous permits obtained from many different agencies. As local and regional stakeholders should be included in all preliminary discussion, therefore, public outreach is as important to the project as is collecting accurate data and developing a technically sound project design. The development of a collaborative technical advisory committee is an important tool in ensuring that project plans, data, and communication are shared and discussed among stakeholders to move the project efficiently through the various technical phases. The HWG composed of highly qualified technical experts representing major stakeholders was developed during the planning and permitting stages of the MPWSP project and has worked closely to plan investigations and understand and interpret project data.

Prior to the construction process, a close collaboration with the drilling contractor is essential. Reviewing well designs and step-by-step methods to be used at every phase of well drilling and construction should be prioritized before equipment is mobilized. The Geohydrologist should satisfy themselves that appropriate equipment is selected and set apart for the project. Equipment should be inspected prior to mobilization, and the contractor should ensure that timely replacement of high-quality parts, which inevitably require replacement during the operation will occur. Figs. 6.8 and 6.9 show the design of the MPWSP test slant well.

Fig. 6.8 MPWSP Test Slant Wellhead and Casing with "Half Moon" Design.

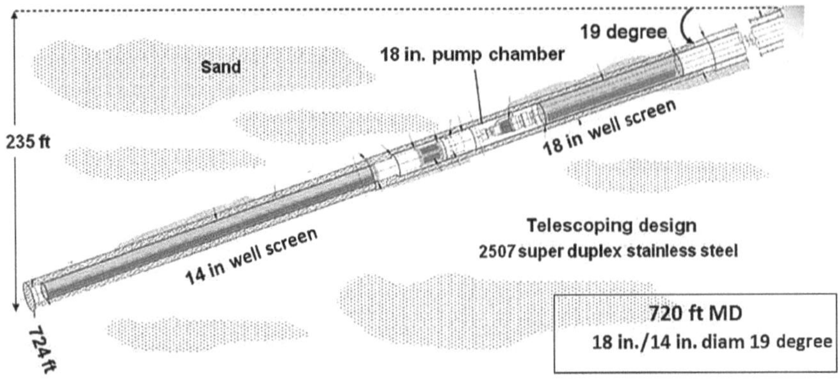

Fig. 6.9 MPWSP Test Slant Well Showing Telescoping Design.

6.2.5 Subsurface Infiltration Galleries

6.2.5.1 Long Beach Pilot Study, California—Beach Gallery

The Long Beach Water Department (LBWD) began a pilot study in 2001 to establish the design criteria for a full-scale facility planned for 2015. The project aims at developing an alternative method for seawater intake and brine discharge for the operation of a prototype desalination plant. In order to avoid impingement and entrainment issues and to provide a natural pretreatment for the feedwater, LBWD decided to use a subsurface intake. Originally, a Ranney well collection system was planned, but flow modeling results indicated that a surf zone infiltration gallery filled with an engineered sand would provide the highest capacity [23]. Feedwater will originate from a series of pipes under the beach sand around the high tide line, and brine will also discharge through a similar infiltration gallery. Permitting was obtained to test a temporary infiltration gallery to investigate its potential as an intake for the future desalination facility.

A 3100-ft^2 infiltration gallery intake system was constructed 5 ft deep with 6-in PVC collection pipes and backfilled with engineered sand to allow greater infiltration than the local fine silty sand [24]. Two objectives of the pilot test were to determine the maximum sustainable intake and discharge rates and to determine the long-term water quality performance of the intake system as pretreatment for the desalination process [25]. Preliminary results showed that the intake system alone may not be sufficient as a pretreatment for RO and that there were frequent system shut downs with an infiltration rate of 0.15 gpm/ft^2 (465 gpm). This may be due to the system design, air binding from tidal fluctuations, or local sand build-up. The results also show that the contribution of infiltration from the side is significant, which can cause inaccuracies in the calculation of the system capacity for the future design [25]. LBWD wants to continue to test the current infiltration gallery to determine the exact limitations to the system's capacity.

6.2.5.2 *Fukuoka, Japan*

To meet increased water demand, the Fukuoka District Waterworks Agency in Fukuoka, Japan, began designing the Uminonakamichi Nata seawater desalination center in 1999. Construction started in 2001 and it began operating in 2005. The total cost for the project was around $400 million USD (40.8 billion Yen) [26]. To ensure high-quality feedwater with a low SDI value and eliminate a negative visual impact, the plant uses an infiltration gallery positioned 2132 ft (650 m) offshore in the Genkai sea at a depth of about 38 ft (11.5 m) [27]. Seawater infiltrates into the gallery pipes and flows into a seawater holding tank, which is situated below sea level. This allows water to be drawn in slow enough for effective pretreatment without entraining sand in the system. The system takes up 216,300 ft^2 and can provide feedwater of reliable quality in quantities of 103,000 m^3/d (about 27.2 mgd) [27].

According to director Akira Shimokawa and operations manager Taketo Tanaka, virtually no maintenance of the infiltration gallery is needed [27]. What maintenance is necessary is performed in the winter months when water demand diminishes [28]. Taketo also confirmed that head loss across the system remains almost unchanged since operation began and the seabed and piping has never required any cleaning. It is believed that the current action keeps the surface of the sand relatively clean. Despite this, most of the information about the Fukuoka plant infiltration gallery is anecdotal and unattributed [27].

6.3 SUBSURFACE INTAKE SYSTEMS—ADVANTAGES AND DISADVANTAGES

The advantages and disadvantages of subsurface intakes are described below, and their applicability is summarized in Table 6.2.

6.3.1 Advantages

- Subsurface intakes have no impacts to the ocean environment in the form of entrainment or impingement to marine life.
- Beach or offshore infiltration galleries have no impacts to riparian habitat, wetlands, or onshore groundwater resources.
- HDD and slant wells will have little to no impacts to riparian habitat, wetlands, or onshore groundwater resources.

Table 6.2 Applicability of subsurface feedwater supply systems

	Subsurface intake type					
Evaluation criteria	**Vertical beach wells**	**Ranney collector wells**	**HDD wells**	**Slant wells**	**Beach subsurface infiltration gallery**	**Offshore subsurface infiltration gallery**
Constructability	Technically feasible	Technically feasible	Not technically feasible unless artificially filter packed	Technically feasible	Technically feasible	Technically feasible but may be difficult to construct
Need for pretreatment	No	No	No	No	Unknown	Unknown
Environmental acceptability and permit ability	May be acceptable depending on the number and location of wells	May be acceptable depending on the number and location of wells	May be acceptable depending on the number and location of wells	Acceptable if wellheads off beach area	Not acceptable	May be difficult to permit
System capacity/production potential—able to meet feedwater supply requirements (100 mgd)	Uncertain as to maximum capacity. Requires pilot testing and modeling	Uncertain as to maximum capacity. Requires pilot testing and modeling	Uncertain as to maximum capacity. Requires pilot testing and modeling	Uncertain as to maximum capacity. Requires pilot testing and modeling	Yes assuming at least a 16–20 acre footprint	Yes assuming at least a 16–20 acre footprint
On-going maintenance required to sustain feedwater supply	Requires on-going periodic rehabilitation	Requires on-going periodic rehabilitation	Requires on-going periodic rehabilitation	Requires on-going periodic rehabilitation	May be difficult to maintain once clogged	May be difficult to maintain once clogged

	Subsurface intake type					
Evaluation criteria	Vertical beach wells	Ranney collector wells	HDD wells	Slant wells	Beach subsurface infiltration gallery	Offshore subsurface infiltration gallery
Impacts to onshore riparian habitat, wetlands or groundwater resources	No impact	No impact	No impact	No impact	No impact	No impact
Potential for entrapment and impingement of marine life	No impact	No impact	No impact	No impact	No impact	No impact
Erosion/storm impacts	Wellhead locations need to account for coastal erosion and 100 yr flood event	Wellhead locations need to account for coastal erosion and 100 yr flood event	Wellhead locations need to account for coastal erosion and 100 yr flood event	Wellhead locations need to account for coastal erosion and 100 yr flood event	Need to account for coastal erosion 100 yr flood event high storm events	May be susceptible to high storm events
Proven technology and track record	Technology well understood and proven for smaller systems	Technology well understood but not proven for large feedwater supply systems in a beach environment	Technology not proven for long artificially filter packed HDD wells	Technology proven and still evolving for large feedwater supply systems	Few systems in the world—technology still evolving	Few systems in the world—technology still evolving

- The need for pretreatment (in the case of well feedwater supply) is significantly reduced or eliminated as SDIs are typically well below the 3–5 design requirements to prevent RO membrane fouling.
- Water well technology is a very proven technology developed over the last 70 years. Application of this technology to vertical wells, Ranney wells, HDD wells and slant wells ensures success of supply and ability to maintain the supply. Well supplies yield low turbidity water that is free from entrained sand with low turbidity and SDIs.
- Beach wells, Ranney wells, HDD wells, and slant wells have no ocean construction impacts.
- Beach wells, Ranney wells, HDD wells, and slant wells tapping subsea or near-shore aquifers provide natural filtration of organic matter and suspended sediment, particularly during storm surges and periods of heavy precipitation and surface runoff.
- Both slant and vertical well intake systems can be completely buried such that there are no visible structures at the land surface, offering an acceptable solution to environmental concerns.
- Proven subsurface intake technology can be proven by scientific exploration and pilot testing.
- Subsurface intakes can be modular with a combination of systems providing the required feedwater supply.

6.3.2 Disadvantages

- Research has shown that subsurface intake systems are site specific and one system, which may work in one environment (and country), may or may not be applied to other areas. As local geologic, hydrologic, oceanographic, and environmental conditions vary widely, success or failure of a particular system may also vary widely.
- Use of subsurface Intake systems is an emerging technology with a relatively short history.
- All subsurface intake systems (wells and galleries) are in a minority as compared to open ocean intakes. As such, only a handful of successful subsurface intakes are in operation worldwide.
- In some coastal areas, obtaining permits is the number one constraint in constructing a near-shore or offshore subsurface feedwater supply. For example, in California, permits to construct the Dana Point Test Slant Well on Doheny State Beach required the following permits:
    - California Environmental Quality Act (CEQA)
    - National Environmental Policy Act (NEPA)

- California Coastal Commission Coastal Development Permit
- California State Lands Commission Lease
- California State Parks Right of Entry Permit
- San Diego Regional Water Quality Control Board
- NPDES Permit No. CAG919002
- United States Army Corps of Engineers Nationwide Permit Number
- San Diego Regional Water Quality Control Board Clean Water Act §401
- Orange County Health Care Agency Well Construction Permit
- California Department of Fish and Game Streambed Alteration Agreement

- Many of the operational intakes do not have the same stringent environmental requirements as required in the United States and especially along the coast of California.
- Depending on site conditions, there may need to be some pretreatment initially to remove iron and manganese until old marine groundwater is removed. However, this can be verified by pilot testing. Also, considering the large feedwater requirement (100 mgd), and based on Dana Point pilot testing, old marine groundwater containing iron and manganese would most probably be removed in a relatively short time period.
- During construction, visual impacts may include unsightly facilities on the beach or in near-shore areas where recreational or other high use occur.

6.4 SUMMARY OF FINDINGS

6.4.1 Findings Related to Subsurface Intakes in General

- Based on the literature review, six general types of subsurface desalination intake systems were reviewed for potential use in the Huntington Beach area.
- Subsurface feedwater supply technology is still evolving.
- All of the subsurface intake systems reviewed are relatively few in number worldwide compared to open ocean intakes.
- Impacts to offshore, onshore, surface and subsurface resources and habitat are important considerations in considering subsurface intakes.
- All subsurface intake systems reviewed would not impact marine life (i.e., impingement and entrainment).
- Subsurface sediments that make up the subsea aquifer system provide natural filtration of organic matter and suspended sediment, particularly during storm surges and periods of heavy precipitation and surface runoff.

6.4.2 Findings Related to Subsurface Intakes Using Wells

- Both offshore slant wells and near-shore vertical wells can provide desalination feedwater supplies in permeable near-shore and offshore subsea aquifer systems. The maximum capacity of these systems depends on site-specific conditions.
- The need for pretreatment in feedwater supplied from wells is significantly reduced or eliminated as SDIs are typically below the 3–5 requirement for RO membrane fouling.
- Shallow-angled slant wells obtain almost all of the feedwater from vertical leakage through the sea floor and lateral inflow from subsea aquifers.
- Unless HDD wells are artificially filter packed, they will not work as a subsurface intake system due to sand clogging of the well screen area. Experience in the water well industry show that prepacked filter systems tend to clog over time and are difficult to clean unlike a conventional artificial filter pack, which can be successfully redeveloped if necessary.
- Proven well-designed methods developed for vertical water supply wells over the past 70 years may be applied to desalination feedwater supply wells.
- Sustainability of the large feedwater supply requirement (100 mgd) would require a large number (~25–30) of 1000-ft slant wells spaced at the edge of the beach. The maximum number of slant wells would have to be determined based on pilot testing and interference modeling
- In order to maintain feedwater production, planned rehabilitation of the vertical, Ranney or slant well subsurface supply will be necessary. The frequency between rehabilitation depends on both site-specific conditions and operational schedules.
- Drawdown (i.e., groundwater-level changes) in areas of sensitive vegetation, fish habitat, or other wildlife may also restrict placement of the wells or hinder construction and maintenance.
- Slant wells constructed using the Dual Rotary method enable placement of artificial filter packs in subsea and near-shore aquifers.

6.4.3 Findings Related to Infiltration Galleries

- Difficult permitting and construction problems are expected with the Beach and offshore infiltration galleries.
- Maintaining sustainability of supply for beach and offshore subsurface infiltration galleries may be difficult due to clogging.
- Based on an infiltration rate of 0.1 gpm/ft^2, both beach galleries and offshore galleries would have to have a minimum footprint of at least 16–20 acres to achieve the 100 mgd.

REFERENCES

[1] Missimer TM, Ghaffour N, Dehwah AH, Rachman R, Maliva RG, Amy G. Subsurface intakes for seawater reverse osmosis facilities: capacity limitation, water quality improvement, and economics. Desalination 2013;322(2013):37–51.

[2] GHD. Desalination Portfolio. 2012. Available from: http://www.ghd.com/PDF/Desalination_experience_document.pdf.

[3] California State Water Resources Control Board (US). Amendment to the Water Quality Control Plan for Ocean Waters of California, Draft Staff Report. 2014.

[4] Williams DE. Design and construction of slant and vertical wells for desalination intake. In: Proceedings IDA World Congress on Desalination and Water Reuse 2011. Perth, Western Australia: IDA; 2011 Sep. Ref IDAWC/PER11-050.

[5] Hunt H. American experience in installing horizontal collector wells. In: Ray C, Melin G, Linsky RB, editors. Riverbank filtration: improving source-water quality. New York: Kluwer Academic Publishers and NWRI; 2003. [Chapter 2].

[6] Babac D, Babac P. wells with horizontal drains: theory, practice, calculation examples. Belgrade: Balby International; 2009.

[7] Whittman Hydro Planning Associates, Inc. 2006.

[8] Wampler D. Well performance and casing selection for horizontal wells. In: The Basics of Environmental Remediation Utilizing Horizontal Wells. Jackson Creek Enterprises L.C. Available from: http://www.jackson-creek.com/soil-remediation/remediation.

[9] Neodren. Catalana de Perforacions SA; 2003. Available from: http://www.catalanadeperforacions.com/en/neodren.

[10] Williams DE. Results of drilling, construction, development, and testing of Dana Point Ocean Desalination Project Test Slant Well. Desalination and Water Purification Research and Development Program Report, US Department of the Interior, Bureau of Reclamation, Technical Service Center, Water and Environmental Services Division, Water Treatment Engineering Research Team, Washington, DC; 2009.

[11] Schwartz J. Beach well intakes for small seawater reverse osmosis plants. Project 97-BS-015 of the Middle East Desalination Research Center, 2000.

[12] Oceaneering Technologies. Intake/outfall structure for proposed desalination facility—draft final report, phase I—preliminary design. Prepared for CCSD 1994.

[13] Driscoll FG. Groundwater and Wells. 2nd ed. Saint Paul, MN: Johnson Screens; 19861089.

[14] Pankratz T. In: Seawater desalination technology overview. Presentation for Georgia Joint Comprehensive Desalination Study Committee, St. Simons Island, Georgia; 2006. p. 22–3.

[15] Jamaluddin AT, Hassan AM, Reweli AA, Saeed MO, Bakhit LM, Al-Amri MM. Operation of Al-Birk plant introducing beachwell intake system. Technical Report No. TR 3807/APP02005, Saline Water Desalination Research Institute, Al-Birk SWRO Plant; 2005.

[16] Saeed MO, Jamaluddin AT, Tisan IA, Lawrence DA, Al-Amiri MM, Chida K. Biofouling in a seawater reverse osmosis plant on the Red Sea coast, Saudi Arabia. Desalination 2000;128(2000):177–90.

[17] Voutchkov NSWRO. Desalination: on the beach—seawater intakes. Filtr Sep 2005;10(2005):24–7.

[18] Fariñas M, López LA. New and innovative seawater intake system for the desalination plant at San Pedro del Pinatar. Desalination 2007;203(2007):199–217.

[19] Charette MA. In: Natural isotope tracer study: test slant well phase 3 extending pumping test—South Orange Coastal Ocean Desalination Project. Prepared for the Municipal Water District of Orange County by Coastal Groundwater Consulting; 2012.

[20] California Coastal Committee. Establishing development setbacks from coastal bluffs. 2003. Memorandum W11.5.

[21] California Ocean Protection Council. Resolution of the California Ocean Protection Council on sea-level rise. 2011.
[22] California Ocean Protection Council. State of California sea-level rise guidance document. 2013.
[23] Wang S, Allen J, Tseng T, Cheng R, Carlson D, Henson J. In: Design and performance update of LBWD's under ocean floor seawater intake and discharge system. Black & Veatch presentation at the Alden Desalination Intake/Outfall Workshop, October 16; 2009.
[24] Tseng TJ, Cheng RC, Wattier KL. In: Pilot and demonstration testing of subsurface filtration for seawater desalination. LBWD Presentation for the American Water Works Association 2008 CA-NV Section Fall Conference, October 22; 2008.
[25] Zhang Y, Tseng TJ, Andrews-Tate C, Cheng RC, Wattier KL. In: LBWD's experience with subsurface intake as pretreatment for seawater desalination. Presentation for the 2012 Water Quality Technology Conference November 6; 2012.
[26] Fukuoka District Waterworks Agency. Available from: http://www.f-suiki.or.jp/english/seawater/.
[27] WDR. WDR visits Fukuoka. Water Desalination Report, 50(2):2014.
[28] Shimokawa A., Desalination plant with unique methods in Fukuoka. Fukuoka District Waterworks Agency, Available at: https://www.niph.go.jp/soshiki/suido/pdf/h21JPUS/abstract/r9-2.pdf.

CHAPTER 7

Brine Disposal and Management—Planning, Design, and Implementation

Ozeair Abessi
Babol Noshirvani University of Technology, Babol, Iran

7.1 INTRODUCTION

Today 60% of world's population lives in coastal areas (200 km from the sea) [1]. Because of the expansion in international trade, natural attraction, and population growth in coastal cities, this ratio increases each year. Most of this population also lives in arid areas that intensely suffer from water shortage. It is accepted worldwide that the desalination technology can be a reliable way of answering this growing demand [1a]. Due to technological advances nowadays fresh water can be produced out of seawater in large volumes to meet the requirements of domestic and municipal supplies. Desalination plants use a variety of technological processes, e.g., thermal processes such as Multistage Flash (MSF) plants, or membrane processes such as Reverse Osmosis (RO) to produce desalinated water removing salts from brackish or seawater. Desalination plants are taking out the seawater into inland using marine intakes and utilize large amounts of energy to remove salt while producing concentrate as byproduct [2,3]. The main environmental impact of desalination projects is actually associated with the disposal of the concentrated brine into adjacent environment. Here in this chapter, different methods developed for brine disposal and management for coastal desalination plants were addressed and recent developments in modeling and design are discussed.

7.2 DESALINATION TECHNOLOGY

Desalting seawater goes back to long time ago in history to Greek civilization and Rome Empire [4]. The techniques of desalination, such as evaporation and distillation, were commonly used in the mid-1900s.

Sustainable Desalination Handbook
https://doi.org/10.1016/B978-0-12-809240-8.00007-1

In the 1940s during the Second World War, thermal desalination process was one of the major sources to supply potable water to troops in arid areas [5]. Regarding the rises in population especially in coastal cities around the world, commercial desalting units, mostly thermal processes, were developed to provide potable water demand raise after World War II [6]. As a result of lower production rate of thermal methods, osmotic processes to desalinate water attracted a lot of attention after 1970s. In this era, development efforts mainly focused on commercial membrane processes, such as Reverse Osmosis (RO), to make these technologies more efficient and cost effective. Desalination technology became a fully commercial investment by the 1980s, and many important cities especially in the Middle East and Southern Europe become dependent on desalination plants for their municipal water supplies by the end of 1990s [3]. It is worth noting that during the 2000s the major desalination processes were membrane and thermal technologies; however, proving capability of osmotic processes, RO technology took over and currently is approximately 60% of installed capacity of costal desalination plants around the world [7]. Between thermal technologies multistage flash and among membrane processes, Reverse Osmosis has been used more in desalination industries all these years. MSF thermal plants work with small conversion rates (10%–20%), so they need greater amounts of feed water to produce the same amount of distilled water [1a]. The MSF plants need more water intake, larger marine structures, and much more energy and chemical additives [8]. The byproducts of this process are greenhouse gases and wastewater, which is highly polluted with chemical additives, corrosion products, hot temperature and salt, hardly affecting water quality of the receiving water body. The wastewater would be lighter or even heavier than the receiving ambient water. It will affect the process of mixing and entrainment while disposing into adjacent water body. The visual impact is also significant because of the large amount of piping, tanks, and chimneys associated with such plants [8a]. RO plants work with higher conversion rates, i.e., about 40%–50% and require lower amount of feed water. Energy consumption is high but still lower than in MSF plants. The byproducts are greenhouse gas and concentrate water, which has no thermal pollution but still has chemical additives and is particularly hypersaline, which makes it denser than seawater. RO plants do not include combustion processes resulting in no direct air pollution but have unfavorable sound pollution due to the pump operations. Its visual impact is less because the plants are usually compacted. However, an additional solid waste is generated by RO plants compared to those of MSF, since membranes need to be changed at a certain frequency and at the moment they are not reusable [8a]. RO plants also require high level of pretreatment due to the nature of osmosis process. So variety of chemicals and additives are used during the desalination process. Additives are chemicals used for biofouling control (e.g., chlorine), scale control (antiscalants), foam reduction, and corrosion inhibition and finally discharge into the coastal waters as contaminants [9]. Higher chemical content and extremely dense brine (more than twice salty as seawater) produces in these plants make the environmental impacts of RO plant even more problematic when it comes to brine discharge in aquatic sensitive coastal ecosystems.

7.3 ENVIRONMENTAL CONSIDERATIONS AND RULES

During last decade, developments in desalination technologies are mainly focused on reducing energy consumption and production costs. Becoming more economical and versatile, desalination industry is now paying a great deal of attention to environmental considerations, as environmental consequences have become increasingly important in the siting and construction of new plants [3]. New technologies are being used successfully to decrease energy consumption and lessen disruptions to marine life during the intake and outfall processes. As the future trend of installation confirms, RO is becoming a dominant technology all around the world, which besides the fresh water, dramatically produce brine that needs to be discharged back to the environment [9]. As a result, environmental safeguards become more associated to impacts of desalination brine on marine environment. Because it is denser than seawater, brine produced in RO desalination plants upon disposing into the sea, spreads out over the sea floor and it increases the risk of negative effects on stenohaline benthic ecosystems. The long-term effects of hypersaline stress on *Cymodocea nodosa* and *Posidonia oceanica* seagrass meadows in Mediterranean region and the Canary Islands are reported by Procaccini et al. [10] and Portillo et al. [11]. They argued that the impact of the hypersaline plume of desalination discharge caused the loss of these habitats in large area. Similar damages are also reported for other benthic organisms in different regions [12,13].

7.3.1 Siting of Costal Desalinations

Siting is the first step in minimizing the environmental impact of coastal desalination plants. Environmental and hydrogeological factors such as bathymetry, waves, currents, depth of the water column, sediment, and tidal

Table 7.1 Sensitivities of marine habitats to desalination plants

Sensitivities	Marine habitats
1	High-energy oceanic coasts, rocky or sandy, with coast-parallel current
2	Exposed rocky coast
3	Mature shoreline (sediment mobility)
4	Coastal upwelling
5	High-energy soft tidal coast
6	Estuaries and estuary-similar
7	Low energy sand-, mud-, and beach rocks-flats
8	Coastal sabkhas
9	Fjords
10	Shallow low-energy bay and semienclosed lagoon
11	Algal (cyanobacterial) mats
12	Seaweed bay and shallows
13	Coral reefs
14	Salt marsh
15	Mangal (mangrove flats)

regime are the characteristics of the marine environment that needs to be considered in a siting. These factors would determine the extent of the mixing of the brines, the quality of intake water, and the impact of costal structures on marine biota and define the extent of the vulnerability of the marine environment from the place to place [14]. This is measured by the nature of the marine habitat, e.g., coral reef, rocky beach, or sandy surfaces and by the origin of the surrounding organism. Hopner and Windelberg [14a] divide the global marine habitats into 15 categories according to their sensitivities to the effects of desalination plants, shown in Table 7.1. Maximum distances from the sensitive habitats are recommended.

Desalination plants should also be away from tourism developments and populations. Coastal zones, because of their landscape, importance for fishery, tourist housing and recreation, are always been on favor. Any activity, constructional and operational, for the purpose of desalination of sea water, based on the environmental regulation of many countries is permitted only after strict Environmental Impact Assessment (EIA) studies. Safe distance from Eco-Sensitive Zones (ESZs), inland, or in aquatic ecosystem should be considered. Environmental management and monitoring programs should also be utilized widely in air, soil, and aquatic media.

7.3.2 Brine Disposal

Brine is by-product of almost all types of desalination plants. The amount of brine and its concentration varies depending on the desalting method, salinity of feed water, and conversion rates. World's largest desalination plants are coastal; to use the infinite source of sea water, while inland desalination plants are smaller in size and attracted less attention, seemingly is not the future trend of big desalination plants. Desalination process produces 10%–50% of the total feed water flow as the brine, depending on the technology used [15]. The concentration of produced brine, based on the process, varies from 10% to twice of the feed water concentration. The produced brine essentially needed to be discharged back to the environment from the perspective of its economic and environmental priority compared to the other methods (evaporation ponds, deep well injection, or disposing to sewage system) [16]. Adjacent water bodies, e.g., rivers, estuaries, and ocean are the main receiving environment that by nature can tolerate great amount of hot temperature and salt due to their high capacity of assimilation [17]. So if adequately diluted in the receiving water body, brine can be safely discharged back to the environment. Efficient mixing of brine can be obtained by marine outfalls. The dilution that results can rapidly reduce concentrations of possibly toxic materials contained in the effluent to a safe level [18]. When discharging flow into ambient water, hydrodynamically three regions can be considered near to the source [19] (Fig. 7.1). First near field in the vicinity of the discharge point that is characterized by high dilution rate, mainly depends on the discharge configuration and the properties of effluent and ambient water. The turbulence that was created by the shear layer along the near field controls jet initial mixing, and it decreases the concentration of discharge in a short distance from the source. Secondly, far-field region, further away from the discharge point, where mixing depends on the ambient conditions and background turbulences of the environment. The stratification of ambient water, solid boundaries, currents, and waves is influential in the far field and dilution rate is small in this region. And finally, far from the source, the region of large-scale flushing where the long-time flushing causes accumulation of

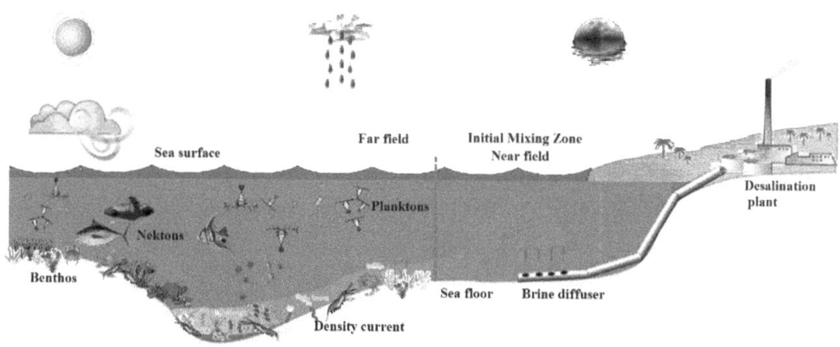

Fig. 7.1 Schematic of a coastal desalination plant and brine disposal system.

pollutants and weak transport of substances, which are controlled by the net flow-through, tidal cycles, up and down wellings [20].

Brine discharge system especially for larger desalination plants typically is a direct disposal system in shore bank that extends up to some distance into the sea in gravel or sandy beaches or rocky shores. The brine outfall system can be a surface shoreline channel, overflow spill from cliff, discharge on a breakwater, submerged single jet, multiport diffuser, or multiple rosette discharges. In Fig. 7.1, the schematic of brine disposal system for a coastal desalination plant is shown.

The outfall, therefore, is a hydraulic structure designed to minimize discharge environmental impacts by optimal mixing while minimizing the construction costs too. The outfall structures are significant components of the overall cost of desalination plants, so it is desirable to keep them as short as possible while meeting the dilution requirements. The factors affecting the choice of a disposal method for desalination plants are brine flow rate and concentration, imposed by the size of the plant and desalination process and also the condition of receiving marine ecosystem, currents, waves, depth, regional flushing, and transport features [19]. The design of the outfall system determines the degree of brine dilution accessible in the near field by engineering control of the flow momentum and buoyancy and discharge configuration, to accomplished ambient water quality standard at the end of regulatory mixing zone. The objective is rapid dilution and mixing of the concentrate to reduce salinity to near background levels. Water quality requirements for desalination concentrate discharges and the literature on the environmental impacts of brine were recently reviewed in SCCWRP [21]. There are not many regulations, standards, or guidelines for brine disposal around the world. Based on SCCWRP [21], some international regulations for brine discharges and the point of compliance for each are listed in Table 7.2.

7.4 BRINE DISCHARGE MODELING AND DESIGN

To environmentally assess the efficiency of discharge and satisfaction of quality standards, it is essential to simulate the behavior of flow in various conditions. For the design purposes, experimental and mathematical models (numerical and integral) are commonly used for brine discharges. There are several water quality models developed to simulate the behavior of brine flows, more based on numerical methods. These models are less expensive and if correctly calibrated can be used on any type of discharge

Table 7.2 International regulations for brine discharge

Region/authority	Salinity limit	Compliance point (relative to discharge)	Source
Iran DOE	Increment ≤10%	200 m	Iran Environmental Law [21a]
Oman	Increment ≤2 ppt	300 m	Sultanate of Oman [22]
Abu Dhabi	Increment ≤5%	Mixing zone boundary	SCCWRP [22a]
US EPA	Increment ≤4 ppt	–	US EPA
Carlsbad, CA	Absolute ≤40 ppt	1000 ft	San Diego Regional Water Quality Control Board 2006
Huntington Beach, CA	Absolute ≤40 ppt salinity (expressed as discharge dilution ratio of 7.5:1)	1000 ft	Santa Ana Regional Water Quality Control Board 2012
Western Australia guidelines	Increment <5%	–	–
Perth, Australia/ Western Australia EPA	Increment ≤1.2 ppt at 50 m and ≤0.8 ppt at 1000 m	50 m and 1000 m	SCCWRP [22a]
Sydney, Australia	Increment ≤1 ppt	50–75 m	SCCWRP [22a]
Gold Coast, Australia	Increment ≤2 ppt	120 m	GCD Alliance [23].
Okinawa, Japan	Increment ≤1 ppt	Mixing zone boundary	Okinawa Bureau for Enterprises

configuration or ambient water scenarios [1a]. Numerical modeling had an important role in the advancement of our understanding of physical processes over the last few years and currently is essential prediction tool for designing brine discharges, controlling the mixing, and optimizing the dilution [24].

Experimental study was indeed the first attempt that lit up the complicated dynamics of turbulent jet and plumes [24a]. Laboratory experiments on scaled physical models, e.g., smaller scale of the prototype can be carried out on any effluent, discharge configuration, and ambient conditions [25]. Experimental studies are essential to identify the physical behavior of flow and can be used to develop and validate the mathematical models too [18]. Experimental studies also can be used to develop empirical equations and standards for design. There are also computer models developed based on the empirical equations that are simple to use and accurate enough for engineering design purposes.

7.4.1 Numerical Models

Water quality models mathematically solve the equations of transport and transformation for the effluent discharge. The equations represent the physical and chemical mechanisms and are solved to determine the behavior of discharge into the various conditions of receiving water. For the pollutant in the brine, i.e., mainly salt, the mechanism of transformation is not distinctive; however, regarding the flow turbulences, diffusion due to fluctuation of turbulence is the fundamental mechanism. There are several models developed to solve the hydrodynamic and transport equations while adapted to a negatively buoyant effluent [1a]. The equations can be set up in Lagrangian or Eulerian framework and can be solved with different numerical methods. Using the enormous range of simplifying assumptions to the governing partial differential equations of flow, e.g., Mass conservation, Momentum conservation and Transport equations, can be solved in a simple up to an extremely difficult way [26,27].

Based on the analytical simplification assumed, there are two different types of numerical models developed for the brine discharge, i.e., Models based on integration of differential equations along the cross section of flow (Integral Models) and Hydrodynamic numerical models (Computational Fluid Dynamics models).

7.4.1.1 Integral Models

Integral models solve the hydrodynamic governing equations of the flow while integrated over the cross section, assuming the cross-sectional distribution is Gaussian priori. The model then transforms the equations into simple ordinary differential equations that can be easily solved with Runge Kutta formula (a solution method of numerically integrating of differential equations). The model solves the equations stepwise along the general trajectory of each individual jet. The receiving fluid requires assuming unbounded ambient environment; consequently, boundary effects cannot be modeled. The unlimited receiving water body can be non or uniformly stratified under stagnant or steady sheared current conditions [28]. Integral models are based on analytical simplification of very complex turbulent mixing phenomena; hence, they usually employ a semiempirical entrainment closure approach with the coefficients obtained from the experiments, able to simulate the three-dimensional hydrodynamics of the mixing process. These models predict steady-state plume trajectories up to the regulatory mixing zone, end of initial mixing zone or closest boundaries and cannot provide details of the mixing behavior [29]. There are several models developed for the marine discharge modeling based on this approach, some more commercial with the prediction engine of either Lagrangian or Eulerian framework.

CORJET developed by Gerhard Jirka in 1990s is a Fortran program based on Eulerian approach and constitutes turbulence closure coefficients for the complex turbulent mixing process that is obtained experimentally [28,30]. CORJET provide a mechanistically sound, accurate, and reliable model for complex mixing processes of buoyant jet. The CORJET model is imbedded as a computational engine into the CORMIX, which is an expert-system user-interface water quality model developed by Doneker and Jirka [31], for the design and prediction of single and multiport diffuser discharges (CORMIX.info). CORMIX has been supported by the United State Environment Protection Agency (USEPA) Center for Exposure Assessment Modeling (CEAM) for a long time and contains three major subsystems, e.g., CORMIX1, CORMIX2, and CORMIX3 developed for submerged single port, submerged multiport, and positively buoyant surface discharges, receptively (available in www.mixzon.com). D-CORMIX is the extension of CORMIX that modified, improved, and validated for water quality modeling of dense materials and dredge disposal plumes [32]. Several discharge configurations are considered in D-CORMIX, including submerged, surface, and above-surface discharges. D-CORMIX using a hydrodynamic classification scheme and the mirror images of positively buoyant flow regime simulate the behavior of brine discharge in various conditions [32,32a]. There are studies seriously questioned the accuracy of D-CORMIX classification especially for surface discharges [33,34].

UM3 is a three-dimensional integral model that was developed based on UM [35] and UMERGE models [36] for simulating single- and multiport submerged discharges. UM3 is a Lagrangian model that features the Projected-Area-Entrainment (PAE) hypothesis [37] for the initial mixing calculation in near field. UM3 is coded in Delphi Pascal and fully integrated with Visual Plumes package that is coded with the same language [38]. Besides UM3, The DKHW model; the Eulerian integral method for single- and multiport submerged discharges; the PDS; another Eulerian model for the surface discharge; and NRFIELD, which is an empirical length scale model for multiport diffusers, are embedded in Visual plumes [26].

This software is a free access software developed by the USEPA and with its submodels is capable to simulate positively, negatively, and neutrally buoyant discharges into stagnant and flowing ambient water. Even though the Visual Plumes' integral submodels do not simulate the interaction of the flow with the boundaries, the software is not strictly limited to near-field region and Brooks far-field algorithm developed for far-field simulation, although the accuracy of this has been doubted [24].

JETLAG is an integral model developed based on Lagrangian entrainment formulation by Lee and Cheung [39] to simulate the mixing and three-dimensional behavior of buoyant jets in ambient water. The model simulates the key physical processes represented in the governing equations using the Projected-Area Entrainment (PAE) hypothesis (suggested by [37]). VISJET is a Windows-based flow visualization tool with JETLAG as modeling engine. The model was developed by Lee and Cheung [39] at the University of Hong Kong intended to facilitate and ensure the outfall design process. It is claimed that the model has been tested extensively against theory, laboratory observation, field verification studies and applications [39a]. The model is limited to near-field region modeling and does not simulate the interaction of the flow with boundaries [24].

These commercial integral models are extensively used in marine outfall design, while most of them were initially developed for the positively buoyant discharges and later adapted for negatively buoyant flow. Hence, the common shortcoming of these models is the lack of validation studies for negatively buoyant jets, which make the use of these models doubtful for brine discharges. Regarding the higher level of simplification and plane mathematical works for the equation solving, these models are quite easy and simulation needs few input data and no calibration is usually applied.

Recently, a comprehensive validation study was carried out by Palomar et al. [24] to estimate the limitation and the advantages of these software for desalination brine discharge modeling. It is argued that for single-port discharges, CORJET, UM3, and JetLag integral models lead to better results rather than the CORMIX1 and CORMIX2 subsystems, as they use very simplified formulas and have not been validated for negatively buoyant discharges. Palomar and her colleagues also do not recommend these models for the discharge angles under 30° and over 75°, since COANDA and re-entrainment are not taken into account. It observed that CORJET, UM3, and JetLag underestimate the geometric features and dilution of the brine jet in all cases. CORJET, in general, gives better result for the geometrical properties of dense jet, while UM3 provides the greatest differences.

For jet mixing and entrainment, dilution calculation at impact point is greatly underestimated (up to half time) by all the models in all cases. Integral models are also limited in their ability to characterize the dynamic interaction between jets from multiple ports in diffusers although some progress has been reported recently [40]. So even though these models are simple, fast, and accurate enough for the engineering design for positively buoyant discharges, their applications for brine discharge is under serious doubt. For more detailed analysis, see Palomar et al. [24].

7.4.1.2 Computational Fluid Dynamics Models

Computational Fluid Dynamics (CFD) is the art of solving a set of algebraic equations using digital computers. CFD is an interdisciplinary research area that uses a mathematical method to solve physical problems and only due to recent progress in computer performances it became possible for engineering problems. CFD numerically solves the partial differential equations of the fluid motion, which represent conservation laws for the mass, momentum, and energy to provide a quantitative prediction of systems involving fluid flow [41]. It enables scientists and engineers to perform numerical experiments of the problem without need to build physical models or consider high simplifying assumptions for the governing equations as is common for the Integral models (mentioned earlier). There are a number of different solution methods used in CFD, e.g., finite volume, finite difference, finite element, meshless techniques, etc. These methods are used to solve the set of partial differential equations of fluid motion known as the Navier-Stokes equations together with the equations describing other processes, i.e., turbulent dispersion and chemical reactions [42]. Using these techniques, several software were developed with advanced solver and powerful pre- and postprocessing capabilities for steady-state and transient flows, in both laminar and turbulent conditions. Discharges of desalination brine like many other engineering flows are turbulent. Robust and accurate modeling of turbulence is vital for the simulation of brine flow mixing and entrainment in near field. Complex processes of turbulence are often modeled using empirical relationships, while there are some sophisticated models developed based on the turbulence modeling theories.

A turbulence model is a computational procedure to close the system of mean flow equations [43]. The turbulence models provide broad range of applicability, accuracy, and robustness. There are several models developed for this purpose. Classical models that were developed based on the Reynolds Averaged Navier-Stokes (RANS) equations are generally accepted for engineering purposes [44]. These models categorize into Zero-equation, one-equation, and two-equation models, while Direct Numerical Simulation (DNS) and Large Eddy Simulation (LES) existed too. Direct Numerical Simulation (DNS) as it sounds directly resolves the whole spectrum of turbulent scales and is the most accurate method available. Due to the high computational cost especially for high Reynolds number, DNS is not still affordable for practical engineering problems. Large eddy simulation (LES) solves large eddies directly, while small eddies are simulated with a subgrid model [43]. LES is attracting more attention these days [45]. Between various turbulence models, no single model is universally accepted for all problems due to the capabilities and limitations of each. How to choose an appropriate turbulence model for a given problem strongly depends on the physics of the problem, the level of accuracy required, the available time and computational resources, and many other factors.

CFD software like Ansys FLUENT, CFX, and Openfoam are the most advanced commercial models available today for the fluid flow simulations. All these models can be used for the modeling of complex three-dimensional features of brine flow in near field. The application of CFD tools to predict the behavior of dense discharge has been reported by Vafeiadou et al. [46]. They employed the commercial computational fluid dynamics package (CFX-5) to simulate inclined negatively buoyant jets discharge at 45°, 60°, 70°, 80°, and 90° to the horizontal. For turbulence closure, the Shear Stress Transport (SST) turbulence model was used. They concluded that the model underestimated slightly the terminal rise height; however, the return point calculated was considerably lower than experimental data. They overall suggested that this model can properly predict the behavior of inclined dense discharges. Other scholars have also investigated the application of CFD models for the brine discharge simulation. Oliver et al. [47] compared the predictions of k-ε model with the experimental data for inclined dense jet. The k–ε model as a part of CFX package has run for the standard and the calibrated forms (where the turbulent Schmidt number is modified) to predict the behavior of 60° inclined dense jet in stationary water. They concluded that the overall effect of the calibrated approach on the quality of the predictions is not tangible and both approaches predict fairly good the trajectory of jet. However, mean-integrated concentrations at the centerline maximum height were overpredicted.

Gildeh et al. [50] also applied a CFD model, OpenFOAM, to simulate 30°- and 45°-inclined dense jets along the near field. OpenFOAM is an open-source CFD code that is developed based on Finite-Volume Method for complex fluid dynamics. Two Reynolds-Averaged Navier–Stokes turbulence models, i.e., the realizable k-ε and the Launder Reece Rodi model were used to evaluate the accuracy of the numerical predictions for these discharges. Reasonable agreement was found between numerical results and previous experimental studies. Later, the same authors (2015) investigated the accuracy of five different turbulent models: two Linear Eddy Viscosity Models (RNG k–ε, and realizable k–ε), one Nonlinear Eddy Viscosity Model (nonlinear k–ε), and two Reynolds Stress Models (LRR and Launder–Gibson) for brine discharge through 30°- and 45°-inclined nozzle. Based on the comparison analysis of their results with the experimental and analytical studies, it has been reported that the LRR and realizable k–ε turbulence model predicts the flow geometry and dilution better among the other models. Davis et al. [48] also utilized FLUENT, famous commercial software for fluid dynamic simulation, to model mixing zone of a marine diffuser and tried to demonstrate the promises that existed for applying CFD models for initial mixing zone studies.

Computational fluid dynamics has always been an expensive tool to be widely used in engineering applications. It consumes high computational resources and time and needs extensive information and data; it also requires experts for modeling and is not completely independent from the experimental data. Recent advancements in numerical methods and computing systems make it more feasible to use CFD software armed with the complex turbulent models for the prediction of flow initial mixing in near field. Recent works illuminated the challenges engineers are faced with the application of right turbulence model for brine discharge modeling [49]. However, the full potential of CFD models as a powerful engineering tool for the simulation of complex discharge problems is yet to be explored [49]. These difficulties, however, make CFD tools a better option for final stages of engineering design and optimization of near-field processes [50]. So experimental equations and Integral models still seem reasonable tools for brine discharge modeling with adequate accuracy and computation cost for engineering applications.

7.4.2 Experimental Models

Experimental studies have formed the basis for much of our knowledge of the dynamics of plumes and jets. They are also essential in validating and developing mathematical models [25]. First serious attempts in laboratory studies started in early 1960s to provide better understanding of jet and plume behavior [51–53]. Conducting experiments in variety of initial and discharge conditions ultimately provided an empirical basis for the developing length scale and semiempirical dilution models for marine outfall design system. Most of current outfalls are designed using these models for near-field analysis. The evolution of three-dimensional CFD models is changing this trend and the CFD models are becoming more popular tool for optimization of engineering design for near field. To model far field, CFD models show even more robustness and comfort.

In spite of the recent advent of progressive CFD tools, empirical and integral models (i.e., RSB, VisualPlumes, CORMIX and VISJET) are still used for initial design stages and have exhibited to be accurate enough for final design applications of positively buoyant effluents. However, their accuracy and proportion for the modeling of negatively buoyant discharges is still in doubt. These models were developed based on analytical simplification of mixing process and huge data of laboratory observations.

In experimental studies, physical models of smaller scale of marine outfalls were tested on any effluent, discharge configuration, and ambient conditions. Geometric and kinematic similarities are guaranteed by scaling magnitudes, and dynamic similarity is considered to be achieved when the Densimetric Froude number remains the same between the model and the prototype [24]. The results are then non dimensionalized and plotted in the form that covers a wide range, typical of those expected in reality or field. Engineers can use these nondimensional graphs to predict near-field mixing and the geometry and dilution of waste field. So far various measurement techniques have been used for the flow visualization in laboratories. Visualization started with simple process of adding dye to discharge to illustrate the mixing processes and later expanded to application of on-time thermal and conductivity probes. More recently, optical techniques and digital image processing have been used by the scholars for the visualization. The accuracy of captured length scale is up to camera's resolution and time scale is down to the speed of frame capturing (frame rate). Therefore, after a short history of using dye and probes, currently optical and digital techniques such as Light Attenuation (LA), Light Intensity (LI), and Laser-induced Fluorescence (LIF) are generally accepted to be used for this process. These techniques facilitated and refined the experimental works and are accurate enough to be touchstone for the comparison of computer model results. LA is based on the attenuation of light passing through a dyed solution while discharging in front of a white-light source. This technique was developed by Cenedese and Dalziel [54] and widely used by Zhang and Chu [55], Gaskin et al. [56], and Kikkert et al. [57]. LI is based on the analysis of light intensity of dyed solution captured in gray scale by a digital camera in front of a white-light source in dark room. It has been used by Abessi et al. [58,59] and Saeedi et al. [34]. However, the most sophisticated and versatile technique for marine outfall studies is Laser-induced Fluorescence (LIF). In LIF technique, a fluorescent tracer dye is added to the flow and a laser causes the dye to fluoresce and emit light, which is captured by the camera. Quantitative concentration for tracer can be obtained only after a process of calibration. Detailed descriptions of LIF are given in many publications [19,59a]. This technique is mostly used in 2D or Planar Laser-Induced Fluorescence (PLIF) where only a plain in jet centerline is illuminated. However, it was upgraded to Three-Dimensional Laser-Induced Fluorescence (3DLIF) by Tian and Roberts [60]. In 3DLIF, a laser sheet is swept horizontally at high speed through the flow and sequential LIF images captured. Then it is visualized to show three-dimensional distributions by computer graphics methods [61]. Currently 3DLIF is the only visualization technique available to illustrate the three-dimensional mixing processes of flow in near field. The ability of this system to disclose the 3D complex behavior of some regimes that have never been reported before is discussed in Tian and Roberts [60,62] and Abessi and Roberts [18].

The methods of brine discharge vary from submerge (single-port, multiport, and rosette-shaped diffusers) up to surface and also over surface discharges. Following, experiments on each, in various conditions of ambient water, are reported. The results of the experiments have led to some de facto and global standards for the design of brine discharge system.

7.4.2.1 Single-Port Discharges

The experiment on the dynamics and mixing of dense flow goes back to about 50 years ago [63,64]. Upwardly inclined jets were discovered to be a common way to dispose brine concentrate back into the sea from the coastal desalination plants [64]. The simplest case is the single vertical jet called fountain but vertical jet falls back onto itself when discharged into a stationary environment, resulting in lowered dilutions [65]. When the jet is discharged vertically at some intermediate angle to the horizontal, negative buoyancy causes the jet to reach a terminal rise height and then fall back on the lower boundary where it spreads as a density current (Fig. 7.2). Inclined jet so results in lower dilution compared to vertical jets and clears waste field from the source site; hence, it has become accepted for brine discharge due to its advantages [19]. Different aspects of dense jet have been investigated during last decades. Researchers have been trying to find the

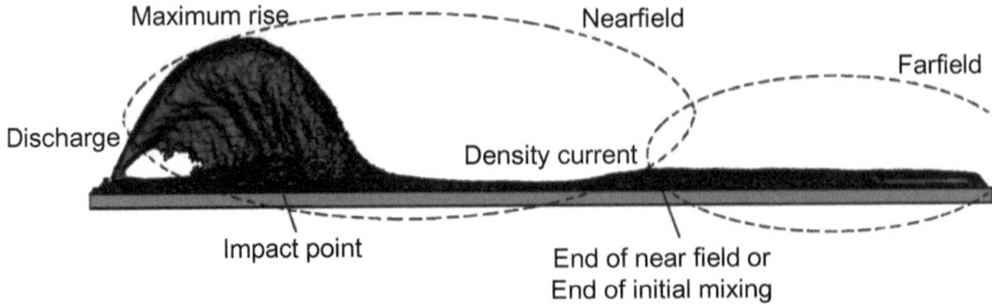

Fig. 7.2 Near-field and far-field regions on a single inclined dense jet in stationary water.

optimal angle of inclination for single-port dense discharges. Cederwall [63] commented that 45° has to be the optimal. Zeitoun et al. [64] measured dilution for various inclinations and concluded that a 60° angle gives the longest trajectory and highest dilution compared to 30°, 45°, and 90° discharges. Thereafter, the 60° nozzle became accepted as the de facto standard for dense discharges, although this has been questioned in later studies by Jirka [66] and Lai and Lee [67]. Recent experiments on the prediction of nozzle angle effect have been performed by Nemlioglu and Roberts [68], Kikkert et al. [57], Jirka [66], and Papakonstantis et al. [69,70], Lai and Lee [67]. Conducting a theoretical study, Jirka [66] challenged the previously accepted standard for optimal angle of nozzle for dense discharges. He argued that the existing design practice of 60° is based on limited and outdated laboratory data for dilutions at the level of maximum rise and is not reliable. Based on parametric application of the numerical model CORJET, Jirka suggested that a flatter discharge angle of about 30–45° above horizontal may be the optimal design. In other studies also, low dependency of dilution on nozzle inclination over the range 45–75° is reported [67–70], without a clearly defined optimal design angle. So even though the relatively steep discharge angle of 60° is accepted as the standard, there was no universal agreement up to Abessi and Roberts [71,72] recent works.

Most of these experiments investigated flow properties up to the return point or boundary impact points. There are few measurements reported beyond the impact point, to consider the end of near field where water quality standards actually must be met and there are no measurements of the effect of nozzle angle on near-field dilution too. Abessi and Roberts [71,72], in part of an extensive program of systematic experiments on brine discharge, reported the flow properties as a function of nozzle angle, for 15–85° in increments of 5° (Fig. 7.3). They used the sophisticated 3DLIF techniques to study flow geometry and dilution for inclined dense jets up to the near field. Normalized expressions were derived and plotted along with

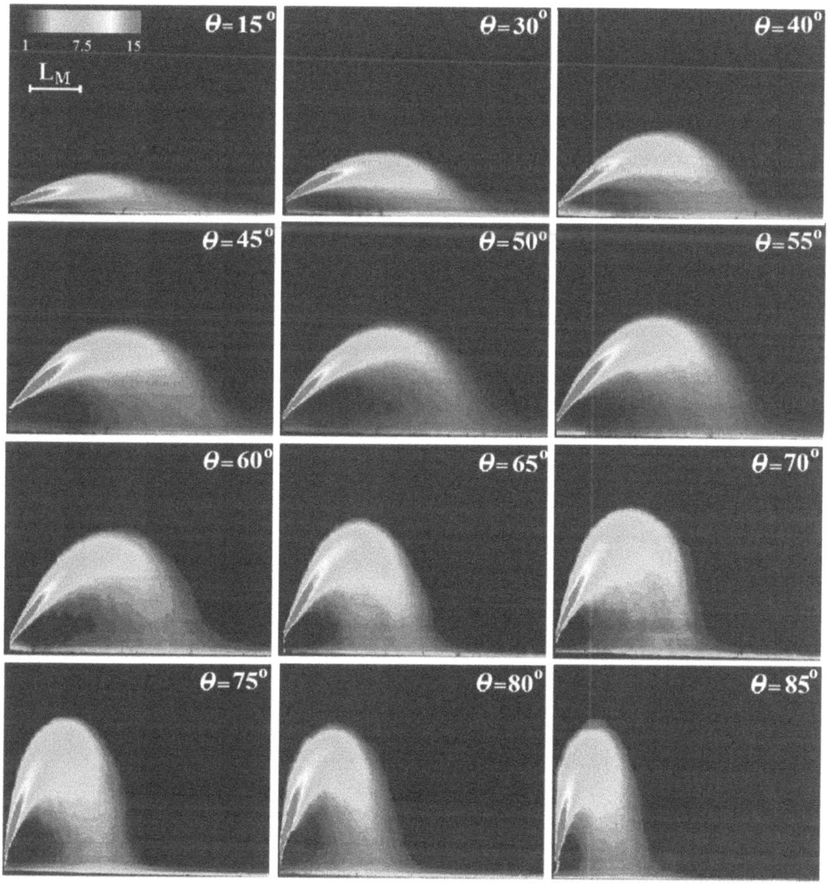

Fig. 7.3 Tracer concentrations (μg/L) at central plane for various nozzle angles of 15–85°.

previously reported data. The length of the near field and its dilution were also measured for various nozzle angles.

Based on the dimensional analysis, known for experimental analysis [19,73], the geometrical distances (χ) for a single dense jet can be nondimensionalized as:

$$\frac{x}{dF} = Con. \tag{7.1}$$

And dilutions as:

$$\frac{s}{F} = Con. \tag{7.2}$$

where d is nozzle diameter, F is discharge densimetric Froude number, and S is dilution defined as $S = c_0/c$, where co is the effluent tracer concentration and c the local time-averaged tracer concentration. The dilution at the location of impact point and the end of near field is plotted in Figs. 7.4 and 7.5 in nondimensional form (S/F) for different angle. As shown, dilution at the location of both near field and impact point was relatively insensitive to nozzle angle over the range of 45–65°; however, the near-field dilution was more sensitive to nozzle angle than impact dilution. The maximum dilution was found to be occurring at an angle of about 60° that confirms the

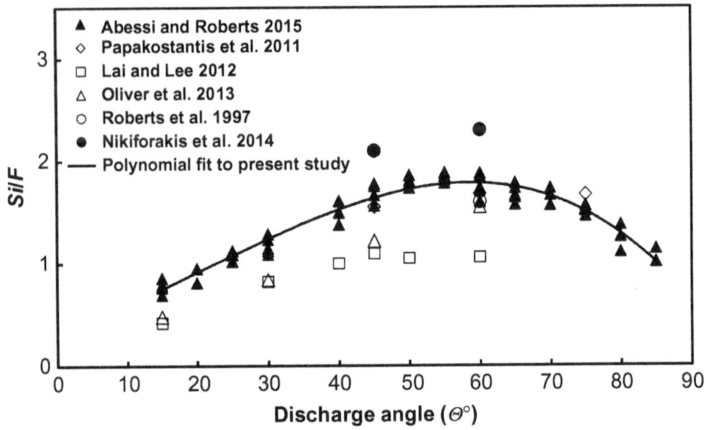

Fig. 7.4 Impact point dilution for various nozzle angles, comparison with previous studies.

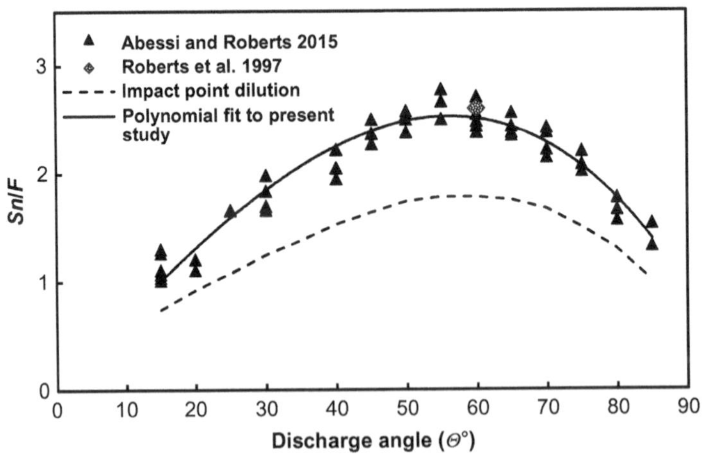

Fig. 7.5 Near-field dilution for various nozzle angles, comparison with Roberts et al. [19].

Table 7.3 Coefficients reported for dense jet properties in various nozzle angles based on Abessi and Roberts [18]

Parameter	Normalized Eq.	Nozzle angle		
		30°	45°	60°
Terminal rise height	y_t/dF	1.18	1.80	2.25
Centerline maximum height	y_c/dF	0.79	1.33	1.65
Location of maximum height	x_t/dF	2.13	2.25	1.90
Minimum Dilution at maximum height	S_t/F	0.55	0.58	0.61
Impact point location	x_i/dF	3.50	3.60	2.75
Impact point dilution	S_i/F	1.20	1.60	1.65
Near-field location	x_n/dF	10.3	11.0	9.50
Near-field dilution	S_n/F	1.85	2.50	2.60

generally accepted design practice. The values of constants for various nozzle angles are summarized in Table 7.3 based on Abessi and Roberts studies.

7.4.2.2 Multiport Diffuser

Most of the studies reported on the dynamics and mixing of inclined dense jets have been related to single jets. Multiport diffusers have also been used for brine disposal and are becoming a more common tool, especially for RO desalination plants, as the brine produced in these plants is more concentrated and salty, due to their higher efficiency and conversion rates. Examples are Perth, Sydney, and Melbourne, Australia. The nozzles may discharge perpendicular or angled to the diffuser axis and be uniformly distributed along either one or both sides of the diffuser, or the nozzles may be clustered in rosettes each having multiple ports. The Perth diffuser has 40 ports distributed uniformly along one side of the diffuser at four-meter intervals [74], the Sydney diffuser has two rosette risers spaced 25 m apart, and each riser has four ports [75]. Schematic of a brine diffuser (multiports distributed along both sides) and a rosette diffuser with 4 nozzles clustered in each riser are exhibited in Figs. 7.6 and 7.7.

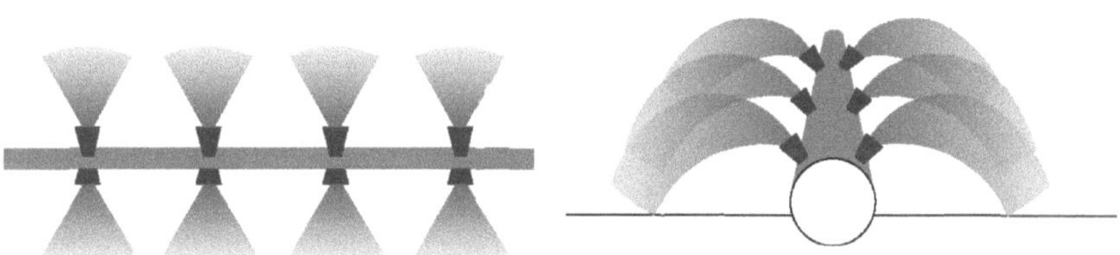

Fig. 7.6 Schematic of a multiport diffuser with nozzles distributed along both sides for brine discharge.

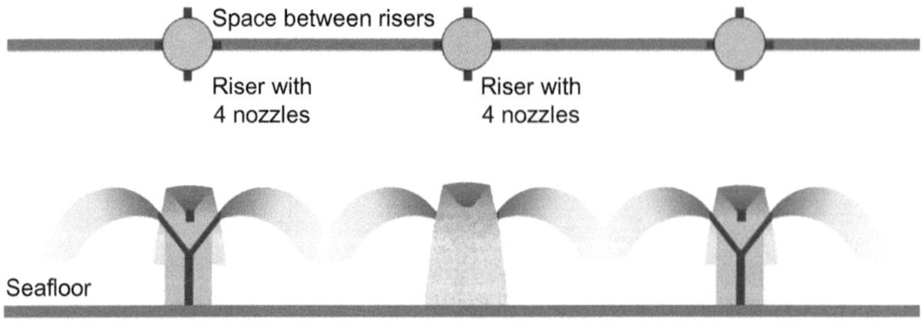

Fig. 7.7 Schematic of brine rosette diffuser with 4 nozzles clustered in each riser.

Although some hydraulic model tests of specific multiport diffusers have been made [76,77], there are few systematic studies that have been reported for dense discharge from multiport diffusers. So most of the desalination diffusers, in service, are designed based on the formulas derived for single jets or specific study on scaled models. Miller et al. [76] report physical model studies for the Perth diffuser and Tarrade et al. [75] reported similar study on Sydney and Melbourne rosette diffusers.

Roberts and Abessi [78] in their experiments on various configurations of brine discharge have used 3DLIF to determine dense jet behavior on multiport diffusers with discharge from one and both sides and multiple rosette diffusers. The experiments were conducted with and without ambient currents. Even though the experimental study was rich for multiport diffusers of positively buoyant effluent [40,79,80], similar study was relatively rare for negatively buoyant flow and their study was actually the first systematic experiments reported on dense discharge from multiport diffusers. No studies on the effect of port spacing have been reported before that.

Abessi and Roberts [18] investigated the effect of port spacing on multiport diffusers with zero current. The nozzles were from one or both sides and inclined upwards at 60°. The three-dimensional visualization of the averaged results for dense flow from two-sided diffuser is shown in Fig. 7.8. The port spacing and other parameters were varied to cover conditions expected to be typical of oceanic diffusers. Geometrical properties and dilutions at jet maximum rise height, location of impact point, and the end of near field were investigated. Normalized expressions were derived for each and plotted versus s/dF. The dimensionless parameter s/dF where s is the port spacing, d the nozzle diameter, and F the jet densimetric Froude number have been used to describe the effect of port spacing. It was found that for $s/dF > \sim 2$ the single jets did not merge and the geometrical and dilution results followed the expected asymptotic results for single jets.

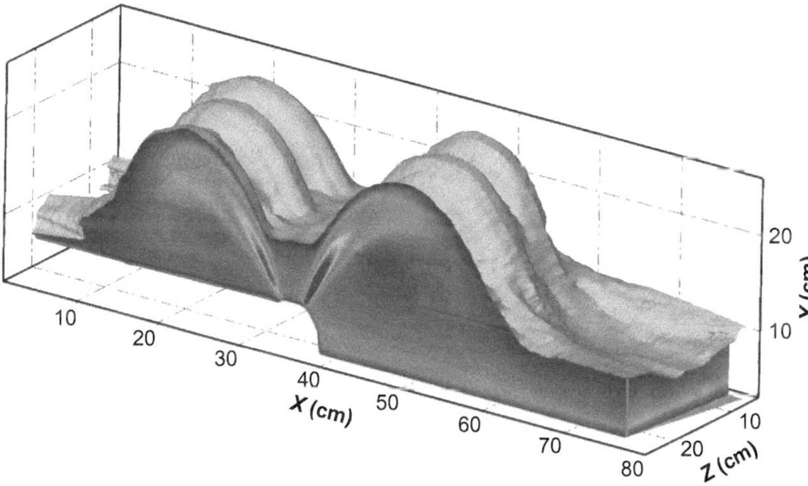

Fig. 7.8 Three-dimensional time-average image of flow from a multiport diffuser with nozzle in both sides.

For $s/dF<\sim 2$ the jets merged, but did not follow the expected line source solutions. Instead of the rise height and other geometrical variables increasing with decreased port spacing, they actually decreased. The dilutions decreased, as expected; however, they decreased much more rapidly than predicted as the spacing decreased. This unpredicted behavior was found to be attributed to Coanda effects combined with merging. The Coanda effect caused an under pressure on the interior jet surfaces, which caused their trajectories to curve inwards more sharply and contract. This caused re-entrainment of the descending jets by their ascending portions. The results for near-field dilution are plotted in the dimensionless forms and are shown in Fig. 7.9 for diffusers with nozzles in one and two sides. For zero current speed ($u_rF=0$), equations of best fit become functions of s/dF only. For this case, Abessi and Roberts [18] suggested the following equations for one- and two-sided discharges for narrow ($s/dF<\sim 2$) and wide ($s/dF>\sim 2$) port spacings, which can be used for the design process on multiport diffusers:

$$\text{narrow}\left(s/dF <\sim 2\right)\text{and wide}\left(s/dF >\sim 2\right)$$

$$\text{Dilution at the end of near field}: \frac{S_n}{F} = 6\left(\frac{S}{dF}\right)^{1/2} \text{ and} \tag{7.3}$$

$$\frac{x_n}{F} = 9 \tag{7.4}$$

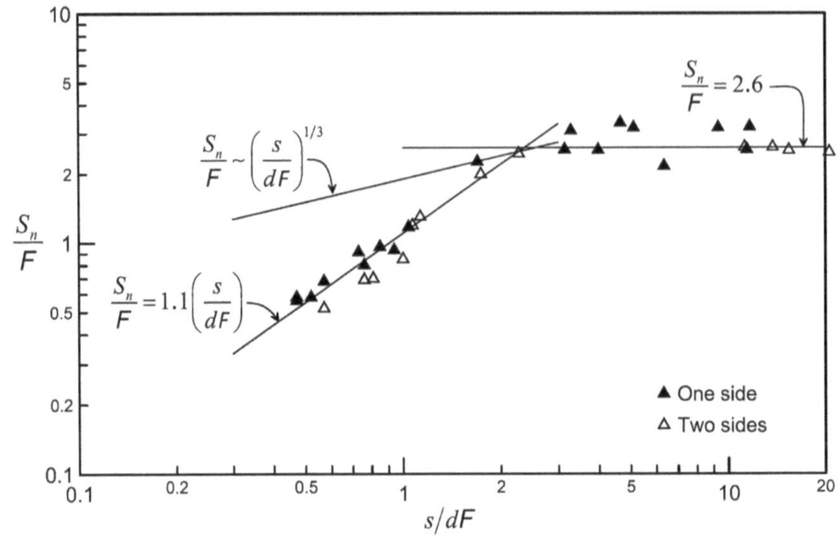

Fig. 7.9 Near-field dilution for multiport diffuser nozzles in one and two sides.

$$\text{Location of near field}: \frac{S_n}{F} = 1.1\left(\frac{S}{dF}\right) \text{ and} \tag{7.5}$$

$$\frac{S_n}{F} = 2.6 \tag{7.6}$$

$$\text{Impact dilution}: \frac{S_i}{F} = 0.9\left(\frac{s}{dF}\right) \text{ and} \tag{7.7}$$

$$\frac{S_i}{F} = 1.6 \tag{7.8}$$

$$\text{Impact location}: \frac{x_i}{dF} = 2.0\left(\frac{s}{dF}\right)^{1/2} \text{ and} \tag{7.9}$$

$$\frac{x_i}{dF} = 2.4 \tag{7.10}$$

$$\text{Rise height}: \frac{y_t}{dF} = 1.9\left(\frac{s}{dF}\right)^{1/2} \text{ and} \tag{7.11}$$

$$\frac{y_t}{dF} = 2.2 \tag{7.12}$$

The result of the experiments on multiport diffusers with zero current is reported. Although stagnant water is usually considered to be the worst-case scenario, the effects of ambient currents should also be considered for assessment under more realistic oceanic conditions. Abessi and Roberts [80a] investigated the diffusers discharging from one side into co- and counter-flowing currents and from both sides. Jet rise height, impact location, and impact point dilution were investigated in flowing water using 3DLIF techniques and the results were reported in nondimensional forms. This is also the only study found in the literature for diffuser dense discharges in flowing ambient bodies. The nozzles were inclined upwards at 60°, uniformly distributed and perpendicular to the diffuser axis either to one or both sides. The results were plotted in dimensionless form (Fig. 7.10) as functions of port spacing $\left(\frac{S}{dF}\right)$ and current speed parameters $\left(U_rF = \frac{u_a}{\sqrt{g'_0 .d}}\right)$ in which $U_r = \frac{u_a}{u_0}$ is the ratio of the ambient current speed to the jet velocity, F is Froude number, and U_rF itself is a type of Froude number based on the crossflow velocity. Empirical equations were fitted to the data to aid in using them in outfall and diffuser design. As shown in Fig. 7.10, for slow current speeds, the jets were slightly bent downstream. A counterflow can cause the jets to fall back on themselves. As the current speed increases, the

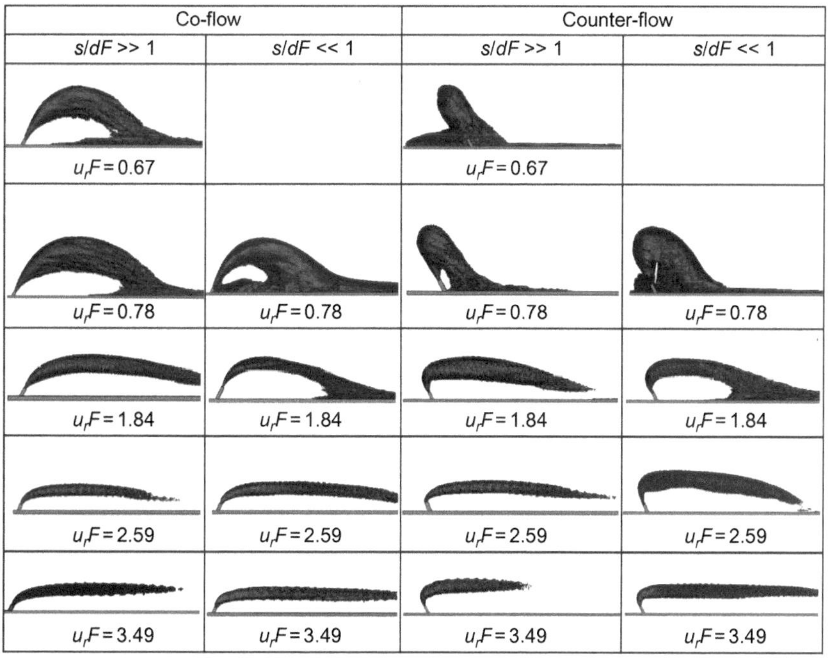

Fig. 7.10 Dense jets in co- and counterflow currents with wide and narrow jet spacings.

impact location is swept farther downstream, the rise height decreases, and impact dilution increases. Dilutions are generally greater than diffusers with similar port spacings but stationary; at low current speeds, dilutions may occasionally be lower. When the jets are closely spaced, they quickly merge, preventing entrainment into the core region between the rising and falling jets. This can cause the flow fields to be significantly different for narrow and wide port spacings, especially at lower current speeds. In contrast to experiments with stationary currents, where the results became independent of port spacing for $s/dF > 2$, the results did not asymptote to constant values for wide-port spacings. It also observed that the widely spaced jets are deflected more by the current than narrowly spaced jets, so they impact the lower boundary farther downstream.

Dilutions at the impact point *Si* are shown in Fig. 7.11 along with the equations for stationary environments, ($\frac{S_i}{F} = 0.9\left(\frac{s}{dF}\right)$ for narrow ($s/dF < \sim 2$) and $\frac{S_i}{F} = 1.6$ for wide $s/dF > 2$ port spacings). It is observed that dilution increases with increased current speeds, as expected. In general, dilutions increase with increased port spacing, also as expected, but do not asymptote to constant values as occurs for zero current. Although the data are scattered, it is convenient to fit the results by empirical equations of the general form:

$$\frac{S_i}{F} = A\log\left(\frac{s}{dF}\right) + B \qquad (7.13)$$

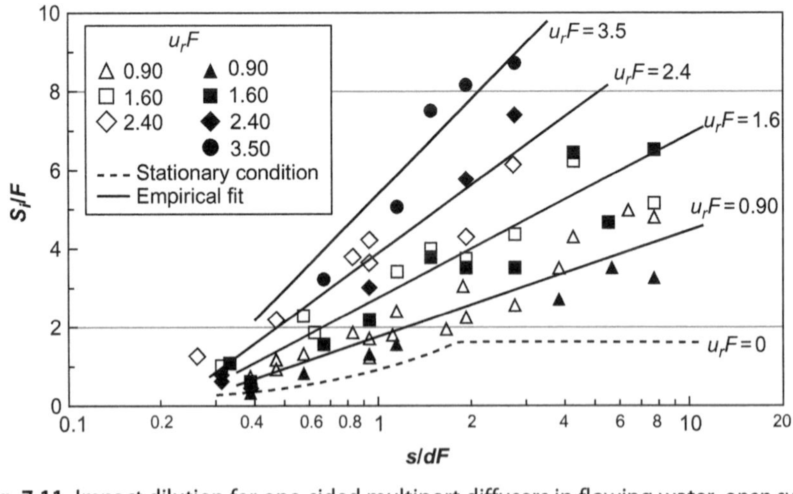

Fig. 7.11 Impact dilution for one-sided multiport diffusers in flowing water, *open symbols* are coflow; *solid symbols* are counterflow.

Table 7.4 Summary of coefficients for empirical equations[a]

	Equation	a_1	a_2	b_1	b_2
Impact dilution, one-sided diffusers	$\frac{S_i}{F} = A\log\left(\frac{s}{dF}\right) + B$	2.1	0.86	1.4	0.49
Terminal rise height, one-sided diffusers	$\frac{y_t}{dF} = A\log\left(\frac{s}{dF}\right) + B$	−0.07	0.31	−0.33	2.1
Impact dilution, two-sided diffusers	$\frac{S_i}{F} = A\log\left(\frac{s}{dF}\right) + B$	1.4	2.0	0.52	0.03
Impact location, two sided diffusers[b]	$\frac{x_i}{dF} = A\log\left(\frac{s}{dF}\right) + B$	−0.71	3.7	4.1	1.4
Terminal rise height, two-sided diffusers	$\frac{y_t}{dF} = A\log\left(\frac{s}{dF}\right) + B$	0	1.5	−0.35	2.3

[a] $A = a_1(u_rF) + a_2$ and $B = b_1(u_rF) + b_2$.
[b] Blocked only, $u_rF < 2.16$.

where A and B are constants that depend on u_rF in the form:

$$A = a_1\left(u_rF\right) + a_2 \quad \text{and} \quad B = b_1\left(u_rF\right) + b_2 \tag{7.14}$$

Similar equations were fitted to the other variables and the values of the coefficients are summarized in Table 7.4 for one- and two-sided diffusers to aid the design process.

For flow from two-sided diffusers in current, the upstream jets are swept back directly onto their downstream counterparts and merge with them before impacting the floor. The impact location is swept considerably downstream for higher current speeds and the rise height decreases. Dilution increases with increasing current speed compared to the stagnant case and does not asymptote to constant values with increasing port spacing (as observed for dense diffuser in stagnant water) [18]. Fig. 7.12 shows the change in impact point dilution $\left(\frac{S_i}{F}\right)$ versus port spacing $\left(\frac{S}{dF}\right)$ for different current speeds (U_rF). Dilutions are generally measured higher than for zero currents, except for narrow spacings ($s/dF<\sim 2$) at low current speeds ($u_rF<\sim 1$). In those cases, the upstream jets merge with their downstream counterparts (see Fig. 7.12) as though the discharge per port is doubled. So the dilutions for two-sided diffusers are significantly lower than for corresponding single-side discharges. The presence of the gap below the diffuser is predicted to change the flow field substantially. Abessi and Roberts [80a]

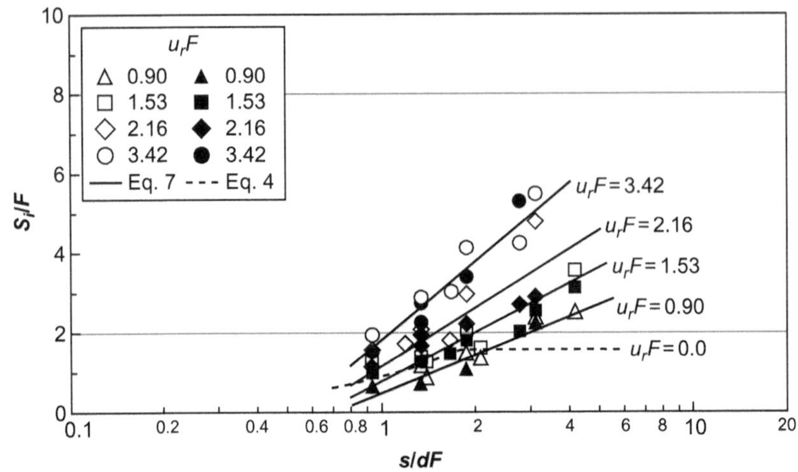

Fig. 7.12 Impact dilution for two-sided diffusers.

also conducted experiments to investigate the effects of raising the diffuser above the bed, e.g., a gap between the diffuser and bed, and some experiments with no gap. These two conditions are referred to as unblocked and blocked condition. It was found that this gap can affect the observed flow patterns significantly, by allowing entrained flow to pass beneath the diffuser. Fig. 7.13 shows the flow field from two-sided diffusers by the presence or absence of the gap. Despite visual differences between blocked and unblocked flows, the dilution was not significantly affected or was within the experimental scatter. For two-sided diffuser, the impact locations are always farther from the diffuser than for zero currents. But as can be inferred from Fig. 7.13, the impact distances for blocked flows are much less than unblocked condition. The blocked and unblocked results were both approximated by empirical equations of the general form (Eq. 13) with the coefficients in Table 7.4. For more information, check Abessi and Roberts [80a].

Generally for dense discharge in flowing water through multiport diffusers, co- and counterflows did not show major differences, although in counterflow, jets fall back directly onto itself for slow current ($U_rF \leq 0.67$) and dilution is slightly lower. In all counter- and coflowing cases, the jets are ultimately swept downstream by the current and usually no upstream wedge forms. Single-side discharges always give higher dilutions than two-sided. For two-sided discharges, the upstream jets are swept back directly onto their downstream counterparts and merge with them before impacting the floor. This can result in a reduction in dilution compared to the stagnant case. For both cases, one and two-sided diffusers, although blocked rise heights were

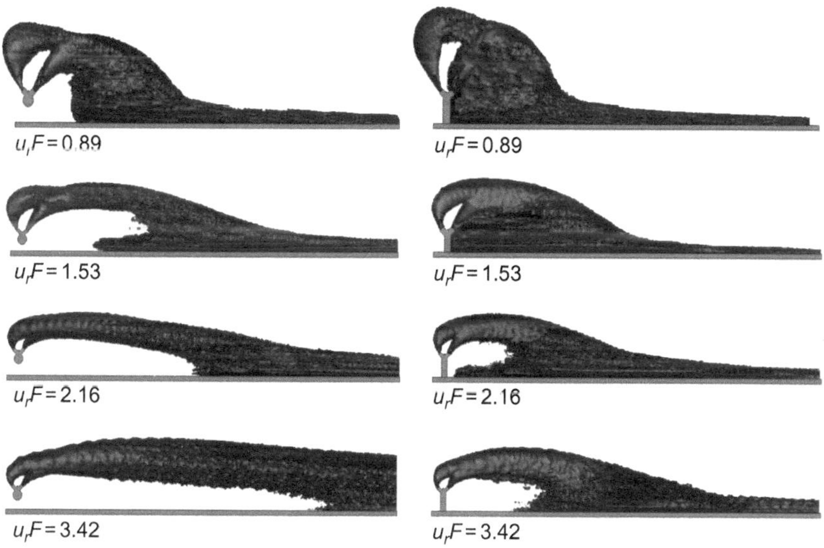

Fig. 7.13 Two-sided multiport diffusers in currents of various speeds with and without blocking, $s/dF = 0.93–3.11$.

sometimes slightly higher than unblocked and the location of impact point closer than unblocked cases, there was no systematic pattern in dilution and it was not significantly affected by the gap underneath the diffuser.

As discussed, multiport diffusers (uniformly distributed or clustered in rosettes) are now a common way to discharge out the brine to sea for large coastal desalination plants. Examples are the diffuser on the outfall at Perth, Australia, which has ports uniformly distributed along one side, whereas the outfalls at Sydney and Melbourne, Australia, are tunneled with diffusers consisting of risers that extend upwards to multiport rosettes on the seabed. The Sydney diffuser has four risers, each with four ports, 370-mm diameter spaced 25 m apart. It discharges 65 ppt concentrate back to the ocean with a target dilution of 30:1 to reduce the salinity increment to 1 ppt above background at the end of the near-field mixing zone [76]. The Melbourne outfall initially considered a design consisting of four rosette risers spaced 100 m apart, but a later design change resulted in risers spaced 50 m apart with 9 or 12 ports per riser [81]. Physical model tests have been used for the part of design for these rosette diffusers. Apart from these physical model tests, the basis for design of many multiport and rosette diffusers has mostly been the studies conducted on the dynamics and mixing of single inclined dense jets and there have been relatively few studies on rosettes and no systematic experiments of jet interaction or the effect of riser spacing.

An extensive program of experiments on various configurations of multiple rosette diffusers consisting of clusters of four ports for desalination brine discharge has been reported by Abessi et al. [81a] and Abessi and Roberts [80a]. The study has done on single risers and also on the role of riser spacing for multiple risers in stationary and flowing water. Schematic of the configuration of multiple rosette diffuser is similar to those used in Sydney brine disposal system exhibited in Fig. 7.7. The rosettes each had four nozzles pointing upwards at 60° and arranged 90° apart in platform. Abessi et al. [81a] found that the effect of riser spacing for multiple rosette risers depends on the dimensionless parameter of s_r/dF, where s_r is the riser spacing, d the nozzle diameter, and F the jet densimetric Froude number. For $s_r/dF > \sim 2$, the results were independent of riser spacing; however, the near-field dilution was lower (up to 30%) than a single jet. The three-dimensional flow visualizations of brine discharge in stagnant water through rosette diffuser that developed by 3DLIF to map tracer concentration fields are shown in Fig. 7.14. Experimental results showed that near-field dilution for rosette discharges is less dependent on riser spacing than multiport diffuser. However, it seems due to less construction cost and simplicity, as it has its own priorities for brine discharge.

Dense disposal into flowing currents as a more realistic case of what actually happening for rosette diffusers has also been investigated by Abessi

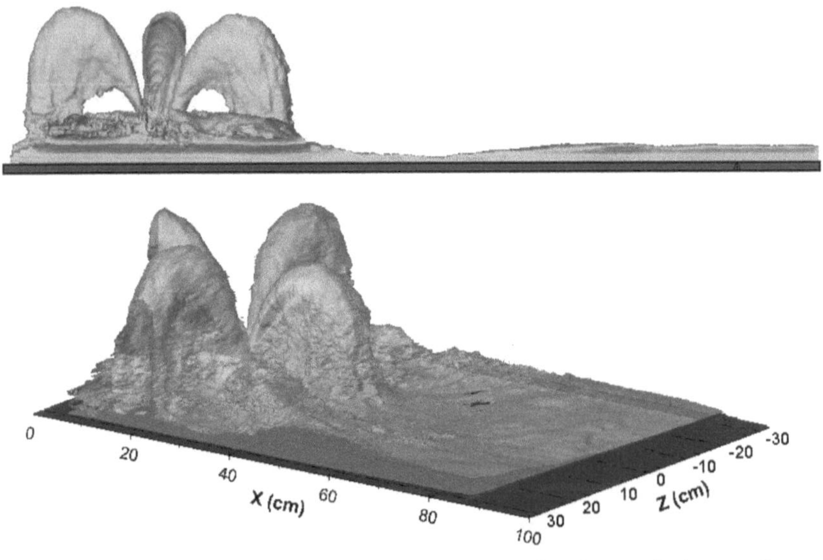

Fig. 7.14 The three-dimensional flow visualizations for dense discharge through rosette diffuser in stagnant water.

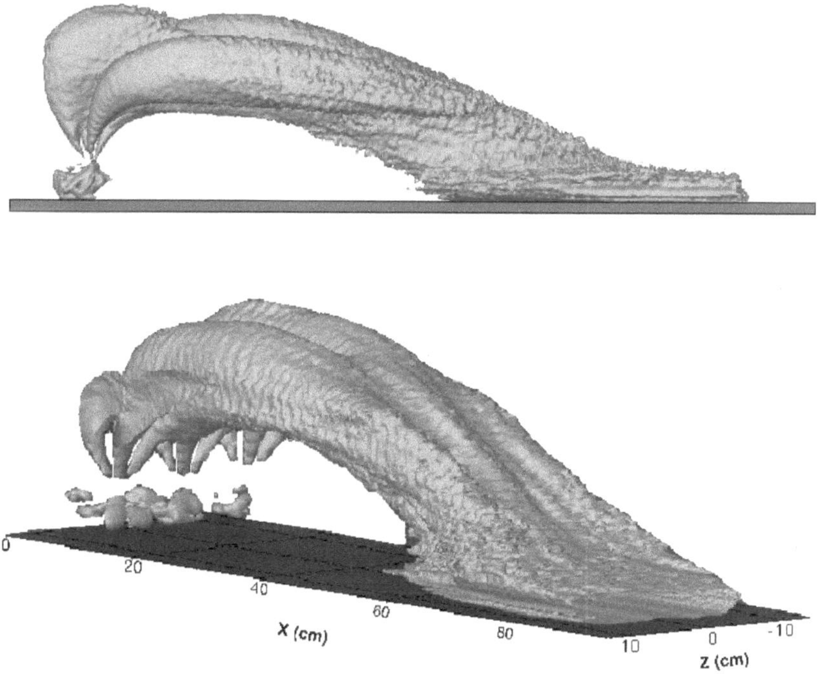

Fig. 7.15 The three-dimensional flow visualizations of rosette discharge in flowing current.

and Roberts [81b]. It observed that for rosette diffusers as the current speed increases, dilution and impact distance increase and rise height decreases. The results also show that the rosette diffusers generally have higher dilutions than equivalent multiport diffusers, for the same diffuser length and number of ports. The three-dimensional tracer concentration fields of discharge from single and multiple rosette diffusers in flowing current are shown in Fig. 7.15. People who are interested in details and want to study flow characteristics on rosette discharge in flowing water can refer to our papers on rosette diffusers for dense flows, published/publishing in *Journal of Hydraulic Engineering, ASCE* [81a,b].

7.4.2.3 Discharge in Shallow Water

Effect of water depth on dense jet properties has attracted little attention while discharges in deep waters have been quite intensively investigated over the years. When discharging the brine into ambient water, the diffusers should be situated far enough from the shore to reach sufficient depth and rapid currents. In low-slope shore when no deep waters available, it may be

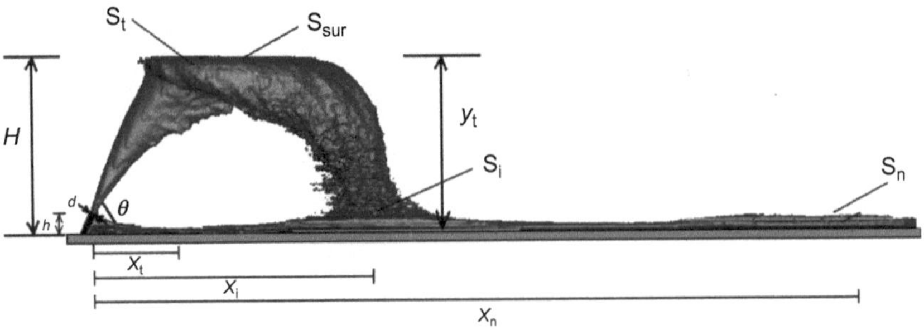

Fig. 7.16 Inclined dense jet in shallow water for 60° nozzle.

still necessary to design an outfall system with high dilution rate. Inclined jet with steeper nozzle angles is common way for dense discharges, but it increases the rise height and may cause surface impact in shallow water for the high discharge velocities. If the jet impacts the surface the water available for dilution may decrease and the flow dynamics will change; it is known as the shallow water condition [71]. In Fig. 7.16, shallow water condition when dense jet impacts the water surface is illustrated for 60° nozzle.

The shallow water condition for vertical positively buoyant jets are previously investigated by Lee et al. (1974), Koester [82], and Roberts and Wright [83]. They experimentally studied the discharge flow in various conditions of shallow water and suggested some criteria to delineate deep from shallow water. However, the first of few studies on the condition of shallow water for dense jet was reported by Jiang et al. [84]. They investigated the effect of water depth on the mixing and dilution of a 45°-inclined dense jet using LIF techniques. The effect of water depth is described by a parameter dF/H where d is the nozzle diameter, F the jet densimetric Froude number, and H the water depth. Jiang et al. [84] measured the minimum surface dilution for various water depths and identified three flow regimes encompassing the full submergence, plume contact, and centerline impingement regimes. Their results show that for $dF/H \approx 0.62$, the top of the jet at 45° contacts the water surface, which delineates the deep water from surface contact regimes where dilution begins to decrease. For 45° jets, the surface contact to shallow water transition values are observed at 0.62 to 0.77. This study was extended by Jiang et al. [85] for 30° and 45° dense jets in limited depth. For 30° jets they proposed $dF/H < 0.95$ as the criterion for deep water, $0.95 \leq dF/H \leq 1.4$ for surface contact, and $dF/H > 1.4$ for shallow water. These results were used by Jiang et al. (2013) to develop a semianalytical solution for the optimal design of noninterfering multiport brine diffusers in deep, surface contact, and shallow waters. The optimal combination of the discharge port diameter, *d*, and number of ports, *N*, was investigated to enable meeting design objectives for all possible conditions.

Abessi and Roberts [71,72] reported a series of experiments on single inclined dense jets on the nozzles oriented at 30°, 45°, and 60° in various Froude numbers and ambient depths (deep to shallow). As shown in Fig. 7.16, the flow characteristics such as surface, impact and near-field dilutions, and the locations of the impact point and the end of near field were investigated. It was the first three-dimensional observations reported on the complex interactions of the jets with the surface. The perspective view of tracer profiles in vertical planes perpendicular to the main flow direction are exhibited at various distances from the nozzle tip for a 60°-dense jet (Fig. 7.17). Similar to Jiang et al. [85], three flow regimes were identified, deep water where the jet is deeply submerged and unaffected by the water surface, surface contact where the top of the jet impacts the water surface, and shallow water where the jet centerline impacts the water surface (Fig. 7.16). For the deep water condition, it is observed that all results

Fig. 7.17 The 3D flow images of 60° jet in shallow water, $dF/H = 0.9$.

followed those previously reported for fully submerged jets. As the depth decreases (or dF/H increases), the top of the jet begins to interact with the water surface. This occurs at $dF/H = 0.80, 0.48$, and 0.42 for nozzle angles of 30°, 45°, and 60°, respectively [71,72]. Above these values, the dilutions begin to decrease. These values obtained by plotting surface minimum dilution (S_{sur}/F) against dF/H while H decreasing. The surface dilution begins to decrease from infinity when the jet approaches the water surface. As the depth is decreased further, the jet centerlines intersect the water surface marking the transition to shallow water condition. This occurs at $dF/H = 1.15, 0.7$, and 0.64 for nozzle angles of 30°, 45°, and 60°, respectively. It has been suggested that dense jets are submerged if the rise height to the jet top is <75% of the water depth to avoid surface effects [82]; however, Abessi and Roberts [71,72] suggested this is overly conservative and the rise can be as much as 90% of the water depth for all angles with no deleterious effect on dilution. These results are summarized in Table 7.5 to help environmental authorities, promoters, and designers to anticipate the behavior of inclined dense jets in shallow water for the design purposes in real applications. For more details on the geometry and dilution value at impact point and the location of near field, check Abessi and Roberts [71,72].

7.4.2.4 Surface Discharge

The preferred method for brine discharge is multiport diffuser due to its higher dilution rate. Co-disposal of brine with power plant cooling water or domestic wastewater that results in significant in-pipe dilution has been reported too. Shoreline surface discharges with codisposal of power plant cooling water have also been employed in Fujairah Power and Desalination Plant. In Ashkelon at the coast of Mediterranean sea, the brine flow produced in their RO plant discharges directly bach into sea through an open channel (Fig. 7.18). In these cases, the mixed effluent could be positively, neutrally, or negatively buoyant that forms discharges with complex hydrodynamics [86]. However, shoreline discharge of raw effluent is discouraged

Table 7.5 Values of dF/H for transition between deep, surface contact, and shallow flow regimes based on Jiang et al. and present study

	Jiang et al. [85]			Present study		
Nozzle angle	Deep	Surface contact	Shallow	Deep	Surface contact	Shallow
30°	<0.95	0.95–1.4	>1.4	<0.80	0.80–1.15	>1.15
45°	<0.62	0.62–0.77	>0.77	<0.48	0.48–0.70	>0.70
60°	–	–	–	<0.42	0.42–0.64	>0.64

due to slow near-field mixing and potentially high exposures of near-shore organisms to elevated salinity [24,87]. Surface discharges, in the form of open channel near or parallel to the free surface of the receiving water, are a common method for the disposal of large volumes of effluent into the sea [32,88]. This is oldest and could be the least costly form of wastewater discharges [88]. It has been reported to be used in many coastal power plants like Hesham power station, United Kingdom (Fig. 7.19) and Bandar Abass, Iran, where the lower dilution rate is required for thermal discharges.

Fig. 7.18 Inshore open channel for the discharge of negatively buoyant brine, Ashkelon RO plant at the coast of Mediterranean.

Fig. 7.19 Protruding surface discharge in Hesham power station outfall, United Kingdom.

For brine discharges, other types of surface and over surface discharge such as overflow spill from cliff or jetties, and surface channels protruding into deep water (in steep shore) can also be considered as an option. Such discharges are different from typical shoreline (diverging) channels. For example, protruding surface discharge is basically a channel on breakwater or mounted on piles that take some distance from the coastline to provide jet free falls to the sea bed [88a]. For the brine flow, discharge creates a jet moving along the sea surface and then plunging down to the deep sea in the form of a negatively buoyant flow. By reaching the sea floor, the brine forms a density current that runs down the bottom slope with entrainment quit slow [88b]. Providing an offshore free fall from the sea surface enhances the chance for brine to entrain and mixed with ambient water before reaching the floor in deep water. So protruding channel configuration reduces the chances of direct shoreline impacts and enhances the initial mixing characteristics of flow, which is the main restrictions of shoreline surface discharges.

In Fig. 7.20, a channel constructed on breakwater to discharge the cooling water of Neka power plant back to the Caspian Sea (northern Iran) is illustrated. Schematic of an off-shore protruding discharge (the channel built on piles) is also shown in Fig. 7.21. Abessi et al. [58,59] suggest it as an economical option for brine disposal from large desalination plants (with low recovery rates), where the effluent flows are very large (5 m^3/s). This is despite the fact that such discharges are less efficient in terms of mixing and entrainment.

Fig. 7.20 Surface discharge as a channel on breakwater at Neka power station, southern Caspian Sea, Iran.

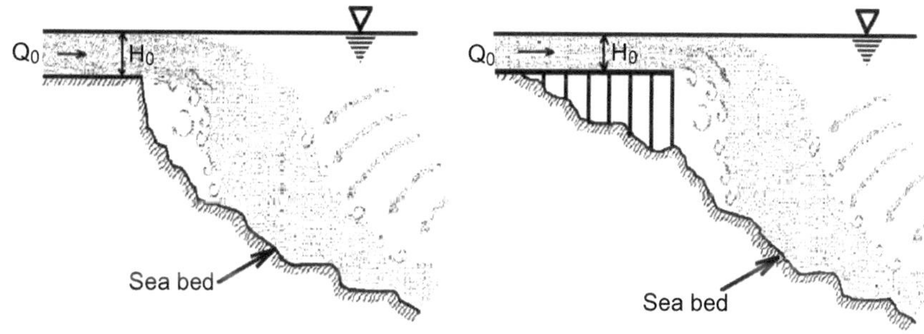

Fig. 7.21 Schematic view of protruding surface dense discharge, channel mounted on piles and dense jet free falls to sea bed.

The negatively buoyant surface discharge is characterized by a plunging motion of the effluent toward the seabed, which makes the behavior of flow distinctively different from the positively buoyant discharges. There is a long history of experimental studies on surface buoyant jet dynamics (Shirazi and Davis 1974, Jirka et al. 1981, [88,89]), while experimental work on surface dense discharge is rare [58,59] and mostly related to diverging channels [90,90a]. Abessi (2012) conducted a comprehensive series of experiments on protruding surface discharge, and investigated dense flow mixing and trajectory in stationary water and ambient current. The jet trajectories have been measured for the wide range of densimetric Froude numbers and plotted in dimensionless form, normalized with L_M [58,59]. They compared the behavior of dense surface discharge with a submerged positively buoyant jet with similar features observed to those of horizontal buoyant jets, considering the mirror effect. The trajectories collapse as expected and are almost identical to data from submerged horizontal buoyant discharges and the theoretical results of jet integral model CorJet [53,91–93]. However, small differences are observed in the flow behavior prior to it plunging away from the free surface. The flow trajectory of surface jets was observed to travel farther along the water surface, before diving downwards. This was discussed to be the effect of suction close to the water surface. The dilutions were measured using conductivity probes and normalized by the densimetric Froude number, plotted versus z/L_M. As expected, the overall dilution is lower when compared to submerged discharges, which is assumed to be due to the interactions with water surface and limited mixing. In Figs. 7.22 and 7.23, trajectory and dilution for dense flow discharging from protruding channel into stationary water are exhibited with the experimental results of submerged horizontal buoyant jet and CORJET model. Some differences were observed and quantified [58,59]. These results provided valuable information for the preliminary design of dense surface discharges where lower dilution is required or limited capital cost is available.

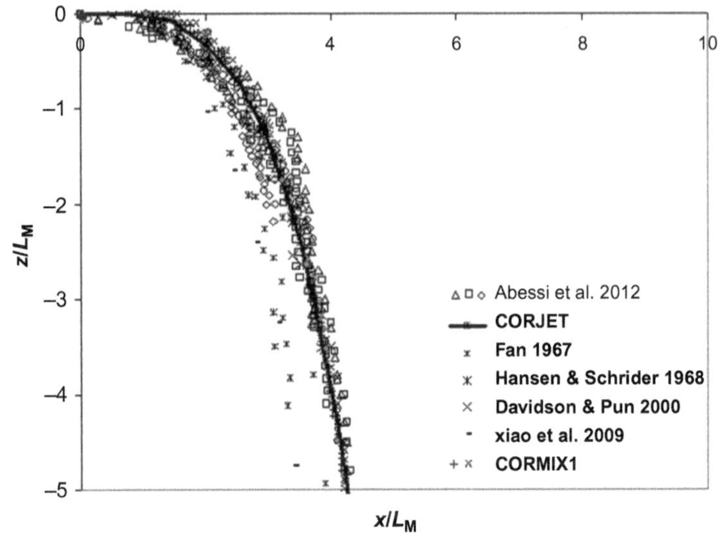

Fig. 7.22 Normalized jet centerline trajectory for negatively buoyant surface jet discharges into stagnant ambient.

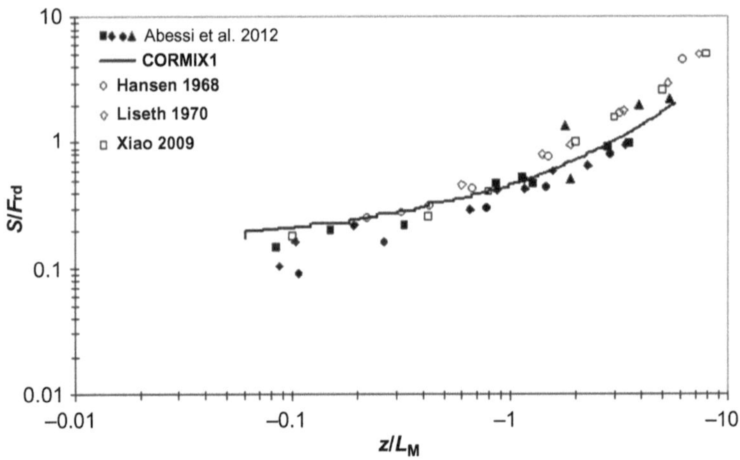

Fig. 7.23 Normalized centerline dilution S/Fr_d, function of water depth z/L_M.

Surface discharge of negatively buoyant effluents into moving water bodies also creates a range of different flow behaviors. Depending on the geometric and dynamic characteristics of the receiving water, some complex hydraulic processes will occur affecting the shape and extent of the resulting plume and its mixing. These complexities develop through the interplay between a discharge's initial fluxes and the motion of the ambient current.

Chu and Jirka [89] conducted the first laboratory experiments on positively buoyant surface jets with flowing ambient and classified the motion of flow into three distinctive regimes. The results of their study, along with the comprehensive laboratory data of previous experiment ([94], and [95]), provided a basis for the development of the tree-like classification scheme for surface discharges in CORMIX3.

Abessi et al. [58,59] investigated the changes of initial fluxes for surface discharge of negatively buoyant flow in ambient current from rectangular and semicircular protruding channels. Such discharges can form a wide variety of flow patterns within the receiving water such as crossflow deflection, dynamic shoreline interaction, and bottom attachment. These interactions can dramatically alter the initial mixing and configuration of the flow. Abessi et al. [58,59] identified three flow regimes for negatively buoyant surface discharges - the free surface jet, the shore-attached surface jet, and the plunging plume - similar to those reported for positively buoyant discharge. However, there are distinct differences in their range of occurrences for positively and negatively buoyant flows (Fig. 7.24). So based on earlier works of Chu and Jirka [89], a new classification scheme was developed by Abessi et al. [59] that can be applied in the flow regime prediction methodology for brine surface discharges [96]. The flow behavior in each

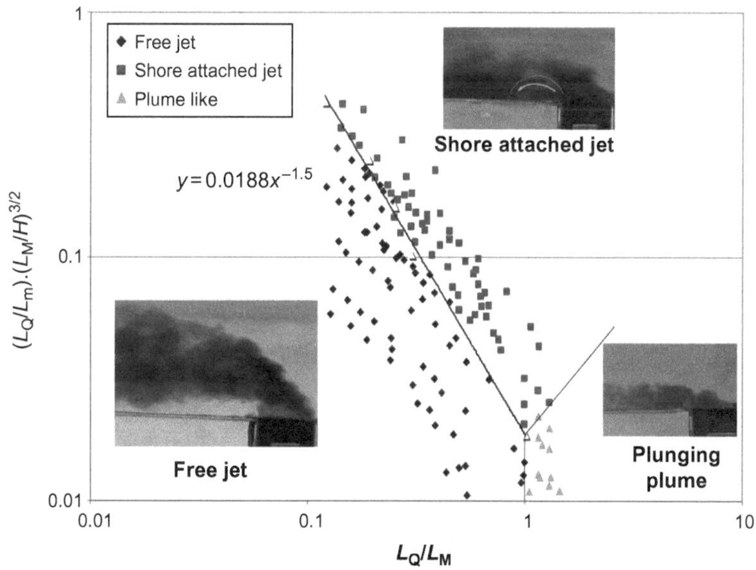

Fig. 7.24 The classification diagram developed for the regime prediction in negatively buoyant surface discharge from the rectangular channels.

regime is quantified. For more information check Abessi (2012). The influence of discharge configuration or the geometry of the channel on mixing and trajectory of surface flow is also investigated recently, as reported by Shacheri et al. [97].

7.5 CONCLUSIONS

Desalination as a practical method to remove salts from saline water has become an alternative way to produce fresh water in many parts of the world, especially coastal areas. The disposal of brine, as the byproduct of desalination process, back into the sea is the main environmental concern associated to these plants. Ocean outfalls are the most widely method used for disposal of brine in coastal plants while evaporation ponds, deep well injection, and disposing to rivers or sewage systems are more consider for inland plants. By designing an outfall system, engineers try to provide rapid mixing and dilution for the brine to reduce salinity to near-background levels and minimize its harm on marine life. The proper discharge makes it possible to accomplish the ambient water quality standard at the point of compliance.

Outfalls can be surface, over surface and submerged, each classified into different categories. Surface discharge can be shoreline diverging channel, protruding channel, overflow spill from cliff or discharge on a breakwater and submerged discharge can be a single jet, multiport diffuser, or multiple rosette diffuser. Knowing the effluent characteristics (density, flow rate, salinity and temperature) and ambient bathymetry, currents, tides, density profiles, etc. are crucial in outfall modeling and design.

The discharged flow in the ambient water can hydrodynamically be classified into Near field and Far field regions. The Near-field region or the zone of initial mixing mainly depends on the brine discharge configuration and the effluent and ambient properties and is responsible for the most of dilution in the environment. The flow in near field is the only region controlled by designer; however, because of the complexities, there are difficulties with simulation and modeling of the near field. Experimental, integral, and numerical models are developed to simulate the flows in this region. Experimental studies, performing laboratory experiments on scaled physical models, were used to clarify the complicated behavior of flow in different conditions. They led to the understanding of the physics of mixing to develop the mathematical models, empirical equations and standards for design. Integral models are based on analytical simplification of complex mixing process and are simple to use and accurate enough for engineering design purposes. Numerical CFD models are complicated and robust models, and using them for engineering purposes has only recently become possible due to advancement in numerical methods and computational technology. CFD models are being increasingly used in hydraulic engineering and their use will certainly increase in the future. If correctly calibrated, CFD model can be used on any type of discharge configuration or ambient water scenarios. Regarding the difficulties in using the CFD models, empirical equations and integral models are used for initial design and scenario planning and using CFD models is becoming more common for final-stage optimizations and far-field modeling.

This chapter, by reviewing the recent works, outlined the discharge methods commonly used for the desalination brine disposal and management. Each method has its own pros and cons. Shoreline surface discharge was introduced as the oldest and simplest way to discharge brine back to the sea. This method is discouraged due to slow near-field mixing and high exposures of near shore to elevated salinity. Surface channel in the form of protruding channel (mounted on breakwater or piles) to provide jet free falls to the seafloor in deep water has also been investigated for its advantages in steep shores. This discharge found to be less expensive compared to other methods however is limited in dilution and mixings. Compared to surface discharges, a longer history of experimental and numerical works was found for submerged discharges. Single jet is the simplest method of submerged discharges that can be used for lower volumes of brine disposal. The single jet that upwardly inclined at 60° was discovered to be the optimal degree with maximum dilution and mixing at the end of near field. However, it was not very sensitive to nozzle angle over the range of 45–65°. Compared to multiport diffuser, single jet provides lower dilution and mixing for the equal flow of brine and it limits its applications, especially for RO desalination plants where the brine is more concentrated and salty. Hence, multiport diffusers are becoming a more common tool for brine disposal where higher dilution is required. Multiple rosette diffusers consisting of clusters of ports at riser have also found to be a favorite method for desalination brine discharge with many examples.

Experiments on multiport and rosette diffusers show that small changes in diffuser design can significantly changes the flow field and associated mixing and dilution. The difficulties in predicting interaction of multiple jets, dynamics of the bottom layer and sloping seabeds, shallow water condition, and boundary interactions cause serious challenges in simulating the physical features of near-field entrainment. It indicates the need for physical models in complex situation design and more realistic look on the abilities of CFD model for engineering designs.

REFERENCES

[1] Population Reference Bureau. Ripple Effects: Population and Coastal Regions. 2003. September 2003, Washington, DC, USA.

[1a] Palomar P, Losada IJ. Impacts of Brine Discharge on the Marine Environment. Modelling as a Predictive Tool. INTECH Open Access Publisher; 2011. https://doi.org/10.1016/10.5772/14880· ISBN 953307311X, 9789533073118.

[2] Raed B, Persson Kenneth M, Magnus L. New criteria for brine discharge outfalls from desalination plants. In: Intakes and Outfalls for Seawater Reverse-Osmosis Desalination Facilities. Switzerland: Springer; 2015. ISBN: 978-3-319-13202-0, 978-3-319-13203-7, https://doi.org/10.1007/978-3-319-13203-7.

[3] Mahad B, Choudri BS, Mushtaque A, Anton P. Recent Progress in desalination, environmental and marine outfall systems. ISBN 978-3-319-19122-5, ISBN 978-3-319-19123-2, Switzerland: Springer; 2015. https://doi.org/10.1007/978-3-319-19123-2.

[4] Burke PA (2013) Understanding desalination: the well that won't run dry. Water Technol Mag 36 (Edition 4), 01 April 2013.

[5] Birkett JD. A brief illustrated history of desalination: from the Bible to 1940. Desalination 1984;50:17–52.

[6] Henthorne L. Desalination: a critical element of water solutions for the 21st century. In: Forare J, editor. Drinking water—sources, sanitation and safeguarding. Sweden: The Swedish Research Council Formas; 2009.

[7] Maalouf S (2014) Planning and design of desalination plants effluent systems. PhD thesis, UCLA Electronic Theses and Dissertations, University of California, USA.

[8] Afgan NH, Al Gobaisi D, Carvalho MG, Cumo M. Sustainable energy management. Renew Sustain Energy Rev 1998;2:235–86.

[8a] Hópner T. A procedure for environmental impact assessment (EIA) for seawaterdesalination plants. Desalination 1999;124:1–12.

[9] Lattemann S, Hopner T. Environmental impact and impact assessment of seawater desalination. Desalination 2008;220(1–3):1–15.

[10] Procaccini G, Gambi MC, Pérez MC, Pergent G, Pergent-Martini C, Romero J. Sea grasses of the western Mediterranean. In: Green EP, Short FT, editors. World atlas of seagrasses. Los Angeles, California: University of California Press; 2003. p. 48–52.

[11] Portillo, E , Ruiz de la Rosa, Louzara M., Quesada G., Ruiz J.M., Mendoza, H. (2014) "Dispersion of desalination plant brine discharge under varied hydrodynamic conditions in the South of Gran canaria and influence on nearby seagrass meadows" Desalin Water Treatm, 52, 1–3, 164–177.

[12] Gacia E, Invers O, Manzanera M, Ballesteros E, Romero J. Impact of the brine from a desalination plant on a shallow seagrass (*Posidonia oceanica*) meadow. Estuar Coast Shelf Sci 2007;72:579–90.

[13] Yoon SJ, Park GS. Ecotoxicological effects of brine discharge on marine community by seawater desalination. Desalin Water Treatm 2011;33:240–7.

[14] Rachel E, Kobi H, Dan P. The footprint of the desalination processes on the environment. Desalination 2002;152:141–54.

[14a] Höpner T, Windelberg J. Elements of environmental impact studies on the coastaldesalination plants. Desalination 1996;108:11–8.

[15] Bleninger, T. 2006. Coupled 3D hydrodynamic models for submarine outfalls: Environmental hydraulic design and control of multiport diffusers. Ph.D. Dissertation, Institute for Hydromechanics. University of Karlsruhe, Germany.

[16] Bleninger T, Jirka GH. Modeling and environmentally sound management of brine discharges from desalination plants. Desalination 2008;221:585–97.

[17] Nadeem A, Baddour RE. A review of sources, effects, disposal methods, and regulations of brine into marine environments. J Ocean Coast Manage 2013;87(2014):1–7.

[18] Abessi O, Roberts WP. Multiport diffusers for dense discharges. J Hydraul Eng, ASCE 2014;140(8). 04014032-1-04014032-12.
[19] Roberts PJW, Ferrier A, Daviero GJ. Mixing in inclined dense jets. J Hydraul Eng 1997;693–9. https://doi.org/10.1061/(ASCE)0733-9429(1997)123:8(693).
[20] Wood IR, Bell RG, Wilkinson DL. Ocean disposal of wastewater. Singapore: World Scientific; 1993.
[21] SCCWRP (2012). "Management of brine discharges to coastal waters recommendations of a science advisory panel." Southern California Coastal Water Research Project, Technical Report 694, March 2012.
[21a] Iran Human Environmental Law, Regulation, Criteria and Standard. Islamic Republic of Iran: Department of Environment; 2012.
[22] Sultanate of Oman. Ministerial Decision No: 159/2005, Promulgating the bylaws to discharge liquid waste in the marine environment. Environment and Water Resources: Ministry of Regional Municipalities; 2005.
[22a] SCCWRP. Management of brine discharges to coastal waters recommendations of a science advisory panel. Southern California Coastal Water Research Project, Technical Report 694; March 2012
[23] Alliance GCD. Material Change of Use Application ERA 16, 19 & 7, Gold Coast Desalination Project 2006, http://www.desalinfo.com.au/Environment.asp#Project.
[24] Palomar P, Lara JL, Losada IJ. Near field brine discharge modeling part 2: Validation of commercial tools. Desalination 2012;290(2012):28–42.
[24a] Fedele F, Abessi O, Roberts JWP. Symmetry reduction of turbulent pipe flows. J Fluid Mech 2015;779:390–410. https://doi.org/10.1017/jfm.2015.423.
[25] Roberts PJW. Near field flow dynamics of concentrate discharges and diffuser design. In: Intakes and Outfalls for Seawater Reverse-Osmosis Desalination Facilities Innovations and Environmental Impacts. Switzerland: Springer; 2015. ISBN 978-3-319-13202-0 ISBN 978-3-319-13203-7 https://doi.org/10.1007/978-3-319-13203-7.
[26] Frick WE. Visual Plumes mixing zone modeling software. Environ Modell Software 2004;19:645–54. Elsevier.
[27] VISJET 2.0 User Manual. The University of Hong Kong. Principal Investigators: Prof. JHW Lee and Prof. WP Wang, 2010.
[28] Jirka GH. Integral model for turbulent buoyant jets in unbounded stratified flows. Part 1: The single round jet. Environ Fluid Mech 2004;4:1–56.
[29] Jirka GH. Integral model for turbulent buoyant jets in unbounded stratified flows. Part 2: Plane jet dynamics resulting from multiport diffuser jets. Environ Fluid Mech 2006;6:43–100.
[30] Jirka GH, Doneker RL, Hinton SW. In: User manual for CORMIX: A hydrodynamic mixing zone model and decision support system for pollutant discharges into surface waters. U.S. Environmental Protection Agency, Techn. Rep., Environmental Research Lab., Athens, Georgia, USA; 1996.
[31] Doneker RL, Jirka GH. CORMIX1: An Expert Systems for Hydrodynamic Mixing Zone Analysis of Conventional and Toxic Submerged Single Port Discharges. Technical Report, De Frees Hydraulic Laboratory, School of Civil and Environmental Engineering, Cornell University, 1990.
[32] Doneker LR, Jirka GH. D-CORMIX continues dredge disposal mixing zone water quality model laboratory and filed data validation study. Washington, DC: U.S. Environmental Protection Agency, 1997.
[32a] Doneker Robert L, Jirka Gerhard H. CORMIX-GI systems for mixing zone analysis of brine wastewater disposal. Desalination 2001;139:1–3, 263–274.
[33] Abessi O. Analysis of surface discharge of negatively buoyant effluents into non-stratified water bodies, PhD thesis, School of Civil Engineering. Tehran, Iran: Iran University of Science and Technology; 2011.

[34] Saeedi MA, Farahani Abessi A, Bleninger OT. Laboratory studies defining flow regimes for negatively buoyant surface discharges into crossflow. J Environ Fluid Mech 2012;12:439–49. https://doi.org/10.1007/s10652-012-9245-4.
[35] Baumgartner DJ, Frick WE, Roberts PJW. Dilution models for effluent discharges. 3rd ed. Newport, OR: Pacific Ecosystems Branch, ERL-N; 1994. EPA/600/R-94/086, 189 pp.
[36] Muellenhoff, W.P., Soldate, Jr., A.M., Baumgartner, D.J., Schuldt, M.D., Davis, L.R., Frick, W.E., 1985. Initial mixing characteristics of municipal ocean outfall discharges, volume 1. Procedures and Applications (EPA/600/3-85/073a, Nov 1985).
[37] Frick WE. Non-empirical closure of the plume equations. Atmos Environ 1984;18(4):653–62.
[38] Frick WE, Roberts PJW, Davis LR, Keyes J, Baumgartner DJ, George KP. Dilution models for effluent discharges. 4th ed; 2003. (Visual Plumes)" US Environmental Protection Agency, National Exposure Research Laboratory, EPA/600/R-03/025.
[39] Lee JHW, Cheung V. Generalized Lagrangian model for buoyant jets in current. J Environ Eng ASCE 1990;116(6):1085–105.
[39a] Cheung SKB, Leung DYL, Wang W, Lee JHW, Cheung V. VISJET—a computerocean outfall modelling system. In: Proceedings of the International Conference on Computer Graphics (CGI '00.), IEEE Computer Society, Washington, DC, USA; 2000.
[40] Lee JHW. Mixing of Multiple Buoyant Jets. J Hydraul Eng 2012;138(12):1008–21.
[41] ANSYS Fluent Theory Guide. ANSYS, Inc 2013, http://www.ansys.com.
[42] ANSYS CFX-Solver Modeling Guide. ANSYS, Inc 2013, http://www.ansys.com.
[43] Lars D. An Introduction to Turbulence Models. Sweden: Chalmers University of Technology Goteborg; 2003.
[44] OpenCFD Limited. OpenFOAM - Programmer's Guide. Version 21 2012;1.
[45] Shuai Z, Law Adrian WK, Mingtao J. Large eddy simulations of 45 and 60 inclined dense jets with bottom Impact. J Hydro-environ Res 2017;15(2017):54–66.
[46] Vafeiadou P, Papakonstantis I, Christodoulou G (2005) Numerical simulation of inclined negatively buoyant jets. Proc 9th Int Conf Environ Sci Tech A, pp 1537–1542.
[47] Oliver CJ, Davidson MJ, Nokes RI. k-ε Predictions of the initial mixing of desalination discharges. Environ Fluid Mech 2008;8:617–25. https://doi.org/10.1007/s10652-008-9108-1.
[48] Davis L, Davis A, Frick W. In: Computational fluid dynamic application to diffuser mixing zone analysis—Case studies. Proceedings of 3rd International Conference on Marine Waste Water Discharges and Marine Environment, Catania, Italy; 2004.
[49] Tang HS, Paik J, Sotiropoulos F, Khangaonkar T. Three-Dimensional Numerical Modeling of Initial Mixing of Thermal Discharges at Real-Life Configurations. J Hydraul Eng 2008;134(9):1210–24.
[50] Gildeh KH, Mohammadian A, Nistor I, Qiblawey H. Numerical modeling of 30° and 45° inclined dense turbulent jets in stationary ambient. Environ Fluid Mech 2015;15:537–62. https://doi.org/10.1007/s10652-014-9372-1.
[51] Abraham G. Jet Diffusion in Stagnant Ambient Fluid. Delf Hydraulic Lab. Publ.; 1963. [No. 29].
[52] Cederwall K. The Initial mixing of jet disposal into a Recipient. Tech Report 14 and 15, Div. of Hydraulics, Geteborg, Sweden: Chalmers Institute of Technology; 1963.
[53] Fan LN. Turbulent buoyant jets into stratified or flowing ambient fluids. In: Rep KH-R-18, W.M. Keck Laboratory for Hydraulics and Water Resources. Pasadena, Calif: California Institute of Technology; 1967.
[54] Cenedese C, Dalziel S. Concentration and depth fields determined by the light transmitted through a dyed solution; 1998 [Private Communication].
[55] Zhang JB, Chu VH. Shallow turbulent flows by video imaging method. J Eng Mech 2003;129(10):1164–72.

[56] Gaskin S, McKernan M, Xue F. The effect of background turbulence on jet entrainment: an experimental study of a plane jet in a shallow coflow. J Hydraul Res 2004;42(5):531–40.
[57] Kikkert GA, Davidson MJ, Nokes RI. Inclined negatively buoyant discharges. J Hydraul Eng 2007;545–54. https://doi.org/10.1061/(ASCE) 0733-9429(2007)133:5(545).
[58] Abessi O, Saeedi M, Bleninger T, Davidson M. Surface discharge of negatively buoyant effluent in unstratified stagnant water. J Hydro-environ Res 2012;6:181–93. https://doi.org/10.1016/j.jher.2012.05.004.
[59] Abessi O, Saeedi M, Davidson M, Hajizadeh Zaker N. Flow classification of negatively buoyant surface discharges in an ambient current. J Coast Res 2012;28(1A):148–55. ISSN: 1551-5036, https://doi.org/10.2112/JCOASTRES-D-10-00131.1.
[59a] Crimaldi JP. Planar laser induced fluorescence in aqueous flows. Exp Fluids 2008;44:851–63. https://doi.org/10.1007/s00348-008-0496-2.
[60] Tian X, Roberts PJW. A 3D LIF system for turbulent buoyant jet flows. Exp Fluids 2003;35(6):636–47.
[61] Tian X. 3DLIF and its applications to studies of the near-field mixing of wastewater discharge. In: PhD thesis, School of Civil and Environmental Engineering. Atlanta: Georgia Institute of Technology; 2002.
[62] Tian X, Roberts Philip JW. Mixing in Water Storage Tanks. I: No Buoyancy Effects. J Environ Eng 2008;134(12):974–85.
[63] Cederwall K. Hydraulics of marine waste water disposal. Rep. No. 42, Gothenburg, Sweden: Chalmers Institute of Technology; 1968.
[64] Zeitoun MA, Reid RO, McHilhenny WF, Mitchell TM. Model studies of outfall system for desalination plants. Research and Development Progress Rep. 804, Office of Saline Water, U.S Washington, DC: Dept. of Interior; 1970.
[65] Gungor E, Roberts PJW. Experimental studies on vertical dense jets in a flowing current. J Hydraul Eng 2009;935–48. https://doi.org/10.1061/(ASCE)HY.1943-7900.0000106.
[66] Jirka GH. Improved discharge configurations for brine effluents from desalination plants. J Hydraul Eng 2008;116–20. https://doi.org/10.1061/(ASCE)0733-9429(2008)134:1(116).
[67] Lai CKC, Lee JHW. Mixing of inclined dense jets in stationary ambient. J Hydro-Environ Res 2012;6(1):9–28.
[68] Nemlioglu S, Roberts PJW. In: Experiments on dense jets using three-dimensional laser-induced fluorescence (3DLIF). MWWD 2006—4th Int. Conf. on Marine Waste Water Disposal and Marine Environment Antalya. Istanbul, Turkey: MWWD Organization; 2006.
[69] Papakonstantis IG, Christodoulou GC, Papanicolaou PN. Inclined negatively buoyant jets. 2: Concentration measurements. J Hydraul Res 2011;49(1):13–22.
[70] Papakonstantis IG, Christodoulou GC, Papanicolaou PN. Inclined negatively buoyant jets. 1: Geometrical characteristics. J Hydraul Res 2011;49(1):3–12.
[71] Abessi O, Roberts WP. Effect of nozzle orientation on dense jets in stagnant environments. J Hydraul Eng, ASCE 2015. https://doi.org/10.1061/(ASCE;), HY.1943-7900.0001032, 06015009.
[72] Abessi O, Roberts WP. Dense jet discharges in shallow water. J Hydraul Eng, ASCE 2015. ISSN 0733-9429/04015033(13).
[73] Fischer B, List JE, Imberger J, Brooks NH. Mixing in inland and coastal waters. San Diego: Academic Press; 1979.
[74] Marti CL, Antenucci JP, Luketina D, Okely P, Imberger J. Near-Field Dilution Characteristics of a Negatively Buoyant Hypersaline Jet Generated by a Desalination Plant. J Hydraul Eng 2011;137(1):57–65.
[75] Tarrade L, Rayner DS, Miller BM. Physical modeling of brine diffuser; 2009. Technical Rep. 2009/17, Univ. of New South Wales, Water Research Laboratory, Sydney, Australia.

[76] Miller BM, Cunningham IL, Timms WA. Physical modeling of the sea water concentrate diffuser for the Sydney desalination study, 2007. University of New South Wales, Water Research Laboratory Technical Report 2007/04.

[77] Miller BM, Tarrade L. In: Design considerations of outlet discharges for large seawater desalination projects in Australia. 6th International Conference on Marine Wastewater Discharges, MWWD 2010; 2010. Langkawi, Malaysia.

[78] Roberts PJW, Abessi O. Optimization of desalination diffusers using three-dimensional laser-induced fluorescence. In: Agreement Number R11 AC81 535. Atlanta: United States Bureau of Reclamation, School of Civil and Environmental Engineering Georgia Institute of Technology; 2014.

[79] Roberts PJW, Snyder WH, Baumgartner DJ. Ocean outfalls I. submerged wastefield formation. J Hydraul Eng 1989;115(1):1–25.

[80] Tian X, Roberts PJW, Daviero GJ. Marine wastewater discharges from multiport diffusers. IV: Stratified flowing water. J Hydraul Eng 2006;132(4):411–9.

[80a] Abessi O, Roberts WP. Multiport diffusers for dense discharge in flowing ambient water. J Hydraul Eng ASCE 2017. https://doi.org/10.1061/(ASCE)HY.1943-7900.0001279.

[81] Miller BM, Tarrade L. In: Design considerations of outlet discharges for large seawater desalination projects in Australia. 6th International Conference on Marine Wastewater Discharges. Malaysia: Langkawi; 2010. MWWD-IEMES 201025–29 October, 2010.

[81a] Abessi O, Roberts WP, Gandhi V. Rosette diffusers for dense effluents. J Hydraul Eng ASCE 2016. https://doi.org/10.1061/(ASCE)HY.1943-7900.0001268.

[81b] Abessi O, Roberts WP. Rosette diffusers for dense effluents in flowing currents. J Hydraul Eng ASCE 2017(in press).

[82] Koester GE. Experimental study of submerged single-port thermal discharges. S.M. dissertation, Cambridge, MA: Dept. of Civil Engineering, Massachusetts Institute of Technology; 1974.

[83] Roberts PJW, Wright SJ. Vertical round buoyant jets in shallow water. J Hydraul Eng 1983;109:490–4.

[84] Jiang B, Law AW-K, Lee JH-W. Mixing of 45° inclined dense jets in shallow coastal waters: Surface impact dilution. In: 3rd International Symposium on Shallow Flows, IIHR-Hydroscience and Engineering. Iowa City, IA: Univ. of Iowa; 2012.

[85] Jiang B, Law AWK, Lee JHW. Mixing of 30° and 45° inclined dense jets in shallow coastal waters. J Hydraul Eng 2014;241–53. https://doi.org/10.1061/(ASCE)HY.1943-7900.0000819.

[86] Bleninger T, Jirka GH. Near- and far-field model coupling methodology for wastewater discharges. In: Lee JHW, Lam KL, editors. Proc. 4th International Symposium on Environmental Hydraulics and 14th Congress of Asia and Pacific Division. Hong Kong, China: International Association of Hydraulic Engineering and Research; 2004.

[87] Palomar P, Lara L, Losada I, Rodrigo M, Alvarez A. Near Field brine discharge modeling. Part 1: Analysis of commercial tools. Desalination 2012;290(2012):14–27.

[88] Jones G, Nash D, Doneker L, Jirka H. Buoyant surface discharge into water bodies. I: flow classification and prediction methodology. J Hydraul Eng 2007;133:1010–20.

[88a] Abessi O, Saeedi M, Zaker NH. Flow characterization dilution in surface discharge of negatively buoyant flow in stagnant and non-stratified water bodies. J Water Waste Water 2012;80:71–90. ISSN: 1024-5936.

[88b] Abessi O, Saeedi M, Zaker NH. Waste field characteristics, ultimate mixing and dilution in surface discharge of dense jets into stagnant water bodies. J Water Waste Water 2012;81:2–16. ISSN: 1024-5936.

[89] Chu VH, Jirka GH. Surface buoyant jets. In: Encyclopedia of fluid mechanics. 1986. Gulf, Houston, Chap. 25.

[90] Gholamreza-Kashi S, Martinuzzi RJ, Baddour RE. Mean flow field of a nonbuoyant rectangular surface jet. J Hydr Eng 2007;133(2):234–9.
[90a] Kassem JAM, Khan JA. Three-dimensional modeling of negatively buoyant flow in diverging channels. J Hydraul Eng 2003;129(12):936–947.
[91] Davidson MJ, Pun KL. Locating discharge trajectories in still and moving ambient fluids. J Hydr Eng 2000;126(7):513e524.
[92] Hansen J, Schroder H. Horizontal jet dilution studies by use of radioactive isotopes. Acta Polytech Scand, Civil Eng Build Constr Ser 1968;49:24.
[93] Xiao J, Travis JR, Breitung W. Non-business integral model for horizontal turbulent buoyant round jets. Sci Technol Nucl Install 2009. https://doi.org/10.1155/2009/862934.
[94] Abdelwahed, M.S.T., Chu, V.H., 1981. Surface jets and surface plumes in cross-flows. Technical Report No. 81e1. Fluid Mechanics Laboratory, McGill University, Montreal.
[95] Nash JD, Jirka GH. Buoyant surface discharges into unsteady ambient flows. J Dyn Atmos Oceans 1996;24(1e4):75–84.
[96] Saeedi M, Abessi O, Aliabadi Farahani A. Prediction of flow pattern in surface discharge of negatively buoyant wastes comparing the results from CORMIX model. J Env Sci Tech 2016;18(2). ISSN: 1563-4809.
[97] Shacheri F, Abessi O, Saeedi M, Samani J. Flow characteristics of surface dense discharge from rectangular and trapezoidal channels into stagnant water. 2017 (under review).

FURTHER READING

[1] Bleninger T, Jirka GH, Weitbrecht V. Optimal discharge configuration for brine effluents from desalination plants. Deutsche Meerwasser Entsalzung, Berlin: Proc. DME-Congress; 2006.
[2] Cheung S.K.B. Leung D.Y.L. Wang W. Lee J.H.W. Cheung V. (2011) "VISJET—A Computer Ocean Outfall Modelling System" Computer Graphics International Proceedings, Geneva, Switzerland, 19-24 June 2000, p. 75-80. DOI: https://doi.org/10.1109/CGI.2000.852322.
[3] Jiang B, Law AWK. Non-interfering multiport brine diffusers in shallow coastal waters. J Appl Water Eng Res 2013;1(2):148–57.
[4] Isaacson MS, Koh RCY, Brooks NH. Plume dilution for diffusers with multiple risers. J Hydraul Eng 1983;199–220. https://doi.org/10.1061/(ASCE); 0733–9429(1983) 109:2(199).
[5] Lee JHW, Valiant C, Kuang CP, Choi DKW, Wang WP, Tu CH, et al. VISJET & VISFLOOD: Software for Environmental Hydraulic Modeling and Visualization. In: 7th Asian Symposium on Visualization; 2003.
[6] Tian X, Roberts Philip JW. Experiments on marine wastewater diffusers with multiport rosettes. J Hydraul Eng 2011;137(10):1–13.

CHAPTER 8

Post-Treatment of Desalinated Water—Chemistry, Design, Engineering, and Implementation

Liat Birnhack, Ori Lahav
Faculty of Civil and Environmental Engineering, Technion, Haifa, Israel

NOMENCLATURE

AI	Aggressiveness Index
Acd	CO_3^{2-} acidity
Alk	$H_2CO_3^*$ alkalinity
AWW	American Water Works Association
BW	brackish water
CCPP	calcium carbonate precipitation potential (mg/l as $CaCO_3$)
C_T	total inorganic carbon
DFI	driving force index
DiaNF	diananofiltration
DO	dissolved oxygen
HRT	hydraulic retention time
EC	electrical conductivity
IX	ion exchange
LSI	Langelier saturation index
LR	Larson ratio
NTU	nephelometric turbidity units
PT	post treatment
RCI	Riddick corrosion index
SW	seawater
TDS	total dissolved solids (mg/l)
TH	total hardness
WHO	World Health Organization

8.1 INTRODUCTION

The quality of unconditioned desalinated water can differ significantly with respect to TDS and specific species' concentrations (e.g., boron), depending both on the type of desalination technology used for salt separation (Reverse Osmosis (RO), ED or thermal desalination techniques) and on

Sustainable Desalination Handbook
https://doi.org/10.1016/B978-0-12-809240-8.00008-3

the specific design and operation of the desalination process [1]. However, in almost all cases, desalination permeates are slightly acidic, contain very low buffering capacity, and are very soft (i.e., contain very low calcium and magnesium concentrations). Water with such quality cannot be supplied directly for either domestic consumption and/or irrigation, for a variety of reasons [2]. Delivery of unconditioned desalinated water to consumers may result in unwanted repercussions throughout the usage cycle: in the distribution system, to the consumers, to downstream wastewater treatment facilities, and in some arid or semiarid locations also to consumers reusing the downstream treated wastewater for agriculture. For these reasons, post treatment (PT) is invariably applied to adjust the water quality, as required according to local considerations and potential uses.

Surveying the literature reveals that (a) water quality standards, specific for desalinated water, which take into account these issues were not set anywhere in the world, until recently, when Israel and Cyprus have adopted such standards. In 2007 the Israeli Ministry of Health adopted a set of water quality criteria specific for desalinated water [3,4]. Thereafter, an almost identical set of quality parameters were adopted by the Water Development Department in Cyprus [5]. Since hardly any official guidelines exist, practiced PT processes around the world lead to varying product water qualities, which are frequently inadequate for meeting final user needs (i.e., maintain the integrity of the distribution system, protect public health, and be appropriate for plant irrigation purposes). For example, a survey among 25 desalination plants (mainly from the USA) showed that many desalination plants still apply a post-treatment sequence comprising merely pH adjustment with NaOH, blending with groundwater, and dosage of corrosion inhibitor to the water. The calcium concentration appeared not to be addressed at all, and consequently, so was the $CaCO_3$ precipitation index, which is perceived by many to be essential for minimization of corrosion and red water phenomena. As a result, locations served by such plants often report on red water events, enhanced distribution-system corrosion, and problems related to lack of pH stability [6]. In addition, the issue of minimum Ca^{2+} concentration required for public health reasons is not addressed. Additionally, $CaCO_3$-precipitation indices for assessing water's corrosivity potential are frequently used incorrectly, due to what appears as misinterpretation of the fundamental principles by which these indices should be implemented in addition to misconception of their limitations [7]; (b) There are various differences between the alternative PT methods, which have been only recently addressed [8].

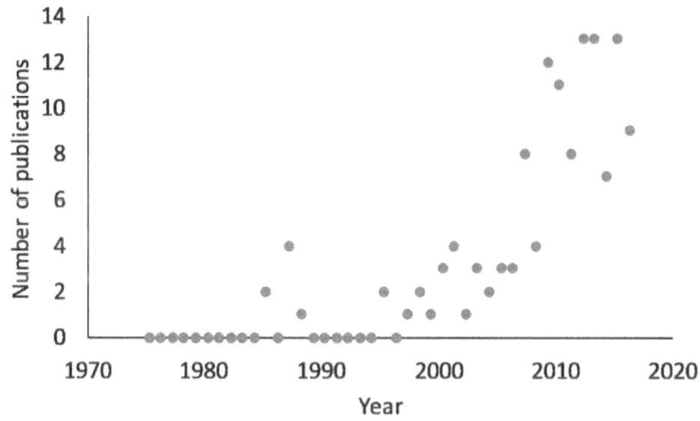

Fig. 8.1 Number of publications dealing with PT issues in the last four decades.

A literature survey reveals that the importance of the subject related to conditioning of desalinated water appeared first when the first large desalination plants were built in the 1980s. However, the subject received very limited attention in the next two decades, as shown in Fig. 8.1, which presents the number of publications dedicated to PT of desalinated water over the years. In recent years, since the adverse effects of ineffective PT have become apparent, professionals began focusing on this subject. (c) Finally, the knowledge available on the subject is sometimes misused or misunderstood, as mentioned, for example in the context of corrosion control strategies, by Imbrogno et al. [9].

The limited understanding of the alternatives available for post-treatment of desalinated water and the general lack of knowledge in this field has often led to unfavorable results. Examples include corrosion and other product water-quality-related problems in a number of projects worldwide, e.g., USA (Boston) [10], Israel [11], USA (Florida) [12], Cyprus [13], Sweden [14], and USA (Arizona) [15].

The aim of this chapter is to provide a detailed overview of the existing knowledge on PT processes. The chapter encompasses the theoretical background, the advantages/disadvantages, and main challenges of commonly practiced PT processes as well as of newly developed water conditioning technologies. The chapter is concluded with description of knowledge gaps and recent research trends.

8.1.1 Considerations Associated With Desalinated Water Quality

In order to determine the required quality of desalinated water (following the PT step) for a given use, the following aspects should be considered:

8.1.1.1 Interaction of the Water With the Distribution System

The need to stabilize the water so that it would not enhance metal corrosion and concrete dissolution has been recognized for decades. Corrosion of system pipes is associated with health hazards due to unwanted release of metal ions such as lead, copper, and zinc into the drinking water [7,16–21]. Moreover, several works report on serious red water cases owing to source water switch, such as in South California, Tucson Arizona, Tampa Florida in the United States, and China [15,22]. More recently (2015), one can also mention the lead contamination event in Flint, Michigan [23]. These works reveal that alternation of the water quality may undermine the fragile balance between the water and the corrosion scale. Corrosion-related problems are associated with jeopardize of distribution system integrity, longevity, and with maintenance costs [7,18,20,23–25]. In 1999, for example, the AWWA estimated that the cost inflicted on US water utilities would be $325 billion over the following 20 years to upgrade damaged water distribution systems [25]. This figure may perhaps be used to explain why most PT technologies and scientific research focus primarily on controlling the corrosivity of desalinated water (by its chemical stability), while often overlooking/neglecting other important water quality aspects. "Red water" events (sudden detachment of corrosion products from the pipe internal sidewalls and their release to the water) are by product of previously occurred corrosion. It should however be referred to and treated as a separate problem, being an outcome of a combination of conditions, which result in a different effect than "crawling" corrosion [20,26]. Red water events and crawling corrosion are two of the most common adverse water quality effects occurring as a result of ineffective desalinated water post treatment [6].

8.1.1.2 Public Health

Two water quality aspects may be directly related to the consumer's health: (1) Biostability of the water in the distribution system, i.e., the efficiency of water disinfection by chlorine species, as well as minimization of bacterial regrowth. (2) Deficiency of important minerals, such as iodide, calcium, and magnesium ions: an important issue, with a potential to directly affect consumers' health. Iodide, which is not supplied with desalinated water but invariably appears in natural drinking waters, is essential for proper functioning of the thyroid gland [27]. Calcium and magnesium ions are minerals of key importance for both human health and agricultural and horticultural water uses. These minerals are notably absent from unconditioned desalinated water.

8.1.1.3 Suitability for Irrigation Purposes

Many countries already use desalinated water directly for agricultural irrigation, e.g., Spain, Italy, Kuwait, and Saudi Arabia, while other countries will probably use desalinated water for irrigation purposes in the near future [28]. In these cases, the mineral content of the water is essential (mainly the concentrations of Ca^{2+}, Mg^{2+} and SO_4^{2-}). On the other hand, low boron, sodium and chloride concentrations are considered beneficial for agricultural uses.

8.1.1.4 Possible Detrimental Effects on Downstream Wastewater Treatment Plants

This issue relates to biological nitrogen removal within wastewater treatment plants. When complete removal of nitrogen species is required, high alkalinity in the wastewater is necessary in order to prevent pH drop and maintain biological process stability [3,29]. The minimum amount of alkalinity needed in the wastewater to sustain stable nitrification - denitrification process may not be available if water with too low alkalinity (below 100 mg/L as $CaCO_3$) is released by the desalination plant and distributed without proper conditioning [29]. Another issue related to WWTP is the settling of the sludge: low total hardness concentrations (i.e., low Ca^{2+} and Mg^{2+} concentrations), or, more specifically low ratio between divalent to monovalent cation concentrations, as may be encountered in wastewater originating from desalinated water, is associated with poor sludge settling. It has been shown that high divalent to monovalent cation concentration ratios improved floc properties, reduced sludge index volume, effluent chemical oxygen demand, and effluent total suspended solids [30].

8.1.1.5 Effect on the Quality of Reclaimed Water Used for Agricultural Irrigation

Reclaimed water (i.e., highly treated secondary or tertiary effluent from wastewater treatment plants) is often used for agricultural irrigation in the same locations where desalinated water constitutes a major source of freshwater (e.g., Spain, Australia, and Israel) and in the future more countries are expected to follow, e.g., Florida and California in the USA [4]. Certain quality characteristics of reclaimed water are directly dependent on the quality of the original desalinated water. From this aspect, it is advantageous to enrich the desalinated water with Ca^{2+}, Mg^{2+}, and SO_4^{2-} ions and in some cases also with alkalinity [29] and reduce the boron concentration to minimum.

8.1.2 Water Quality Parameters

The following parameters are usually used to determine the required water quality of soft waters with respect to their interaction with the distribution system, desalinated water included: [Ca^{2+}], alkalinity, pH, TDS, and the Calcium Carbonate Precipitation Potential (CCPP) (or another $CaCO_3$ precipitation related index such as LSI). Clearly, different values for these parameters can and have been set in different locations, depending on the local conditions, e.g., the quality of other water sources with which the desalinated water may be blended; the portion of desalinated water out of the overall water supply; the average daily Ca^{2+} and Mg^{2+} intake from food, etc. Between 2005 and 2007, a comprehensive work was carried out in Israel, with the aim of determining a set of target water quality criteria, specific for desalinated water. The work was performed under the auspices of the Committee for the Update of Water Quality Standards, appointed by the Israeli ministry of Health. It relied partly on various worldwide water quality standards and guidelines for corrosion control of naturally occurring soft waters. Based on this work, the following specific water quality criteria for desalinated water were derived and are now being enforced by the Israeli Ministry of Health: Alk > 80 mg/l as $CaCO_3$; $80 < [Ca^{2+}] < 120$ mg/l as $CaCO_3$; $3 < CCPP < 10$ mg/l as $CaCO_3$ and LSI > 0; and pH < 8.5. A similar set of criteria was also set in Cyprus, as reflected, for example in the updated Larnaka desalination plant contract from 2010 [5]. The considerations that led to the establishment of these new criteria are summarized in the following paragraphs [8]:

- *Ca^{2+} concentration.* The minimum Ca^{2+} concentration required for adequate protection of human health is 50–60 mg/L as $CaCO_3$ [31]. An upper recommended value was also set in these criteria due to economic reasons attributed to the need to supply water that is not excessively hard. On the other hand, high calcium concentration is required for establishing a positive CCPP value at pH close to 8 (or not higher than 8.5, as required for biostability reasons, see later). Combined together, the required Ca^{2+} concentration was set between 80 and 120 mg/l as $CaCO_3$ (i.e., $32 < [Ca^{2+}] < 50$ mg/l).
- *Alkalinity* (with $H_2CO_3^*$ as reference species). It is widely accepted that relatively high alkalinity values are advantageous in desalinated water for various reasons including increased water buffer capacity (e.g., [7]); prevention of metal ion release in the water distribution system (e.g., [32,33]), i.e., for the minimization of corrosion and red water events (except when lead and copper form a part of the water distribution

systems [7]); and for minimizing the potential negative effects of blending of desalinated water with other water sources within the distribution system [34]. In case the desalinated water is used for agricultural irrigation, a relatively high alkalinity value may be also needed for maintaining a stable pH when acidic or basic fertilizers are added through the irrigation system. Finally, a target minimum content of alkalinity in the drinking water may be needed to allow for stable operation of downstream wastewater treatment plant, as mentioned before. Accordingly, the alkalinity requirement was set at >80 mg/l as $CaCO_3$.

- *$CaCO_{3(s)}$ precipitation potential.* CCPP is the quantitative measure of the precise potential of a solution to precipitate (or dissolve) $CaCO_{3(s)}$. As such, it constitutes an unambiguous parameter that can be used in the context of guidelines or regulations without invoking misunderstanding [9]. Maintaining a high CCPP has been associated with a decrease in the rate of corrosion and potential for red water events occurrence. However, an upper CCPP value should also be set in order to prevent possible buildup of excessive $CaCO_3$ scales on pipes and pumping stations [7]. Note that if desalinated water is supplied only to nearby agricultural areas and its retention time in the distribution system is relatively short, a less stringent minimum threshold of CCPP can be imposed.
- *Langelier saturation index (LSI).* Langelier's saturation index was developed in 1936 to answer the question whether a certain solution will tend to precipitate or dissolve $CaCO_{3(s)}$ or whether it is exactly at saturation with respect to this solid. LSI is a qualitative parameter. As such, it does not give any information about the potential amount of $CaCO_{3(s)}$ that would precipitate or dissolve, and thus cannot be used in a quantitative context [18,25,35]. Before the computer era, LSI was a convenient technique for determining the saturation state. To date, determining the saturation state by comparing between Q (the quotient) and K'_{sp} is easily attained using a simple computer algorithm and the LSI procedure has become much less significant. Moreover, computation of the value of CCPP is also possible by a variety of computer software. Nevertheless, LSI is still mentioned in many text books and papers as well as in desalination post-treatment guidelines.
- *pH.* Within the range of pH values attained after PT (between ~6 and ~8), a lower pH value results in a higher chlorine-based disinfection efficiency [36,37] and higher buffer capacity. Maintaining an appropriate buffering capacity in drinking water is essential because several

processes (either deliberate or not) that occur in the distribution system can potentially affect the pH value, possibly resulting in changing the CCPP value from positive to negative. However, a pH value between 8.0 and 8.5 may decrease lead corrosion [38]. Examples of deliberate water treatment processes that result in a decrease in the pH value are disinfection with Cl_2 gas and dosage of fluoride ions by the addition of H_2SiF_6 (fluorosilicic acid). Unintentional processes include organic matter oxidation by aerobic and anaerobic bacteria, biological nitrification of ammonia present due to chloramination practices, and iron oxidation and precipitation.

- *Mg^{2+} concentration.* Only very few desalination plants worldwide add a substantial Mg^{2+} concentration to desalinated water in the PT step. Nevertheless, the World Health Organization (WHO) has recently concluded that "There is a growing consensus among epidemiologists that the epidemiological evidence, along with clinical and nutritional evidence, is already strong enough to suggest that new guidance should be issued" [39]. The WHO has thus recommended in its recent publications to maintain a minimum Mg^{2+} concentration in all drinking waters [16,39]. Following this recommendation, and due to the acknowledged importance that Mg^{2+} ions have on both crop irrigation [11] and the public health [16,21,31,39–41], the Israeli Ministry of Health decided to add a minimum requirement of 20 mg Mg/l to the desalinated water quality criteria in 2009 (pending a pilot study) [42].

8.1.3 Chemicals Used for Corrosion Minimization in Water Distribution Systems

Addition of chemicals described as "corrosion inhibitors" (e.g., polyphosphates, monophosphates, zinc orthophosphate, and silicates) is a common technique for conditioning of soft- and low-buffering-capacity waters (e.g., in the USA [6]). This practice is typically used in addition to post-treatment but sometimes also as a standalone corrosion protection alternative [18,25,43–46]. The recognition of the harmful effect of zinc and phosphate loading on wastewater treatment plants has curtailed the usage of zinc-phosphate [25]; however, this corrosion inhibitor is still used in desalination plants [6]. Many studies claim that dosing polyphosphates and silicates to water has the potential to reduce iron release and visible appearance of released iron particles, i.e., red water events [16,47]. However, the interaction between iron scales and these chemicals is still unclear [25,43,44]. Moreover, the use of corrosion inhibitors has often been criticized because of their apparent environmental adverse effects. It has been shown that even small phosphorus dosages are correlated with an increase in the total microbial count in water sampled from drinking water distribution systems [48–50].

It is important to note that in the presence of inhibitors iron is released to the water in the form of stabilized colloid particles [25]. Therefore, the concentration of dissolved iron is naturally reduced when polyphosphates and orthophosphates are dosed to the water, while the total iron (dissolved and suspended) concentration either remains constant or may even increase. Thus, iron release (hence corrosion rate) does not necessarily decrease when the so-called "corrosion inhibitors" are dosed to the water.

It is actually possible that phosphate inhibitors actually enhance iron corrosion rate, as shown in several studies (e.g., [16,25,51]) and as mentioned, this application has been associated with further adverse effects. As a result, in some locations, this application was stopped and corrosion control is maintained solely by adjusting the water quality with respect to alkalinity, Ca^{2+}, and pH to result in slightly positive CCPP value.

8.1.3.1 Post-Treatment Nomenclature

PT processes are sometimes termed recarbonation, remineralization, or potabilization processes. Recarbonation can be defined as the elevation of the total inorganic carbon concentration (C_T) of the water while "remineralization" can be defined as the introduction of minerals to the water [1]. An appropriate PT process aims at both recarbonizing and remineralizing the water. "Potabilization" is defined as a process by which the water becomes suitable for drinking. In desalination plants, it usually refers to enriching the water with minerals, stabilizing it (i.e., recarbonating it), and also disinfecting it [52]. It is noted that these terms are often used interchangeably. In this chapter, the expression "post treatment" is used to describe the method by which the quality of the desalination process product is improved, in order to comply with drinking water regulatory requirements in terms of alkalinity, TH, and corrosivity.

8.1.4 Subjects Not Included in the Scope of this Chapter: Disinfection, Fluoridation, Boron Removal, and Aeration

Disinfection, fluoridation, boron removal, and aeration are occasionally referred to as a part of the PT process. However, these issues are not discussed in this book chapter, which solely addresses PT processes designed to control the water's buffering capacity, TH (i.e., Ca^{2+} and Mg^{2+} concentrations) and corrosivity indices. Nevertheless, these issues are not completely ignored in this chapter, and indirect effects of the PT process on them are discussed and explained.

8.2 BASIC CHEMICAL PRINCIPLES

8.2.1 The Carbonate System

The carbonate system is the main (often the only) weak acid system in natural fresh waters and also in desalinated product waters. The carbonate system is a diprotic weak acid system, comprising three species: $H_2CO_3^*$, HCO_3^-, and CO_3^{2-}. The sum of the concentrations of the species is often denoted C_T or TIC (total inorganic carbon) concentration.

In actual fact, there are four carbonate species present in the aqueous phase since $H_2CO_3^*$ is a fictitious species, representing the sum of two "real" species: $H_2CO_{3(aq)}$ and $CO_{2(aq)}$, which are in equilibrium according to the following reaction:

$$CO_{2(aq)} + H_2O \leftrightarrow H_2CO_{3(aq)}, \quad K = 1.7 \times 10^{-3} \tag{8.1}$$

Note that according to the equilibrium constant of Eq. (8.1), the concentration of the species $CO_{2(aq)}$ is three orders of magnitude greater than the concentration of the species $H_2CO_{3(aq)}$.

Both H_2CO_3 and CO_2 can release a proton to the water according to its own equilibrium constant. However, for simplicity sake, the sum of the concentration of the two species, i.e., $H_2CO_3^* = CO_{2(aq)} + H_2CO_3$, is usually considered instead of the individual concentrations. A theoretical equilibrium constant K $H_2CO_3^*/HCO_3^- = 10^{-6.35}$ was established to describe the tendency of the virtual species $H_2CO_3^*$ to release a proton:

$$H_2CO_3^* \leftrightarrow HCO_3^- + H^+, \quad K_{C1} = 10^{-6.35} \tag{8.2}$$

The second protonation reaction of the carbonate system is:

$$HCO_3^- \leftrightarrow CO_3^{2-} + H^+, \quad K_{C2} = 10^{-10.33} \tag{8.3}$$

Fig. 8.2 describes the distribution of the carbonate system species as a function of pH, according to the quoted equilibrium constants. Note that at pH values lower than ~5, the dominant species is $H_2CO_3^*$. Low pH values prevail, for example, in water pumped into calcite ($CaCO_3$) dissolution

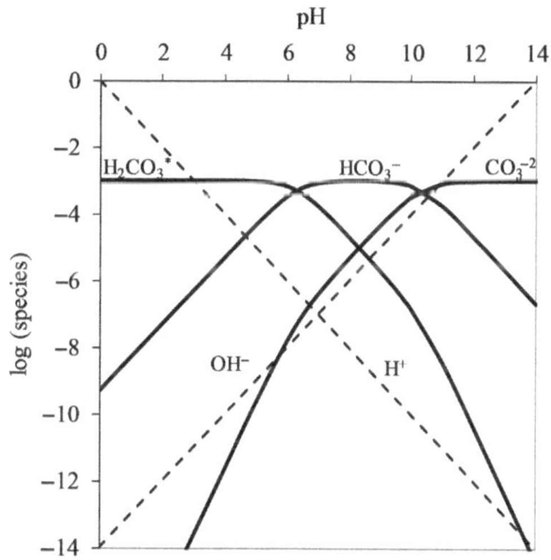

Fig. 8.2 pH log species representation of the carbonate system. $C_T = 10^{-3}$ M. The concentration of the species H^+ and OH^- appears in dashed lines.

reactors. Similarly, at the range $7.5 < pH < 9.5$, the dominant species is HCO_3^-. This is usually the case in the product water of a desalination plant, following the PT step.

Understanding the characteristics of the carbonate system forms the basis for understanding the terms alkalinity, CCPP, and buffer capacity, and for the design of most PT processes.

8.2.2 Aqueous—Gaseous Phase Interaction (CO_2 Saturation State)

The carbonate system has an important component in the gaseous phase, i.e., carbon dioxide ($CO_{2(g)}$). $CO_{2(g)}$ diffuses into (or out of) the aqueous phase to attain equilibrium with dissolved carbon dioxide ($CO_{2(aq)}$), which is the main (~99.7%) constituent of the species $H_2CO_3^*$ (see Eq. (8.1)). The equilibrium between the gaseous and dissolved species is described by Henry's law:

$$CO_{2(aq)} \leftrightarrow CO_{2(g)}, \quad H = \frac{P_{CO_2(g)}}{\left[CO_{2(aq)}\right]} \tag{8.4}$$

where H is Henry's constant for CO_2 (30.2 bar M^{-1}), $P_{CO2(g)}$ stands for the partial pressure of $CO_{2(g)}$ (bar), and [ ] stands for concentration (M).

The partial pressure of $CO_{2(g)}$ in the atmosphere is unfortunately crawling upwards in the last decades, currently (November 2016) amounting to ~4.0×10^{-4} bar. Thus, according to Henry's law, if equilibrium between the two phases is attained, then the concentration of $CO_{2(aq)}$ would be ~0.59 mg/L at 25°C:

$$\left[CO_{2(aq)}\right] = \frac{P_{CO_2(g)}}{H_{CO_2}} \cdot M_{WCO_2} = \frac{3.9 \times 10^{-4}\ \text{bar}}{30.2\left(\frac{\text{bar} \times \text{L}}{\text{mol}}\right)} \cdot 44{,}000\frac{\text{mgCO}_2}{\text{mol}} \quad (8.5)$$
$$= 0.59\frac{\text{mgCO}_2}{\text{L}}$$

When the water is in contact with the atmosphere, from the practical point of view, there are two initial water qualities of interest: (1) water characterized by a dissolved CO_2 concentration greater than 0.59 mg/L, i.e., oversaturated with regard to atmospheric CO_2. Such water will tend to release CO_2 gas; and (2) The dissolved CO_2 concentration is lower than 0.59 mg/L. In this case, the solution is undersaturated with regard to atmospheric CO_2 and consequently, CO_2 will tend to be absorbed from the gaseous phase to the solution.

When CO_2 is released from solution, the pH value increases and the C_T concentration decreases. Similarly, absorption of CO_2 into solution decreases pH and increases C_T. The alkalinity ($H_2CO^*_{3alk}$) concentration remains constant when CO_2 is released or absorbed into solution (see Section 8.2.3).

Several steps in the PT processes include deliberate or unintentional emission/absorption of CO_2. Understanding these processes is essential for making the right decisions on design and operation of PT processes.

8.2.3 $H_2CO_3{}^*{}_{\text{alkalinity}}$

The term "Alkalinity" in drinking water typically refers to the value of $H_2CO_3^*$ alkalinity, which is defined as the proton-accepting capacity of the water with respect to $H_2CO_3^*$ as reference species. Such alkalinity value is typically determined by strong acid titration to a pH value close to 4.5 (i.e., close to the $H_2CO_3^*$ equivalence point, where all the proton-accepting species are converted to $H_2CO_3^*$, see Fig. 8.2), or, more accurately, by applying the Gran titration technique [53]. The mathematical expression of this value is given in Eq. (8.6):

$$\text{Alkalinity}_{\left(H_2CO_3^*\right)} = 2\left[CO_3{}^{2-}\right] + \left[HCO_3{}^{-}\right] + \left[OH^{-}\right] - \left[H^{+}\right] \quad (8.6)$$

where alkalinity is expressed in units of meq/L, and [ ] stands for concentrations in mmol/L.

The concentration of each of the species of the carbonate system is a function of C_T, pH, and the equilibrium constants of the carbonate system: K_{C1} and K_{C2} (see Eqs. (8.2) and (8.3), respectively). Therefore, Alk is a function of C_T and pH. For a given pH value, a higher C_T value results in a higher Alk concentration. Out of the advantages gained from maintaining a high Alk value, attaining a sufficient buffer capacity and CCPP values are the most significant, as mentioned before.

Note that the Alk concentration of a given solution (Eq. 8.6) is completely unrelated to its TH value and these terms should not be used interchangeably.

8.2.4 Buffer Capacity

Buffer capacity is defined as the ability of water to withstand changes in pH when a strong base or a strong acid are added to it. The buffer capacity, β, of natural water can be calculated from the knowledge of the alkalinity value (or C_T) and the pH using Eq. (8.7):

$$\beta = \frac{2.303 C_T K_{C1} \left(H^+\right)}{\left(K_{C1} + \left(H^+\right)\right)^2} + \frac{2.303 C_T K_{C2} \left(H^+\right)}{\left(K_{C2} + \left(H^+\right)\right)^2} + 2.303\left(\left(H^+\right) + \left(OH^-\right)\right) \quad (8.7)$$

From Eq. (8.7) it can be seen that for a given C_T value, a higher buffer capacity (sometimes termed buffer intensity) is attained when the pH is close to the pK values of the carbonate system, i.e., at pH values close to either 6.3 or 10.3. In addition, for a given pH, a higher C_T value (corresponding to a higher alkalinity value) results in higher buffer capacity. Despite these relationships, it should be pointed out that Alk, C_T, and buffer capacity are obviously different and independent parameters.

When a strong acid, for example, is dosed to a water containing a weak acid (e.g., the carbonate system), the proton-accepting species of the weak acid (i.e., CO_3^{-2} or HCO_3^-, in this case) reacts with the protons to form the more acidic species of the weak acid system (Eqs. (8.2) and (8.3))'; thus, the extent of pH drop is decreased. Similarly, when strong base is dosed to such a solution, pH elevation is minimized by the reaction of the proton-donating species ($H_2CO_3^*$ and HCO_3^-) with the hydroxide ions.

Supplying well-buffered water ensures that the pH value would remain fairly constant, regardless of processes the water might be subjected to in water treatment facilities, in the distribution system, and in downstream wastewater treatment plants. Examples of the importance of this parameter include the fact that high buffer capacity inhibits the potential of pH drop in distribution systems as a result of chlorination and fluoridation thereby minimizing the consequent drop in CCPP/LSI. In the context of PT, buffer capacity has an important effect on the extent of dissolution of minerals such as dolomite and calcite (see Sections 8.2.6 and 8.2.7), as well as on the required dosage of NaOH necessary for final pH adjustment in these processes.

8.2.5 pH

pH, the negative logarithm of the active H^+ concentration, supplies a general indication on the acidic or basic nature of the solution. However, it should be borne in mind that the pH value alone does not provide sufficient information on the acid-base characterization of a given water sample, since when a pH value is accompanied by very low buffering capacity, the result may well be a significant downstream pH change due to a variety of possible reactions. Thus, information on the alkalinity or the C_T concentration in the water, in addition to the value of pH, is essential in order to define the acid-base characteristics of the water.

8.2.6 $CaCO_3$ Solubility

Thermodynamically, the dissolution of metal carbonates can be described by the following parallel reactions, which occur at the solid/water interface [54]:

$$MeCO_3 + H^+ \xrightarrow{k_1} Me^{2+} + HCO_3^- \tag{8.8}$$

$$MeCO_3 + H_2CO_3^* \xrightarrow{k_3} Me^{2+} + 2HCO_3^- \tag{8.9}$$

$$MeCO_3 \overset{k_4}{\leftrightarrow} Me^{2+} + CO_3^{2-} \tag{8.10}$$

where Me^{2+} represents Ca^{2+} or Mg^{2+}.

Note that as a result of the dissolution process either bicarbonate (as described in Eqs. (8.8) and (8.9)) or any other species of the carbonate system can form, as a function of pH and not directly of the dissolution path. In other words, although Eqs. (8.8) and (8.9) describe the formation of bicarbonate, if solution pH is lower than 6, the dominant species will be $H_2CO_3^*$.

Water is defined as over- or undersaturated with respect to $CaCO_{3(s)}$ by comparing the product of Ca^{2+} and carbonate (CO_3^{2-}) activities present in the water (denoted Q) to $K_{sp(CaCO3)}$. In case $Q > K_{sp}$, the water is oversaturated, and in case $Q < K_{sp}$ the water is undersaturated.

In most raw desalinated waters, the influence of ionic strength and formation of ion pairs can be neglected. Accordingly, the dissolution/precipitation propensity of $CaCO_3$ (denoted also calcite, limestone or marble, and consists mainly of the calcite mineral) can be considered to be an exclusive function of the concentrations of Ca^{2+} and carbonate (CO_3^{2-}) ions. The concentration of the latter is a function of pH and C_T (or any other two independent carbonate-system parameters, such as alkalinity and acidity). A low pH value corresponds to a high $CaCO_3$ dissolution potential (high Calcium Carbonate Dissolution Potential (CCDP)), even when the C_T is very low. This can be explained as follows: dissolution of $CaCO_3$ releases Ca^{2+} and CO_3^{2-} to the water. At low pH values (e.g., pH~2), the CO_3^{2-} ions react with H^+ ions (present at a high concentration) to form HCO_3^- and thereafter $H_2CO_3^*$, according to Eqs. (8.2) and (8.3) (reacting from right to left), respectively. Thus, only if a sufficiently high amount of $CaCO_3$ is dissolved, the pH value becomes sufficiently high to result in a high CO_3^{2-} concentration (see Fig. 8.2), leading to a significant increase in the value of Q, thus resulting in a significant decrease in the dissolution potential. Therefore, reducing the pH of desalinated water by dosage of a strong acid is a common method for elevating the dissolution propensity of the water with respect to $CaCO_3$. Another option for enhancing $CaCO_3$ dissolution is by dissolving $CO_{2(g)}$ in the water. In this alternative, the pH is also lowered, but to less acidic values relative to strong acid addition. On the other hand, since the dissolution of CO_2 elevates the value of C_T, this, in turn, contributes to the dissolution capacity of the water because a higher C_T value results in higher buffering capacity. Thus, a significant amount of $CaCO_3$ may be dissolved, before pH and consequently the concentration of CO_3^{2-}, are increased and the dissolution potential exhausted.

From the stoichiometric standpoint, $CaCO_3$ dissolution results in addition of Ca^{2+} and C_T to the water at a 1:1 molar ratio. The ratio between Ca^{2+} and Alk added to the water is also 1 to 1, in meq/l units.

8.2.7 Dolomite Solubility

Busenberg and Plummer [55] suggested that the dissolution of the mineral dolomite under acidic conditions (pH < 5) is a two-step process, the first step described by Eq. (8.11):

$$MgCa\left(CO_3\right)_2 + H^+ \rightarrow MgCO_3 + Ca^{2+} + HCO_3^- \tag{8.11}$$

The second step, which is the slower and thus the rate-limiting step, is described by Eq. (8.12):

$$MgCO_3 + H^+ \rightarrow Mg^{2+} + HCO_3^- \tag{8.12}$$

The stoichiometric addition of Ca^{2+}, Mg^{2+}, and C_T to the water as a result of dissolution of a pure dolomite mineral is at a molar ratio of 1:1:2.

Kinetics wise, at similar water quality and operational conditions, the dissolution of dolomite is much slower than that of calcite [55–57]. For example, Liu et al [56] reported that the dissolution rate of dolomite is lower by a factor of 3–60 relative to calcite. Due to the slow dissolution kinetics of dolomite, it is seldom practiced in the context of water treatment.

8.2.8 $CaCO_3$ Dissolution Indices

As corrosion in distribution systems poses a significant economic and health problem, determining the potential corrosivity of given water is perceived important, as reflected by the large number of aggression/corrosion-related indices developed over the years. The most commonly used and investigated indices are listed in Table 8.1.

It is commonly thought that deposition of a thin layer of $CaCO_{3(s)}$ on the pipe's wall acts to minimize the rate at which metal pipes are corroded [9]. Thus, supplying water that has a slight tendency to precipitate

Table 8.1 Important properties of principal water corrosion indices

Index	$CaCO_{3(s)}$ solubility based on thermodynamics	Other considered parameters	Source
LSI	Yes	Ionic strength and temperature	
CCPP	Yes	Ionic strength and temperature	[58]
Riddick Corrosion Index	No	(Cl^-), (SiO_2), (DO) TH, and nitrate	Cited in [32]
Driving Force Index	Yes		Cited in [32]
Aggressiveness Index	No	pH, TH	[59]
Larson Index (LI)	No	(Cl^-), (SiO_2), (HCO_3^-)	Cited in [32]
Singley index	No	Buffer capacity, EC	[60]

$CaCO_{3(s)}$ is often considered a corrosion control strategy. For this reason, many of the so-called "ggression indices" or "corrosion indices" are in fact indices that predict the tendency of calcite to precipitate/dissolve [25]. There are several problems associated with the use of each index, and many studies showed rather poor correlation between the indices' prediction and actual corrosion rate (e.g., (25,32,35,58–61)). Moreover, as mentioned by Seacord et al. [7], the indices are too often applied incorrectly, due to what seems to be lack of understanding of the fundamental principles by which they were developed. The AWWA manual on corrosion, for example, states explicitly that the use of Langelier Saturation Index (LSI) should be abandoned [61]. Rossum and Merrill [35] compared a large number of indices. They concluded that CCPP is the best-suited index to describe the water's saturation state with respect to calcite; Seacord et al. [7] arrived at a similar conclusion. CCPP is the only quantitative $CaCO_{3(s)}$ precipitation index [7,32]. However, it should be kept in mind that CCPP is a pure calcite-related index, and as such it has limited capability in predicting corrosion phenomena unrelated to calcite dissolution. For example, it does not take into account the corrosive nature of high concentrations of Cl^-.

8.2.9 Main Gaps in Knowledge

In the context of PT processes, the kinetics of the dissolution of both calcite and dolomite in packed bed reactors have been satisfactorily quantified lately [62–64]. However, the kinetics, as well as the engineering aspects of dissolving powdered calcite or dolomite ($D < 100\,\mu m$), have been thus far only partially investigated. Consequently, the precise water quality at the outlet of such dissolution reactors as a function of the reactor's operational parameters cannot be predicted with reasonable accuracy, to date.

Another subject not addressed thus far is the development of ingenious PT techniques (beyond mere dosage of chemicals) for introduction of other components, such as iodine and potassium, to desalinated water, as well as expanding the understanding on the importance of these components or lack thereof, on the public health.

Finally, it should be noted, that surveying the published PT literature shows that even the knowledge that is already well established, is often overlooked or misunderstood. The aim of this chapter is to bring together all the relevant knowledge including the relevant theoretical background, the advantages and drawbacks of commonly practiced PT processes as well as PT processes, which are still under development.

8.3 DESALINATION POST-TREATMENT METHODS: STATE OF THE ART

8.3.1 Direct Dosage of Chemicals

Direct dosage refers to direct injection of chemicals to the water. The chemicals may be either in a slurry form (usually hydrated lime, $Ca(OH)_2$, as discussed in Sections 8.3.1.1 and 8.3.1.2), dissolved in solution (e.g., calcium salts, as detailed for example in Section 8.3.1.3), or in a condensed, liquid form ($CO_{2(l)}$) that transforms into $CO_{2(g)}$ and dissolves in the water (as discussed in Sections 8.3.1.1 and 8.3.1.4). Typical direct dosage treatments combine the addition ofTH and alkalinity to the water. However, as a function of the required product water quality, in some cases only alkalinity is supplied (as demonstrated in Section 8.3.1.4). This is, in most likelihood, an outcome of inadequate understanding of the inverse impact of soft water consumption by humans.

The main advantages of applying direct chemicals dosage are simplicity, the relatively low capital cost, and required footprint and the flexibility with regard to product water quality, i.e., the wide range of water qualities that can be attained. Table 8.2 summarizes the chemicals, which are usually applied in post-treatment units that are based on "direct dosage" methods and the associated ions and other water quality parameters (i.e., C_T and Alk) added to the water as a result of the dissolution of 1 mol of each of the chemicals. Clearly, almost any required water quality combination can be achieved by choosing appropriate chemical dosages. However, the disadvantages associated with this method include high operational costs due to the high cost of the chemicals and/or their on-site preparation [65] and the unavoidable addition of unwanted counter ions. The practical chemical

Table 8.2 The increase in main water quality parameters as a result of the dissolution of 1 mol of each of the chemicals applied via "direct dosage" methods

	Added quantity				
	Na^+	Cl^-	C_T	Alkalinity	Ca^{2+}
Chemical dissolved (1 mol)	Equiv	Equiv	mol	Equiv	Equiv
CO_2 [a]	0	0	1	0	0
$NaHCO_3$	1	0	1	1	0
Na_2CO_3 [a]	2	0	1	2	0
$Ca(OH)_2$ [a]	0	0	0	2	2
$CaCl_2$ [a]	0	2	0	0	2
NaOH	1	0	0	1	0

[a] 1 mol of chemical equals 2 equivalents.

combinations typically dosed to desalinated water and the associate changes in water quality are listed in Sections 8.3.1.1–8.3.1.4. Note that direct dosage of each of these chemicals can be applied as a complementary practice to any other PT approach.

8.3.1.1 Dosage of $Ca(OH)_2 + CO_2$

Lime ($Ca(OH)_2$) enriches the water with both TH and alkalinity (at a 1:1 ratio, in equivalent units); however, carbonate alkalinity is not added, as shown in Table 8.2. Thus, mere lime addition does not contribute to the (carbonate) buffer capacity of the water. Moreover, in order to efficiently dissolve hydrated lime, the water must be acidified. Therefore, CO_2 is dissolved into the water following the addition of hydrated lime. Hydrated lime is dosed as a slurry, and CO_2 is applied in a condensed, liquid form ($CO_{2(l)}$), which transforms to $CO_{2(aq)}$ in the water, and immediately thereafter converted to HCO_3^-, due to the high pH value caused by the lime dosage. This method has several drawbacks. For example, the use of hydrated lime slurry is relatively complex from the engineering point of view [1,66–68], especially if the permeate is warm, which reduces the solubility of lime [7]. In addition, the approach may raise water turbidity to values higher than 5 Nephelometric Turbidity Units (NTU) [1,7,13,65,69]. Finally, the lime-CO_2 method has been associated with control problems, i.e., difficulties in maintaining consistent product water pH [1]. On the other hand, in this alternative, no unwanted counter ions are added to the water (Table 8.2). In spite of its known drawbacks, this method is widely applied [68,69]. For example, in a survey of approximately 100 Swedish municipalities, it was found that 73% used this method [14]. Fig. 8.3 shows a photo of the lime feed system serving in the Tampa Bay SW desalination plant.

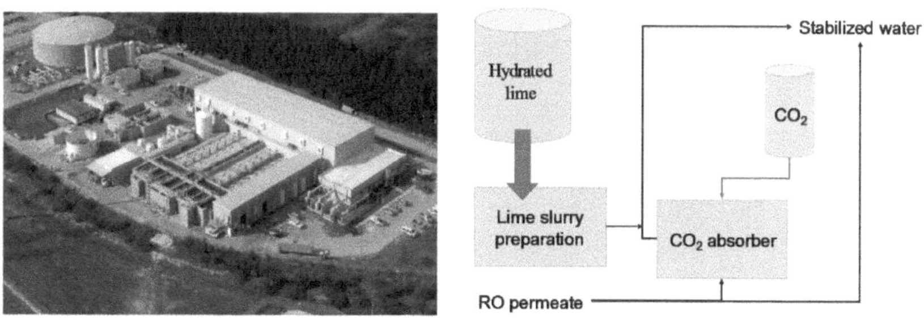

Fig. 8.3 Tampa Seawater Desalination Plant (Left-hand side, lime feed system shown in upper left corner); right-hand side: schematic of the CO_2 and lime direct dosage process.

8.3.1.2 Dosage of $Ca(OH)_2 + Na_2CO_3$ or $Ca(OH)_2 + NaHCO_3$

Using Na_2CO_3 or $NaHCO_3$ as the C_T source instead of CO_2 (as in the previous alternative) results in an elevated pH value, since these salts consist of the basic species of the carbonate system. Consequently, the dissolution potential of lime decreases considerably, and more significantly, even at relatively low final alkalinity and Ca^{2+} concentrations (~30 mg/L as $CaCO_3$ and ~10 mg/L, respectively) the resulting pH is excessively high (pH > 10.5), making this approach impractical as a PT of conventional RO permeates. However, as Withers [1] stated, this method is appropriate to raw waters, which contain a certain initial alkalinity and relatively low pH values. In such cases, the raw water contains a substantial amount of buffer capacity as compared to RO permeates, and thus the dissolution capacity of lime is increased.

Comparing the dissolution of this combination of chemicals to the one described in Section 8.3.1.1 (lime and CO_2 dissolution) shows that attaining the same C_T and Ca^{2+} concentration in both methods would result in lower product water alkalinity if CO_2 is used, as can be concluded from Table 8.2. Finally, the methods based on dissolving $NaHCO_3$ and Na_2CO_3 result in unwanted elevation of the Na^+ concentration.

8.3.1.3 Dosage of $CaCl_2 + NaHCO_3$

This process is based on simple dissolution of chemicals, and no handling of slurries or gases is required. Hence, from the engineering point of view, it is a straightforward method. However, $CaCl_2$ is often a more expensive source of Ca^{2+} than lime (and quarry calcite) and the method involves the introduction of unwanted Cl^- and Na^+ ions to the water. Withers [1] stated that further pH adjustment is often required when applying this alternative. This statement is somewhat puzzling, since dissolution of $NaHCO_3$ results in a pH value around 8.3 (the HCO_3^- equivalent point), and the addition of $CaCl_2$ can lead to practically any required CCPP value. It is correct, however, that at low Ca^{2+} concentrations ($Ca^{2+} < 80$ mg/l as $CaCO_3$) pH elevation would be required to attain positive CCPP (or LSI).

Note that in the literature the following combinations are also mentioned: $CaCl_2 + Na_2CO_3$ and $CaCl_2 + NaHCO_3 + Na_2CO_3$ (2). However, to the writers' understanding, these methods are impractical since these combinations of chemical additions would render product water pH (and thus also CCPP) excessively high.

8.3.1.4 Dosage of $Na_2CO_3 + CO_2$ or $NaOH + CO_2$

These alternatives are also mentioned in the literature and even practiced in a few desalination plants, although they do not involve the addition of

Ca^{2+} or TH to the water, and thus, no $CaCO_3$ dissolution index can be satisfied. Accordingly, these alternatives can be considered only for elevating the carbonate alkalinity and the pH value of acidic water, rather than as PT methods for desalinated water.

8.3.2 Blending Desalinated Water With Other Water Sources

Blending desalinated water with either SW or BW may be considered a low-cost method to increase the concentration of some desired ions in desalinated water product. However, such practice invariably adds other, undesired species, to the water. The concentrations of all the introduced ions are a function of the blended water composition and the blending fractions, making control over the product water quality limited. Consequently, blending is not recommended for water intended for domestic use [65] or agricultural use [70]. When desalinated water is destined for irrigation, blending with seawater (SW) or brackish water has both negative environmental and economic implications such as elevation in the concentrations of boron, chlorides, and sodium ions. Nevertheless, the literature continues to address the option of blending either as a sole remineralization technique, e.g., when groundwater is blended with permeate [71–73], or as a complementary alternative [66]. Since it is often recognized that blending alone introduces mainly TH and TDS to the water, in case blending is practiced as a PT approach, it is in many cases followed by pH correction. Such a combination is still commonly applied in BW desalination plants in the USA [6]. Or, for example, in the City of Abu Dhabi, which receives practically all its water from a desalination plant, the PT approach is to blend the permeate with SW at a ratio of 1:500 followed by final pH elevation by NaOH dose [74].

8.3.3 Calcite Dissolution Processes

Unlike direct dosage, calcite and also dolomite dissolution processes are conducted in reactors, in which the retention time is on the order of minutes (typically >15 min). These reactors are often mistakenly termed "filter beds" although their exclusive purpose is the introduction of Ca^{2+} and CO_3^{2-} (and also Mg^{2+} in case dolomite is dissolved) to the water through dissolution, and not filtration. In order to enable rapid dissolution of a high $CaCO_3$ concentration, the CCDP of the desalinated water should be considerably high, this is commonly practiced via pH reduction the introduction into the dissolution reactor (as explained in Section 8.2.6). It is stressed that in any dissolution reactor thermodynamic equilibrium is in practice not attained, due to kinetic limitations. In other words, the CCPP of the water leaving the calcite dissolution reactor is always slightly negative. Thus, although the dissolution of $CaCO_3$ results in elevated pH (pH~6.5) it must be further increased, both for achieving a more appropriate pH value for drinking water and mainly for elevating the $CaCO_3$ precipitation potential to a positive value, i.e., to produce water that is chemically stable within the distribution system. Fig. 8.4 shows the calcite addition system in the Ashkelon SWRO desalination plant in Israel.

8.3.3.1 Acidifying Agents Used to Enhance Calcite Dissolution

Two acidic substances are typically used to lower pH to enhance calcite dissolution: H_2SO_4 and $CO_{2(g)}$. Unless an unconventional low-cost source of CO_2 is available (see elaboration in Section 8.3.5), using a strong acid such as H_2SO_4 is a less costly and simpler practice. Lehmann et al. [63] examined the dissolution of calcite in dissolution reactors fed with H_2SO_4-acidified water. They found that more than 70% of the calcite dissolution potential was realized after merely 2 min of hydraulic retention time inside the reactor. That is, after around 30 cm in the reactor, 70% of the calcite dissolution potential has already been attained. Note that once the majority of calcite has dissolved, and pH has been raised to between pH4.5 and pH6, the $CO_{2(aq)}$ concentration may become very high, and in fact, acid/base conditions closely resemble those prevailing in the entrance to a CO_2-based dissolution reactor. Therefore, the conditions prevailing in the rest of the reactor (i.e., after the first ~30 cm) resemble the conditions prevailing in a CO_2-based dissolution process. Lehmann et al., 2013's [63] results also strengthen the conclusion that CO_2-based dissolution is much slower than H_2SO_4-based dissolution. Accordingly, H_2SO_4-based dissolution was chosen, for example, as the PT method in the Ashkelon (130 Mm^3/y) and Palmachim (45 Mm^3/y) [75] desalination plants in Israel. The main advantage of this approach is that, due to the high CCDP values that can be attained when a strong acid is introduced to the water and the rapid dissolution rate of calcite under these conditions, it is possible to dissolve a significant amount of calcite in the water that passes through the reactor, and let the majority of the flow (between ~75% and ~90% of the total flow rate of the plant, in case the product water alkalinity should be ~80 mg/L as $CaCO_3$), bypass the reactor, a fact that renders the reactor considerably less costly [76]. The limiting factor in determining the practical maximal concentration of calcite that can be dissolved in this method is the potential precipitation of gypsum ($CaSO_{4(s)}$). The percentage of treated water out of the total desalinated water flow rate is denoted in this chapter "% split flow" (see Fig. 8.4).

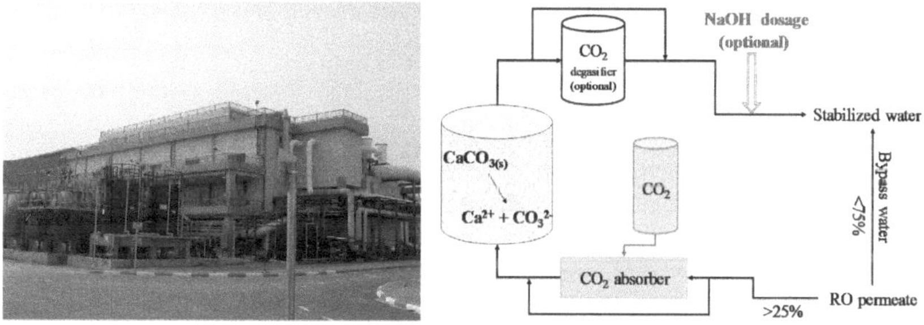

Fig. 8.4 Calcite Contact Tanks for the Ashkelon SWRO Plant in Israel (left-hand side); Schematic of the CO_2-based calcite dissolution process (right-hand side). Final pH adjustment can be achieved either by CO_2 stripping or by NaOH dosage.

From the water-quality standpoint, the main difference between applying H_2SO_4 and CO_2 is that the H_2SO_4-based calcite dissolution process results in dissolved calcium to alkalinity concentration ratio that is always equal to or higher than 2 to 1 (in equivalent per liter units), while the alternative process, i.e., CO_2-based calcite dissolution, results in a ratio of approximately 1 to 1. This statement is corroborated by the characteristics of the water produced in PT plants that apply this method [77,78]. The ratios obtained in the two dissolution alternatives are discussed in detail in [3,8]. The 2 to 1 ratio can be explained by the data shown in Table 8.3 and Fig. 8.5, which presents the alterations in $H_2CO_3^*$ alkalinity (Alk), CO_3^{2-} acidity (Acd), and Ca^{2+} concentrations as a result of applying the H_2SO_4-based calcite dissolution process. The process ends when the pH is raised (by NaOH addition) to around 8.3, i.e., close to the HCO_3^- equivalent point. At this point, the Alk value exactly equals the Acd value. Substituting the expressions for the final Alk and Acd (given in the last row of Table 8.3) yields:

$$\text{Alk}_{\text{final}} = \text{Acd}_{\text{final}} \Rightarrow y + z - x = x - z$$

Thus,

$$y = 2(x - z)$$

and thus:

$$\text{Alk}_{\text{f}} = y - (x - z) = 0.5y, \text{i.e., exactly half the } Ca^{2+} \text{ concentration.}$$

The 1:1 ratio attained in the CO_2-based process can be simply derived from the data given in the first row of Table 8.2 and the second row of Table 8.3.

Table 8.3 Alterations in water quality parameters as a result of the H_2SO_4-based calcite dissolution process steps

Alteration in component value	Dosage	Chemical added in the process		
$[Ca^{2+}]$	Acd	Alk		
eq/L	eq/L	eq/L	eq/L	
0	$+x$	$-x$	x	H_2SO_4
$+y$	0	$+y$	y	$CaCO_3$
0	$-z$	z	z	NaOH to ~pH 8.3
y	$x-z$	$y+z-x$	Product water	

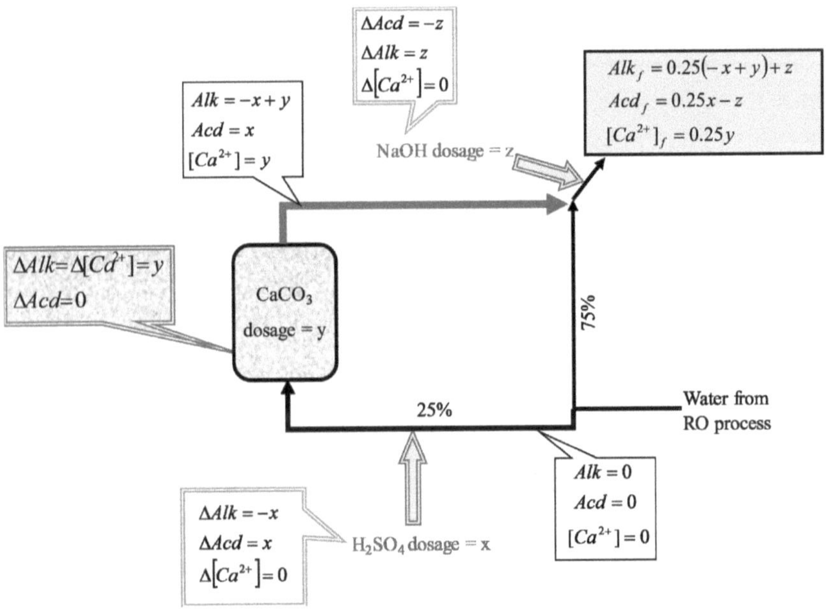

Fig. 8.5 Alterations in water quality parameters as a result of the H_2SO_4-based calcite dissolution process steps.

Calcite dissolution enhanced by a combined dosage of CO_2 and H_2SO_4 is also theoretically possible. In this case, the resultant water quality (e.g., from the Alk to Ca^{2+} ratio and the CO_2 concentration perspectives) will fall in between the qualities obtained when either CO_2 or H_2SO_4 are applied as the sole acidifying agent.

8.3.3.2 Final pH Adjustment

As mentioned, calcite dissolution results in pH values lower than pH7.0 (e.g., (64)). Arriving at the final pH value in drinking water applications typically involves controlled addition of NaOH (commonly practiced) or $Ca(OH)_2$ (not often practiced, due to product water turbidity problems. However, this process is applied in the 130 Mm^3/y Hadera desalination plant, Israel). In some cases, pH elevation can also be carried out by controlled CO_2 stripping (also less commonly practiced than NaOH addition). pH elevation by CO_2 stripping can be practiced only in cases in which the solution coming out of the dissolution reactor is significantly supersaturated with respect to atmospheric $CO_{2(g)}$, since this technique is based on emitting excess CO_2 and approaching aqueous—gas phase equilibrium with respect to CO_2. As a result, this approach can be applied only when CO_2 is used as the acidifying agent. The maximal final pH value that can be theoretically achieved in this method is primarily a function of the extent of supersaturation, i.e., the concentration of CO_2 that can be emitted before equilibrium is approached. In practice, thermodynamic equilibrium can never be attained because of kinetic limitations. However, the final distance from equilibrium can be controlled by engineered parameters, such as the applied air flow rate, the hydraulic retention time, and the average air bubbles size.

Often, following the stripping step, the pH and CCPP values are still not sufficiently high. In such cases, further pH elevation is realized by NaOH dosage, as practiced, for example, in Kuwait [77] and Qatar [78].

8.3.3.3 Unintentional $CO_{2(g)}$ Stripping

CO_2-based $CaCO_3$ dissolution reactors are normally operated at pressures higher than atmospheric (i.e., sealed reactors). The loss of CO_2 to the atmosphere in such reactors is minimal. On the other hand, H_2SO_4-based $CaCO_3$ dissolution reactors are often open to the atmosphere, since unsealed reactors are less costly and easier to maintain. In such reactors, after a significant amount of calcite has dissolved, the water is characterized by low pH values (~pH5) and a relatively high C_T concentration, which is comprised almost entirely of $CO_{2(aq)}$. In other words, the water is highly supersaturated with respect to atmospheric CO_2. At the outlet of the reactor, the C_T concentration is higher, but the pH is also higher, making the dissolved CO_2 concentration lower (see Fig. 8.2). Nevertheless, $CO_{2(aq)}$ supersaturation still prevails. According to this description, it is clear that both upflow and downflow operation of H_2SO_4-based $CaCO_3$ dissolution open reactors can result in loss of CO_2 to the atmosphere. It is noteworthy that CO_2 emission does not affect the calcium or Alk concentrations; on the other hand, it reduces the Acd value and elevates pH. As a result, a decreased dosage of NaOH is required in order to reach the same final pH (~8.3); thus, the elevation of the Alk concentration as a result of NaOH dosage is also lower. In this case, the ratio between dissolved calcium and alkalinity concentrations in the product water is higher than 2, as observed in the water quality produced, for example, in the Ashkelon plant: Alk = 45–50 mg/l as $CaCO_3$ and $[Ca^{2+}]$ = 90–110 mg/l as $CaCO_3$ [72].

8.3.4 Dolomite Dissolution

Dissolution of quarry dolomite ($CaMg(CO_3)_2$) as means of supplying carbonate alkalinity, Mg^{2+} and Ca^{2+} ions to the water as part of the PT step suffers from several drawbacks: (1) The dissolution rate of dolomite (at the neutral pH range) is slow relative to that of calcite, requiring larger reactors, as compared with equivalent calcite dissolution reactors; (2) Dolomite dissolution practically stops at relatively low pH values (the magnesium part of the rock stops dissolving at ~pH5.5). Therefore, using a strong acid as the acidifying agent results in low alkalinity concentrations at the outlet of the dissolution reactor [62,79]. Consequently, the water leaving the dissolution reactor has a highly negative CCPP value (i.e., high CCDP value) and low buffering capacity. Consequently, the increase in pH due to NaOH dosage is steep, and excessively high pH values may be obtained at relatively low Alk values. On the other hand, using a weak acid, i.e., CO_2, leads to slower dissolution, and results at relatively high buffer capacity, due to the high C_T, and low pH and CCPP (high CCDP) [79]. Therefore, NaOH dosage required for stabilizing the water is substantial; (3) The composition of quarry dolomite (i.e., the percentage of calcite within the excavated rock) varies from location to location and often also within the same quarry. Thus, the Ca^{2+} and Mg^{2+} concentrations in the product water are expected to vary in somewhat uncontrolled manner; (4) Engineering know-how related to dissolving dolomite for water remineralization purposes is lacking; (5) Problems associated with parallel dissolution of two minerals ($CaCO_3$ and $MgCO_3$). These include both time fluctuation in water quality ($[Ca^{2+}]$, $[Mg^{2+}]$, pH, and the Alk concentration) and operational difficulties. Because of these negative aspects of dolomite dissolution, it was concluded that it is probably impractical to generate water quality that complies (for example) with the criteria set for desalinated water in Israel or in desalination bids in Cyprus, using this alternative. Evidently, this approach is very rarely used. However, similar to the blending method, it is also widely mentioned in the scientific literature as a PT alternative (e.g., [24,46,66,74,79,80]). According to reports, dolomite dissolution is practiced for soft waters stabilization by ~9% of the Swedish municipalities [14].

8.3.5 Sources of CO_2 Used in the PT Step

Different post-treatment methods require formation of dissolving conditions in the water, which can be created by dissolving $CO_{2(g)}$ in the water. A possible source of CO_2 in the permeate water could be the carbonate species present in the raw water fed to the desalination plant (i.e., in seawater or brackish water). However, in order to use these in the post-treatment step, the raw water should be acidified, to shift the bicarbonate (which is the main carbonate species in seawater) toward the CO_2 species (see Eq. (8.2) and Fig. 8.2), which are known to readily pass through RO membranes. Such a technique is already practiced in some SW desalination plants. For example, in the Laayoune desalination plant in Morocco, acid is dosed to the seawater prior to the RO stage, to lower the pH. As a result, the RO permeate contains approximately the same total carbonic species concentration as the seawater; however, it is all present as dissolved CO_2. Consequently, no further CO_2 dosage is required for the PT step (e.g., [69]). However, it should be noted that a high percentage of the desalinated water should be treated in the PT stage in this case, since the CO_2 concentration in the permeate is not high enough to enable treating only a smaller portion of the water and blend it thereafter with untreated water. Similarly, brackish water often contains a considerable CO_2 concentration even without acidification. For example, Oren et al. [81] showed that the permeate of the BW fed to the Kantor desalination plant in Ma'agan Michael (Israel) naturally contains CO_2 concentration of 59 mg/L. Oren et al. [81] also showed that for treating only 50% of the permeate (i.e., blending at a ratio of 1:1 with untreated water), 82 mg/L of HCl should be dosed to the BW fed to the RO plant, manifested in ~150 mg/L CO_2 in the permeate, which in turn enables dissolving ~250 mg/L $CaCO_3$ in the calcite reactor [81]. Finally, note that an inexpensive technical-grade strong acid can be dosed to the feed BW, since this acid is fed upstream the membrane.

8.4 INNOVATIVE POST-TREATMENT PROCESSES FOR ATTAINING MAGNESIUM IN THE PRODUCT WATER

Two parallel trends led to recent developments of new PT methods: (1) the increasing acknowledgment regarding the need for efficient PT solutions capable of producing water complying with new and more stringent quality criteria; and (2) the recently accumulated evidence on the importance of threshold concentrations of Ca^{2+} and Mg^{2+} in drinking water (e.g., [31,39,41,82–85]), water used for agricultural irrigation [11,86,87], as well as aquaculture.

In Section 8.3, currently applied PT methods are described. From this evaluation, it is clear that so far no PT process is practiced that supplies Mg^{2+} in a cost-effective manner, and which also results in low addition of unwanted components to the water. Two recently developed PT methods aimed at addressing these issues are described in Sections 8.4.1 and 8.4.2. In addition, the two newly developed methods described in Sections 8.4.3 and 8.4.4 are complementary steps, aimed at adding Mg^{2+} to any post-treated water.

8.4.1 Calcite Dissolution Combined With an Ion Exchange Step (IX) (The Calcite Dissolution-IX Process)

The method is shown schematically in Fig. 8.6. It is based on dissolving $CaCO_3$ to supply carbonate alkalinity and Ca^{2+} ions and subsequently replacing a part of the dissolved Ca^{2+} concentration with Mg^{2+} ions, which originate from the raw water supplied to the desalination plant (either SW or BW), or from BWRO brine. The Mg^{2+} rich solution is denoted in this process "loading solution" since it is used for loading an IX resin with Mg^{2+}. Separating the Mg^{2+} ions from the loading solution is performed by a specific cationic IX resin, characterized by a high affinity toward Mg^{2+} and Ca^{2+} and a very low affinity toward Na^{+} ions. A detailed description of the process and pilot-scale operation results can be found elsewhere (e.g., [76,88,89]).

The calcite dissolution-IX process can be operated with either SW or BW as the loading solution. In case SW is available, it is preferred over BW, since both the Mg^{2+} concentration and the Mg^{2+} to Ca^{2+} concentration ratio are higher, resulting in higher efficiency of the Load step, which manifests itself in lower costs and higher flexibility, especially with respect to the product water Mg^{2+} concentration. A detailed description of the application of this alternative appears in Birnhack and Lahav [88]. Alternatively, when SW is not available (i.e., inland locations), BW can be used. However, the use of BW characterized by Mg^{2+} to Ca^{2+} concentration ratio lower than 1:1 was found impractical when H_2SO_4 was the acidifying agent considered for enhancing calcite dissolution prior to the IX step [90].

The use of different loading solutions and different acidification alternatives, i.e., CO_2, H_2SO_4, or combination of both acids (resulting in different

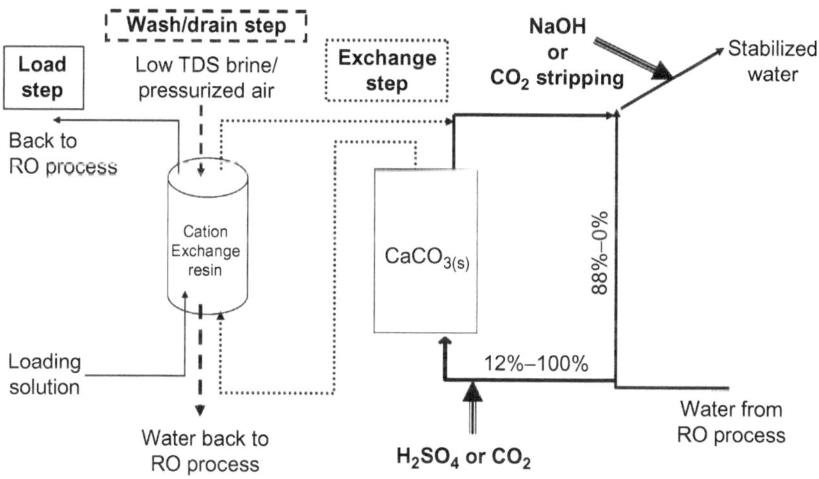

Fig. 8.6 Schematic of the post-treatment process based on calcite dissolution followed by IX

possible %split flow values—see Fig. 8.6) determines the length of the exchange and load steps, for a given required Mg^{2+} concentration. Since there is a minimum practical ratio between the length of the exchange step and the load step, applying operational conditions that increase this ratio (e.g., using CO_2 as the acidifying agent or SW as the loading solution) enables supplying higher Mg^{2+} concentration.

The use of a combination of resins may be needed in case a stringent restriction on the TH is imposed in addition to a demand for a substantial SO_4^{2-} concentration (required for agricultural uses). Under such water quality requirements, the use of CO_2 is not possible, since it results in a zero SO_4^{2-} concentration, while the use of H_2SO_4 is not a viable solution because of the high resultant TH to Alk ratio. The use of a combination of the acids may produce water that complies with the criteria. However, a combined requirement of low TH and high SO_4^{2-} concentration cannot be achieved in this way. An alternative solution is to use two different IX resins: the first resin is aimed at replacing the excess Ca^{2+} ions by Mg^{2+} ions and the second one is aimed at replacing Ca^{2+} with Na^+ [91].

8.4.2 Dolomite Dissolution Combined With Calcite Dissolution

As mentioned in Section 8.3.4, H_2SO_4-enhanced dolomite dissolution is not feasible as a sole PT method. Birnhack et al. [62] described a process based on dolomite and calcite dissolution in series, which is capable of producing water able to meet a wider range of quality criteria.

As mentioned before, the effluent of a dolomite dissolution reactor is characterized by low pH, very negative CCPP but also by a high $CO_{2(aq)}$ concentration and a high buffering capacity. The logic behind the new alternative is that the water leaving the dolomite reactor can be used to further dissolve calcite to increase alkalinity, pH, and CCPP. Such a strategy is superior over NaOH dosage from both water quality and cost standpoints. Birnhack et al. [62] showed that the pH and C_T values in the effluent of the dolomite reactor, which were found unsuitable for further dolomite dissolution, were indeed appropriate for dissolving further several meq/l of calcite. Thus, the alkalinity and pH values obtained in the product water in this alternative were significantly higher than those obtained when the dolomite was dissolved alone. Using this method, it was possible to produce water characterized by an alkalinity value that is slightly lower than the required threshold in the Israeli criteria (i.e., 75 instead of 80 mg/L as $CaCO_3$), but the CCPP, pH value, and Ca^{2+} concentration complied with the criteria, and in addition 12.4 mgMg/L were supplied. On the down side, the use of this alternative resulted in relatively high TH concentration in the product water.

8.4.3 A Complementary Step for Addition of Mg^{2+}, SO_4^{2-} (and Ca^{2+}) Through Dosage of Seawater Nanofiltration Brine

In the heart of this complementary step stands a process for separating Mg^{2+} ions from seawater, which applies a nanofiltration membrane. The process, described in Birnhack et al. [92], aims at maximizing the Mg^{2+} concentration in the target solution, while at the same time minimizing the addition of the unwelcome seawater species (Cl^-, Na^+, B, Br^-, etc.) to the receiving desalinated water. The Mg^{2+} separation method is based on passing seawater through suitable nanofiltration (NF) membranes (e.g., Lentech DL-4040-F1021 Stinger) in a way that: (1) enhances the separation between divalent and monovalent ions, i.e., rejects the divalent ions, while allowing monovalent ions to pass the membrane; and (2) generates brine with the highest Mg^{2+} concentration possible. Since the solution dosed to the desalinated water is the brine of the NF step, pathogens initially present in seawater will remain in the seawater NF brine. Thus, in order to prevent the contamination of the desalinated water, the NF brine is passed through an ultrafiltration membrane, prior to its introduction to the desalinated water.

As the brine is produced, food-grade antiscalant (e.g., Genesys CAS, manufactured by Genesys International, http://www.genesysro.com/membrane-anti-scalants-antiscalant-chemicals.php) is dosed to the feed water in order to prevent $CaCO_3$ and $CaSO_4$ precipitation and subsequent chemical fouling of the membrane.

Birnhack et al. [92] showed that the overall quality of the desalinated water is hardly impaired by the addition of the brine. With respect to chloride and sodium ions, they concluded that their additions are not high (<40 and <12 mg/L, respectively, assuming the water is enriched with 10 mg Mg/L). The additions of other components were calculated to be ~3 mg/L of calcium, ~0.06 mg/L of strontium, 0.005 mg/L of boron, and 0.025 mg/L of food-grade antiscalant.

8.4.4 A Complementary Step for Mg^{2+} Enrichment by Dosage of the Brine Produced Through Seawater Nanofiltration, Followed by Diananofiltration Step

This process was developed in order to reduce the addition of ions invariably present in the brine of the NF process (described in Section 8.4.3 in this chapter). The technique (described in detail in [93]) comprises nanofiltration of seawater (or first-pass SWRO brine) followed by a DiaNanofiltration step, applied on the nanofiltration brine. DiaNanofiltration is a procedure involving dosage of low-TDS water to the nanofiltration feed solution, to improve the "wash out" of unwanted components. The purpose of the preceding nanofiltration step is to produce low-volume brine retaining most of the Mg^{2+}, Ca^{2+}, and SO_4^{2-} from the seawater, thereby minimizing the cost of the DiaNanofiltration diluting solution.

In the heart of this complementary step is a method termed "DiaFiltration," which is used in the current context with an NF membrane; therefore, it is denoted DiaNF. A fundamental NF-DiaNF process and some modifications of it were examined by Nativ et al., 2016 [93]. The modifications of the process were developed in order to reduce the costs associated with the technique while at the same time attaining the best water quality. In this respect, the water quality was determined as the ratio between Mg^{2+} and the unwanted components in the produced brine.

The fundamental version of the developed technique is presented schematically in Fig. 8.7. Seawater (the Mg^{2+} and Ca^{2+} source) is first acidified in order to circumvent $CaCO_3$ precipitation in the next steps; the acidified SW is then nanofiltered at a recovery ratio ranging from 40% to 80%. The recovery ratio is limited by the potential of gypsum ($CaSO_4$) to precipitate in the brine. Two main goals are achieved by first applying a conventional NF step: (a) Elevation of the divalent ion concentrations in the brine while maintaining the concentrations of the monovalent ions approximately

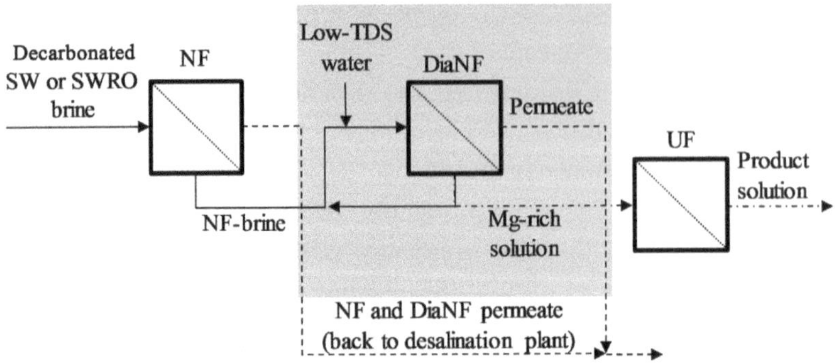

Fig. 8.7 Schematic of the fundamental NF-DiaNF process for selective separation of Mg^{2+} from SW or SWRO brine. The number of DiaNF cycles applied determines the ratio between Ca^{2+}/Mg^{2+} and the unwanted species (Na^+, Cl^-, B, etc.).

constant; and (b) Decreasing the volume of the costly diluting solution (the low-TDS water) to be fed to the DiaNF step, i.e., to the following separation step. Once the brine of the NF step has been produced, the DiaNF operation is initiated. The eventual overall cost of the Mg^{2+} addition to the desalinated water depends largely on the total volume of the diluting solution and on the final Mg^{2+} mass retained in the final brine volume. When a single DiaNF cycle is performed (or for that matter the last DiaNF cycle), the dilution solution comes almost at no cost, since exactly the same volume of water is returned, for example to the desalination plant product, when the Mg^{2+}-rich solution is dosed back to the product water.

From the water quality standpoint, it was shown that the lowest Cl^- to Mg^{2+} concentration ratio that can be attained in the produced solution of this process is ~1.52 (w:w). A cost assessment (operational and capital expenses) showed the cost of all examined scenarios to be lower than $0.01 for supplying desalinated water with 10 mg/l of Mg^{2+} [93].

Tang et al. [94] developed a process based on the hypothesis that the passage of chloride ions through the membrane in the NF-DiaNF process was inhibited due to the lack of electroneutrality of the divalent ions. In order to further reduce the ratio between magnesium and chlorides in the produced solution (for example, in order to comply with the Israeli regulations), Tang et al. [94] suggested utilizing a different NF membrane to treat the seawater as a preceding step to the NF-DiaNF process: a relatively "open" nanofiltration membrane characterized by dense negative surface charge. As a result, SO_4^{2-} was rejected better than Mg^{2+} and Ca^{2+}, and thus a retentate was characterized by a TH to SO_4^{2-} ratio that is lower than the original ratio in seawater was attained (~1.3 M/M). This retentate was then subjected to the

NF-DiaNF steps, in which chloride ions easily pass to the permeate side. A dosage of 20 mg Mg^{2+}/l to desalinated water using the retentate of the NF-NF-DiaNF process was estimated at \$0.006/m^3 and \$0.017/m^3 for addition of 32 and 12 mg Cl^-/l along with the Mg^{2+}, respectively [94].

8.5 COMPARISON BETWEEN PT METHODS

In this section, the engineering, water quality, and economic aspects of several PT processes are compared. The PTs considered in this comparison are direct dosage, calcite dissolution (CO_2-based, H_2SO_4-based, or based on a combination of the acids), blending, in-series operation of dolomite and calcite dissolution reactors, calcite dissolution followed by IX, nanofiltration of SW, and NF of seawater followed by DiaNF.

8.5.1 Water Quality

The water qualities attained in the various PT processes are similar in the sense that they are all characterized by a certain Ca^{2+} and carbonate alkalinity concentrations (in the following comparison, it was assumed that all the methods aim at reaching a similar Ca^{2+} concentration of 30 mg/L). However, they differ from each other in terms of other parameters. The following evaluation focuses on the key differences between the discussed PT processes with respect to six parameters. Explanation on the importance of each of the parameters is given in Section 8.1.4.

8.5.1.1 TDS Addition

Attaining a Ca^{2+} concentration of 30 mg/L through blending SW with the desalinated water would result in the addition of ~2400 mg/L of TDS to the desalinated water, out of which ~1200 mg/L Cl^- and ~700 mg/L Na^+. As a result, blending of RO permeate and pretreated SW cannot be considered a viable method to post-treat desalinated water. Direct dosage of $CaCl_2$ at a concentration that adds 30 mg/L of Ca^{2+} to the permeate leads to the addition of 50 mg/L Cl^-. Similarly, direct dosage of $NaHCO_3$ and Na_2CO_3 results in an increase of 23 mg/L of Na^+ for each meq/L of alkalinity supplied to the water. Dissolution of limestone or dolomite rocks results in a negligible elevation of the TDS due to impurities in the rock, and an additional slight elevation in the Na^+ concentration (around 10 mg/L) due to final pH adjustment carried out via NaOH dosage. In case CO_2 stripping is applied for final pH adjustment (feasible only in case CO_2 is used as the acidifying agent), the dissolution leads to a negligible amount of unwanted

TDS. The calcite dissolution-IX process results in a very small increase in TDS. This is the consequence of a very small volume of loading solution (SW, in the worst scenario) that remains in the bed after the drain step and is incorporated in the product water in the following "exchange" step [88].

To conclude, from the TDS addition point of view, the processes can be graded as follows: blending SW $\ll$ direct dosage of $CaCl_2$, $NaHCO_3$ and Na_2CO_3 < calcite dissolution followed by IX < calcite/dolomite dissolution.

Note that addition of Mg^{2+} through dosage of NF brine also results in elevation of TDS. Addition of Mg^{2+} through dosage of NF-DiaNF brine results in a smaller addition of Na^+ and Cl^-.

8.5.1.2 SO_4^{2-} Supply

Any PT that relies on H_2SO_4-based calcite (or dolomite) dissolution results in the addition of SO_4^{2-} to the water. A supply of around 32 mg/L SO_4 as S is attained when H_2SO_4 is used to dissolve around 170 mg/L of calcite (all units are per liter of product water) (e.g., [90]). Such a SO_4^{2-} concentration lies within the recommended concentration for irrigation water [4,11]. Clearly, the water is not enriched with SO_4^{2-} when CO_2 is used for dissolving calcite. Direct dosage of $MgSO_4$ also results in a considerable concentration of SO_4^{2-}. Blending SW for attaining a required Ca^{2+} concentration of 80 mg/l as $CaCO_3$ results in the addition of more than 60 mg/l SO_4 as S. The amount of SO_4^{2-} added when BW is blended with the permeate depends on the BW composition. When using both the NF and NF-DiaNF methods (assuming enrichment of the water with 10 mg/L Mg^{2+}), the SO_4^{2-} addition would be around 30 mg/L.

8.5.1.3 Mg^{2+} Supply

As explained before, most of the currently implemented PT processes do not enrich the water with Mg^{2+}. Nevertheless, a considerable concentration of Mg^{2+} can be supplied by the following alternatives: direct dosage of $MgCl_2$ or $MgSO_4$; dolomite dissolution; calcite dissolution followed by IX; dosage of the nanofiltration brine, or the brine produced vie diananofiltration; and blending of seawater.

The concentration of Mg^{2+} added by direct dosage of Mg salts is almost unlimited. Similarly, when using both the NF-based methods, any concentration of magnesium can be attained, although with some unavoidable addition of TDS, more pronounced in the NF-based process than in the NF-DiaNF-based process. In the dolomite-calcite dissolution process, it is a function of the amount of dolomite dissolved, which, in turn, impacts the amount of calcite dissolved and the overall product water quality; in the calcite dissolution followed by IX process, it is a function of (1) the availability of SW, (2) the concentration of Ca^{2+} (and also the alkalinity) since in this process the Mg^{2+} ions are exchanged with Ca^{2+}, which is supplied through calcite dissolution. Thus, there is a certain linkage between the carbonate alkalinity and the Ca^{2+} and Mg^{2+} concentrations.

8.5.1.4 Buffering Capacity

The buffer capacity attained in the water varies significantly between the PT alternatives and also (to a certain degree) within each PT. Thus, comparing the methods from the buffer capacity perspective is tricky. As stated at the beginning of this section, the baseline water quality of all compared PTs was $[Ca^{2+}]$ of 30 mg/L. However, this constraint is not sufficient for the current evaluation since each PT can result in a variety of buffer capacities, while maintaining the same $[Ca^{2+}]$, which itself has no influence on the buffer capacity. Therefore, another constraint was defined for this comparison, i.e., product water pH ~ 8.2. Under these two constraints, the PT processes can be ranked by the buffer capacity parameter as follows (from the highest to the lowest): (1) CO_2-based calcite dissolution followed by NaOH dosage; (2) CO_2-based calcite dissolution followed by CO_2 stripping; (3) $Ca(OH)_2 + CO_2$; (4) H_2SO_4-based calcite dissolution followed by NaOH dosage; (5) H_2SO_4-based dolomite dissolution followed by NaOH dosage; (6) direct dosage of $CaCl_2 + NaHCO_3$; (7) blending SW; and (8) the lowest buffer capacity is attained when $CaCl_2 + Na_2CO_3$ are dosed, since the addition of Na_2CO_3 elevates the pH dramatically, and consequently, C_T remains very low. Note that the addition of an IX stage to the calcite dissolution processes, under the mentioned assumptions, results in a higher C_T (because more calcite is dissolved), and thus, also in a higher buffer capacity, as compared with the same dissolution process in the absence of the IX stage.

Note that the addition of lime ($Ca(OH)_2$) for achieving the required Ca^{2+} concentration, results in a very high pH value (>11). Thus, as explained in Section 8.3.1.2, the combined addition of lime and $NaHCO_3$ or Na_2CO_3 is impractical, since the final pH value is excessively high. Accordingly, this alternative was not further evaluated.

Dosage of NF brines has no direct influence on the buffer capacity of the water. However, it elevates the Ca^{2+} concentration. Therefore, if the extent of Ca^{2+} added through a conventional PT (e.g., calcite dissolution) is reduced, for attaining the same final Ca^{2+} concentration, the final buffer capacity might be reduced.

8.5.1.5 *Flexibility*

When desalinated water is blended with other water sources (either SW or BW), there is practically no flexibility since it is impossible to attain control over more than one quality parameter. In the other discussed processes, the flexibility is higher. Nevertheless, there are several restraints on the quality of the water that can practically be attained in each of the processes. Table 8.4 shows the three main product water quality limitations of the considered processes.

8.5.2 Practical Aspects

8.5.2.1 *Percentage of Treated Water*

The percentage of treated water is a significant parameter in comparing PT processes. Operating the PT plant in bypass mode is favorable from both operational and capital cost points of view [82]. With respect to processes that rely on calcite dissolution, the minimum possible percentage of treated water out of the total desalinated water flow rate (%split flow) is a function of two parameters: (1) the available acid concentration in the influent to the dissolution reactor (either CO_2 or H_2SO_4) and the resultant CCDP of the water fed to the reactor. Higher CCDP values enable treating lower volumes of water, i.e., lower %split flow. (2) The required product water quality, in particular, the required Ca^{2+} and alkalinity concentrations. A lower requirement for Ca^{2+} and alkalinity concentrations enables operating the plant with a lower %split flow. However, for a given required water quality, the H_2SO_4 dosage is limited by the propensity of gypsum ($CaSO_4$) precipitation. The combined high concentrations of SO_4^{2-} (from the added acid) and Ca^{2+} (from the dissolved calcite) may lead to unwanted gypsum precipitation. Lehmann et al. [63] showed that at H_2SO_4 dosage higher than 1400 mg/L the precipitation potential of gypsum becomes positive; therefore, higher acid dosage is not recommended.

Generally speaking, as the addition of 1 meq/l of CO_2 results in a less negative CCDP than the addition of 1 meq/l H_2SO_4 and higher pressure is needed for the application of CO_2, it can be concluded that H_2SO_4-based calcite dissolution would result in lower practical %split flow values, despite the gypsum precipitation limitation.

A literature survey on the actual %split flow used in desalination plants resulted in only a handful of examples: A PT plant in Ras Laffan (Qatar) is designed to dissolve 220 mg/l $CaCO_3$ using CO_2 as the acidifying agent [78]. Thus, in order to supply product water containing >80 mg/l as $CaCO_3$ alkalinity, a split flow of at least 37% has to be applied. According to de-Souza et al. [68], who also used CO_2 for dissolving 45 mg/l as $CaCO_3$ of alkalinity to

Table 8.4 The resultant water quality attained in the various PTs

PT process		Ca^{2+} to Alk ratio (eq/eq)	TH to Alk ratio (eq/eq)	Mg^{2+} to Ca^{2+} ratio (eq/eq)
1	CO_2+CaCO_3+NaOH	Slightly above 1	Slightly above 1	Not relevant
2	$CO_2+CaCO_3+CO_2$ stripping	Precisely 1	Precisely 1	Not relevant
3	$H_2SO_4+CaCO_3+NaOH$	~2	~2	Not relevant
4	H_2SO_4+Dolomie+NaOH	Depends on the ratio in the rock (>1)	~2	As the ratio in the rock (<1)
1a	CO_2+CaCO_3+IX+NaOH	Depends on the IX stage, very flexible ≤1	Slightly above 1	Depends on the IX stage, very flexible (usually <1)
2a	$CO_2+CaCO_3+IX+CO_2$ stripping	≤1	1	
3a	$H_2SO_4+CaCO_3$+IX+NaOH	≤2	2	
5	$H_2SO_4+CaCO_3$+Dolomite+NaOH (in series operation)	Depends on the ratio between dissolved calcite and dolomite ~2	Depends on the NaOH dosage >2	Depends on the ratio between dissolved calcite and dolomite <1
6	$Ca(OH)_2+Na_2CO_3$[a]	Very flexible	Very flexible	Not relevant
7	$CaCl_2+NaHCO_3/Na_2CO_3$	Very flexible	Very flexible	Not relevant
8	$CO_2+Ca(OH)_2$	1	1	Not relevant
9	Blending SW	8	49	5
10	Blending BW	Depends on the BW composition		
11	NF brine dosage	Slightly elevated	Elevated, flexible	Elevated
12	NF-DiaNF brine dosage	Slightly elevated	Elevated, flexible	Elevated

[a] This alternative is impractical (see Section 8.3.1.2).

the product water, it was feasible to allow up to 80%–90% of the flow rate to bypass the treatment plant, i.e., apply a %split flow of 10%–20%. Accordingly, for attaining an alkalinity concentration of >80 mg/l as $CaCO_3$, a split flow of around 17%–35% is required. The 130 and 45 million m^3/y Ashkelon and Palmachim plants in Israel use H_2SO_4 as the acidifying agents. Both plants are operated with %Split flow in the range 18%–25%.

8.5.2.2 Reliability

Limestone dissolution is a reliable process, providing constant alkalinity without overdose risks [66,95] and without encountering high turbidity in the product water [63]. On the other hand, it was also recognized that lime injection is potentially associated with handling and injecting problems [66], and somewhat high water turbidity.

8.6 RECENT FULL-SCALE PROJECT EXPERIENCE

At present, the most commonly used post-treatment system in desalination plants worldwide is the sequential addition of lime and carbon dioxide. The most frequent challenge with the operation of such remineralization systems is maintaining low turbidity in the finished drinking water because lime can cause turbidity increase, in some cases up to 5 NTU.

Another remineralization process, which at present is fairly common for small- and medium-sized plants and is gaining more attention and popularity for larger desalination plants over the past 10 years, is the use of calcite dissolution reactors (often termed "calcite contactors"), as elaborated in Section 8.3.3 of this chapter. In comparison to the lime-based post-treatment systems, calcite dissolution systems are usually less costly in terms of both capital and chemical expenditures, and typically produce lower-turbidity finished water. However, in many locations worldwide, high-quality food-grade calcite is not as readily available as lime, which is one of the main reasons why this technology has not been used as frequently as lime/carbon dioxide conditioning. In addition, the desalination industry has limited experience with the use of calcite contactors for large desalination plants—most of this experience today is in Israel, where all large SWRO desalination plants in operation at present use this technology for permeate conditioning. Outside Israel, the largest operational SWRO plants using calcite contactors are in Barcelona, Spain (200,000 m^3/d), Larnaka, Cyprus (64,000 m^3/d), and Carlsbad, California (190,000 m^3/d).

The following section describes a case study of a post-treatment process that relies on calcite dissolution, in which the source of acid used to enhance the dissolution is the inherent $CO_{2(aq)}$ present in the brackish water fed to the desalination plant.

8.6.1 The Kantor Desalination Plant

The Kantor plant is a small brackish water RO desalination plant located in Ma'agan Michael, Israel, whose current aim is to supply drinking water to the small Kibutz it serves. It was originally designed to serve as a pilot plant to a much larger desalination plant, which is now operative beside it. The plant has an annual production of ~800,000 m^3, an operational recovery ratio of 0.73, and typical rejection for Cl^-/Na^+ of 98.5%. Accordingly, the typical Cl^- concentration in the product water is ~30 mg/l.

8.6.1.1 Characteristics of the Brackish Water Fed to the Kantor Plant

The composition of the ground BW fed to the Kantor desalination plant is shown in Table 8.5. The water is pumped from the Carmel Coast Pleistocene aquifer in Northern Israel (typical water table depth: 4–6 m; temperature: 22–24°C). Based on the EC, temperature, pH, and alkalinity values shown in Table 8.5, the $CO_{2(aq)}$ concentration in the well water is ~59 mg/l.

8.6.1.2 Operation of the Kantor Desalination Plant Post-Treatment Step

The plant applies a PT step aimed at meeting the following water quality: pH value between 7.5 and 8.3, dissolved calcium concentration between 80 and

Table 8.5 Typical composition of the BW fed to the Kantor desalination plant

Parameter	Units	Value	Parameter	Units	Value
pH	–	6.90	K^+	mg/l	35.5
Turbidity	NTU	0.2	Ca^{2+}	mg/l	195.5
Alkalinity	mg/l as $CaCO_3$	305	Mg^{2+}	mg/l	146.0
TDS	mg/l	3890	NH_4^+	mg/l	0.1
EC	μS/cm	6810	Fe (total)	μg/l	12.0
F^-	mg/l	0.5	Mn^{2+}	μg/l	<3
Cl^-	mg/l	1970	Co^{2+}	μg/l	5.0
NO_3^-	mg/l	6.0	Ba^{2+}	μg/l	112
HCO_3^-	mg/l	371.8	Zn^{2+}	μg/l	60
SO_4^{2-}	mg/l	251.0	Al^{3+}	μg/l	5.0
Hardness	mg/l as $CaCO_3$	1088	Na^+	mg/l	1014
H_2S	mg/l	0			

120 mg/l as $CaCO_3$, and alkalinity higher than 80 mg/l as $CaCO_3$ (according to Annex 6 of the Israeli ministry of health regulations of drinking water quality, 2013, http://www.health.gov.il/Subjects/Environmental_Health/drinking_water/Documents/Briut47-Eng.pdf). Note that according to these regulations, the CCPP (or LSI) threshold is not mandatory for brackish water plants producing less than 5000 m^3 water per day, such as the Kantor plant. The PT is based on $CaCO_3$ dissolution using the inherent CO_2 concentration in the permeate water, resulting from the naturally high $CO_{2(aq)}$ concentration in the groundwater (see Table 8.5). The entire water path from the well to the permeate tank is almost hermetically sealed from the atmosphere; thus, the CO_2 remains dissolved in the permeate stream, resulting in low pH and a negative CCPP value. The CO_2 concentrations in the permeate stream closely simulate the CO_2 concentrations in the feed water (~50 mg/l). The 20-m^3 permeate tank, despite being maintained under atmospheric pressure, is not well ventilated, and consequently, (based on a mass balance) only a small mass of CO_2 is emitted from the water. As a result, the dissolved CO_2 concentration in the water at the calcite reactor's inlet is high (~50 mg/l); thus, calcite dissolution is enabled with no need for external acidic agent dosage.

The entire permeate flow rate is introduced to the calcite dissolution reactor at a downflow velocity of ~17 m/h. The hydraulic retention time in the reactor is between 7 and 10 min. The average calcite particle size is 1 mm. Every ~20 days the reactor is refilled with calcite, brought from a quarry in Northern Israel. Differences between winter and summer, with respect to the water composition, are negligible.

The composition of the water flowing out of the calcite dissolution reactor is detailed in Table 8.6. As explained in Section 8.3.3 of this chapter, due to kinetic limitations, the CCPP of the effluent of calcite dissolution reactors is negative. In order to raise the pH and produce water with

Table 8.6 Composition of the water at two points along the PT process in the Kantor desalination plant

Parameter	Units	Calcite reactor effluents	Product water (after NaOH dosage)
Turbidity	NTU	0.2	0.2
TDS	mg/l	200	205
pH	–	7.4	8.1
Ca^{2+}	mg/l	36	36
Alkalinity	mg/l as $CaCO_3$	95	104
CCPP	mg/l	−10	1

positive CCPP (as required), NaOH dosage (~6 mg/L) is applied at the end of the post treatment, prior to releasing the water to the consumers.

8.7 SUMMARY

This chapter starts by developing the basic chemistry concepts required for designing and operating post-treatment processes for desalinated water. Based on these fundamentals, the main state-of-the-art post-treatment methods are then reviewed, i.e., direct dosage of a variety of chemicals, blending of the desalination permeate with external water sources, and processes that are based on calcite and dolomite dissolution. Out of these, calcite dissolution (in its various manifestations) is likely the most often used method, worldwide. The detailed explanation on calcite dissolution in this chapter includes the effect of applying strong acid (i.e., H2SO4) vs. the effect of applying weak acid (i.e., CO2) for enhancing dissolution, as well as application of alternative sources of CO2. Similarly, two options for final pH adjustment are reviewed: CO2 stripping and strong base (e.g., NaOH) dosage. It is stressed that calcite and lime dissolution, the most practiced post-treatment processes, do not supply minerals to the water, except for calcium.

The chapter further provides a review on post-treatment processes, which are still under research, that focus on attaining magnesium in the product water. The chapter includes a discussion on the advantages and disadvantages of each one of the processes, from the engineering, cost, and water-quality perspectives.

The chapter also brings a case study showing an irregular operation of post-treatment process, to demonstrate how creative solutions can be applied in the industry.

In conclusion, the post-treatment (remineralization) step in desalination plants can be operated in a variety of modes and approaches. The choice of the applied post-treatment scheme results in the production of different water qualities and a different cost. Overall, the post-treatment step has a significant impact on the final water quality, although its cost is typically only ~10% of the overall cost. It is therefore imperative to design and operate this step based on profound knowledge and understanding of the processes that govern it.

It is noted that this chapter does not include details related to boron removal, disinfection, fluoridation, and aeration, although these are sometimes considered to be a part of the post-treatment step in desalination plants. In addition, processes aimed at supplying other required species in the final product water, such as iodide and selenium, were not covered either.

REFERENCES

[1] Withers A. Options for recarbonation, remineralisation and disinfection for desalination plants. Desalination 2005;179(1–3 SPEC. ISS):11–24.

[2] Delion N, Mauguin G, Corsin P. Importance and impact of post treatments on design and operation of SWRO plants. Desalination 2004;165(suppl):323–34.

[3] Lahav O, Birnhack L. Quality criteria for desalinated water following post-treatment. Desalination 2007;207(1–3):286–303.

[4] Martínez-Alvarez V, Martin-Gorriz B, Soto-García M. Seawater desalination for crop irrigation—a review of current experiences and revealed key issues. Desalination. 2016;381:58–70.

[5] Philippou J. Quality considerations from integrating desalinated water into existing water infrastructure. Desalin Water Treat [Internet] 2015;55(13):3519–26. Available from: https://doi.org/10.1080/19443994.2014.1002990.

[6] Duranceau SJ, Wilder RJ, Douglas SSA. survey of desalinated permeate post-treatment practices. Desalin Water Treat 2012;37(1–3):185–99.

[7] Seacord TF, Singley E, GVN J. In: Post-treatment Concepts for Seawater and Brackish Water Desalination. American Water Works Association Proceedings, Membrane Technology Conference, Atlanta, Georgia; 2003.

[8] Birnhack L, Voutchkov N, Lahav O. Fundamental chemistry and engineering aspects of post-treatment processes for desalinated water-A review. Desalination 2011;273(1):6–22.

[9] Imbrogno J, Keating JJ, Kilduff J, Belfort G. Critical aspects of RO desalination: A combination strategy. Desalination [Internet]. 2016; Available from: https://doi.org/10.1016/j.desal.2016.06.033.

[10] Sung W, Huang X, Wei IW. Treatment and distribution system effects on chloramine decay, pH, nitrification, and disinfection by-products: case study. J Water Resour Plan Manag 2005;131(3):201–7.

[11] Yermiyahu U, Tal A, Ben-Gal A, Bar-Tal A, Tarchitzky J, Lahav O. Rethinking desalinated water quality and agriculture. Science (80-). 2007;318(5852):920–1.

[12] Franchi A, Jousset S, Brodeur T. Study and prevention of "Red Water" occurrences cause by water chemistry changes. In: Water quality conference and exhibition. Seattle, WA: American Water Works Association; 2002.

[13] Marangou VS, Savvides K. First desalination plant in Cyprus—Product water aggressivity and corrosion control. Desalination 2001;138(1–3):251–8.

[14] Berghult B, Hedberg T, Broo AE. Drinking water distribution: Corrosion control in Swedish municipalities. Aqua 1999;48(2):44–52.

[15] Price F.T. S. J. Corrosion control strategies for changing water supplies in Tucson, Arizona. J New Engl Water Works Assoc, Vol. 111 1997.

[16] Guidelines for Drinking—Water Quality 2008 2nd addendum to 3rd Ed [Internet]. World Health Organization. 2008 [cited 2016 Nov 30]. Available from: http://www.who.int/water_sanitation_health/dwq/secondadden-dum20081119.pdf.

[17] Karalekas PC, Ryan CR, Taylor FB. Control of lead, copper, and iron pipe corrosion in Boston. J Am Water Works Assoc 1983;75:92–5.

[18] Maddison LA, Gagnon GA, Eisnor JD. Corrosion control strategiesfor the Halifax regional distribution system. Can J Civ Eng 2001;28(2):305–13.

[19] Plottu-Pecheux A, Democrate C, Houssais B, Gatel D, Cavard J. Controlling the corrosiveness of blended waters. Desalination 2001;138(1–3):237–49.

[20] Sarin P, Snoeyink VL, Bebee J, Kriven WM, Clement JA. Physico-chemical characteristics of corrosion scales in old iron pipes. Water Res 2001;35(12):2961–9.

[21] WHO World Health Organization. Expert committee meeting on health effects of calcium and magnesium in drinking-water. DC: Washington; 2006.

[22] Yang F, Shi B, Gu J, Wang D, Yang M. Morphological and physicochemical characteristics of iron corrosion scales formed under different water source histories in a drinking water distribution system. Water Res 2012;46(16):5423–33.
[23] Bellinger DC. Lead contamination in flint—an abject failure to protect public health. N Engl J Med 2016;374(12):1101–3.
[24] Ginocchio JC. Protection against corrosion in drinking water distribution systems. Anti-Corros Meth Mater 1985;32(8):14–6.
[25] Mcneill LS, Edwards M. Iron pipe corrosion in distribution systems. J Am Water Work Assoc 2001;93(7):88–100.
[26] Rodriguez AD, Reiber S, Black B, Graff NA. Predicting colored water episodes when integrating new source waters into your distribution system: a bench scale protocol. In: Water quality technology conference. Neshville, Tennessee: American Water Works Association; 2001. p. 2316–31.
[27] Ovadia YS, Gefel D, Aharoni D, Turkot S, Fytlovich S, Troen AM. Can desalinated seawater contribute to iodine-deficiency disorders? An observation and hypothesis. Public Health Nutr 2016;19(15).
[28] Barron O, Ali R, Hodgson G, Smith D, Qureshi E, McFarlane D, et al. Feasibility assessment of desalination application in Australian traditional agriculture. Desalination 2015;364:.
[29] Lew B, Cochva M, Lahav O. Potential effects of desalinated water quality on the operation stability of wastewater treatment plants. Sci Total Environ 2009;407(7).
[30] Higgins MJ, Tom LA, Sobeck DC. Case study I: Application of the divalent cation bridging theory to improve biofloc properties and industrial activated sludge system performance—direct addition of divalent cations. Water Environ Res 2004;76(4).
[31] Kozisek F. Health significance of drinking water calcium and magnesium [Internet]. [cited 2016 Nov 30]. p. 29. Available from: http://www.szu.cz/uploads/documents/chzp/voda/pdf/hardness.pdf; 2003.
[32] Imran SA, Dietz JD, Mutoti G, Taylor JS, Randall AA, Cooper CD. Red water release in drinking water distribution systems. J Am Water Work Assoc 2005;97(9).
[33] Sontheimer H, Kolle W, Snoeyink VL. Siderite model of the formation of corrosion-resistant scales. J Am Water Work Assoc 1981;73(11).
[34] Lahav O, Salomons E, Ostfeld A. Chemical stability of inline blends of desalinated, surface and ground waters: the need for higher alkalinity values in desalinated water. Desalination 2009;238(1–3).
[35] Rossum JR, Merrill DT. Evaluation of the calcium carbonate saturation indexes. J Am Water Work Assoc 1983;75(2).
[36] Barbeau B, Huffman D, Mysore C, Desjardins R, Prévost M. Examination of discrete and confounding effects of water quality parameters during the inactivation of MS2 phages and Bacillus subtilis spores with free chlorine. J Environ Eng Sci 2004;3(4).
[37] Gyürék LL, Finch GR, Belosevic M. Modeling chlorine inactivation requirements of Cryptosporidium parvum oocysts. J Environ Eng 1997;123(9).
[38] Sheffer M. Guidelines for drinking-water quality. 4th ed. 2011. [Internet]. Geneva. Available from: http://apps.who.int/iris/bitstream/10665/44584/1/9789241548151_eng.pdf.
[39] BJ CJ, editor. Calcium and magnesium in drinking water: Public health significance [Internet]. 1st ed. Spain: World Health Organization; 2009. 194 p. Available from: http://apps.who.int/iris/bitstream/10665/43836/1/9789241563550_eng.pdf.
[40] Catling LA, Abubakar I, Lake IR, Swift L, Hunter PRA. Systematic review of analytical observational studies investigating the association between cardiovascular disease and drinking water hardness. J Water Health 2008;6(4).

[41] Monarca S, Donato F, Zerbini I, Calderon RL, Craun GF. Review of epidemiological studies on drinking water hardness and cardiovascular diseases. Eur J Cardiovasc Prev Rehab. 2006;13(4).
[42] Sinclair, M.; and Hayes P. Desalination Decision for Israel. Water Qual Res Aust [Internet]. 2011;61:4–37. Available from: www.wqra.com.au.
[43] McNeill LS, Edwards M. Phosphate inhibitors and red water in stagnant iron pipes. J Environ Eng 2000;126(12).
[44] Alshehri AA, Duranceau SJ, Taylor JS, Stone ED. Investigating iron release in distribution systems with blend variations of source waters and phosphate inhibitors. Desalin Water Treat 2009;8(1–3).
[45] Volk CJ, LeChevallier MW. Impacts of the reduction of nutrient levels on bacterial water quality in distribution systems. Appl Environ Microbiol 1999;65(11).
[46] Ludwig H, Hetschel M. Treatment of distillates and permeates from seawater desalination plants. Desalination 1986;58(2).
[47] Lytle DA, Sarin P, Snoeyink VL. The effect of chloride and orthophosphate on the release of iron from a cast iron pipe section. J Water Supply Res Technol -AQUA 2005;54(5).
[48] Fang W, Hu J, Ong SL. Effects of phosphorus on biofilm disinfections in model drinking water distribution systems. J Water Health 2010;8(3).
[49] Chu C, Lu C, Lee C. Effects of inorganic nutrients on the regrowth of heterotrophic bacteria in drinking water distribution systems. J Environ Manage 2005;74(3).
[50] Hozalski RM, Elizabeth E-A, Chen CF. Comparison of stannous chloride and phosphate for lead corrosion control. J Am Water Work Assoc 2005;97(3).
[51] Morton SC, Edwards M. Reduced phosphorus compounds in the environment. Crit Rev Environ Sci Technol 2005;35(4).
[52] Khawaji AD, Wie J-M. Potabilization of desalinated water at Madinat Yanbu Al-Sinaiyah. Desalination 1994;98(1–3).
[53] Gran G, Johansson A, Johansson S. Automatic titration by stepwise addition of equal volumes of titrant. Part VII. Potentiometric precipitation titrations. Analyst 1981;106:.
[54] Pokrovsky OS, Golubev SV, Schott J. Dissolution kinetics of calcite, dolomite and magnesite at 25 °C and 0 to 50 atm pCO2. Chem Geol 2005;217(3–4 SPEC).
[55] Busenberg E, Plummer LN. The kinetics of dissolution of dolomite in CO2-H2O systems at 1.5 to 65°C and 0 to 1 atm PCO2. Am J Sci 1982;282(1).
[56] Liu Z, Yuan D, Dreybrodt W. Comparative study of dissolution rate-determining mechanisms of limestone and dolomite. Environ Geol 2005;49(2).
[57] Morse JW, Arvidson RS. The dissolution kinetics of major sedimentary carbonate minerals. Earth-Science Rev 2002;58(1–2).
[58] Holm TR, Schock MR. Computing SI and CCPP using spreadsheet programs: Spreadsheets can automate any chemical equilibrium calculation that can be performed by hand. J Am Water Work Assoc 1998;90(7).
[59] Millette JR, Hammonds AF, Pansing MF, Hansen EC, Clark PJ. Aggressive water: assessing the extent of the problems (USA). Am Water Work Assoc J 1980;72(5).
[60] Pisigan Jr. RA, Singley JE. Technical note: experimental determination of the calcium carbonate saturation states of water systems. J Am Water Work Assoc 1985;77(10).
[61] Internal corrosion of water distribution systems. 2nd ed. Denver, CO: American Water Works Association Research Foundation; 1996, 586 p.
[62] Birnhack L, Fridman N, Lahav O. Potential applications of quarry dolomite for post treatment of desalinated water. Desalin Water Treat 2009;1(1–3).
[63] Lehmann O, Birnhack L, Lahav O. Design aspects of calcite-dissolution reactors applied for post treatment of desalinated water. Desalination 2013;314.
[64] Hasson D, Bendrihem O. Modeling remineralization of desalinated water by limestone dissolution. Desalination 2006;190(1–3).

[65] Fritzmann C, Löwenberg J, Wintgens T, Melin T. State-of-the-art of reverse osmosis desalination. Desalination 2007;216(1–3).
[66] Gabbrielli E. A tailored process for remineralization and potabilization of desalinated water. Desalination. 1981;39(C).
[67] Glade H, Meyer J-H, Will S. The release of CO2 in MSF and ME distillers and its use for the recarbonation of the distillate: A comparison. Desalination 2005;182(1–3).
[68] PF de Souza G du Plessis and G Mackintosh. An evaluation of the suitability of the limestone based sidestream stabilization process for stabilization of waters of the Lesotho highlands scheme. In: Water SA - Special Edition, Wisa Proceedings [Internet]. Stellenbosch, South Africa: Water SA; 2002. p. 10–5. Available from: http://www.wrc.org.za/Knowledge Hub Documents/Water SA Journals/Manuscripts/2002/05/WaterSA_2002_05_4.pdf.
[69] El Azhar F, Tahaikt M, Zouhri N, Zdeg A, Hafsi M, Tahri K, et al. Remineralization of Reverse Osmosis (RO)-desalted water for a Moroccan desalination plant: Optimization and cost evaluation of the lime saturator post. Desalination 2012;300:.
[70] Ben-Gal A, Yemiyahu U, Cohen S. Fertilization and blending alternatives for irrigation with desalinated water. J Environ Qual 2009;38(2).
[71] Glueckstern P, Priel M, Kotzer E. Blending brackish water with desalted seawater as an alternative to brackish water desalination. Desalination 2005;178(1–3 SPEC).
[72] Dreizin Y, Tenne A, Hoffman D. Integrating large scale seawater desalination plants within Israel's water supply system. Desalination 2008;220(1–3).
[73] Ghermandi A, Messalem R. Solar-driven desalination with reverse osmosis: The state of the art. Desalin Water Treat. 2009;7(1–3).
[74] Shams El Din AM. Three strategies for combating the corrosion of steel pipes carrying desalinated potable water. Desalination 2009;238(1–3).
[75] Hermony A, Sutzkover-Gutman I, Talmi Y, Fine O. Palmachim Seawater desalination plant—seven years of expansions with uninterrupted operation together with process improvements. Desalin Water Treat 2015;55(9).
[76] Birnhack L, Oren S, Lehmann O, Lahav O. Development of an additional step to current CO_2-based $CaCO_{3(s)}$ dissolution post-treatment processes for cost-effective Mg^{2+} supply to desalinated water. Chem Eng J 2010;160(1).
[77] Al-Rqobah HE, HA-M A. A recarbonation process for treatment of distilled water produced by MSF plants in Kuwait. Desalination 1989;73(C).
[78] Migliorini G, Meinardi R. 40 MIGD potabilization plant at Ras Laffan: Design and operating experience. Desalination 2005;182(1–3).
[79] Greiserman M, Hasson D, Semiat R, Shemer H. Kinetics of dolomite dissolution in a packed bed by acidified desalinated water. Desalination 2016;396.
[80] Rooklidge SJ, Ketchum Jr. LH. Corrosion control enhancement from a dolomite-amended slow sand filter. Water Res 2002;36(11).
[81] Oren S, Birnhack L, Lehmann O, Lahav O. A different approach for brackish-water desalination, comprising acidification of the feed-water and $CO_{2(aq)}$ reuse for alkalinity, Ca^{2+} and Mg^{2+} supply in the post treatment stage. Sep Purif Technol 2012;89.
[82] King DE, Mainous III AG, Geesey ME, Woolson RF. Dietary magnesium and C-reactive protein levels. J Am Coll Nutr 2005;24(3).
[83] Rowell C, Kuiper N, Shomar B. Potential health impacts of consuming desalinated bottled water. J Water Health 2015;13(2).
[84] Shlezinger M, Amitai Y, Goldenberg I, Shechter M. Desalinated seawater supply and all-cause mortality in hospitalized acute myocardial infarction patients from the Acute Coronary Syndrome Israeli Survey 2002–2013. Int J Cardiol 2016;220.
[85] Nriagu J, Darroudi F, Shomar B. Health effects of desalinated water: Role of electrolyte disturbance in cancer development. Environ Res 2016;150.

[86] Lahav O, Kochva M, Tarchitzky J. Potential drawbacks associated with agricultural irrigation with treated wastewaters from desalinated water origin and possible remedies. Water Sci Technol 2010;61(10).
[87] Díaz FJ, Tejedor M, Jiménez C, Grattan SR, Dorta M, Hernández JM. The imprint of desalinated seawater on recycled wastewater: Consequences for irrigation in Lanzarote Island, Spain. Agric Water Manag 2013;116.
[88] Birnhack L, Lahav O. A new post-treatment process for attaining Ca^{2+}, Mg^{2+}, SO_4^{2-} and alkalinity criteria in desalinated water. Water Res 2007;41(17).
[89] Birnhack L, Penn R, Oren S, Lehmann O, Lahav O. Pilot scale evaluation of a novel post-treatment process for desalinated water. Desalin Water Treat 2010;13(1–3).
[90] Birnhack L, Shlesinger N, Lahav O. A cost effective method for improving the quality of inland desalinated brackish water destined for agricultural irrigation. Desalination 2010;262(1–3).
[91] Penn R, Birnhack L, Adin A, Lahav O. New desalinated drinking water regulations are met by an innovative post-treatment process for improved public health. Water Science and Technology: Water Supply 2009;9.
[92] Birnhack L, Nir O, Lahav O. Establishment of the underlying rationale and description of a cheap nanofiltration-based method for supplementing desalinated water with magnesium ions. Water (Switzerland) 2014;6(5).
[93] Nativ P, Birnhack L, Lahav O. DiaNanofiltration-based method for inexpensive and selective separation of Mg^{2+} and Ca^{2+} ions from seawater, for improving the quality of soft and desalinated waters. Sep Purif Technol 2016;166.
[94] Tang SCN, Birnhack L, Nativ PLO. Highly-selective separation of divalent ions from seawater and seawater RO brines. Sep Purif Technol 2017;175:460–8.
[95] Kettunen R, Keskitalo P. Combination of membrane technology and limestone filtration to control drinking water quality. Desalination 2000;131(1–3):271–83.

FURTHER READING

[1] Feigenbaum C, Gal-or L, Yahalom J. Scale protection criteria in natural waters. Corrosion 1978;34(4).
[2] Singley JE. Search for a corrosion index. J Am Water Work Assoc 1981;73(11).

CHAPTER 9

Desalination Concentrate Management and Valorization Methods

Antonio Santos Sánchez*, Ângelo Paggi Matos†
*Federal University of Ouro Preto, Ouro Preto, Brazil
†Federal University of Santa Catarina, Florianópolis, Brazil

9.1 INTRODUCTION

Brine reject, hereafter desalination concentrate (DC), can pose a serious environmental threat due to its salinity and also due to containing residues of pretreatment and cleaning chemicals, together with their reaction by-products. Therefore, an adequate discharge of the DC is a key part of the environmental management of desalination plants. DC is a waste effluent generated in many industrial processes: the most common are produced water from the primary treatment of fluids in the oil and gas industry and desalination of seawater or brackish waters for the production of freshwater. Irrigation practices can eventually produce significant volumes of DC, especially in arid climates where the water is abducted from overexploited aquifers. The process of salinization due to irrigation occurs when the salts dissolved in the water get accumulated in the soil. Since soil salinity makes it more difficult for plants to absorb soil moisture, these salts must be removed out of the plant root zone by applying an additional amount of water: this excess of irrigation is called the leaching fraction [1]. This subsurface drainage from agricultural fields can also produce large volumes of DC [2–4]. Other sources of DC are geothermal waters; effluents from mines, metallurgy, industrial processes and power plants; marinating brine waste; and aquaculture wastewater. The discharge of these DCs without any treatment can cause severe ecological impacts. In coastal desalination plants, discharge through submarine outfalls is the most common practice. They must be adequately modeled and projected in order to assure proper dilution as brine discharge poses a potentially serious threat to marine ecosystems [5].

Sustainable Desalination Handbook
https://doi.org/10.1016/B978-0-12-809240-8.00009-5

In inland desalination plants, diluting the DC in a water stream is not a feasible option. Direct brine discharge in the soils would lead to its degradation by salinization, even if the DC is only at slightly above recommended values [6]. Other strategies must be considered, such as impervious evaporation ponds or the valorization of the brine through aquaculture or agricultural uses [7–11]. Pervious evaporation ponds allowing percolation [12,13] as well as the injection in shallow wells could pollute groundwater. In addition, salts may rise by capillary transport from a salt-laden water table and then accumulate in the soil surface due to evaporation. This problem is of particular importance in arid regions where there is already an anthropogenic cause of salinization due to irrigation or due to an extensive use of potassium as fertilizer [14–16].

There are two main approaches for the management of DC. The first one focuses on its valorization through agriculture, aquaculture, and algaculture. Around the world, systems have been developed aiming the use of the DC evaporation ponds for fish or algae farming. Cyanobacteria species such Spirulina, and algae species such as *C. vulgaris* and *N. gaditana* can be grown in open raceway ponds. These species are a source of proteins, fatty acids, and also lipids from which biodiesel can be obtained. Other schemes that have been assessed consider the growth of fish in the ponds, particularly from the Tilapia genus, together with the irrigation of salt-tolerant forage shrubs. Halophyte plants such as Atriplex genus have been widely tested due to their ability to tolerate DC from brackish desalination plants and to act as fodder for goats [17]. Those schemes have been developed for inland desalination plants located in arid regions, but are valid as well for coastal areas worldwide. Aquaculture wastewater, typically the effluent originated from fish and prawn farms, offers good possibilities for further use as a substrate for algae cultivation. Meanwhile, the DC that is generated in reclamation and reuse plants is typically loaded with nitrogen compounds. For that case, the production of microalgae using the DC blended with wastewater, which acts as a source of phosphorus, is of special interest. Following that strategy, blending treated urban wastewater (20%–30%) with DC from seawater in reverse osmosis (RO) plants (80%–70%) can also produce a final effluent with uses for agricultural irrigation [18]. Blending the DC with less saline water can also open opportunities for renewable energy generation through osmotic energy recovery, especially in those DC with higher salinity [19,20]. Overall, the composition of a DC would be determinant in the assessment of its potential beneficial uses. For example, geothermal waters can have a use for agricultural irrigation after boron removal [21]. Other process wastewater, besides its salinity content, can contain toxic chemicals making the DC unsuitable for its use for agricultural/aquaculture purposes. RO concentrates from wastewater treatment plants can contain pharmaceuticals compounds that are potentially toxic to aquatic organisms. Some of these compounds can be removed with the use of specific treatments. In the case of beta blockers and other toxic compounds, the application of moderate ozone doses was found to be sufficient for its removal, allowing the reuse of the DC [22,23].

The second approach of the management of DC focuses on a series of treatments that reduce the volumetric flow rate to be discharged. Salt concentration increases as the volumetric flow rate reduces. Therefore, it is easier to precipitate dissolved salts and other compounds, which turns the effluent less hazardous and also opens the possibility of mineral extraction from the DC. If the DC carries dissolved toxic compounds, a convenient approach may be the volume minimization of the brine followed by selective adsorption of the compounds or alternatively, trying to precipitate/neutralize them in a solid form for proper disposal. Ideally, a volume minimization process would achieve a ZLD (Zero Liquid Discharge). An ideal ZLD process would generate, from one input (the DC), two different outputs: value-added solids (salts and other chemicals) and clean reclaimed water. Table 9.1 shows the set of volume minimization technologies that is currently available.

Solar evaporation, which requires a significant amount of land, and Wind-Aided Intensified Evaporation (WAIV) are among the simplest volume minimization methods. The latter is a less land-intensive method that reduces DC volumes through the use of the drying power of the wind without generating small droplets that can cause salt drift [79].

More sophisticated technologies allow higher efficiencies and compact units for the treatment of DC but require more energy input. This is the case of membrane distillation or pervaporation (aka pervaporative separation) and Forward Osmosis (FO). Membrane distillation is as alternative for the processing of highly concentrated aqueous solutions that allows very compact installations. Vacuum Membrane Distillation (VMD) is an evaporative technology that uses a membrane to support the liquid–vapor interface. Meanwhile, FO utilizes a highly concentrated solution generally referred to as the draw solution to generate an osmotic pressure differential across the membrane, thus resulting in the transport of water from the less concentrate feed stream to the highly concentrated draw solution. A wide range of draw solutions can be used: sulfur dioxide, aluminum sulfate, fructose, ammonium

Table 9.1 Volume minimization approaches for DCs

Volume minimization technique	Reference
Two-step RO with intermediate chemical precipitation	[24,25]
Seeded Slurry Precipitation and Recycle Reverse Osmosis (SPARRO)	[26,27]
Two-step RO with intermediate biological treatment	[28]
RO with softening pretreatment and high pH	[29]
Two-step nanofiltration	[30]
Membrane distillation/pervaporation (pervaporative separation)	[31]
FO (forward osmosis)	[4,32–37]
Evaporation–crystallization processes	[38–45]
Vibratory shear enhanced processing (VSEP)	[46,47]
Electrodialysis (ED)	[38,48–60]
Electrodialysis with Bipolar Membranes (EDBM)	[61–63]
Electrodialysis Reversal (EDR)	[64,65]
Electrodialysis Metathesis (EDM)	[66]
Wind-aided intensified evaporation (WAIV)	[67–69]
Vacuum Membrane Distillation (VMD) or Air Gap Membrane Distillation (AGMD)	[70–72]
Brine evaporative cooler/concentrator (BECC)	[73]
Ohmic evaporator	[74]
Multiple Effect Distillation (MED)	[75–77]
Mechanical vapor compressor (MVC)	[75,78]

bicarbonate, etc. [79]. The main advantage of FO is to have lower energy requirement than RO.

As said, the ZLD technique may be of interest in those cases where the brine has heavy metals or other toxics that are wanted to be captured in the resulting solid product [80,81]. One solution for ZLD is evaporation–crystallization processes [38,39]. The evaporation–crystallization technology combines these two techniques to produce a solid residue (essentially calcium sulfate salts) from a liquid effluent, but the energy consumption of this technique is high. Therefore, it may be economically viable only in some particular cases, when associated with a vapor recovery system or waste heat. The use of solar energy could be of special interest for the increase in the viability of this technique.

In sum, there are two main alternatives for the management of DC. The composition of the DC (level of salinity, presence of contaminants) would determine which alternative is more convenient to choose. Both of them allow beneficial uses of DC that can recover resources or produce value-added byproducts. Fig. 9.1 shows the two alternative routes, namely, valorization (with optional blending of the DC with wastewater) and volume minimization.

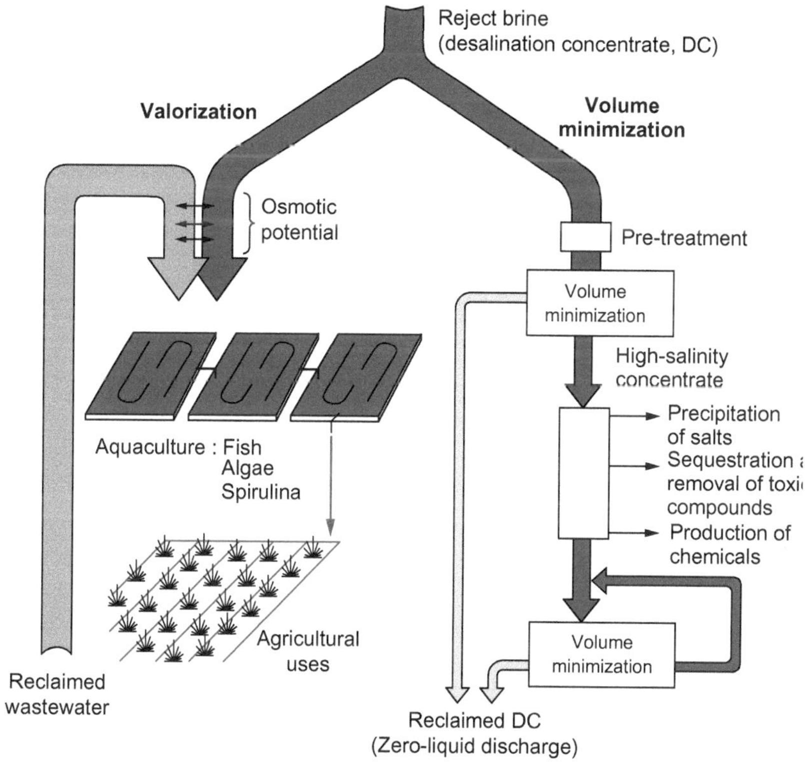

Fig. 9.1 Valorization or volume minimization? Two main approaches for the management of DC.

Previous reviews on DC management would help the reader to get an overview of all the stages involved, from modeling to the different treatment and discharge possibilities [8,11,27,79,82–88]. Practically all of those works mention the possibility of resource recovery from the DC as a convenient option prior to its discharge. This chapter of the handbook identifies and describes the beneficial uses of DC. Among the two mentioned routes that aim to turn this waste effluent into a resource asset, the focus of this chapter is set more on valorization rather than on volume minimization.

Hereafter, each one of the different beneficial uses of DC will be assessed in detail.

9.2 BENEFICIAL USES OF DESALINATION BRINES/CONCENTRATES

9.2.1 Renewable Energy Production

Osmotic power, salinity gradient power, or blue energy is the energy available from the difference in the salt concentrations between two liquid streams [89]. Two main methods can be used to harvest this renewable energy source: pressure-retarded osmosis [19,90] and reverse electrodialysis [20]. Both processes rely on osmosis with ion specific membranes. The energy source to be harvested is the salinity gradient between the two liquids, that is, their osmotic pressure (OP) difference. OP can be defined as the chemical potential of concentrated and diluted solutions of salt: the effluent with the higher concentration of salt has higher OP. Since 2009, there is a pilot osmotic power plant operating in Norway driven by the salinity gradient between the sea and a fiord [91]. The technology chosen was pressure-retarded osmosis in which seawater is pumped into a pressure chamber. The chamber contains the two liquid streams, separated by a semipermeable membrane. The pressure of that chamber is lower than the difference between the pressure of the two water streams (seawater and fresh water). As the salt water pulls fresh water through the membrane, fresh water moves in the membrane and increases its volume in the chamber. Around 90% of fresh water is transferred by osmosis across the membrane across the membrane into the pressurized seawater. Pressure increases even more on the salt water side, and is enough to power a turbine and generate electricity. Operating pressure for such system is in the range of 11–15 bars [91], which is the equivalent to a water head of 100–145 m in a conventional hydropower plant. Fig. 9.2 shows a scheme of this process:

The membrane is the heart of the process. The development of highly efficient membranes at affordable cost would determine the profitability of this renewable energy source in the next years [92]. Regarding brine management, the generation of osmotic power could be a feasible option if the

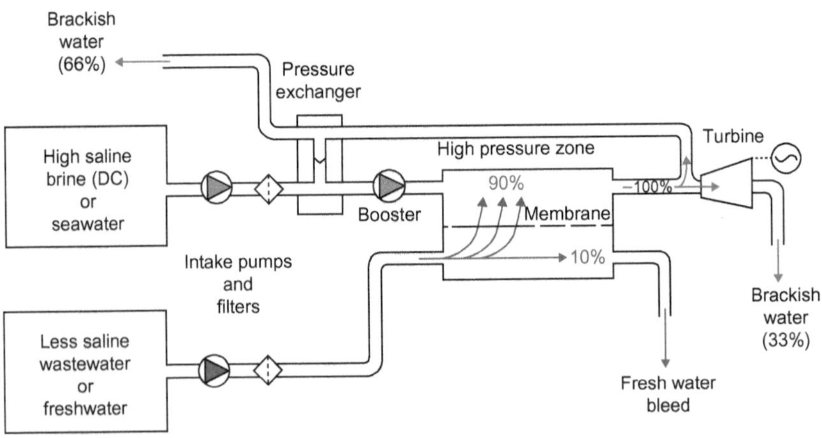

Fig. 9.2 Extraction of osmotic power through pressure-retarded osmosis.

DC has a high content of salts and there is another liquid stream available locally so enough salinity gradient can be created [88]. Blending the DC with low-salinity reclaimed wastewater could be an interesting option [19,93]. After mixing both streams and extracting osmotic energy, the waste effluent is brackish water that could still have a use in agriculture or aquaculture.

9.2.2 Salts Recovery and Acids/Bases Production

Various types of inorganic salts can be produced from the DC of desalination plants: gypsum, sodium chloride, magnesium hydroxide, calcium chloride, calcium carbonate, and sodium sulfate [40]. Magnesium concentration can reach values up to 30–40 kg m^{-3} in the DC, which is 20–30 times that of typical seawater [94]. Other minerals whose extraction from seawater or DC has been indicated as potentially attractive are potassium, lithium, bromine, boron, and uranium [41]. Table 9.2 shows the typical mineral contents of seawater, DC, and permeate:

A "salt recovery" approach aims to treat the concentrate stream, at the tail end of the desalination system, to recover valuable salts as byproducts. Within that approach, both the salt load and the volume of the concentrate are reduced, which minimizes the requirements for effluent discharge, getting close to a ZLD (Zero-Liquid-Discharge). The preconcentration of the DC before its treatment is a key factor for the viable recovery of salts and the production of value-added chemicals. Solar or wind-aided evaporation, FO, VMD, and other evaporation techniques (see Table 9.1) reduce the volume of the DC and increase the load of dissolved salts, which can later be recovered through a series of emerging technologies. The following ones are highlighted:

- Combination of evaporation and crystallization
- Membrane distillation + direct-contact membrane crystallization

Table 9.2 Typical salt concentration in RO concentrate and permeate with 45% recovery and operating pressure of 900 psi [95]

	Seawater (PPM)	Concentrate (PPM)	Permeate (PPM)
Sodium, Na	10,967	19,888	64
Potassium, K	406	736	3
Magnesium, Mg	1306	2372	2
Calcium, Ca	419	761	0.5
Bicarbonate, HCO_3	109	194	0.9
Chloride, Cl	19,682	35,771	105
Sulfate, SO_4	2759	5014	1.5
Total dissolved salts	35,666	64,771	176
pH	7.8	7.7	6.1

- Liquid–liquid extraction with organic solvents
- Electrodialysis (ED)

An example of salt recovery by combining evaporation and crystallization is the SAL-PROC™ process. Appropriate for some inorganic saline streams, it is based on simple closed-loop processing and fluid flow circuits. Depending on the chemical composition of the saline feed water, the process route may involve one or more steps of reaction, evaporation and cooling supplemented by mineral and chemical processing. The produced salts and chemical compounds come in crystalline, slurry, and liquid form. The electricity consumption of the process is low: the main usage of electricity is for the operation of pumps and agitators in the chemical reactors and fluid transfer circuits [40]. However, an energy source is required for salt crystallization in the cooling circuits or for powder production in the drying circuits. That is the main disadvantage of this system, but it could be a viable option if a local source of waste heat is available. In addition, this type of facilities requires land for establishing various ponds to preconcentrate (recycle or enhanced evaporation techniques) and store the DC, and also to crystallize and harvest the sodium chloride salt (NaCl).

An integrated scheme that treats a saline effluent and produces salts as byproducts typically comprehends one or several RO units followed by evaporation and crystallization stages. Another configuration is based on a RO tandem (primary and secondary RO systems) with an intermediate stage of lime softening. After the secondary RO system, several stages of evaporation-crystallization are located. In the intermediate stage of lime-soda treatment, a precipitate containing magnesium hydroxide (51%–58%) is obtained together with two impurities: calcium carbonate (19.5%–23.3%) and calcium sulfate (1.3%–7.8%). Calcium sulfate can also be obtained as a byproduct, much more concentrated (92%), same as calcite (95%) [42]. The evaporation and crystallization step that follows the secondary RO system produces a mixture of sodium sulfate (86%–88%) and NaCl (5%–14%). The final evaporation to dryness of the left effluent results in a mixture of ionic compounds: sodium, sulfate, chloride, and potassium (35%–40%, 17%–33%, 25%–41%, 1%–5%, respectively) [43]. The combination of RO with evaporation and crystallization has also been applied to the saline effluent of combined-cycle power plants [44] and coal mines [45]. In those cases, the goal was achieving zero liquid discharge instead of recover the salts, so the crystallized compounds arise as a filter press cake that must be disposed of.

Another option for salts recovery is an integrated process composed of reverse osmosis + membrane distillation + direct-contact membrane crystallization. The major advantage of the membrane distillation coupled with crystallization (MDC) process is its capability to concentrate the salt up to the supersaturated state, which allows its crystallization [79]. MDC has been recently applied for the treatment of shale-gas-produced water [96]. Such systems have proven their stability in the treatment of saturated solutions for the recovery of high-quality epsomite crystals [97] and NaCl [39]. MDC has also succeeded in recovering sodium sulfate, in the form of Thenardite, from industrial wastewater [98].

Liquid–liquid extraction consists in transferring one or more solutes contained in a feed solution (the DC) to another immiscible liquid (the solvent). Separation of divalent salts from brines using extraction with an organic solvent (for example octanoic acid [99]) is technically feasible, but enhancements are still needed to make the process economically viable. Several researches have evaluated the recovery of valuable metals from concentrates by means of liquid–liquid extraction [100–102]. Rubidium, whose compounds have various chemical and electronic applications, has been successfully recovered from DC using extractants at a recovery ratio higher than 80% [101]. Liquid–liquid extraction has been proposed to recover materials from the concentrated brine of RO plants [102]. For that type of DC the following treatment process, divided into several extraction stages, has been projected: phosphates extraction using a blend of iron and aluminum sulfates; liquid–liquid extraction to recover cesium, indium, and gallium; rubidium extraction; evaporation of the remaining brine to recover carnallite crystals; and, as a final step, extraction of germanium and magnesium.

Lithium ions are another interesting material that can be recovered from some DCs. Lithium is the lightest metal element and is widely used in various fields, including electronics, aerospace, ceramics, polymers, and metal additives. Its demand is increasing significantly due to the development of lithium batteries and electric vehicles. Lithium resources exist naturally in two forms: as a mineral and as a dissolved salt. Onshore liquid lithium resources contain around 90% of the world's recoverable lithium [103]. In particular, continental brine from salt lakes is the biggest resource (59%) for lithium occurrence [48]. Lithium ions can be recovered from salt brine via a number of different processes, including precipitation, ion exchange, adsorption, solvent extraction, and supported liquid membrane [48–55]. The use of solvent extraction is viable but requires strongly acidic conditions, which results in the generation of large amounts of acidic wastewater and corrosion problems in the equipment. Those disadvantages could be overcome with the development of alternative solvents [56]. Recently, the technologies of fractional crystallization [57] and electrodialysis for brines with a high magnesium/lithium ratio [103] have also been proposed for extracting lithium from salt lake brines.

Finally, applying electrodialysis (ED) to brine effluents is another interesting option. ED pilot plants combined with a gypsum precipitation step have already been tested [58]. ED using ion exchange membranes for the separation of nutrient ions and organic compounds from salts has been tested for agroindustrial wastewaters achieving high overall recovery [59,60]. Electrodialysis with bipolar membranes (EDBM) is a promising technology that valorizes DC by producing acids and bases. Nanocomposite anion exchange membranes are capable to produce good-quality acids and bases, such as HCl and NaOH [61]. EDBM has been also applied to RO concentrates from wastewater treatment plants, converting them into mixed acid and mixed base with concentrations as high as 0.2 N [62]. Meanwhile, diffusion dialysis (DD) utilizes concentration difference as the driving force and has the advantages of simple operation, lower energy consumption than crystallization, RO and ED, and lower installation and operating cost. The recovery of salts from DC such as NaCl and sodium acetate is difficult due to their low diffusivity. Sodium acetate is present in some industrial salt wastewaters. For those effluents, DD using polyelectrolyte complexes/PVA membranes can be an appropriate option [63].

9.2.3 Irrigation and Fertigation

When considering the water from an irrigation source, salt concentration starts posing a high hazard at levels above $1.6\,mS\,cm^{-1}$ ($1\,g\,L^{-1}$). High levels of soluble salts can induce physiological drought in the plant as the roots are unable to absorb the water due to osmotic pressure, even if they have an adequate water supply. Ions of Na^+ and Cl^- can be absorbed through roots or leaves if sprinkler irrigation is used. These ions can cause direct root injury as they accumulate in shoot tissues and cause shoot toxicity problems, or cause direct foliar toxicity on plant leaves. However, a low-saline DC can be a valid source for irrigation if adequately managed. The key point is to determine the flow of saline water that can be effectively supplied. An adequate irrigation scheme should comprise using appropriate and resistant plant species, blending the DC with less saline waters to a lower final salt content, and periodically monitoring the salt content of the soil. Applying the DC into crops or halophytes at tolerable levels will result in the phytoextraction of some of its salt. That is, plants will assimilate mainly ions of Na^+ and Cl^- from the DC. Some crops have higher salt tolerance (halotolerance) than others, as they are capable to develop adaptive responses to salinity stress. Plants take up water from soils of low water accessibility by adjusting osmotically. That is, by accumulating ions and synthesized organic compounds inside their cells. The most commonly used ion for osmotic adjustment is K^+, but in saline situations Cl^- and Na^+ can also be used, provided these ions are sequestered into cell vacuoles, which avoid their toxic effect on metabolism [104]. This is the main mechanism for saline adaptation used by halophytes and the more salt-tolerant crops.

Halophyte plants are capable of growing in saline deserts, swamps, or seashores by using the phytoaccumulation of cations (K^+, Na^+, Cl^-) within their aerial plant parts. Differently, the Brassica genus (mustard family) uses among its mechanism for salt tolerance the retention of K^+ within the roots [105]. In addition, some species may develop secondary strategies under drought/salinity stress such as early flowering. That is the case of wheat and barley, which through early flowering can eventually avoid high salinities of the soil solution during terminal drought [104]. Some important crops such as barley and the date palm can tolerate about $5\,g\,L^{-1}$ ($7.8\,mS\,cm^{-1}$) of total dissolved salts in irrigation water, and can be considered as marginal halophytes [106]. Meanwhile quinoa's most tolerant varieties are able to cope with salinity levels as high as those present in sea water ($34.6\,g\,L^{-1}$ or $54\,mS\,cm^{-1}$) [107,108].

Soil texture and structure may have a significant role in the salt stress process under saline irrigation. Thus, salt tolerance of plants can be enhanced by an adequate type of soil. For instance, the performance of salt-tolerant plants *Medicago arborea* and *Atriplex lenformis* is better in clay than in sandy soils [109]. Another mechanism to enhance salt tolerance of plants is the modification of specific genes in crops. An example of that is the transformation of commercial crops such as tomato by adding genes that allow the synthesizing of glycine betaine, which allows high potassium levels and low sodium levels to be maintained in transgenic plants under salt stress condition [110]. Salt-tolerant varieties of crops can be developed using techniques that range from the simple selection of the plants with desirable characteristics for propagation, to precision breeding of hybrid crops through genomics and molecular tools [111]. Plant breeders have focused on identifying crops which will ensure crops perform under drought and salt stress conditions. In a near future, genetic engineering of ion transporters and their regulators holds promise for further development of salt-tolerant crops [112].

Crop growth is inhibited as the salinity of the irrigation water increases, until eventually the plants die when surpassing their tolerable range of salinity. Table 9.3 shows the performance of different crops when irrigated with waters of different level of salinity.

Results from Table 9.3 indicate that the cultivation of some salt-tolerant crop varieties is viable using brackish waters, preferably under hydroponic conditions and mixed with fresh water if available. Appropriated cultivation techniques and the use of liquid manure can improve the performance of the yield grown with brackish waters. However, periodical leaching is necessary in order to remove the accumulated salts and conserve an acceptable salt balance of the soil. Meanwhile, the direct use of DC from desalination plants to irrigate crops does not show to be a convenient option due to the adverse effects detected in the plants (extreme growth inhibition) [17]. Other irrigation strategies have proven to be successful such as diluting brackish waters with less saline water [113] or blending brackish waters/DC with wastewater [18]. Biologically treated secondary effluent (from sewage or agroindustry) is being used increasingly as a source of water for irrigation purposes [114]. Nanofiltration and RO can remove most of the pharmaceutical and personal care products from urban sewage effluent, which protects soil and ground water from those potential hazards [115]. In addition to using this type of effluent for irrigation, it is necessary to previously decrease the content of biodegradable organic matter. Treatments such as biodegradation by microorganisms and adsorption in activated carbon, followed by microfiltration, end up producing an effluent with a high reuse potential for irrigation purposes [116]. After neutralizing the organic content of the effluent, the salt content may still be high for long-term irrigation. Depending on the final salt content, it may be necessary to blend the treated effluent with RO permeate or other source of freshwater. Microfiltered, biologically treated sewage effluent generally has high sodium adsorption ratio and sodium (Na^+) and chloride (Cl^-) concentrations, so it cannot be directly used for irrigating sensitive crops [115]. In semiarid or desert regions, the blending of treated secondary effluent with brackish waters or DC can provide a vital resource for local irrigation. Israel is an example of this, as around 85% of the treated municipal effluents are recycled for irrigation, accounting for about one-half of the total water demand for agriculture [114].

While the irrigation of commercial crops with DC has its limitations and is not viable for unblended saline effluents, halophytes are a valid alternative for the valorization of higher salinity brines. They are the preferred option for both the irrigation of soil [117–121] and the discharging of the DC into

Table 9.3 Yield reduction (inhibition of the growth) of different crops irrigated with saline waters

Species	Type of cultivation	Range of salinity ($mS\,cm^{-1}$)	Average growth inhibition for each increasing $mS\,cm^{-1}$ within the range
Quinoa—Pandela rosada variety	Soil cultivation, under NPK fertilization	0–36.5	0.48%
Quinoa—Huallata variety	Soil cultivation, under NPK fertilization	0–36.5	1.44%
Sunflower—Red Sun variety	Hydroponic	0.5–16.4	5.8%
Sunflower	Soil cultivation	0.5–5.0	10.5%
Sunflower	Soil cultivation, under nitrogen fertilization	0.5–4.9	7.6%
Olive tree—Picholine variety	Soil cultivation	9.7–14.0	0.8%
Olive tree—Picholine variety	Soil cultivation	0–9.7	2.0%
Olive tree—Meski variety	Soil cultivation	0–14.0	3.5%
Olive tree—Ascolana variety	Soil cultivation	9.7–14.0	3.6%
Olive tree—Ascolana variety	Soil cultivation	0–9.7	8.0%
Feijoa	Soil cultivation	2.8–11.1	20.0%
Tomato	Hydroponic	2.1–10.1	2.6%
Tomato—Miller Astona variety	Soil cultivation	1.8–8.8	6.0%
Cotton	Soil cultivation, under nitrogen fertilization	0.7–8.7	9.5%
Cotton	Soil cultivation	0.5–8.5	10.0%
Sugar cane—My 5514 variety	Soil cultivation	3.2–6.0	10.0%
Lettuce	Hydroponic	1.3–5.2	13.0%
Beetroot	Hydroponic	1.2–5.2	8.0%
Papaya	Soil cultivation	1.2–4.8	7.2%

Sources: Sánchez AS, Nogueira IBR, Kalid RA. Uses of the reject brine from inland desalination for fish farming, Spirulina cultivation, and irrigation of forage shrub and crops. Desalination. 2015;364: 96–107; Bader B, Aissaoui F, Kmicha I, Salem AB, Chehab H, Gargouri K et al. Effects of salinity stress on water desalination, olive tree (Olea europaea L. cvs 'Picholine', 'Meski' and 'Ascolana') growth and ion accumulation. Desalination. 2015; 364: 46–52; Adolf VI, Shabala S, Andersen MN, Razzaghi F, Jacobsen SE. Varietal differences of quinoa's tolerance to saline conditions. Plant Soil. 2012;357(1–2): 117–129.

constructed wetlands [122]. Halophytes can act in the phytoremediation and reclamation of polluted soils or saline wastelands. They can provide a use for the DC sequestrating a part of the salts and producing biomass, which can later be used as animal fodder or even as biofuel feedstock [123].

Nonhalophytes plants can also be used for treating DC that have low salinity but carry other types of pollutants. For instance, large perennial grasses from the genus Phragmites and Typha have some advantages for phytoremediation. *T. latifolia*, for example, has the advantage of having some salt tolerance, as naturally occurs in slightly brackish marshes. Meanwhile, *P. australis* has lesser salt tolerance but shows good efficiency in treating wastewaters polluted with metals. Constructed wetlands planted with *P. australis* and *T. latifolia* have been used to treat DC generated from iron and steel companies, resulting that 70% of metals carried by the DC (K, Ca, Na, Mg, Zn, Cu) were accumulated in the aboveground of *P. australis* plants [124].

On the other hand, halophyte plants such as the Atriplex genus (commonly known as saltbush) are appropriate for the treatment of DC from the desalination of brackish groundwater [17,120] but its capacity of removing heavy metals from wastewater still needs more assessment. *A. halimus* L. and *A. hortensis* L. have been tested for phytodesalination of road runoff polluted by deicing salts (NaCl is used to maintain roads safe for traffic under winter conditions) and also by trace metals from road pollutants: Cd, Cr, Cu, Ni, Pb, and Zn. While succeeding in accumulating sodium, no remarkable results in the remediation of heavy metals were found [125]. Differently, other works have highlighted *A. halimus* species, commonly known as the Mediterranean saltbush, as appropriate for the phytostabilization of metals in mine soils contaminated with Cu, Zn, and Cd [126]. Other research addressing the uptake of metals by the Atriplex genus has concluded that these plants accumulate the metals in the roots and therefore are able to exclude metals from the aboveground portions [127,128]. That characteristic makes some of these species valuable for the production of animal forage when irrigated with DC or wastewater, as the metals would accumulate in the roots and not pass to the hay. In particular, *A. nummularia* irrigated with DC has been used as fodder for goats in arid regions, although its hay contains moderate levels of salt (NaCl), which do migrate from the soil to the leaves. Results from more than one decade of continuous use have proven Atriplex hay irrigated with DC to be a valuable nutritional complement for goats [17,129]. This halophyte species can tolerate saline waters with up to $31.2\,mS\,cm^{-1}$ [118] and has the capacity of effectively removing salts from the soil. Therefore, it is a convenient option for saline soils to replace

Table 9.4 Nutritional comparison between Atriplex (*nummularia, amnicola, lentiforme*) and other common hays

Hay	Nutrients		
	Dry matter, %	**Crude protein, %**	**Ash, %**
A. nummularia	99.1	20.2	28.3
A. amnicola (leaves)	44.7	23.2	26.6
A. amnicola (stems)	59.7	7.0	12.0
A. lentiformis	98.0	12.2	20.7
Alfalfa	89.70	16.0	9.9
Cynodon hay (Bermuda grass)	91.1	12.3	8.8
Bluegrass	92.7	6.4	5.1
Sudan grass	88.9	10.8	7.6
Agrostis stolonifera	87.4	6.1	6.4

Sources: Eissa MA, Ahmed EM, Reichman, SM. Production of the forage halophyte Atriplex amnicola in metal-contaminated soils. Soil Use Manage. 2016;32(3): 350–356; Meneses R, Varela GI, Flores H. Evaluating the use of Atriplex nummularia hay on feed intake, growth, and carcass characteristics of creole kids. Chilean J Agric Res. 2012;72(1): 74–79; Joint United States-Canadian Tables of Feed Composition: Nutritional Data. Washington DC: National Academy of Sciences; 1964.

traditional forage plants as alfalfa (*Medicago sativa*), which has much lower tolerance to salinity (1.3–4.0 $mS\,cm^{-1}$) [17]. Table 9.4 offers a comparative chemical analysis of three Atriplex species and other hays.

A combination of fodder shrubs based on *A. nummularia* and spineless cactus (*Opuntia* spp.) was also studied in Argentina and proved to be useful for the feeding of goat herds in arid environments [129,130]. The threshold limit for salt in livestock feeding (cows and goats) is 10 g salt/100 g of dry mass [117]. This is one of main limitations to the use of saltbush forage as it has to be mixed with other nourishment. Despite this inconvenience, it offers a high content of proteins and other nutrients. There is a slight difference in metabolizable energy between *A. nummularia* and alfalfa hay as the latter contains 10% more energy. Conversely, *A. nummularia* reveals to be an interesting source of proteins, with a higher content of them than alfalfa. The limiting factor about the use of *A. nummularia* is the content of sodium, whose percentage is 36 times higher than in alfalfa hay.

A. nummularia has been used to valorize the DC originating in inland RO desalination plants located in the semiarid region of Brazil [17]. In Argentina and Chile, it has been cultivated in desert landscapes using brackish waters as irrigation [129,130]. In North America, the *A. lentiformis* (quailbush), native from the U.S.–Mexican Sonoran Desert, has also been used specifically for the brine disposal of RO desalinization plants [119,120].

In Egypt, *A. amnicola* has been used in metal-contaminated soils irrigated with untreated sewage effluent 129]. In South America and in North Africa, those experiences have been oriented to the production of fodder for goats, while experiences in North America lack of that focus as the only goal has been to provide a viable strategy for disposing saline water in arid zones [119]. However, the experiences in North America also identified the potential of *A. lentiformis* as a forage crop, specially its leaves and fruits. While their high content of sodium and other minerals limits the proportions that can be used in animal diets, fruits and seeds from *A. lentiformis* have protein levels within the range of alfalfa and other high-quality forages. Moreover, they have relatively low levels of acid-detergent fibers and high levels of starch and soluble carbohydrates [119]. The main conclusion of the experiences with Atriplex is that its leaves and fruits (rather than its woody stems) are valuable as a mineral and protein supplement and forage replacement in feedlot diets for ruminants, but its value as fodder is limited by containing excess sodium and antinutritional compounds such as saponins substances and non-protein nitrogen sources that can be difficult to digest. Table 9.5 summarizes the results of these experiences, in terms of fresh biomass yield per hectare and year.

In sum, the Atriplex genus is indicated for the phytostabilization of metals and the reclamation of already-saline soils and is interesting for DC management as it allows the valorization of the brine-producing fodder that can be used by local herds. Table 9.6 recaps the possible uses identified in the most studied Atriplex species.

Throughout this section, it has been shown that halophytes and some salt-tolerant varieties of crops can be cultivated under saline conditions, producing acceptable yields of biomass and hence valorizing saline effluents. Now it will be assessed how efficient are these halophytes, particularly Atriplex, in preventing the accumulation of salts in a soil irrigated with DC. Table 9.7 shows the salt removal rate of *A. nummularia* after one year of irrigation. In the location of the study, the average annual rainfall is 433 mm and the evapotranspiration is 1520 mm. Four different irrigation flows were applied, in order to determine which one maximized the biomass yield while minimizing the salt accumulation in the soil.

The experiences shown in Table 9.7 revealed that the optimal performance of *A. nummularia*, in terms of yield, was obtained through the application of 225 l of DC per plant/week, considering an average of 625 plants per ha [132]. At lower volumes of DC applied on the fields, the salinity of the soil got the greatest increase: despite low volumes of the DC fed

Table 9.5 Performance of *A. nummularia* (South America), *A. lentiformis* (U.S.A.), and *A. amnicola* (North Africa) irrigated with saline water

Location	Average annual precipitation (mm)	Atriplex Species	Type of water used for irrigation	Electrical conductivity ($mS\,cm^{-1}$)	Yield ($ton\cdot ha\cdot year^{-1}$)
Argentina—Mendoza	175	*A. nummularia*	Rainfed agriculture and a little of brackish water from a saline aquifer	3.5–5.0	0.47 (after year 3) 1.88 (year 6 onwards)
Brazil—Semi Arid region	433	*A. nummularia*	RO reject brine	6.0–11.4	14.5–14.9 (after the first year)
U.S.A.—Sonoran Desert	260	*A. lentiformis*	RO reject brine	3.5	16.2–22.0 (after the first year)
Egypt—banks of the Nile	5	*A. amnicola*	Untreated sewage effluent	2.7	3.6–4.0 (after the first year)

Table 9.6 Possible uses of Atriplex species irrigated with DC or planted in degraded soils

Atriplex species	Uses/remarks	Ref.
A. nummularia	DC from desalination of brackish groundwater. Usable as forage	[17,129,130]
A. canescens	Reclamation of degraded areas. Usable as forage	[131]
A. acanthocarpa	High salt tolerance due to accumulation of sodium in leaves. Reclamation of degraded areas. Usable as forage	[131]
A. lentiformis	DC from desalination of brackish groundwater	[109,119,120]
A. amnicola	Untreated sewage wastewater, metals. Usable as forage	[127]
A. rosae	Untreated sewage wastewater, metals	[128]
A. halimus L.	Phytostabilization of metals in mine soils	[126]
A. hortensis L.	Phytodesalination of road runoff polluted by deicing salts	[109,125,128]

and good salt removal rates, the plants generated low content in biomass, which resulted in less salt effectively removed. Greater volumes of DC fed resulted in lesser salt removal rates, but on the other hand the halophytes achieved better yields, which resulted in more total salts removed from the soil. For this Atriplex species and this soil conditions, the optimum flow of DC was around 225 l·plant^{-1}·week^{-1}. That flow maximized the yield while minimizing the salinity of the soil. For flows above that value, the yield of forage tends to decrease while the salinity of the soils gets stagnated around 4.60 mS cm^{-1}. These results exemplify the importance of adjusting the flow of DC to the local conditions of soil and plant species to be irrigated. However, even if the flow of DC is adjusted to levels that minimize the salt deposition in the soil, studies assessing long-term effects of irrigation halophytes with DC suggest that a progressive salinization of the soil may occur, particularly when dealing with high-salinity brines such as DC from desalination plants [17]. Blending the DC with other streams of less saline waters (brackish water or sewage effluent) seems like the only alternative to prevent an excessive salinization of the soil. Thus, measurements of 5 years of continuous irrigation and drainage of Atriplex fields with DC (8.3 mS cm^{-1}) recommend caution when using the brine directly in the soil. Starting with a salinity level at 0.90 m depth of 0.6 mS cm^{-1} in 2000, salinity progressively increased to 6.7 mS cm^{-1} (2004), 7.2 mS cm^{-1} (2005), and 8.2 mS cm^{-1} in 2006 [133]. Although the biological quality of the soils

Table 9.7 Application of DC to an *A. nummularia* field using four different amounts of flow of this effluent, during one year

Flow of DC (liters·plant^{-1}·week^{-1})	Total dissolved salts of the DC (mS cm^{-1})	Salinity of the soil before (mS cm^{-1})	Salinity of the soil after (mS cm^{-1})	Salt removal rate (%)	Annual yield of dry biomass (ton ha^{-1})	Annual yield of forage (ton ha^{-1})
75	8.3	0.40	8.02	13.8	9.7	5.5
150	8.3	0.40	6.09	10.9	12.3	7.1
225	8.3	0.40	4.97	8.0	14.5	8.3
300	8.3	0.40	4.60	5.2	13.8	7.9

Source: Porto ER, Amorim MD, Dutra MT, Paulino RV, Brito LDL, Matos AN. Behavior of saltbush (Atriplex nummularia) irrigated with effluents from tilapia raised in brackish water. Rev bras eng agríc ambient. 2006;10(1): 97–103.

along with its fertility was increased, the cultivation of *A. nummularia* was not able to prevent the progressive salinization of the soil.

9.2.4 Aquaculture (Fish/Prawn Farming)

Aquaculture, or the cultivation of populations of fish, crustaceans, mollusks, or algae under controlled conditions, is among the best alternatives to valorize the DC. Harvest stagnation in wild fisheries and overexploitation of popular marine species, combined with a growing demand for high-quality protein, has encouraged aquaculture during last decades. Depending on the type of pollutants present on the saline effluent, the DC may undergo different pretreatments and then be used for aquaculture. For instance, the DC that is rejected from an RO desalination plant can be directly discharged into breeding ponds of tilapia.

Tilapia is the common name given to cichlids from three genera: Oreochromis (*Oreochromis sp.*), Sarotherodon, and Tilapia. They are endemic to Africa and the Middle East and, besides being originally freshwater fishes, they adapt very well to saline waters [134]. At salinities of 34 mS cm^{-1} and above, red strains of tilapia are preferred. In intermediate levels of salinity (17–34 mS cm^{-1}), various usually nonred strains or hybrids of *O. mossambicus, O. aureus, or O. urolepis h.* are used [135]. Nowadays, tilapiines have become the third most important fish in aquaculture after carp and salmon because they are among the easiest and most profitable fish to farm due to their omnivorous diet, high protein content, tolerance of high stocking density, large size, and rapid growth (6–7 months to grow to adequate size). Unlike salmons, which rely on high-protein feeds based on fish or meat, commercially important tilapiine species eat a vegetable- or cereal-based diet. Tilapia cultivated in open raceway ponds are considered safe for the environment, since their waste and disease is contained and not spread to the wild. However, tilapiines have acquired notoriety as being among the most serious invasive species in many subtropical and tropical parts of the world. Tilapia farming schemes that use the DC rejected from desalination plants have been in operation for years [17]. In the semiarid region of Brazil, the Koina tilapia, native of Malaysia, with an attractive pink color, which favors its marketing and a good-quality meat without bones, has been grown in DC with a salinity between 7.0 and 11.4 mS cm^{-1} [132]. The production system for tilapia developed in that region consists in the use of the DC from RO desalination units (typically small plants with a reject flow of 1.0–2.5 $m^3 h^{-1}$), which flows into impervious hatchery ponds. The ponds had a capacity of 330 m^3 and were populated with a density of 4 fries for each m^3. The installation of mechanical facilities for oxygenation, which are common in commercial aquaculture farms, could have allowed a higher population density. However, yields proved to be acceptable even for high-saline DC. Thus, for a brine reject with an electrical conductivity of 11.4 mS cm^{-1}, the fish yield was 1.96 kg for each m^3 of cultivation pond. At that salinity, the average weight of each fish was 518.7 g for a rearing period of 153 days and a survival rate of 94.7% [136]. The waste generated in the processing of the tilapia (gutting and filleting) can be further processed into value-added byproducts. Guts and organic waste can be used for the production of oil. This oil is can be further processed for the production of animal feed and soap. In the semiarid region of Brazil, the production of tilapia grown in DC ponds close to desalination plants and also in freshwater reservoirs was enough to make viable a biodiesel plant that processes the waste fish oil. After around two years of continuous operation, that biodiesel production facility was eased at the end of 2016 due to a strategic decision of its owner company Petrobras to leave the biofuel market and focus on oil and gas production. Since then, the fish oil production has been diverged mainly to animal feed uses, but the experience showed the possibilities of inserting biodiesel production within the value chain of aquaculture in DC. Another valuable byproduct that can be produced is collagen, which can be extracted from tilapia skin [137] and has a potential use in the cosmetic industry.

Worldwide, the polyculture of Tilapia and shrimp is rapidly being adopted as a primary management practice [135]. The polyculture of tilapia with shrimp encourages the development of microbial communities (bacteria, algae, and fungi) that are closer to the natural biota found in estuaries and other native shrimp habitats. Therefore, the shrimp-tilapia polyculture can improve survival rates and feed utilization efficiency. Given the high salt tolerance of both species, at ranges compatible with that of many DC effluents, the polyculture of tilapia in combination with shrimp seems a promising option for DC valorization. Among the different shrimp families, the marine penaeid shrimp is the preferred one for aquaculture in saline waters. Brown shrimp species show a preference for salinities from 28.2 to 54.1 mS cm^{-1} [138]. The white shrimp can be grown well in salinities up to 67.5 mS cm^{-1} [139]. For another important commercially valuable species, the Caledonian blue shrimp, seawater salinity (53 mS cm^{-1}) is optimum during larval stages while for postlarval stages, lowering salinity at 42 mS cm^{-1} leads to a better growth [140].

Tilapia and the mentioned species of shrimp are harvested, processed, and commercialized for human food. A different kind of aquaculture involves the cultivation of small species of nondecapod crustaceans, which are also commonly referred to as shrimp, to serve as food for fish farmed in other breeding ponds. This type of fish food is easy for operation, rich in nutrients, and has small size, which makes it suitable for feeding fish and crustacean larvae live or after drying. These crustaceans typically eat microscopic planktonic algae but when cultivated in ponds can be fed with different types of particulate foods including yeast, wheat flour, soybean powder, or egg yolk. Some genera of crustaceans have high tolerance to saline waters and thus may be suitable for being cultivated in ponds filled with DC. In particular, the brine shrimp *Artemia* spp. can tolerate levels of salinity from 40 to 400 $mS\,cm^{-1}$, with a physiological optimal range of 47.2–54.1 $mS\,cm^{-1}$ [141]. Fish farm owners search for a cost-effective, easy-to-use, and available food that is preferred by the fish, which has led the Artemia genus to be extensively used in aquaculture due to its ability to produce dormant eggs, known as cysts. The cysts may be stored for long periods and hatched on demand to provide a convenient form of live feed for larval fish and crustaceans. Once cysts eclode, they can readily be used to feed fish and crustacean larvae just after one-day incubation.

9.2.5 Growth of Cyanobacteria (Spirulina)

Another alternative for the valorization of the DC is the cultivation of *Spirulina* in shallow brine ponds. Spirulina is a form of cyanobacterium and also known as blue-green algae comprising mainly two species, *Arthrospira platensis* and *Arthrospira maxima*. The common name, *Spirulina*, refers to the dried biomass of the first one. With impressive nutrition characteristics, *Spirulina* was already cultivated by the Aztecs who, back in 16th century, used to dry the harvested algae and then commercialize the dehydrated product in the form of cakes. Nowadays, *Spirulina* has gained popularity as a "super-food" and is considered the most nutritious, concentrated whole food known to humankind [142]. This microorganism that can be consumed by humans and other animals is cultivated worldwide and has a consolidated market. The processes of harvest and dry of its biomass are simple. Its cultivation can be carried out in open artificial ponds and also in closed photobioreactors with an artificial illumination and constant temperature. Its main uses are for human food or animal feed. It can be either a substitute or a support for fish farming as it offers valuable nutrients for tilapia [143–145]. Up to 2% *Spirulina* per day in aquaculture feed can be safely recommended [146]. Also, there is the possibility of using *Spirulina* as a fertilizer, replacing chemical fertilizers in the treatment of depleted soils [147].

Spirulina uses a low volume of water per kilo of protein (approximately 2100 l kg^{-1} protein) as the brackish water fed into the evaporation ponds can be recycled and the only significant water loss is through evaporation. In Chile, a productivity of 30 tons ha^{-1} $year^{-1}$ has been reached in a medium with sodium bicarbonate, ammonia phosphate, saltpeter and as much as 30% of sea water, which is mixed with the waters of a coastal river. Around 30% of the total cost of production corresponds to the cost of nutrients, mainly CO_2 [148]. Furthermore, *Spirulina* has a high tolerance to salt. In particular, *A. platensis* has an optimum growth between 32 and 112 mS cm^{-1}. Experiences have shown that a content of 5%–15% sea water in the culture medium only causes a decrease of biomass yield in 15%–20% as compared with a Zarrouk's culture medium [149]. In fact, some *Spirulina* strains cultured with seawater accumulate more bioactive substances and therefore have a higher nutritive value [150]. Alkaline waters, at a pH around 8.5 or more and a temperature around 30 °C, are the preferred ones. In addition to its high salt tolerance, *Spirulina* has a unique quality to detoxify (neutralize) or to chelate toxic minerals and heavy metals [151] and is also capable of removing calcium and salts from inland saline waters [152].

In sum, *Spirulina* is a valuable product with a consolidated market and low growing requirements, capable of tolerate high-saline brines. For all the mentioned characteristics, its cultivation seems as an interesting option for the valorization of any DC. Among the possible saline effluents that could be discharged in the *Spirulina* cultivation ponds are DC from RO inland desalination facilities [17,153], wastewater with high organic loads [154], vinasse from the sugarcane industry [155], and aquaculture wastewater [147]. Regarding the latter type of DC that was mentioned, it is noticeable that *Spirulina* cultivation gives an adequate treatment to the tail-end effluent of aquaculture farms, while at the same time produces one of the most costly inputs for that industry: quality food for the fishes. When treating aquaculture effluent, *Spirulina* has shown to be adequate for ammonia and nitrate removal, but low efficient for nitrite removal [147]. But perhaps what turns *Spirulina* more interesting for the valorization of DC is cost-effective method for heavy metal removal from wastewater. This is a recurrent problem in the effluents from industries such as mining, metallurgy, fertilizers, leather, photography, and electric appliances manufacturing. Research has shown that *A. platensis*, due to its high adsorption potential, can be efficiently used for removing lead from industrial wastewater. In particular, *Spirulina* has been tested with wastewater from battery manufacturing industry. That effluent was polluted with Pb^{+2} (lead), Cd^{+2} (cadmium), and

Cu^{+2} (copper), was quite acid (pH 5.5) and had 1.1 g L^{-1} of total dissolved salts (1.7 mS cm^{-1}). After increasing the pH to 7, a contact time of 60 min was enough to obtain a lead adsorption efficiency of 84.3% [156]. In addition, using *Spirulina* as a biosorbent for copper has also shown good results, indicating a maximum removal of 85.0% [157].

9.2.6 Integrated Systems: Irrigation and Aquaculture

When considering the valorization of DC, the different possibilities can be combined in order to give a more efficient treatment that, at the same time, maximizes both the biomass yield and the revenues. The different possibilities (aquaculture, *Spirulina* cultivation, and irrigation) can form an integrated system where the different production stages complement each other. For a DC with moderate salinity and nonpolluted with heavy metals, the first production stage may be fish farming in open ponds. The polyculture of tilapia and some shrimp strains seems an appealing option. Both tilapia and shrimps are valuable and its coculture has mutual benefits and enhances the overall productivity of the ponds. Alternatively, the brine shrimp *Artemia* spp. (used as fish feed) can be cultivated in this first stage. The effluent originated from aquaculture farms offers good possibilities for further use as a substrate for algae cultivation. Therefore, *Spirulina* cultivation can be inserted within a fish/shrimp farm scheme as it gives an adequate treatment to the effluent while producing a source of proteins for the fishes. As a final stage, the DC that served as the cultivation media for tilapia/shrimps and for algae or *Spirulina* can irrigate halophyte fields. If the halophytes considered for cultivation are from the Atriplex family, they can be harvested as hay to supplement the diet of goats. Differently, if the DC to be treated has higher salinity (up to 112 mS cm^{-1}) or is polluted with heavy metals, the preferred option may be the cultivation of *Spirulina* in ponds followed by further irrigation of halophytes. That high level of salinity would not be viable for growing tilapia but, more importantly, the heavy metals could accumulate in the fish/shrimp meat. In contrast, *Spirulina* acts as an effective biosorbent and has some capacity of neutralizing heavy metals and other minerals such as calcium and salts. The DC could be used for irrigation after leaving the *Spirulina* ponds. The capacity of some Atriplex species to phytostabilizate metals would provide an adequate final stage of treatment to that effluent. Besides, the Atriplex genus accumulates the metals in the roots and not in the aboveground portions [127,128] so that integrated Spirulina/Atriplex production system could also produce forage from metal-polluted wastewater. For a DC that is free of heavy metals or other pollutants and has

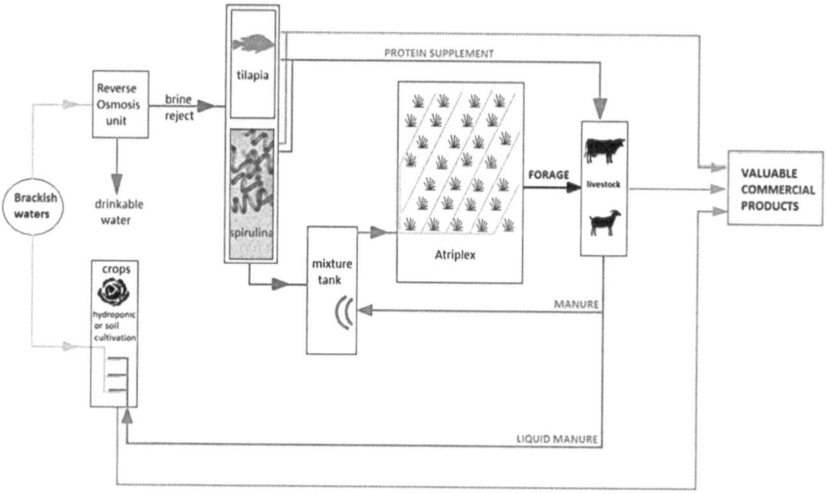

Fig. 9.3 Integrated production system for the valorization of the DC from inland brackish desalination plants [17].

moderate salinity, a wide range of options and combinations is possible. The DC from inland brackish desalination plants is an example of such saline effluent. Fig. 9.3 illustrates a production system that aims to valorize the DC by integrating different irrigation/aquaculture options.

Integrated systems that valorize DC through aquaculture and irrigation have produced interesting results after more than one decade of continuous operation. Particularly, in the semiarid region of Brazil, such a scheme was tested for small- and medium-sized inland desalination plants. An integrated production scheme was conceived that aims to add as much value as possible to the DC while at the same time solves the environmental problem of this effluent. This production scheme has its intake in brackish groundwater and produces, together with fresh drinkable water, fishes for human consumption and fodder for the local cattle. In a similar way to what is depicted in Fig. 9.3, the DC is first discharged into tilapia raceway ponds. In a forward production stage, the DC that leaves the ponds is mixed with organic fertilizer (manure from the local livestock) and then pumped for the irrigation of halophyte forage shrub plantations (*A. nummularia*), located in the surroundings of the desalination plant facilities. The Atriplex fodder is finally used as dietary supplementation for the local livestock, formed mostly by goats. In addition, byproducts of the fish process can also be transformed into oil and used as a protein source for cattle while at the same time the manure produced by the livestock can be used as a fertilizer, mixed with the brine used in irrigation. Hence, unproductive lands and saline water ponds turn to produce valuable commercial products together with forage during the whole year. Since this solution was adopted, several villages of the semiarid region of Brazil have been benefited with both a drinkable water supply and an increase in the local economy [17]. In contrast, the phytoextraction of salt from the DC in the Atriplex fields was not enough, so a progressive salinization of the soil occurred right after the first years of operation [133]. Following, Table 9.8 shows the results of that integrated system in terms of average biomass productivity.

9.2.7 Algae Cultivation Using DC

Microalgae have emerged as very promising candidates as a source of biomass, bioproducts, and biofuels, offering significant advantages such as rapid growth, low demand for land area, and high yield of lipid- and carbohydrate-rich biomass per acre [158,159]. Although microalgal cultivation does not require arable land, it is associated with a high demand for large quantities of water, fertilizers, and solar radiation, which considerably limits the availability of suitable locations [160]. A series of strategies is being developed to circumvent these issues, such as the use of marine species [161], cultivation coupled with wastewater treatment plants [162,163], and mixotrophic cultivation using readily available low-cost or industrial-waste-derived carbon sources [164,165].

Several recent studies have analyzed the growth of microalgae under a variety of wastewater conditions, in particular growth in multiple municipal, industrial, and agricultural wastewaters [162,163]. These studies have mainly been focused on evaluating the ability of algae to remove N and P, and in some instances metals, from wastewater. However, more recently studies have extended the feasibility of algal systems in combination with desalination concentrate (DC) for simultaneous N and P removal with biomass, protein, and lipid production [166–168]. Furthermore, DC is abundant in inorganic minerals, such as Cl^-, Na^+, and Ca^{+2}, which can lead to salt stress in microalgae cultivation, especially due to the effect of salinity on biochemical composition [166]. According to Kumar et al. [169], the salt concentration influences algae via effects on osmotic stress, salt stress, and cellular ionic ratios. For marine algae, salinity is expected to have some effect on the metabolism, especially on important cell membrane constituents, such as lipids [170] and fatty acids composition [171].

Experimental studies on the use of DC as an alternative medium for microalgae cultivation, particularly studies that also assess variables that

Table 9.8 Productivity of an integrated system that products tilapia + *Atriplex nummularia* + goats

			Atriplex		
Salinity of the DC (mS cm^{-1})	**Flow of the DC ($m^3 h^{-1}$)**	**Tilapia production**	**Annual yield of dry biomass (ton ha^{-1})**	**Annual amount of forage saltbush given to goats (ton ha^{-1})**	**Weight increase of the livestock (g day^{-1})**
11.38	1.0–2.5	648.4 kg in 153 days	14.9	1.5	138

Sources: Sánchez AS, Nogueira IBR, Kalid RA. Uses of the reject brine from inland desalination for fish farming, Spirulina cultivation, and irrigation of forage shrub and crops. Desalination. 2015;364: 96–107; Porto ER, Amorim MCC, Paulino RV, Matos ANB. Sistema de produção usando o rejeito da dessalinização de água salobra no semi-árido brasileiro. Proceedings of the XIII Brazilian Congress of Groundwater. Cuiabá: Embrapa; 2004.

determine maximal algal biomass production and biochemical composition, will have significant benefits for the evaluation of wastewater-grown microalgae. As an example of these studies, the following topics summarize the main results of studies on the use of DC as a potential substrate for three microalgae (*Chlorella vulgaris*, *Spirulina platensis*, and *Nannochloropsis gaditana*) cultivation at laboratory stage using photobioreactors) and at pilot scale using raceway ponds realized in Brazil with a focus on the physicochemical characteristics of DC, the impact of this medium on algal growth, biomass productivity, and biochemical composition [172,173].

9.2.7.1 Using DC for Algae Cultivation Under Laboratory Conditions

Several studies carried out in Brazil have addressed the use of DC as a potential substrate for microalgae cultivation under laboratory conditions [166–168,174]. Desalination concentrate is rich in inorganic minerals, such as Cl^-, $Na^{+,}$ and Ca^{2+}, contains other nutrients (N and P), and trace elements necessary for microalgae growth, including Si, K^+, Mg^{+2}, Fe^{+3}, and electrical conductivity normally ranges from 4.0 to 5.5 mS cm^{-1} (Table 9.9).

Table 9.9 Typical compositions of groundwater and desalination concentrate (DC) obtained from the Community of Uruçu (semiarid region of Paraíba state, Brazil)

Parameters	Brackish groundwater	DC
pH	7.3 ± 0.5	8.5 ± 0.3
Electrical conductivity (EC, mS cm^{-1})	3.0 ± 0.8	4.5 ± 1.0
Salinity (M NaCl)	~0.02	~0.04
TDS (mg L^{-1})	1823.8 ± 94	2190.5 ± 78
Cl (mg L^{-1})	720.7 ± 35	1691.3 ± 114
Na (mg L^{-1})	464.0 ± 66	987.5 ± 110
$CaCO_3$ (mg L^{-1})	506.5 ± 97	985.2 ± 112
Ca (mg L^{-1})	74.0 ± 10	126.5 ± 21
Mg (mg L^{-1})	77.2 ± 21	80.7 ± 10
SiO_2 (mg L^{-1})	48.2 ± 11	82.4 ± 24
SO_4^{2-} (mg L^{-1})	15.6 ± 0.6	138.0 ± 14
K (mg L^{-1})	7.0 ± 2	47.0 ± 8
NO_3^- - N (mg L^{-1})	8.2 ± 0.1	30.0 ± 5
NH_4^+ (mg L^{-1})	0.59 ± 0.3	1.35 ± 0.7
PO_4^{3-} - P (mg L^{-1})	0.20 ± 0.1	0.70 ± 0.3
Fe (mg L^{-1})	0.05 ± 0.1	0.13 ± 0.1

Sources: Matos AP, Feller R, Moecke EHS, Sant'Anna ES. Biomass, lipid productivities and fatty acids composition of marine Nannochloropsis gaditana cultured in desalination concentrate. Bioresour Technol 2015;197:48–55; Matos AP, Moecke EHS, Sant'Anna ES. The use of desalination concentrate as a potential substrate for microalgae cultivation in Brazil. Algal Res 2016.

To be able to assess the applicability of DC as a culture medium for algal cultivation, the DC needs to be mixed with a conventional medium, since some microalgae cannot grow in DC due to the high mineral salts strength. For example, Matos et al. [167] cultivated freshwater *C. vulgaris* in dilutions of 25%, 35%, 45%, and 55% DC mixed with Bold Basal Medium (BBM) and demonstrated that incubating *C. vulgaris* with the higher DC concentrations of 35%–55% resulted in growth-salt stress accompanied with lower biomass concentration and lipid production in comparison to control (BBM). Thus, the optimal DC concentration for *C. vulgaris* growth is 25% (~1.6 mS cm^{-1}) and at higher concentrations growth is reduced, probably due to the high Na^+ and Cl^- concentrations in DC. In addition, small quantities of ammonia and sulfate have been detected in DC that may affect *C. vulgaris* growth, while DC can contain residues of pretreatment and cleaning chemicals, and their reaction byproducts, which may also have adverse effects on aquatic organisms [175].

When DC was tested in the same way mixed with Paolettic Synthetic Medium as an alternative culture medium to grow *Spirulina platensis*, it was found that the optimal DC concentration for *S. platensis* growth is 50% DC (~2.5 mS cm^{-1}) [144,166]. On the other hand, marine *N. gaditana* performed well when exposed to a high DC concentration of 75% (~8.0 mS cm^{-1}) [168]. In addition, *N. gaditana* has a tendency toward the synthesis of lipid and higher saturated fatty acids ($C_{16:0}$ palmitic and $C_{18:0}$ stearic acids) when increasing DC concentrations in the culture medium. These examples of the influence of DC on algal growth suggest that DC is a good medium for marine algae strains such as *N. gaditana*, but that the ionic concentration of DC is too high for the cultivation of the freshwater *C. vulgaris* and *S. platensis* [172]. The results of studies on the optimal DC concentration for cultivation of three microalgae species are summarized in Table 9.10.

Different salinities in terms of NaCl concentration in the optimum DC medium for three algae species have been reported in the literature, for *C. vulgaris* (0.01 M NaCl) [167], *S. platensis* (0.05 M NaCl) [144], and *N. gaditana* (0.17 M NaCl) [168]. Among the three microalgae species tested in optimum DC concentration, marine *N. gaditana* has better ability to cope with osmotic changes and alterations in the salt ratios than freshwater algae species, because it has the capability to tolerate a broader range of DC salinities with a satisfactory growth. For example, when marine *N. gaditana* was cultured in F/2-seawater medium (0.5 M NaCl) and induced to a hypotonic salinity, i.e., to brackish DC medium (0.17 M NaCl), biomass concentration and lipids enhanced from 0.71 to 0.96 g L^{-1} and 4.5% to 12.6%,

Table 9.10 Optimal physicochemical parameters of culture media based on DC for three different algae species

	C. vulgaris	*S. platensis*	*N. gaditana*
Culture medium	BBM (fresh water)	Paoletti (fresh water)	F/2 (seawater)
DC proportion (%)	25	50	75
pH	8.5 ± 0.5	9.5 ± 0.5	8.2 ± 0.4
EC (mS cm^{-1})	1.5 ± 0.5	2.5 ± 0.6	9.0 ± 0.1
Salinity (M NaCl)	0.01	0.05	0.18
Reference	Matos et al. [167]	Volkmann et al. [144]	Matos et al. [168]

DC = Desalination concentrate, EC = Electrical conductivity.
Source: Matos AP, Moecke EHS, Sant'Anna ES. The use of desalination concentrate as a potential substrate for microalgae cultivation in Brazil. Algal Res 2016.

Table 9.11 Chemical compositions of three microalgae cultured in optimal DC concentrations (*C. vulgaris* = 25% DC, *S. platensis* = 50% DC, *N. gaditana* = 75% DC)

Component (%)	*C. vulgaris*	*S. platensis*	*N. gaditana*
Moisture	10.2 ± 0.4	6.2 ± 0.3	11.3 ± 0.4
Ash	27.2 ± 2.0	15.0 ± 2.1	22.0 ± 1.2
Total fiber	7.0 ± 0.5	6.0 ± 0.8	15.2 ± 0.6
Total protein	35.5 ± 2.2	55.0 ± 0.6	27.0 ± 1.0
Total lipids	9.7 ± 1.0	4.7 ± 0.6	15.0 ± 0.4
Carbohydrates	10.4 ± 0.5	13.1 ± 0.7	9.5 ± 0.6

Sources: Volkmann H, Imianovsky U, Oliveira JLB, Sant'Anna ES. Cultivation of Arthrospira (Spirulina) platensis in desalinator wastewater and salinated synthetic medium: protein content and amino-acid profile. Braz J Microbiol. 2008;39(1): 98–101; Matos AP, Ferreira WB, Torres RC, Morioka LRI, Canella MHM, Rotta J, Silva T, Moecke EES, Sant'Anna ES. Optimization of biomass production of Chlorella vulgaris grown in desalination concentrate. J Appl Phycol 2014;27: 1473–83; Matos AP, Feller R, Moecke EHS, Sant'Anna ES. Biomass, lipid productivities and fatty acids composition of marine Nannochloropsis gaditana cultured in desalination concentrate. Bioresour Technol 2015;197:48–55; Matos AP, Sant'Anna ES. Avaliação bioquímica da biomassa de microalgas cultivadas em águas residuais dessalinizadas: De escala de laboratório à produção de algas em escala piloto no Semiárido do Nordeste do Brasil. In: Ministério da Ciência e Tecnologia, Brasília – DF; 2017. pp. 117–145; Matos AP, Cavanholi MG, Moecke EHS, Sant'Anna ES. The use of optimum desalination concentrate concentration for biomass, protein and lipid production by the marine alga Nannochloropsis gaditana in response to photoperiod and trophic conditions. Bioresour Technol 2017;224:490–7.

respectively [168]. These results suggest that marine *N. gaditana* production for lipids work well when exposed to suboptimal salinity.

In Table 9.11, it is possible to visualize the chemical compositions of three microalgae (*C. vulgaris, S. platensis,* and *N. gaditana*) cultured in optimal DC under laboratory conditions.

The moisture content in dried microalgae ranged from 6.2% to 11.3%, with *N. gaditana* having the highest content. Ash contents are relatively high in algal biomass (ranges from 15.0% to 27.2%) due to presence of the mineral salts from DC. Protein is the most abundant component in algal biomass, with *S. platensis* showing the highest content (55.0%), followed by *C. vulgaris* (35.5%) and *N. gaditana* (27.0%) of the dry matter. The freshwater algae, i.e., *C. vulgaris* and *S. platensis*, have low content in fiber (6.0%–7.0%) in comparison to marine specie *N. gaditana* (15.2%), which is important for human consumption, since low-fiber values suggest an easily digestible biomass [176]. It is important to note that while the fiber fraction in most terrestrial plants generally comprises cellulose, hemicellulose, and lignin, the fiber in microalgae contains no lignin and has low hemicellulose levels. This makes it easier to digest, and *Chlorella* and *Spirulina* are in fact widely used as a dietary supplement for human consumption. Moreover, *Chlorella* is recognized as safe food ingredient, with GRAS (Generally Recognized as Safe) status by the US FDA (Food and Drug Administration) [177]. While freshwater algae species have a preference in produce protein and carbohydrate when cultured in DC, marine *N. gaditana* has a tendency toward the synthesis of lipid accumulation (15.0%), and this species is gaining attention as potential feedstock for biodiesel production [161,168]. The shift from protein production to accumulation of lipids is a typical response of algae to stress conditions, such as nutrient depletion or salt-osmotic stress (for example [177]).

A recent study by Matos et al. [168] has found that the lipid content in *N. gaditana* cells increased from 5.9% to 12.6% dry matter when the DC concentration increases from 25% to 75% DC in the culture medium. More importantly, at higher DC concentrations, *N. gaditana* tends to synthesize more saturated fatty acids (SFAs, C14:0 and C16:0) accompanied with a lower production of polyunsaturated fatty acids (PUFAs, C18:2, C18:3, and C20:5 ω3). For example, increasing the DC concentration from 0% (control test) to 100% DC in the culture medium for *N. gaditana*, the percentage of palmitic acid (C16:0) increased from 29.4% to 53.4%, and the percentage of eicosapentaenoic acid (EPA, C20:5 ω3) dropped from 12.2% to 0.2% of the total fatty acids. Functionally, increased intracellular lipid content and higher SFAs in *N. gaditana* can be explained by an adaptive osmoregulatory mechanism to cope with rapid or gradual changes in salinity that are associated with algae-cell-membrane permeability [171,178].

In sum, microalgae strains (*C. vulgaris, S. platensis,* and *N. gaditana*) are able to grow in DC, but require different optimum DC concentration. When *N. gaditana* is exposed to optimum DC concentration, which is a "stress" condition, algae decrease the production of protein and increase the synthesis of intracellular lipids and saturated fatty acids. In general, the studies on the use of optimal DC as an alternative culture medium find that freshwater and marine microalgae are virtually devoid of long-chain fatty acids (e.g., C_{20} and C_{22}) and show a trend toward the increased formation of medium-chain (C_{16} and C_{18}) fatty acid methyl esters, which is an ideal lipid source for biodiesel production [159].

9.2.7.2 Using DC for Algae Cultivation at Pilot Scale

Extensive research studies have been conducted using many types of wastewater in combination with microalgae cultivation, particularly in cultures using raceway ponds. As examples, municipal wastewater has been used to grow many types of algae [162], centrate from anaerobic digestion for the cultivation of *Scenedesmus* spp. [179,180], effluent from anaerobic digestion to grow *Spirulina* sp. LEB-18 [181], pig wastewater for the cultivation of *Spirulina* (*Arthrospira*) [182], and dairy wastewater to grow *Chlorella zofingiensis* [183], all studies performed in raceway ponds. In a particular case, DC has been used to grow *C. vulgaris* and *S. platensis* [174], particularly aimed at developing a system, which could be integrated with an inland desalination plant. The current model has been designed in the community of Uruçu, São João do Cariri, Paraíba State—Brazil (Fig. 9.4), as a possible site for algal production due to its favorable climate conditions (low seasonal variation and ample solar irradiance) and geography (local source of brackish groundwater and flat topography).

The outdoor cultivation of *C. vulgaris* in raceway ponds supplemented with 25% DC resulted in an average biomass concentration of 0.23 g L^{-1}

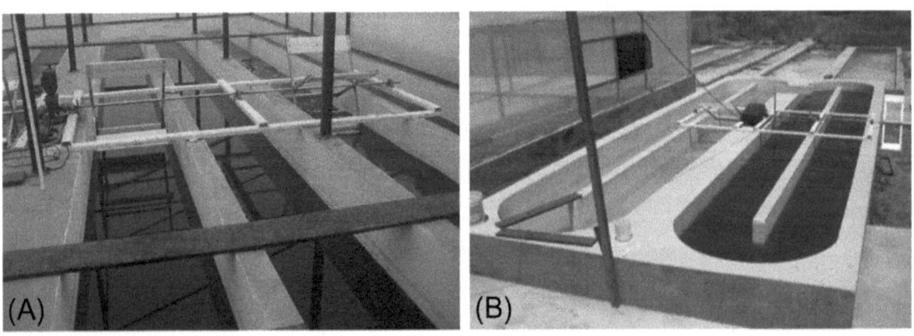

Fig. 9.4 Cultures of algae located in an algal-farm facility in the community of Uruçu, Brazil. Pond measures (5.0-m long, 1.6-m width and 0.5-m depth; totaling 8 m^2 and total capacity of 4000 L/4 m^3 each pond); (A) *S. platensis* cultivation supplemented with 50% DC in raceway ponds that are covered by a greenhouse. (B) *C. vulgaris* cultivation using 25% DC in open raceway ponds.

Table 9.12 Biomass productivity of *C. vulgaris* and *S. platensis* cultured in raceway ponds supplemented with desalination concentrate (DC)

Species	DC (%)	Biomass (g L^{-1})	Areal Productivity (g m^{-2} day^{-1})
C. vulgaris	0 (zero)	0.27 ± 0.2	8.56
C. vulgaris	25	0.22 ± 0.1	6.87
S. platensis	50	1.10 ± 0.3	25.0

Source: Matos AP, Sant'Anna ES. Avaliação bioquímica da biomassa de microalgas cultivadas em águas residuais dessalinizadas: De escala de laboratório à produção de algas em escala piloto no Semiárido do Nordeste do Brasil. In: Ministério da Ciência e Tecnologia, Brasília – DF; 2017. p. 117–45.

and areal productivity of 7.71 g m^{-2} day^{-1}, which is much lower than the biomass concentration (1.10 g L^{-1}) and areal productivity (25.0 g m^{-2} day^{-1}) of *S. platensis* supplemented with 50% DC (Table 9.12) [174]. It is important to note that *S. platensis* has a cell size of 30–70 μm while *C. vulgaris* is 5 μm; therefore, the biomass productivity of *S. platensis* is much higher than *C. vulgaris*. Furthermore, *S. platensis* species is able to grow very well under the semiarid conditions using DC as a substrate. In this case, DC is rich in calcium and carbonates (see Table 9.9), which elevate the pH of the culture medium and favoring the growth of *S. platensis*.

On analyzing the biomass composition of the two freshwater species cultured in raceway ponds and supplemented with DC, *S. platensis* is able to synthesize higher protein content (50.0%) than *C. vulgaris* (20.0%–34.2%) (Table 9.13), indicating that *S. platensis* cultured in DC is a potential source of protein that can be used for human food or animal feed [17]. While the lipid content is almost similar between the species (mean value of 4.6%), *C. vulgaris* is able to produce more fiber + carbohydrate (average of 26.5%) than *S. platensis* (15.7%), and some authors have reported that *Chlorella* sp. (and similar species like *Scendesmus*, *Chlamydomonas*, and *Tetraselmis*) typically produces large amounts of carbohydrate as energy and carbon reserves [184]. As a result, it has been proposed that the use of

Table 9.13 Comparison of chemical composition of *C. vulgaris* and *S. platensis* in raceway ponds supplemented with desalination concentrate (DC)

Specie	DC (%)	Protein (%)	Lipids (%)	Fiber + carbohydrate (%)
C. vulgaris	0 (zero)	34.2 ± 1.0	5.8 ± 0.5	27.3 ± 1.3
C. vulgaris	25	20.0 ± 0.5	3.7 ± 0.2	25.7 ± 1.5
S. platensis	50	50.0 ± 1.2	4.5 ± 0.6	15.7 ± 1.0

Source: Matos AP, Cavanholi MG, Moecke EHS, Sant'Anna ES. The use of optimum desalination concentrate concentration for biomass, protein and lipid production by the marine alga Nannochloropsis gaditana in response to photoperiod and trophic conditions. Bioresour Technol 2017;224:490–7.

carbohydrate-rich algal biomass (for example *C. vulgaris*) as feedstock for bioethanol production may be advantageous over conventional feedstocks by providing increased hydrolysis efficiency, higher fermentable yields, and reduced production costs [185].

Based on the research of using DC as a substrate for microalgae cultivation in semiarid region of Brazil [174], the production cost of *S. platensis* under the natural climatic conditions of the semiarid region has been estimated of US\$ 20.00 kg^{-1} biomass, while for *C. vulgaris* is approximately of US\$ 40–50.00 kg^{-1} biomass. China has become the biggest country in *Spirulina* production in the world and according to Lu et al. [186] Inner Mongolia has become one of the most important bases for commercial production of *Spirulina* in China, with production cost of US\$ 4.00 kg/biomass. This advantage is mainly due to the fact that the climate of western Inner Mongolia is semiarid, the sandy land around of the large number of alkaline lakes, mostly desert, is very cheap to produce *Spirulina*.

Considering that *S. platensis* has been adapted very well under the natural climatic conditions of the community of Uruçu, located inside of the semiarid region of Brazil, with a biomass productivity of around 25.0 $g\,m^{-2}\,day^{-1}$, it is estimated that is possible to produce 72.0 kg of biomass per year in each raceway pond. As there are eight raceway ponds for *Spirulina* production in this experimental farm, this value can reach up 576.0 kg/biomass/year, besides the protein content (50.0%), amino acids, and vitamins of the complex B (thiamine and riboflavin), which makes this algae as a potential source of nutritional supplement.

In summary, the approach using DC as a substrate for microalgae cultivation has been adapted to the community of Uruçu, located in the semiarid region of Brazil, where DC is used for many agricultural purposes [17]. While considerable knowledge has been gained under laboratory and pilot-scale conditions that have demonstrated the potential of DC as a promising alternative culture medium for algae cultivation, the use of DC for algal mass cultivation in semiarid regions of Brazil needs to be validated in further investigations. This region of Brazil has exceptional qualities to produce algae in open raceway ponds on a pilot scale due to its ample solar irradiance, low seasonal variation, extensive arid lands, and abundant brackish groundwater where algae production is proposed [187]. This can be considered as a step forward to the reassessment of saline residues from reverse osmosis desalination plants regarding the valorization of brine management through algal mass cultivation.

9.2.8 Value-Added Product from Algae

Microalgae cultivation has an advancing role in solving some of the future limitations of traditional biomass production and markets (i.e., food, feed, energy, emission mitigation, chemicals, biological material, etc.) [188]. In addition, microalgae are considered to be one of the most promising feedstock materials for developing a sustainable supply of commodities, including food and nonfood products [189]. They also produce natural compounds that could be used as functional food ingredients to enhance the nutritional value of foods [190].

Commercial large-scale production of microalgae for bioproducts began in 1960s and 1970s with *Chlorella* and *Spirulina* and followed in the 1980s with production of β-carotene from *Dunaliella salina* [160]. The ability to grow at high selective environments is the main reason for the successful growth of these species (*Spirulina* = high pH and high HCO_3^-, *D. salina* = high salinity and *Chlorella* = high nutrients) [188].

In the late 1990s, commercial production of the freshwater green alga *Haematococcus pluvialis* as a source of the carotenoid astaxanthin started at Cyanotech in Hawaii [188]. The culture system is a combination of "closed" tower reactors and raceway ponds. *Haematococcus* production by several other small producers commenced in Hawaii in subsequent years using a wide range of, mainly, "closed" culture systems. The astaxanthin from *Haematococcus* is mainly sold as a nutraceutical and antioxidant [191].

A key challenge in mass culturing of microalgae is to find strains that not only produce marketable products or biomass for energy and alternative agriculture, but also grow well under industrially relevant outdoor conditions [160,188]. These may include the need for growth under (i) high (or low) temperatures due to local climatic conditions or bubbling of hot flue gases, (ii) wide temperature variations over a diurnal cycle typically seen in desert conditions, (iii) high light and UV radiation when using solar radiation to drive photoautotrophic growth, (iv) high CO_2 while bubbling flue gases or limiting CO_2 while growing with ambient CO_2 if no source of CO_2 supplementation is available, (iv) local water conditions such as high salt content (e.g., seawater), alkaline or acidic pH, and metal and organic carbon content originating either from local water bodies or from industrial wastewater that needs to be used for algal growth. Some of these can be considered extreme conditions, as most organisms will not survive in such environments. In view of this, extremophilic microalgae have the potential to play an important role in the eventual commercial exploitation of microalgae-based biofuels, bioproducts, and agriculture [160].

Table 9.14 Examples of microalgae that are tolerant to high salt levels and some of their potential products

Species	**Product/application**
Green algae	
Dunaliella salina	β-Carotene (antioxidant)
Dunaliella bardawil	β-Carotene (antioxidant)
Dunaliella sp.	Glycerol
Nannochloropsis salina	Higher level of lipids (up to 28%)
Cyanobacteria	
Aphanocapsa halophytica MN 11	EPS (Exopolysaccharides)/RPS (Released Polysaccharide)
Anabaena sp. ATCC 33047	EPS
Cyanothese sp. ATCC 51142	EPS
Spirulina platensis	Blue pigment phycocyanin

Source: Varshney P, Mikulic P, Vonshak A, Beardall J, Wangikar PP. Extremophilic micro-algae and their potential contribution in biotechnology. Bioresour Technol 2015;184:363–72.

Some algae have developed special mechanisms that allow the cell to grow and thrive under extreme conditions. The most common example is accumulation of glycerol as an osmoregulant in *Dunaliella* due to hypersaline conditions or the accumulation of β-carotene as a protective agent against excess light [192]. Such a phenomenon is the basis for the development of the mass culturing of *Haematococcus pluvialis* for the extraction of astaxanthin [160,190]. A number of cyanobacteria are also capable of growth under high salinities and *Aphanocapsa halophytica* and strains of *Anabaena* and *Cyanothese* produce high levels of extracellular polysaccharides under those conditions (Table 9.14).

Despite their commercial potential having been recognized for over 50 years, the number of microalgal species that have currently produced on a large scale in a sustainable economic process is limited [188]. An ideal microalgae must sustain high productivity under varying environmental and production conditions, and produce commercially sought services and products. As a consequence, much needs to be done in terms of the physiology, molecular biology, metabolic engineering, and outdoor cultivation trials before their true potential is realized [160].

9.3 CONCLUSION AND FUTURE TRENDS

There are two main alternatives for the management of DC that can minimize environmental impacts and produce value-added byproducts as well. The first is volume minimization and the second one is the valorization of the brine. The characteristics of the DC (level of salinity, presence of contaminants) would determine which alternative is more convenient to choose.

Volume minimization of the brine aiming a ZLD (Zero Liquid Discharge) may be a convenient approach if the DC carries dissolved toxic compounds, followed by selective adsorption of the compounds or alternatively, trying to precipitate/neutralize them in a solid form for proper disposal.

Valorization consists in the use of the DC for the production of energy and minerals or for aquaculture and irrigation that can yield food and value-added byproducts. The production of minerals and chemicals is a particular case that requires reducing the volume of the DC to increase the load of dissolved salts. Those salts can later be recovered through a series of emerging technologies that still need further research to decrease costs and energy consumption. Lithium ions stand among the most valuable materials that could be recovered. The other options that aim to valorize the DC can be combined and have interesting synergies among them. If the DC is blended with another effluent such as reclaimed wastewater or less saline waters, some of the salinity gradient power can be recovered in the form of turbine power. The main challenge identified for this technological option is the need to develop highly efficient membranes at an affordable cost.

Diluting the DC with less saline waters to lower the salt content is also crucial if agricultural uses are intended. The studies reviewed throughout this chapter point that even if the irrigation of DC is adjusted to levels that minimize the salt deposition in the soil, a progressive salinization of the soil may occur in the long term, particularly when dealing with high-salinity brines. Among those reviewed studies are results of more than a decade of halophyte cultivation using DC from inland desalination plants in Brazil. Thus blending the DC with other streams of less saline waters (brackish water or biologically treated secondary effluent from sewage or agro-industry) seems like the only alternative to prevent an excessive salinization of the soil.

After blending the DC with a less saline effluent, the resulting dilution can be applied in consecutive cascade-type uses. The first uses would comprehend aquaculture. It was identified that the polyculture of tilapia in combination with shrimp seems a promising option for DC valorization. Another interesting alternative was found in the brine shrimp *Artemia* spp. that has high tolerance to saline waters and produce live feed for larval fish and crustaceans. The effluent leaving the fish/shrimp ponds can further fill cultivation ponds with *Spirulina* and microalgae species such as *Dunaliella salina, Chlorella vulgaris,* or *Nannochloropsis gaditana. N. gaditana* reveals as a promising potential feedstock for biodiesel production due to its high lipid production while *C. vulgaris* yields a carbohydrate-rich algal biomass usable as feedstock for bioethanol. Other extremophilic microalgae have potential uses for biofuel, bioproducts, and agriculture, but improvements are needed in terms of the physiology, molecular biology, metabolic engineering, and outdoor cultivation trials before their true potential is realized. Differently, *Spirulina* is a more viable option that offers synergies for aquaculture farms as it gives an adequate treatment to its tail-end effluent while at the same time produces a source of proteins for the fishes. *Spirulina* is also an effective biosorbent and a cost-effective method for heavy metal removal from wastewater. If the DC is loaded with that type of pollutants, aquaculture should be disregarded and replaced with *Spirulina* cultivation. As a final stage in the valorization of the DC, the effluent could be used for irrigation after leaving the *Spirulina* ponds. Halophytes, especially some Atriplex species, have the capacity to phytostabilizate metals and would provide an adequate final stage of treatment to that effluent. As the Atriplex genus accumulates the metals in the roots and not in the aboveground portions, the integrated *Spirulina*/Atriplex production system could also produce forage from metal-polluted wastewater. For a DC that is free of heavy metals or other pollutants and has moderate salinity, a wide range of options and combinations is possible including the irrigation of nonhalophyte plants. The development of salt-tolerant crops through the breeding of hybrid crops and the use of genetic engineering could increase the range of candidate crops in near future. However, some cautions should be observed. The DC should still be blended with less saline waters and the total salt that can be effectively supplied should be carefully determined. The salt content of the soil should be periodically monitored. Finally, periodical leaching is recommended in order to remove the accumulated salts and conserve an acceptable salt balance of the soil.

In sum, there is not a single standard solution for DC management but a wide range of options are available. Different treatment options can be combined, and the most appropriate combination of technologies would depend on the characteristics of each DC. Treatment and disposal systems for DC can be projected in such a way that allow the sequestration of specific contaminants or make an optimal use of the DC by producing value-added products and maximizing the synergies between the different options. It is possible to obtain beneficial uses of DC and at the same time complain with environmental regulations thus transforming expenses in a potential source of revenue. Desalination concentrates and wastewater from industrial processes are being regarded more and more as opportunities for resource recovery rather than tail-end effluents to be discharged.

REFERENCES

[1] Letey J, Hoffman GJ, Hopmans JW, Grattan SR, Suarez D, Corwin DL, et al. Evaluation of soil salinity leaching requirement guidelines. Agric Water Manag 2011;98(4):502–6.
[2] Siyal AA, Siyal AG, Abro ZA. Salt affected soils: their identification and reclamation. Pak J Appl Sci 2002;2(5):537–40.
[3] Qadir M, Oster JD. Crop and irrigation management strategies for saline–sodic soils and waters aimed at environmentally sustainable agriculture. Sci Total Environ 2004;323(1–3):1–19.
[4] Li J, Niu A, Lu CJ, Zhang JH, Junaid M, Strauss PR, et al. A novel forward osmosis system in landfill leachate treatment for removing polycyclic aromatic hydrocarbons and for direct fertigation. Chemosphere 2017;168:112–21.
[5] Roberts DA, Johnston EL, Knott NA. Impacts of desalination plant discharges on the marine environment: a critical review of published studies. Water Res 2010;44(18):5117–28.
[6] Kathijotes N, Panayiotou C. Wastewater reuse for irrigation and seawater intrusion: evaluation of salinity effects on soils in Cyprus. J Water Reuse Desalin 2013;3(4):392–401.
[7] Brady PV, Kottenstette RJ, Mayer TM, Hightower MM. Inland desalination: challenges and research needs. J Contemp Water Res Educ 2009;132(1):46–51.
[8] Morillo J, Usero J, Rosado D, El-Bakouri H, Riaza A, Bernaola FJ. Comparative study of brine management technologies for desalination plants. Desalination 2014;336:32–49.
[9] Téllez D, Lom H, Chargoy P, Rosas L, Mendoza M, Coatl M, et al. Evaluation of technologies for a desalination operation and disposal in the Tularosa Basin. New Mexico Desalination 2009;249(3):983–90.
[10] DeVito C, Mignardi S, Ferrini V, Martin RF. Reject brines from desalination as possible sources for environmental technologies. In: Ning RY (editor). Expanding Issues in Desalination. InTech; 2011, pp. 978–953.
[11] Giwa A, Dufour V, Al-Marzooqi F, Al-Kaabi M, Hasan SW. Brine management methods: Recent innovations and current status. Desalination 2017;407:1–23.
[12] Arnal JM, Sancho M, Iborra I, Gozálvez JM, Santafé A, Lora J. Concentration of brines from RO desalination plants by natural evaporation. Desalination 2005;182(1–3):435–9.
[13] Ahmed M, Shayya WH, Hoey D, Mahendran A, Morris R, Al-Handaly J. Use of evaporation ponds for brine disposal in desalination plants. Desalination 2000;130(2):155–68.
[14] Alcalá FJ, Custodio E. Using the Cl/Br ratio as a tracer to identify the origin of salinity in aquifers in Spain and Portugal. J Hydrol 2008;359(1–2):189–207.
[15] Duan Z, Hu W. The accumulation of potash in a continental basin: the example of the Qarhan Saline Lake, Qaidam Basin. West China Eur J Mineral 2001;13(6):1223–33.
[16] Vovnyuk SV, Czapowski G. Generation of primary sylvite: the fluid inclusion data from the Upper Permian (Zechstein) evaporites. SW Poland. Special Publications, 285. London: Geological Society; 2007275–84.
[17] Sánchez AS, Nogueira IBR, Kalid RA. Uses of the reject brine from inland desalination for fish farming, Spirulina cultivation, and irrigation of forage shrub and crops. Desalination 2015;364:96–107.

[18] Bunani S, Yörükoğlu E, Yüksel Ü, Kabay N, Yüksel M, Sert G. Application of reverse osmosis for reuse of secondary treated urban wastewater in agricultural irrigation. Desalination 2015;68–74.

[19] Wan CF, Chung TS. Osmotic power generation by pressure retarded osmosis using seawater brine as the draw solution and wastewater retentate as the feed. J Membr Sci 2015;479:148–58.

[20] Wang Q, Gao X, Zhang Y, He Z, Ji Z, Wang X, et al. Hybrid RED/ED system: Simultaneous osmotic energy recovery and desalination of high-salinity wastewater. Desalination 2017;405:59–67.

[21] Samatya S, Köseoğlu P, Kabay N, Tuncel A, Yüksel M. Utilization of geothermal water as irrigation water after boron removal by monodisperse nanoporous polymers containing NMDG in sorption–ultrafiltration hybrid process. Desalination 2015;62–7.

[22] Benner J, Salhi E, Ternes T. vonGunten U. Ozonation of reverse osmosis concentrate: kinetics and efficiency of beta blocker oxidation. Water Res 2008;42(12):3003–12.

[23] Miralles-Cuevas S, Oller I, Agüera A, Llorca M, Pérez JS, Malato S. Combination of nanofiltration and ozonation for the remediation of real municipal wastewater effluents: Acute and chronic toxicity assessment. J Hazard Mat 2017;323:442–51.

[24] Gabelich CJ, Williams MD, Rahardianto A, Franklin JC, Cohen Y. High-recovery reverse osmosis desalination using intermediate chemical demineralization. J Membr Sci 2007;301(1–2):131–41.

[25] Gabelich CJ, Rahardianto A, Northrup CR, Yun TI, Cohen Y. Process evaluation of intermediate chemical demineralization for water recovery enhancement in production-scale brackish water desalting. Desalination 2011;272(1–3):36–45.

[26] Rahardianto A, McCool BC, Cohen Y. Accelerated desupersaturation of reverse osmosis concentrate by chemically-enhanced seeded precipitation. Desalination 2010;264(3):256–67.

[27] Subramani A, Jacangelo JG. Treatment technologies for reverse osmosis concentrate volume minimization: a review. Sep Purif Technol 2014;122:472–89.

[28] Luo W, Phan HV, Xie M, Hai FI, Price WE, Elimelech M, et al. Osmotic versus conventional membrane bioreactors integrated with reverse osmosis for water reuse: Biological stability, membrane fouling, and contaminant removal. Water Res 2017;109:122–34.

[29] Gabelich CJ, Xu P, Cohen Y. Chapter 10 - Concentrate treatment for inland desalting. Sustain Sci Eng 2010;2:295–326.

[30] Mohammad AW, Teow YH, Ang WL, Chung YT, Oatley-Radcliffe DL, Hilal N. Nanofiltration membranes review: recent advances and future prospects. Desalination 2015;356:226–54.

[31] Liang B, Pan K, Li L, Giannelis EP, Cao B. High performance hydrophilic pervaporation composite membranes for water desalination. Desalination 2014;347:199–206.

[32] Tang W, Ng HY. Concentration of brine by forward osmosis: performance and influence of membrane structure. Desalination 2008;224(1–3): 143–153.

[33] McGinnis RL, Hancock NT, Nowosielski-Slepowron M, McGurgan GD. Pilot demonstration of the NH_3/CO_2 forward osmosis desalination process on high salinity brines. Desalination 2013;312:67–74.

[34] Altaee A, Hilal N. High recovery rate NF–FO–RO hybrid system for inland brackish water treatment. Desalination 2015;363:19–25.

[35] Zhao S, Minier-Matar J, Chou S, Wang R, Fane AG, Adham S. Gas field produced/process water treatment using forward osmosis hollow fiber membrane: Membrane fouling and chemical cleaning. Desalination 2017;402:143–51.

[36] Bell EA, Poynor TE, Newhart KB, Regnery J, Coday BD, Cath TY. Produced water treatment using forward osmosis membranes: Evaluation of extended-time performance and fouling. J Membrane Sci 2017;525:77–88.

[37] Zhu S, Li M, El-Din MG. Forward osmosis as an approach to manage oil sands tailings water and on-site basal depressurization water. J Hazard Mat 2017;327:18–27.
[38] Ji X, Curcio E, Al-Obaidani S, DiProfio G, Fontananova E, Drioli E. Membrane distillation–crystallization of seawater reverse osmosis brines. Sep Purif Technol 2010;71(1):76–82.
[39] Gryta M. Concentration of NaCl solution by membrane distillation integrated with crystallization. Sep Sci Technol 2002;37(15):3535–58.
[40] Ahmed M, Arakel A, Hoey D, Thumarukudy MR, Goosen MF, Al-Haddabi M, et al. Feasibility of salt production from inland RO desalination plant reject brine: a case study. Desalination 2003;158(1):109–17.
[41] Loganathan P, Naidu G, Vigneswaran S. Mining valuable minerals from seawater: a critical review. Environ Sci: Water Res Tech 2017;3(1):37–53.
[42] Mohammadesmaeili F, Badr MK, Abbaszadegan M, Fox P. Mineral recovery from inland reverse osmosis concentrate using isothermal evaporation. Water Res 2010;44(20):6021–30.
[43] Mohammadesmaeili F, Badr MK, Abbaszadegan M, Fox P. Byproduct recovery from reclaimed water reverse osmosis concentrate using lime and soda-ash treatment. Water Environ Res 2010;82(4):342–50.
[44] Seigworth A, Ludlum R, Reahl E. Case study: integrating membrane processes with evaporation to achieve economical zero liquid discharge at the Doswell combined cycle facility. Desalination 1995;102:81–6.
[45] Bostjancic J, Rodi L. Getting to Zero Discharge: How to recycle that last bit of really bad wastewater. Technical Paper, GE Water & Process Technologies; 2013.
[46] Subramani A, DeCarolis J, Pearce W, Jacangelo JG. Vibratory shear enhanced process (VSEP) for treating brackish water reverse osmosis concentrate with high silica content. Desalination 2012;291:15–22.
[47] Leong J, Tan J, Heitz A, Ladewig BP. Use of vibratory shear enhanced processing to treat magnetic ion exchange concentrate: A techno-economic analysis. Desalination 2016;383:46–52.
[48] Basudev S. Recovery and recycling of lithium: A review. Sep Purif Technol 2017;172:388–403.
[49] Volkova TS, Rudskikh VV, Orlova VA, Tananaev IG. Recovery of lithium from aqueous solutions by evaporation and precipitation and stability of the resulting compounds in their storage in air. Russ J Appl Chem 2015;88:1388–94.
[50] Xu ZH, Zhang HJ, Wang RY, Gui WJ, Liu GF, Yang Y. Systemic and direct production of battery-grade lithium carbonate from a saline lake. Ind Eng Chem Res 2014;53:16502–7.
[51] Yang LX, Wu SX, Liu XL, He J, Chen WG. Lithium and magnesium separation from salt lake brine by tributyl phosphate under action of co-extraction reagent ClO_4. Chem J Chin Univ-Chin 2013;34:55–60.
[52] Intaranont N, Garcia-Araez N, Hector AL, Milton JA, Owen JR. Selective lithium extraction from brines by chemical reaction with battery materials. J Mater Chem 2014;A2:6374–7.
[53] Hien-Dinh TT, Luong VT, Giere R, Tran T. Extraction of lithium from lepidolite via iron sulphide roasting and water leaching. Hydrometallurgy 2015;153:154–9.
[54] Ooi K, Makita Y, Sonoda A, Chitrakar R, Tasaki-Handa Y, Nakazato T. Modelling of column lithium desorption from Li^+ loaded adsorbent obtained by adsorption from salt brine. Hydrometallurgy 2017;169:31–40.
[55] Marchini F, Rubi D. delPozo M, Williams FJ, Calvo EJ. Surface chemistry and Lithium-Ion exchange in $LiMn_2O_4$ for the electrochemical selective extraction of LiCl from natural salt lake brines. J Phys Chem C 2016;20(29):15875–83.
[56] Shi C, Jing Y, Xiao J, Wang X, Yao Y, Jia Y. Solvent extraction of lithium from aqueous solution using non-fluorinated functionalized ionic liquids as extraction agents. Sep Purif Technol 2017;172:473–9.

[57] Liu W, Xu H, Shi X, Yang X. Fractional crystallization for extracting lithium from Cha'erhan tail brine. Hydrometallurgy 2017;167:124–8.
[58] Korngold E, Aronov L, Daltrophe N. Electrodialysis of brine solutions discharged from an RO plant. Desalination 2009;2428:215–27.
[59] Zhang Y, Ghyselbrecht K, Meesschaert B, Pinoy L, VanderBruggen B. Electrodialysis on RO concentrate to improve water recovery in wastewater reclamation. J Membrane Sci 2011;378(1–2):101–10.
[60] Zhang Y, VanderBruggen B, Pinoy L, Meesschaert B. Separation of nutrient ions and organic compounds from salts in RO concentrates by standard and monovalent selective ion-exchange membranes used in electrodialysis. J Membrane Sci 2009;332:104–12.
[61] Fernandez-Gonzalez C, Dominguez-Ramos A, Ibañez R, Chen Y, Irabien A. Valorization of desalination brines by electrodialysis with bipolar membranes using nanocomposite anion exchange membranes. Desalination 2017;406:16–24.
[62] Badruzzaman M, Oppenheimer J, Adham S, Kumar M. Innovative beneficial reuse of reverse osmosis concentrate using bipolar membrane electrodialysis and electrochlorination processes. J Membrane Sci 2009;326(2):392–9.
[63] Xu C, Xue S, Wang P, Wu C, Wu Y. Diffusion dialysis for NaCl and NaAc recovery using polyelectrolyte complexes/PVA membranes. Sep Purif Technol 2017;172:140–6.
[64] Yen FC, You SJ, Chang TC. Performance of electrodialysis reversal and reverse osmosis for reclaiming wastewater from high-tech industrial parks in Taiwan: A pilot-scale study. J Environ Manage 2017;187:393–400.
[65] Dong Y, Liu J, Sui M, Qu Y, Ambuchi JJ, Wang H, et al. A combined microbial desalination cell and electrodialysis system for copper-containing wastewater treatment and high-salinity-water desalination. J Hazard Mat 2017;321:307–15.
[66] Bond R, Batchelor B, Davis T, Klayman B. Zero liquid discharge desalination of brackish water with an innovative form of electrodialysis: electrodialysis metathesis. Fla Water Resour J 2011;1:4289–317.
[67] Murray B, McMinn D, Gilron J. Waiv technology: An alternative solution for brine management: Results of a full-scale demonstration trial conducted at a location near Roma in Queensland. Water: AWA 2015;42(5):77.
[68] Philip L, Reddy KS, Kumar B, Bhallamudi SM, Kannan A. Performance evaluation of a solar and wind aided cross-flow evaporator for RO reject management. Desalination 2013;317:1–10.
[69] Aravamudan K, Harikumar V, Kumar B, Philip L, Bhallamudi SM, Reddy KS. Simulation of a cross flow wind aided evaporator. Desalination 2014;340:18–29.
[70] Xing YL, Qi CH, Feng HJ, Lv QC, Xu GR, Lv HQ, et al. Performance study of a pilot-scale multi-effect vacuum membrane distillation desalination plant. Desalination 2017;403:199–207.
[71] Hamawand I, Lewis L, Ghaffour N, Bundschuh J. Desalination of salty water using vacuum spray dryer driven by solar energy. Desalination 2017;404:182–91.
[72] Alkhudhiri A, Hilal N. Air gap membrane distillation: A detailed study of high saline solution. Desalination 2017;403:179–86.
[73] Cipollina A, Micale G, Rizzuti L. A brine evaporative cooler/concentrator for autonomous thermal desalination units. Desalin Water Treat 2011;31:269–78.
[74] Assiry AM. Application of ohmic heating technique to approach near-ZLD during the evaporation process of seawater. Desalination 2011;280:217–23.
[75] Onishi VC, Carrero-Parreño A, Reyes-Labarta JA, Ruiz-Femenia R, Salcedo-Díaz R, Fraga ES, et al. Shale gas flowback water desalination: Single vs multiple-effect evaporation with vapor recompression cycle and thermal integration. Desalination 2017;404:230–48.
[76] Qi C, Wang X, Feng H, Lv Q. Performance analysis of low-temperature multi-effect distillation system under different feeding modes. App Thermal Eng 2017;112:1452–9.

[77] Sharan P, Bandyopadhyay S. Solar assisted multiple-effect evaporator. J Clean Prod 2017;142:2340–51.
[78] Ahmadi M, Baniasadi E, Ahmadikia H. Process modeling and performance analysis of a productive water recovery system. App Thermal Eng 2017;12:100–10.
[79] Pérez-González A, Urtiaga AM, Ibáñez R, Ortiz I. State of the art and review on the treatment technologies of water reverse osmosis concentrates. Water Res 2012;46(2):267–83.
[80] Stanford BD, Leising JF, Bond RG. Snyder SA. Chapter 11-inland desalination: current practices, environmental implications, and case studies in Las Vegas, NV Sustain Sci Eng 2010;2: 327–350.
[81] Heijman SGJ, Guo H, Li S. vanDijk JC, Wessels LP. Zero liquid discharge: heading for 99% recovery in nanofiltration and reverse osmosis. Desalination 2009;236(1–3):357–62.
[82] Balasubramanian P. A brief review on best available technologies for reject water (brine) management in industries. Int J Environ Sci 2013;3:2010–8.
[83] Xu P, Cath TY, Robertson AP, Reinhard M, Leckie JO, Drewes JE. Critical review of desalination concentrate management, treatment and beneficial use. Environ Eng Sci 2013;30:502–14.
[84] Leong J, Tan J, Charrois J, Ladewig BP. Review of high recovery concentrate management options. Desalin Water Treat 2014;52(40–42):7609–27.
[85] Joo SH, Tansel B. Novel technologies for reverse osmosis concentrate treatment: A review. J Environ Manage 2015;150:322–35.
[86] Brine GJ. Treatment and High Recovery Desalination. In: Singh R, Hankins N, editors. Emerging Membrane Technology for Sustainable Water Treatment. Amsterdam: Elsevier; 2016. p. 297–324.
[87] Ariono D, Purwasasmita M, Wenten IG. Brine Effluents: Characteristics, Environmental Impacts, and Their Handling. J Eng Technol Sci 2016;48(4):367–87.
[88] Chung HW, Nayar KG, Swaminathan J, Chehayeb KM. Thermodynamic analysis of brine management methods: Zero-discharge desalination and salinity-gradient power production. Desalination 2017;404:291–303.
[89] Cipollina A, Micale G. Sustainable Energy from Salinity Gradients. Cambridge: Woodhead Publishing; 2016.
[90] Wan CF, Chung TS. Energy recovery by pressure retarded osmosis (PRO) in SWRO–PRO integrated processes. Appl Energ 2016;162:687–98.
[91] Gerstandt K, Peinemann KV, Skilhagen SE, Thorsen T, Holt T. Membrane processes in energy supply for an osmotic power plant. Desalination 2008;224(1–3):64–70.
[92] Hickenbottom KL, Vanneste J, Elimelech M, Cath TY. Assessing the current state of commercially available membranes and spacers for energy production with pressure retarded osmosis. Desalination 2016;389:108–18.
[93] He W, Wang J. Feasibility study of energy storage by concentrating/desalinating water: Concentrated Water Energy Storage. Appl Energ 2017;185:872–84. Part 1.
[94] Cipollina A, Bevacqua M, Dolcimascolo P, Tamburini A, Brucato A, Glade H, et al. Reactive crystallisation process for magnesium recovery from concentrated brines. Desalin Water Treat 2015;55(9):2377–88.
[95] Dhawan GK. Seawater Desalination. Water treatment guide—technical database for the water treatment industry. Available from: http://www.watertreatmentguide.com/seawater_desalination.htm [Accessed 4th Feb 2017].
[96] Kim J, Kwon H, Lee S, Lee S, Hong S. Membrane distillation (MD) integrated with crystallization (MDC) for shale gas produced water (SGPW) treatment. Desalination 2017;403:172–8.
[97] Quist-Jensen CA, Macedonio F, Drioli E. Membrane crystallization for salts recovery from brine—an experimental and theoretical analysis. Desalin Water Treat 2016;57(16):7593–603.

[98] Quist-Jensen CA, Macedonio F, Horbez D, Drioli E. Reclamation of sodium sulfate from industrial wastewater by using membrane distillation and membrane crystallization. Desalination 2017;401:112–9.
[99] Sanap DB, Kadam KD, Narayan M, Kasthurirangan S, Nemade PR, Dalvi VH. Analysis of saline water desalination by directed solvent extraction using octanoic acid. Desalination 2015;357:150–62.
[100] Jegatheesan V, Liow JL, Shu L, Kim SH, Visvanathan C. The need for global coordination in sustainable development. J Clean Prod 2009;17(7):637–43.
[101] Jeppesen T, Shu L, Keir G, Jegatheesan V. Metal recovery from reverse osmosis concentrate. J Clean Prod 2009;17(7):703–7.
[102] LeDirach J, Nisan S, Poletiko C. Extraction of strategic materials from the concentrated brine rejected by integrated nuclear desalination systems. Desalination 2005;182:449–60.
[103] Ji ZY, Chen QB, Yuan JS, Liu J, Zhao YY, Feng WX. Preliminary study on recovering lithium from high Mg2+/Li+ ratio brines by electrodialysis. Sep Purif Technol 2017;172:168–77.
[104] Setter TL, Waters I, Stefanova K, Munns R, Barrett-Lennard EG. Salt tolerance, date of flowering and rain affect the productivity of wheat and barley on rainfed saline land. Field Crop Res 2016;194:31–42.
[105] Chakraborty K, Bose J, Shabala L, Shabala S. Difference in root K+ retention ability and reduced sensitivity of K+−permeable channels to reactive oxygen species confer differential salt tolerance in three Brassica species. J Exp Bot 2016;67(15):4611–25.
[106] Glenn EP, Brown JJ, Blumwald E. Salt tolerance and crop potential of halophytes. Crit Rev Plant Sci 1999;18(2):227–55.
[107] Ruiz KB, Biondi S, Martínez EA, Orsini F, Antognoni F, Jacobsen SE. Quinoa–a model crop for understanding salt-tolerance mechanisms in halophytes. Plant Biosystems 2016;150(2):357–71.
[108] Adolf VI, Jacobsen SE, Shabala S. Salt tolerance mechanisms in quinoa (Chenopodium quinoa Willd.). Environ Exp Bot 2013;92:43–54.
[109] Panta S, Flowers T, Doyle R, Lane P, Haros G, Shabala S. Growth responses of Atriplex lentiformis and Medicago arborea in three soil types treated with saline water irrigation. Environ Exp Botany 2016;128:39–50.
[110] Wei D, Zhang W, Wang C, Meng Q, Li G, Chen TH, et al. Genetic engineering of the biosynthesis of Glycinebetaine leads to alleviate salt-induced potassium efflux and enhances salt tolerance in tomato plants. Plant Sci 2017;257:74–83.
[111] Ismail AM, Horie T. Genomics, physiology, and molecular breeding approaches for improving salt tolerance. Annu Rev Plant Biol 2017;68(1).
[112] Chinnusamy V, Jagendorf A, Zhu JK. Understanding and improving salt tolerance in plants. Crop Sci 2005;45(2):437–48.
[113] Himi M, Valdés-Abellán J, Lovera R, Queralt I, Tapias JC, Candela L, et al. ERT Monitoring the Effects of Irrigation with a Blend of Desalinated/Brackish Groundwater on an Instrumental Plot. In: Near Surface Geoscience. 22nd European Meeting of Environmental and Engineering Geophysics. EAGE; 2016; 2016.
[114] Inbar Y. New standards for treated wastewater reuse in Israel. In: Zaidi M, editor. Wastewater reuse-risk assessment, decision-making and environmental security. Netherlands: Springer; 2007. p. 291–6.
[115] Shanmuganathan S, Vigneswaran S, Nguyen TV, Loganathan P, Kandasamy J. Use of nanofiltration and reverse osmosis in reclaiming micro-filtered biologically treated sewage effluent for irrigation. Desalination 2015;364:119–25.
[116] Pramanik BK, Roddick FA, Fan L, Jeong S, Vigneswaran S. Assessment of biological activated carbon treatment to control membrane fouling in reverse osmosis of secondary effluent for reuse in irrigation. Desalination 2015;364:90–5.

[117] Miyamoto S, Glenn EP, Singh NT. Utilization of halophytic plants for fodder production with brackish water in subtropical deserts. In: Squires VR, Ayoub AT, editors. Halophyte as a resource for livestock and for rehabilitation of degraded lands. Proceedings of the International Workshop on Halophytes for Reclamation of Saline Wastelands and as a Resource for Livestock—Problems and Prospects; 1994. p. 43–75.
[118] Miyamoto S, Glenn EP, Olsen MW. Growth, water use and salt uptake of four halophytes irrigated with highly saline water. J Arid Environ 1996;32(2):141–59.
[119] Jordan FL, Yoklic M, Morino K, Brown P, Seaman R. Glenn EP. Consumptive water use and stomatal conductance of Atriplex lentiformis irrigated with industrial brine in a desert irrigation district. Agric For Meteorol 2009;149: 899–912.
[120] Soliz D, Glenn EP, Seaman R, Yoklic M, Nelson SG, Brown PI. Water consumption, irrigation efficiency and nutritional value of Atriplex lentiformis grown on reverse osmosis brine in a desert irrigation district. Agric Ecosyst Environ 2011;140(3–4):473–83.
[121] Glenn EP, Nelson SG, Ambrose B, Martinez R, Soliz D, Pabendinskas V, et al. Comparison of salinity tolerance of three Atriplex spp. in well-watered and drying soils. Environ Exp Bot 2012;83:62–72.
[122] Morteau B. Salt contaminated water phytotreatment by constructed wetland. In: Khan MA, Boër B, özturk M, Clüsener-Godt M, Gul B, et al., editors. Sabkha ecosystems. Volume V: The Americas. Basel: Springer International Publishing; 2016. p. 261–73.
[123] Alassali A, Cybulska I, Galvan AR, Thomsen MH. Wet fractionation of the succulent halophyte Salicornia sinus-persica, with the aim of low input (water saving) biorefining into bioethanol. Appl Microbiol Biot 2017;101(4):1769–79.
[124] Xu J, Zhao G, Huang X, Guo H, Liu W. Use of horizontal subsurface flow constructed wetlands to treat reverse osmosis concentrate of rolling wastewater. Int J Phytoremediation 2017;19(3):262–9.
[125] Suaire R, Durickovic I, Framont-Terrasse L, Leblain JY, De Rouck AC, Simonnot MO. Phytoextraction of Na+ and Cl− by Atriplex halimus L. and Atriplex hortensis L.: A promising solution for remediation of road runoff contaminated with deicing salts. Ecol Eng 2016;94:182–9.
[126] Pérez-Esteban J, Escolástico C, Ruiz-Fernández J, Masaguer A, Moliner A. Bioavailability and extraction of heavy metals from contaminated soil by Atriplex halimus. Environ Exp Botany 2013;88:53–9.
[127] Eissa MA, Ahmed EM, Reichman SM. Production of the forage halophyte Atriplex amnicola in metal-contaminated soils. Soil Use Manage 2016;32(3):350–6.
[128] Kachout SS, Leclerc JC, Mansoura AB, Rejeb MN, Ouerghi Z. Effects of heavy metals on growth and bioaccumulation of the annual halophytes Atriplex hortensis and A. rosea. J Appl Sci Res 2009;5:746–56.
[129] Meneses R, Varela GI, Flores H. Evaluating the use of Atriplex nummularia hay on feed intake, growth, and carcass characteristics of creole kids. Chilean J Agric Res 2012;72(1):74–9.
[130] Guevarata JC, Colomerb JHS, Esteveza OR, Paez JA. Simulation of the economic feasibility of fodder shrub plantations as a supplement for goat production in the north-eastern plain of Mendoza. Argentina J Arid Environ 2003;53(1):85–98.
[131] Mata-González R, Abdallah MA, Trejo-Calzada R, Wan C. Growth and leaf chemistry of Atriplex species from Northern Mexico as affected by salt stress. Arid Land Res Manage 2016;1–14.
[132] Porto ER, Amorim MD, Dutra MT, Paulino RV, Brito LDL, Matos AN. Behavior of saltbush (Atriplex nummularia) irrigated with effluents from tilapia raised in brackish water. Rev bras eng agríc ambient 2006;10(1):97–103.
[133] Silva CMMS, Vieira RF, Oliveira PR. Salinidade, sodicidade e propriedades microbiológicas de argissolo cultivado com erva-sal e irrigado com rejeito salino. Pesq Agrop Bras 2008;43(10):1389–96.

[134] Stickney RR. Tilapia tolerance of saline waters: a review. The Progressive Fish-Culturist 1986;48(3):161–7.
[135] Perschbacher PW, Stickney RR, editors. Tilapia in Intensive Co-culture. 1st ed. Chichester: John Wiley & Sons; 2017.
[136] Porto ER, Amorim MCC, Paulino RV, Matos ANB. In: Sistema de produção usando o rejeito da dessalinização de água salobra no semi-árido brasileiro. Proceedings of the XIII Brazilian Congress of Groundwater. Cuiabá: Embrapa; 2004.
[137] Huang CY, Kuo JM, Wu SJ, Tsai HT. Isolation and characterization of fish scale collagen from tilapia (Oreochromis sp.) by a novel extrusion–hydro-extraction process. Food Chem 2016;190:997–1006.
[138] Doerr JC, Liu H, Minello TJ. Salinity selection by juvenile brown shrimp (Farfantepenaeus aztecus) and white shrimp (Litopenaeus setiferus) in a gradient tank. Estuaries and Coasts 2016;39(3):829–38.
[139] Chong-Robles J, Charmantier G, Boulo V, Lizárraga-Valdéz J, Enríquez-Paredes LM, Giffard-Mena I. Osmoregulation pattern and salinity tolerance of the white shrimp Litopenaeus vannamei (Boone, 1931) during post-embryonic development. Aquaculture 2014;422–423:261–7.
[140] Pham D, Charmantier G, Wabete N, Boulo V, Broutoi F, Mailliez JR, et al. Salinity tolerance, ontogeny of osmoregulation and zootechnical improvement in the larval rearing of the Caledonian Blue Shrimp, Litopenaeus stylirostris (Decapoda, Penaeidae). Aquaculture 2012;362–363:10–7.
[141] Warren JK. Halotolerant life in feast or famine a source of hydrocarbons and a fixer of metals. In: Evaporites: Sediments, Resources and Hydrocarbons. Berlin Heidelberg: Springer; 2006. p. 617–703.
[142] Ghaeni M, Roomiani L. Review for Application and Medicine Effects of Spirulina. Microalgae J Adv AgricTechnol 2016;3(2):114–7.
[143] Olvera-Novoa MA, Dominguez-Cen LJ, Olivera-Castillo L, Martínez-Palacios CA. Effect of the use of the microalga Spirulina maxima as fish meal replacement in diets for tilapia, Oreochromis mossambicus (Peters), fry. Aquacult Res 1998;29(10):709–15.
[144] Volkmann H, Imianovsky U, Oliveira JLB, Sant'Anna ES. Cultivation of Arthrospira (Spirulina) platensis in desalinator wastewater and salinated synthetic medium: protein content and amino-acid profile. Braz J Microbiol 2008;39(1):98–101.
[145] Velasquez SF, Chan MA, Abisado RG, Traifalgar RFM, Tayamen MM, Maliwat GCF, et al. Dietary Spirulina (Arthrospira platensis) replacement enhances performance of juvenile Nile tilapia (Oreochromis niloticus). J Appl Phycol 2016;28(2):1023–30.
[146] Spirulina in combating Protein Energy Malnutrition (PEM) and Protein Energy Wasting (PEW) - A review. J Nutrition Res. 2015;3(1): 62–79.
[147] Wuang SC, Khin MC, Chua PQD, Luo YD. Use of Spirulina biomass produced from treatment of aquaculture wastewater as agricultural fertilizers. Algal Res 2016;15:59–64.
[148] Valderrama G, Gardenas A, Marcovits A. On the economics of Spirulina production in Chile with details on dragboard mixing in shallow ponds. Hydrobiology 1987; 151–152:71–4.
[149] Tambiev AK, Vasilieva SG, Lukyanov AA. Manifestation of Salt Tolerance of Spirulina platensis and Spirulina maxima Cyanobacteria of the Genus Arthrospira (Spirulina). Mosc Univ Biol Sci Bull 2011;66(4):133–7.
[150] Liu C, Li LJ, Wu CY, Guo KN, Li JH. Growth and antioxidant production of Spirulina in different NaCl concentrations. Biotechnol Lett 2016;38(7):1089–96.
[151] Okamura H, Aoyama I. Interactive toxic effect and distribution of heavy metals in phytoplankton. Toxicol Water Qual 1994;9:7–15.
[152] Sandeep KP, Shukla SP, Harikrishna V, Muralidhar AP, Vennila A, Purushothaman CS, et al. Utilization of inland saline water for Spirulina cultivation. J Wat Reuse Desalin 2014;3(4):346–56.

[153] Myint MT, Hussein W, Ghassemi A. Microalgal process for treatment of high conductivity concentrates from inland desalination. Desalin Water Treat 2016;57(10):4313–21.
[154] Çelekli A, Topyürek A, Markou G, Bozkurt H. A multivariate approach to evaluate biomass production, biochemical composition and stress compounds of Spirulina platensis cultivated in wastewater. Appl Biochem Biotechnol 2016;180(4):728–39.
[155] dos Santos RR, Araújo ODQF, de Medeiros JL, Chaloub RM. Cultivation of Spirulina maxima in medium supplemented with sugarcane vinasse. Bioresour Technol 2016;204:38–48.
[156] Malakootian M, Limoni ZK, Malakootian M. The efficiency of lead biosorption from industrial wastewater by micro-algae Spirulina platensis. Int J Environ Res 2016;10(3):357–66.
[157] Prathima B, Rao P, Mahalakshmi MM. Biosorption of copper from wastewater using Spirulina species. In: Mohan BR, Srinikethan G, Meikap BC, editors. Materials, Energy and Environment Engineering. 2017. Singapore: Springer; 2017. p. 129–35.
[158] Foley PM, Beach ES, Zimmerman JB. Algae as source of renewable chemicals: opportunities and challenges. Green Chem 2011;13:1399–405.
[159] Stephens E, Ross IL, King Z, Mussgnug JH, Kruse O, Posten C, et al. An economical and technical evaluation of microalgae biofuels. Nature Biotechnol 2010;28:126–8.
[160] Varshney P, Mikulic P, Vonshak A, Beardall J, Wangikar PP. Extremophilic micro-algae and their potential contribution in biotechnology. Bioresour Technol 2015;184:363–72.
[161] Simionato D, Sforza E, Carpinelli EC, Bertucco A, Giacometti GM, Morosinotto T. Acclimation of Nannochloropsis gaditana to different illumination regimes: Effects on lipids accumulation. Bioresour Technol 2011;102:6026–32.
[162] Pittman JK, Dean AP, Osudenko O. The potential of sustainable algal biofuel production using wastewater resources. Bioresour Technol 2011;102:17–25.
[163] Dickinson KE, Whitney CG, McGinn PJ. Nutrient remediation rates in municipal wastewater and their effect on biochemical composition of the microalga Scenedesmus sp. AMDD Algal Res 2013;2:127–34.
[164] Perez-Garcia O, Escalanate FME, de-Bashan LE, Bashan Y. Heterotrophic cultures of microalgae: Metabolism and potential products. Water Res 2011;45:11–36.
[165] Selvaratnam T, Pegallapati AK, Reddy H, Kanapathipillai N, Nirmalakhandan N, Deng S, et al. Algal biofuels from urban wastewaters: Maximizing biomass yield using nutrients recycled from hydrothermal processing of biomass. Bioresour Technol 2015;182:232–8.
[166] Volkmann H, Imianovisk U, Furlong EB, Oliveira JLB, Sant'Anna ES. Influence of desalinator wastewater for the cultivation of *Arthrospira platensis*. Fatty acids profile. Grasas Aceites 2007;58:396–401.
[167] Matos AP, Ferreira WB, Torres RC, Morioka LRI, Canella MHM, Rotta J, et al. Optimization of biomass production of Chlorella vulgaris grown in desalination concentrate. J Appl Phycol 2014;27:1473–83.
[168] Matos AP, Feller R, Moecke EHS, Sant'Anna ES. Biomass, lipid productivities and fatty acids composition of marine Nannochloropsis gaditana cultured in desalination concentrate. Bioresour Technol 2015;197:48–55.
[169] Kumar K, Mishra SK, Shrivastav A, Park MS, Yang JW. Recent trends in the mass cultivation of algae in raceway ponds. Renew Sustain Energy Rev 2015;51:875–85.
[170] Takagi M, Karseno YT. Effect of salt concentration on intracellular accumulation of lipids and triacylgliceride in marine microalgae Dunaliella cells. J Biosc Bioenerg 2006;101:223–6.
[171] An M, Mou S, Zhang X, Zheng Z, Ye N, Wang D, et al. Expression of fatty acid desaturase genes and fatty acid accumulation in Chlamydomonas sp. ICE-L under salt stress. Bioresour Technol 2013;149:77–83.

[172] Matos AP, Moecke EHS, Sant'Anna ES. The use of desalination concentrate as a potential substrate for microalgae cultivation in Brazil. Algal Res 2016.
[173] Matos AP, dessalinizadas S'AESAbdbdmceár. De escala de laboratório à produção de algas em escala piloto no Semiárido do Nordeste do Brasil. In. Brasília – DF: Ministério da Ciência e Tecnologia; 2017117–45.
[174] Matos AP, Cavanholi MG, Moecke EHS, Sant'Anna ES. The use of optimum desalination concentrate concentration for biomass, protein and lipid production by the marine alga Nannochloropsis gaditana in response to photoperiod and trophic conditions. Bioresour Technol 2017;224:490–7.
[175] Greenlee LF, Lawler DF, Freeman BD, Marrot B, Moulin P. Reverse osmosis desalination: Water sources, technology, and today's challenges. Water Res 2009;43:2317–48.
[176] Rebolloso Fuentes MM, Ácien Fernández GG, Sánchez Pérez JA, Guil Guerrero JL. Biomass nutrient profiles of the microalga Porphyridium cruentum. Food Chem 2000;70:345–53.
[177] Anupama RP. Value-added food. Single cell protein Biotechnol Adv 2000;18:459–79.
[178] Lawton RJ, Nys R, Magnusson ME, Paul NA. The effect of salinity on the biomass productivity, protein and lipid composition of a freshwater macroalga. Algal Res 2015;12:213–20.
[179] Morales-Amaral MM, Gómez-Serrano C, Acién FG, Fernández-Sevilha JM, Molina-Grima E. Outdoor production of Scenedesmus sp. in thin-layer and raceway reactors using centrate from anaerobic digestion as the sole nutrient source. Algal Res 2015;12:99–108.
[180] Tran KC, Mendoza Martin JL, Heaven S, Banks CJ, Acien Fernandez FG, Molina Grima E. Cultivation and anaerobic digestion of Scenedesmus spp. grown in a pilot-scale open raceway. Algal Res 2014;5:95–102.
[181] Borges JA, Rosa GM, Meza LHR, Henrard AA, Souza MRAZ, Costa JAV. Spirulina sp. LEB-18 culture using effluent from the anaerobic digestion. Braz J Chem Eng 2013;30:277–87.
[182] Olguín EJ, Galicia S, Mercado G, Pérez T. Annual productivity of Spirulina (Arthrospira) and nutrient removal in a pig wastewater recycling process under tropical condition. J Appl Phycol 2013;15:249–57.
[183] Huo S, Wang Z, Zhu S, Zhou W, Dong R, Yuan Z. Cultivation of Chlorella zofingiensis in bench-scale outdoor ponds by regulation of pH using dairy wastewater in winter, South China. Bioresour Technol 2012;121:76–82.
[184] Tibbetts SM, Bjornsson WJ, McGinn PJ. Biochemical composition and amino acid profiles of Nannochloropsis granulata algal biomass before and after supercritical fluid CO2 extraction at two processing temperatures. Anim Feed Sci Tech 2015;204:62–71.
[185] Lee OK, YK O, Lee EY. Bioethanol production from carbohydrate-enriched residual biomass obtained after lipid extraction of Chlorella sp. KR-1. Bioresour Technol 2015;196:22–7.
[186] YM L, Xiang WZ, Wen YH. Spirulina (Arthrospira) industry in Inner Mongolia of China: current status and prospects. J Appl Phycol 2011;23:265–9.
[187] Rogers JN, Rosenberg JN, Guzman BJ, Oh VH, Mimbela LE, Ghassemi A, et al. A critical analysis of paddlewheel-driven raceway ponds for algal biofuel production at commercial scales. Algal Res 2014;4:76–88.
[188] Moheimani NR, McHenry MP, Boer K, Bahri PA. Biomass and Biofuels from Microalgae: Advances in Engineering and Biology. Springer International Publishing Switzerland; 2015.
[189] Batista AP, Gouveia L, Bandarra NM, Franco JM, Raymundo A. Comparison of microalgal biomass profiles as novel functional ingredient for food products. Algal Res 2013;2:164–73.

[190] Borowitzka MA. High-value products from microalgae – their development and commercialisation. J Appl Phycol 2013;25:743–56.
[191] Reyes FA, Sielfeld CS, del Valle JM. Effect of high-pressure compaction on superctitical-CO2 extraction of astaxanthin from Haematococcus pluvialis. J Food Eng 2016;189:123–34.
[192] Katz A, Waridel P, Shevchenko A, Pick U. Salt induced changes in the plasma membrane proteome of the halotolerant alga Dunaliella salina as revealed by blue native gel electrophoresis and nano-LC-MS/MS analysis. Mol Cell Proteomics 2007;6:1459–72.

FURTHER READING

[1] Bader B, Aissaoui F, Kmicha I, Salem AB, Chehab H, Gargouri K, et al. Effects of salinity stress on water desalination, olive tree (Olea europaea L. cvs 'Picholine', 'Meski' and 'Ascolana') growth and ion accumulation. Desalination 2015;364:46–52.
[2] Adolf VI, Shabala S, Andersen MN, Razzaghi F, Jacobsen SE. Varietal differences of quinoa's tolerance to saline conditions. Plant and Soil 2012;357(1–2):117–29.
[3] Joint United States-Canadian Tables of Feed Composition. Nutritional Data. Washington DC: National Academy of Sciences; 1964.

PART 2

Issues in Sustainable Desalination

CHAPTER 10

Environmental Regulations—Inland and Coastal Desalination Case Studies

J. Jaime Sadhwani Alonso, Noemi Melián-Martel
University of Las Palmas de Gran Canaria, Las Palmas de Gran Canaria, Spain

10.1 INTRODUCTION

Water desalination in Spain is a consolidated technology market that was born 50 years ago, becoming a global leader in desalination. The first built desalination plants, some already dismantled have contributed to the population as well as to the nature conservation because they have been providing some basic needs for populations, industries, and agriculture where traditional water resources have been scarce.

Desalination has developed in Spain continuously, driven mainly due to the serious imbalance between water resources (motivated by the irregular rainfall in the country) and water consumption supported in certain areas with intensive irrigated agriculture and tourism infrastructure that coincides at the time of lower rainfall.

Currently, the country has over 700 desalination plants, which produce >5.2 million m^3/day and place Spain on the top of the European Union and at the fourth place in terms of desalination after Saudi Arabia, United Arab Emirates, and the United States [1]. It is a leadership kept at our regret, forced by the scarcity of natural resources. Desalination in Spain is mainly reduced to the Spanish Levante (*Cataluña*, *Valencian Community*, *Balearic Islands*, *Andalucía*, and *Murcia*) and the *Canary Islands*.

In Spain, the Canary Islands has been a pioneer in the use of different technologies employed for water desalination, being an important reference point for their experience and knowledge to other communities in the Spanish territory. From the first desalination plant by Multi-Stage Flash (MSF) distillation process installed in 1964 in *Lanzarote* (Canary Islands), today there are >300 desalination plants [2], driven by the breathtaking necessity of water resources for their inhabitants and for the growing tourism industry in some of its most water-challenged coastal areas.

Sustainable Desalination Handbook
https://doi.org/10.1016/B978-0-12-809240-8.00010-1

Among desalination technologies, the largest installed capacity is reverse osmosis, which during the last decades has been rising steeply achieving important goals that affect the user, as is the reduction of costs per cubic meter of desalinated water supply and quality.

However, there has been a lot of discussion in the literature regarding environmental concerns of desalination plants, especially for those in Spain, since the Water National Plan of the country included desalination as one of the alternative technologies for supplying water for human consumption, tourist uses, as well as agricultural or industrial consumption.

Nevertheless, comprehensive environmental monitoring data are not available for the majority of desalination plants that have been constructed around Spain, partly because, many of these plants have been constructed in places and at times when environmental concerns were not at the fore.

This is changing, and the society is concerned to preserve natural resources and sustainability, including the environmental impacts from desalination plants. For that reason, several recently constructed plants have severe monitoring programs, reflected in the Environmental Impact Statement (EIS), to evaluate and minimize the environmental impacts associated with desalination plants.

In this chapter, we analyze the environmental concerns of seawater and brackish water reverse osmosis desalinations plants, focusing on some cases studies, and describing the major impacts identified.

10.2 ENVIRONMENTAL IMPACT REGULATION

The term environmental impact is used in two different but interrelated fields: the scientific and legal-administrative. The first has led to development of methodologies and techniques for the study, identification, and estimation of the potential effects of the implementation of project, work or activity, and enabled an estimation to preserve environmental natural resources and protect them. The second has produced a series of rules and regulations that guarantee that a particular project may be modified or rejected due to its environmental consequences (technical project).

As a general definition, strictly speaking, Environmental Impact Assessment (EIA) means the administrative process followed in certain projects for which, by law, its application is mandatory and include most of the infrastructure projects. In this process, before project approval, an Environmental Impact Study (EsIA) is carried out, in which the potential impacts are anticipated, the most suitable alternative is chosen from the environmental point of view, if any, and preventive, corrective, or compensatory necessary measures are proposed to minimize impacts.

Regulations associated to EIA, pollution prevention and control, protected habitats and environmental liability have made desalination techniques progress in order to achieve the established thresholds and specifications.

10.2.1 Community Regulations

In 1969, the United States the first National Environmental Policy Act (NEPA was introduced for EIA and the need for public participation. This resulted in major changes in planning for the development and management of water resources, emphasizing protection and enhancement of environmental values, and protection of cultural and social values.

While the starting point at the legislative level in Europe emerged in 1985 with the publication of the Directive 85/337/EEC [3] modified by the Directive 97/11/EC [4].

Currently, two community laws regulate the EIA process in Europe. First the newly amended Directive 2014/52/EU [5] on the assessment of the effects of certain public and private projects on the environment (*known as Environmental Impact Assessment—EIA Directive*) that modifies and complete the directive 2011/92/UE and these two directives. Secondly, the Directive 2001/42/EC [6] on the assessment of the effects of certain plans and programs on the environment (*known as Strategic Environmental Assessment—SEA Directive*).

10.2.2 National Regulations

In Spain, the spectacular growth in the number of desalination plants from its beginnings in the 1960s did not come from the hand of a regulated EIA. The Spanish EIA regulation was established in 1986 by the Royal Decree 1302/86 [7] as a basic regulation in the environmental matters by the transpose of the Directive 85/337/ECC [3] into the Spanish law. This regulation defined the stages of EIA and the content of the EIS.

This highlights an environmental gap of nearly 25 years since the first desalination plant installed on the island of Lanzarote (Canary Islands) in 1964.

Later on further regulations were issued; in 2008, the Royal Decree 1/2008 [8] included aspect of public participation, modified aspect of the administrative process and expanded the list of the projects, which are subject inescapably to EIA included in the Annexes I and II. Both Annexes I and II are separated into nine groups (Group 1. Agriculture, forestry, aquaculture, and livestock; Group 2. Food industry, Group 3. Extractive industry; Group 4. Energy industry; Group 5. Iron and steel industry. Production and elaboration of metals; Group 6. Chemical, petrochemical, textile, and paper industry; Group 7. Infrastructure projects; Group 8. Water Engineering and Management Projects; and Group 9. Other projects).

The Law 6/2010 [9] amended the definition of competent authorities; it organized the process in three phases and changed the processing time. Finally, in 2013 the Law 9/2006 (SEA) and the Royal Decree 1/2008 (EIA) [8] were unified into a single regulation the Law 21/2013 [10]. This law repealed the existing environmental assessment legislation and unified in various sections the legal regime applicable to the assessment of plans and projects. The new law includes a range of difficult aspects, such as the definition of the projects subject to assessment, the various procedural stages and the amendments to existing environmental reports, prepared by the so-called environmental authority in the case of plans and programs, but not subject to deadlines.

Nevertheless, this law is only the basic regulation for Spanish, since the different regions (Autonomous Communities) are entitled to produce their own procedures, including the capability to issue their own screening and scoping procedures (establishing the activities subject to environment impact assessment and the terms of reference for the assessments themselves) [11]. This is discussed further in Section 10.2.3.

Fuentes-Barges [12] in a recent paper discussed in detail the analysis of the process of EIA for seawater desalination plants in Spain during the last 12 years. The process of EIA is divided into three phases [12]:

(1) Determining the scope of Environmental Impact Statement (EIS) by the environmental agency.

(2) Preparation of the EIS by the promoter and the process of public information and consultations.

(3) Record of Decisions (RODs) (also called development consent) published by the environmental agency—which determines the environmental viability of the project, as well as the conditions and requirements to be imposed.

A project will have to submit to the EIA process if its activity is in the list of Annex I activities of the regulations effective [10] or if it is on the list of Annex II activities and the environmental agency decides that the assessment process is required. In the case of desalination plants with a new or added capacity over 3000 m^3/day of water production is specified in Annex

II, Group 8, Section E: Water Engineering and Management Projects. Nevertheless, the need for EIA is determined by the own regulations, or because it required corresponding to the European ecological Natura 2000 Network, such as the case of a Place of Community Interest.

In Spain, the reduction of environmental impact of desalination facil ities has been focused on the protection of vital ecosystems from brine discharge like endemic seagrass that are dominants habitats of the Spanish Mediterranean and Atlantic coast with *Posidonia oceanica (L.) Delile* and *Cymodocea nodosa (Ucria) Ascherson* being the most abundant and ecologically relevant species [13]. Those are habitats protected by the Directive 92/43/EEC [14] on the conservation of natural habitats and of wild fauna and flora.

Among the legal instruments to control these brine discharge is the EIA, which is not mandatory in all cases (only if the project is included in Annexes I or II) and the discharge authorization. The discharge authorization for both to the sea and to the public hydraulic domain allows establishing conditions to prevent damage to the environment. However, there are no provisions that establish specific parameters for both salinity and for other substances harmful to the environment that usually accompany brine discharge.

Therefore, it is proposed that the Spanish government adopt specific rules in this respect, where appropriate limits both with respect to the salt concentration, as in other substances whose effect on marine ecosystems is shown to be established, without prejudice to the additional rules safeguarding that could approve the Autonomous Communities.

10.2.3 Regional Regulations

The regional regulation has been developed in parallel to the national, although each integrating its particular characteristics, a fact that has caused sometimes autonomic be more developed, producing differences between them and difficulties in performing projects different regions by the promoters.

Currently, there are four autonomous communities with specific regulations for EIA. *Canary Islands* is one of them, with the regional Law 14/2014 [15] on harmonization and simplification on the protection of land and natural resources, which repeals the Law 11/1990 on Ecological Impact Prevention, which regulated the environmental assessment of projects for nearly 25 years.

The law 11/1990 was characterized by the most demanding environmental assessment of Spanish state regulation. The restructuring of the law does not mean that the level of demand that has characterized our legal system is reduced, but the new regulation reconstructs the system for evaluating the environmental impact of the Canary Islands to adjust state legislation and new socioeconomic needs.

Among the innovations is the creation of two types of EIA of projects, ordinary and simplified in contrast to the previous law that had three types of EIA of projects: (1) basic environmental impact study, (2) detailed environmental impact study, and (3) environmental impact study. With the new Law 14/2014, the EIA is also enforced through accreditation, through associates in this matter, which the promoter may also be able to choose freely between the ordinary and the simplified procedure unless the law is obliged to go to the ordinary, longer, and more complex procedure.

10.3 GENERAL POTENTIAL IMPACTS FROM DESALINATION FACILITIES

There are many potential environmental impacts of desalination process similar to other industry. The EIA for any project include environmental concerns during the construction period such as [12]:

(1) The environmental effect of the land use that starts with the construction and continues during the operation. It has indirect effects as removing habitats, the movement of the wildlife and the visual impact.
(2) Spilling oils, fuel remains, concrete grout, liquids stripping, etc. over the land.
(3) Disturbance of the waters because the earthmoving works generate an increasing quantity of suspended solids and turbidity of the water. This leads to an increase of the possibility of accidental releases of dangerous products, such as oil or petrol, from the machinery used.
(4) Noise generated by the machinery used in the phases of the work.
(5) Potential releases to the air's quality, especially during the earthmoving works, as consequence of dust emission and gas engines from the machinery.

However, there are environmental impacts more specific to desalination plants during the operation period [11,16–18]:

(1) Impacts on the marine environment attributed to the brine and chemical discharges.
(2) Indirect impacts due to the need of intensive energy demand of fossil fuel for the process that leaves behind a relatively high carbon footprint.
(3) Impingement and entrainment of marine organisms due to intake of large quantities of seawater.
(4) Impact of noise generated by desalination process.

(5) Adverse effect of land use and impact on the landscape.
(6) Impact on the aquifer.
(7) Visual disturbance.

Apart from the environmental impacts indicated as follows, the list of possible environmental impacts could be extended, including environmental concerns that are not included in this chapter, such as environmental effects of disposal of used membranes, all kind of containers, waste related to maintenance activities, etc. This factor should be taken into account for an environmental integral evaluation of desalination plants.

In the next sections, the major "direct" environmental impacts will be addressed: brine discharge, energy use, land use, and visual disturbance and impact of noise.

10.3.1 Impact on the Marine Environment Due to the Brine Discharge

An overview on the composition and effects of brine discharge on the marine environment can be found in a WHO document [19], discussed in detail by Lattemann and Höpner [16,20] and Roberts et al. [21].

The reject brine from RO desalination plants represents a brine flow rate of 30%–70% of the feed water flow, which means 1.3–1.7 times the seawater concentration. In addition, the high concentration of salts contain also other additives used during desalination process (biofouling control (biocides), scale control (antiscalants), coagulants, and corrosion inhibitors) and punctual discharges resulting from cleaning membranes (chemical cleaning), which generates highly concentrated flows of suspended solids and detergents [16,18,22].

In nearby coastal desalination plants, this reject water is mainly discharged into the marine environment, which can very often severely damage the receiving environment. Anoxic condition on the seabed, changing light conditions, and especially impact on benthic and stenohaline species [23], such as *Posidonia oceanica* [24] and *Cymodocea nodosa* seagrasses could be generated. Table 10.1 shows salinity thresholds established for these species.

The environmental impact of brine discharge has been minimized through both adequate selections of the alternatives and design of the disposal system and good strategy planning prior to desalination plants building. The following aspects must be taken into account in the EIA:

(1) Preliminary analysis of the marine environment: (i) natural protected areas and presence of priority habitats, (ii) discharge at the estuary of water courses, streams and, in general, degraded areas, and (iii) installation of emissaries on dams, harbors, and ports.

Table 10.1 Sensitivities of *Posidonia oceanica* and *Cymodocea nodosa* to desalination plants and their limits in saline concentration. Salinity in "psu," practical salinity units [24,25]

Species	Critical Salinity Limits
Posidonia oceanica seagrasses	Should not exceed 38.5 psu in >25% of measurements: S25, lim = 38.5
	Should not exceed 40 psu in >5% of measurements: S25, lim = 40
Cymodocea nodosa seagrasses	Should not exceed 39.5 psu in >25% of measurements: S25, lim = 39.5
	Should not exceed 41 psu in >5% of measurements: S25, lim = 41

(Personal compilation by the authors.)

(2) Simulation of the plume discharge: (i) use of behavioral simulation models of the discharge on the marine environment (CORMIX, COHERENS, VISUAL PLUMES, and VISJET MATLAB software are used to model the behavior of the brine), (ii) evaluation of the best design for the discharge system to achieve the highest possible dilution.

(3) Characterization of the receptor environment: physical and chemical environment.

(4) Characterization of the biotic receptor environment.

Although sea disposal of reject brine is a common practice for plants located in coastal areas, it would not be available for inland desalination plants. Deep well injection and surface water discharge, sewer disposal, reuse of concentrates, land applications, or concentration of brines in evaporation ponds is the conventional option for brine disposal from inland desalination plants [26]. Similar options of RO concentrates management have been used in the case of inland brackish groundwater desalination plants. Inland plants have to solve the problem of concentrate disposal without the possibility of their discharge to seawater, so the development of other management options is an urgent demand.

There are many emerging alternative technologies that can be combined to achieve a minimization and valorization of brine with an appropriate management such as solar evaporation, membrane distillation, electrodialysis, and forward osmosis. Pérez-González et al. [27] in a recent paper presented a review on the treatment technologies of water reverse osmosis concentrates aimed at reducing the environmental impact of the disposal of RO concentrates according to their origin, the level of development of the technologies, and the final objective of the treatment. However, in desalination plants already installed, minimization corrective measures, being considered at present, are unviable in most cases. This fact emphasizes the need of developing new effective and economically viable proposals for this residue management, not only for newly constructing plants, but also for those already installed.

10.3.2 Indirect Impact Due to Energy Use

When considering energy consumption of water desalination process, water input characteristics are essential since the operation pressure of RO systems is a function of feed water salinity. Seawater can be desalted and supplied in large quantity, and with a very high quality, but it requires a great amount of thermoelectric energy, which overwhelming majority is obtained from fossil fuels. Less energy is necessary for desalted brackish water. This is due to the lower salt concentration of the input water, which reduces the osmotic pressure to overcome and therefore energy consumption.

Environmental concerns associated with the energy demand and thus indirectly associated with the process of desalination are the emission of air pollutants, from electrical power generation, responsible for greenhouse gas emissions including carbon monoxide (CO), nitric oxide (NO), nitrogen dioxide (NO_2), and sulfur dioxide (SO_2) that further exacerbate global warming.

Modern SWRO plants have energy consumption rate in the range of 3–4 kWh/m^3, but there are some reports of SWRO plants that would consume 2 kWh/m^3 of desalted water [28]. In the case of BWRO desalination plants, the energy consumption is in the rate 0.5–2.5 kWh/m^3. Some of the most relevant airborne emissions produced by the analyzed RO desalination processes (with an energy consumption of 4 kWh/m^3) along all their life cycle are the following: 1.78 kg CO_2/m^3, 3.87 g NO_X/m^3, 10.68 g SO_X/m^3, and 1.10 g NMVOC/m^3 (Nonmethane volatile organic compounds) [29]. In the reference cited, it is shown airborne emissions associated to RO for different energy consumptions where there is a 47% of airborne emissions are reduced when the energy consumptions are reduced from 4 kWh/m^3 until 2 kWh/m^3.

Energy use problem has been significantly reduced through the implementation of advanced technologies that have provided solutions for reduction in energy consumption such as energy recovery system used in the process and through the use of renewable energy and improvements on the efficiency of the process.

10.3.3 Impact of Land Use and Visual Disturbance

Additional questions associated to land use are also considered as a negative impact, however, with a low emphasis.The location of desalination facilities must be governed by rigorous preliminary analysis. The following aspects are taken into account in the selection of alternatives for land infrastructures associated to desalination plants:

(a) Presence of Protected Areas, Priority Habitats, and Natura 2000 Network areas (Area LIC (Place of Community Interest) and area ZEC (Special Area of Conservation)).

(b) Possibility of using existing infrastructures.

(c) Selection of landfill and stockpile areas with a minimum impact on the environment.

Normally desalination plants are located close to the coastal areas. Constructions of buildings containing the characteristics of reverse osmosis desalination facilities are buildings that do not have an eminently industrial character. From the point of view of its visual and landscape impact, they should be easily integrated into their environment, as shown in Figs. 10.2–10.4. Their designs are usually adapted to their surroundings, with large areas to provide for possible expansions, etc.

10.3.4 Impact of Noise

Also to be highlighted is the noise pollution of a desalination plant, which is not mentioned because of its relative remoteness in villages and inhabited areas, which in principle should avoid public nuisances. However, it should be noted especially in small islands or areas with very limited building land, unfortunately is a very common situation in the Levant Spanish, Canary Islands, and Balearic Islands.

Acoustic contamination on seawater reverse osmosis desalination plant, principally, is important, due to the fact high-pressure pumps and energy recovery systems, such as turbines or similar, produce significant level of noise 90 dB (A) [11].Therefore, they should be located far away from populated areas and equipped with appropriate acoustic technology to reduce noise level.

10.4 CASE STUDIES

The large experience on brackish and seawater desalination plants in Spain and particular in Canary Islands is an important starting point to transfer knowledge about adverse environmental effects. For that reason, Spain has been chosen as a case study due to the availability of comprehensive site-specific data and the suitability of its geographical location, climate and low

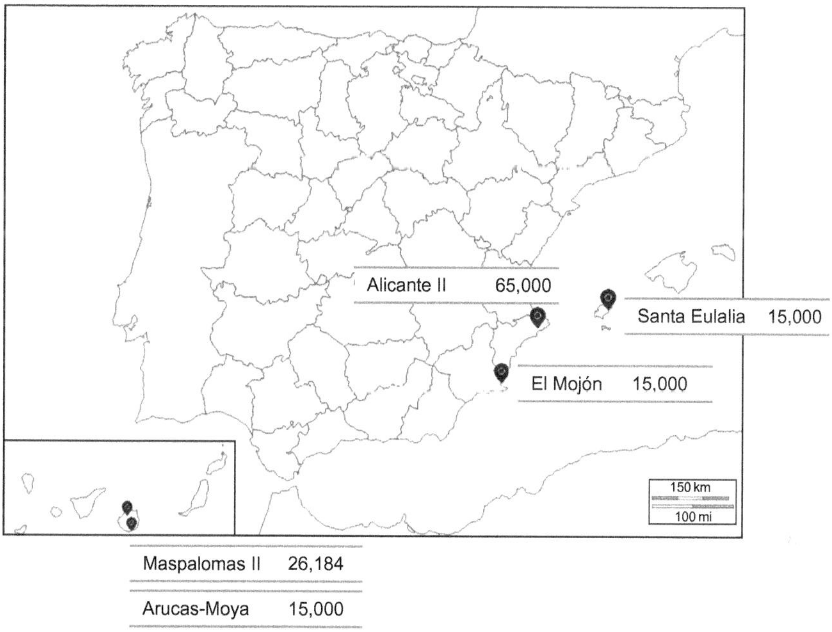

Fig. 10.1 Geographic situation of desalination plants under study. Plant capacity shown in m^3/day. *(Personal compilation by the authors.)*

rainfall, the constant demands of agriculture, and the growing population, based on the tourism to the employment of the desalination technologies.

In this context, we provide an overview of the key environmental concerns of some new and older seawater and brackish water desalination plants in different locations. Five cases are reviewed, providing a brief description of the environmental drawbacks. Each one of these desalination plants has its own environmental features and difficulties; consequently, a "case-by-case" study has been required. Fig. 10.1 shows the location map of case of studies on desalination plants.

The analysis of the environmental concerns of these particular cases is based through an analysis of Records of Decision (RODs) for desalinations plants published in the Official Spanish Magazine.

10.4.1 Case of Study 1: Alicante II SWRO Desalination Plant

The Alicante II seawater reverse osmosis (SWRO) desalination plant is located in the municipality of the city of Alicante (Spain) and more specifically in the vicinity of Salt Marsh *Agua Amarga* and the Alicante I desalination plant (Fig. 10.2). It was designed and constructed by GS Inima Environment for the Commonwealth of Municipalities *Los Canales de Taibilla.*

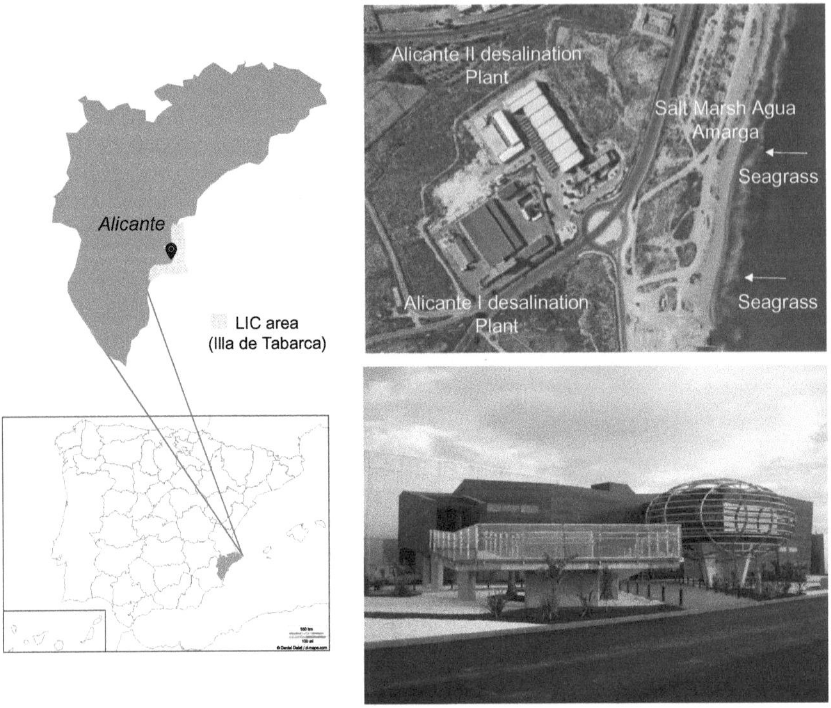

Fig. 10.2 Location of Alicante II SWRO desalination plant (Alicante, Spain) [30]. *(Personal compilation by the authors.)*

Initial operation of the plant began in 2008 with a total capacity of 65,000 m^3/day. The flow rates provided by the Alicante II desalination plant are consumed directly in the municipalities of *Alicante* and *San Vicente del Raspeig*, with a stable population of 350,000 inhabitants, which increases significantly during the summer. The desalination plant is located in a marine area designated as a protected zone area LIC called *Illa de Tabarca* due to the presence of seagrass as can be seen in Fig. 10.2. The most important general data and the technical specifications of Alicante II SWRO desalination plant are summarized in Table 10.2.

As specified in the RODs of Alicante II desalination plant [32], in the environmental impact study, three alternatives for seawater intake were analyzed: beach wells, direct seawater intake, and the alternative deep-water intake by tunnel. Regarding the first, a clear environmental effect on dune ecosystems is specified. Regarding the possible direct seawater intake, the marine coastal environment of *Alicante* has no favorable aptitude for it, for the following reasons: slightly steep bathymetry, poor water quality, intense

Table 10.2 General data and technical specifications of Alicante II SWRO desalination plant [31,32]

General data	
Commissioning	2008
Capacity	65,000 m^3/day
Desalination process	Reverse Osmosis
Raw water type	Seawater
Recovery rate	43%
Product quality (STD)	<400 ppm (domestic consumption)
Technical specifications	
Type of intake	11 horizontal wells + tunnel (104 pipe laterals)
Feed water pumps (N°/unit pot)	7 + 1
Sand/cartridge Filters	18/10
N° High pressure pumps	7 + 1
N° Process line	7
Nª R.O. racks	7
Pressure Vessel	128 × 7
N° of elements per PV	7
RO element type	Hybrid system: 6-SW30HRLE-400i, 1-SWXLE-400i
Energy recovery system	Pressure exchangers, 7 × 12PX-220
Post treatment	Remineralization with limestone beds, and hypochlorite sodium disinfection + CO_2
Product water pumping	2 + 1
Product water tank	3500 m^3
Installed Power	10 MVA
Energy consumption	3.8 kWh/m^3
Brine discharge	(Dilution) Submarine emissary

(Personal compilation by the authors.)

maritime traffic, and possible environmental effect to LIC area *Illa de Tabarca*. For that reason, the seawater flow necessary for Alicante II SWRO desalination plant is extracted by a combination of 11 individual horizontal wells and a deep-water intake by tunnel located in *Agua Amarga* beach. The 1-km-long tunnel was constructed underneath the coastal beach area 14 m below sea level. It is oriented parallel to the shoreline and has a diameter of about 3.5 m. The water intake into the tunnel is provided by 104 pipe laterals, constructed perpendicular to tunnel axis and extending outward.

The raw water is treated by 18 sand filters. The filter seawater then passes through cartridge filters. The pretreatment section involves dosing sulfuric acid, sodium hypochlorite, coagulant, antiscalant, sodium metabisulfite, and sodium hydroxide.

The RO system is based on seven independent and individual first pass train's configuration allows high plant modularity and flexibility to adapt to the water demand with a recovery of water of 43%. Each train is composed of 128 pressure vessels, one centrifuge pump, and one recirculation pump. To optimize and reduce the final energy consumption, each train has implemented 12 energy recovery devices (ERD). These devices allow a high recovery grade (98%) of the mechanical energy contained in the high-pressure discharge brine.

Adjustment of chemical characteristics of potable water is based on a remineralization with calcite filters.

The discharge of brine rejection is carried out in coast, near *La Cala de los Borrachos*, in the municipality of *Alicante*. Brine rejection is evacuated through a shared outfall with Alicante I desalination plant. The brine flow rate amounts to 79,444 m^3/day with a salinity of 69.2 psu. This volume discards management alternatives of the brine such as salt evaporation, since about 2500 ha of free surface water would be needed and as would deep injection into geological storage. Brine rejection into the sea is the only real alternative to brine disposal.

10.4.1.1 Analysis of Significant Environmental Impacts and Corrective Measure

In order to comply with the EIS, the Alicante II SWRO desalination plant has been subjected to implement the following correctives measures in order to avoid or reduce the environmental impact [32]:

(a) *Environmental effect on salt marsh Agua Amarga*. In the environmental impact study, a mathematical modeling with MODFLOW program was carried out to design seawater intake and to study the effects on the environment. To avoid possible environmental effect to nearby wetlands to seawater intake, particularly in the salt marsh *Agua Amarga*, it proceeds to the periodic flooding of certain streams of salt marsh, by opening channels connecting the sea to propitiate renewal of water through natural flood cycles, creating a sheet of water. Thus, maintaining hypersaline flora and fauna of this natural area is favored. Monitoring program ensures control of the evolution of the piezometric levels.

(b) *Impact on habitats and species listed in the Directive 92/43/CEE*. To guarantee salinity limits that guarantee noneffect to marine ecosystems, the project promoter incorporates a solution (2/1) 2 L of seawater/1 L of brine rejected in order to reduce the hypersaline plume extension. Due to dilution, the salinity of the discharge is dissipated down to 38.5 psu salinity concentration within 485 m of the discharge point. This value can be increased to a ratio of 6/1, to ensure no environmental effect to *Posidonia oceanica seagrass*.

(c) *High energy consumption*. Care to minimize the environmental impact is complemented by the use of renewable energy to electricity production through the photovoltaic plant. 1700 MWh is obtained, presenting a savings of 1.87% of the total energy consumption of the desalination plant.

10.4.1.2 Environmental Monitoring Plan

The monitoring plan requirements for Alicante II SWRO desalination plant are the following [32]:

(a) *Protection of Posidonia oceanica seagrass*. Continuous measurement of salinity and temperature is carried out through six sensors. Quarterly to the General Secretary refers a report on the recorded data (specifying maximum, minimum, and average) as well as basic information on the state of the sea near the disposal area. To prevent salinity exceeding the tolerance threshold of *Posidonia oceanica*, a progressive stop protocol of the desalination plant has been established. The protocol of progressive stop in the event of overshooting the limit of is:

(b) *Evolution of Posidonia oceanica seagrass*. A protocol was established to monitor the evolution of *Posidonia oceanica* affected by the discharge of brines, and the following factors should be taken into consideration, at least: structure of the meadow, environmental conditions, and animal community. It shall be at least an annual campaign, coinciding with the plant bloom and shall refer the report considering the most significant aspects to the General Secretariat.

- 38.3 psu in >25% of the observations or 39.5 psu limit at >5% of observations based on monthly data in any of measuring stations.
- 49 psu in >10% of the observations in the discharge point based on weekly data.
- 38.5 in >50% of the observations, based on weekly data in any of measuring stations.

(c) Dispersion control of plume discharge: in order to control salinity and the correct application of the models used to predict the evolution of the plume of the brine discharge, 12 sensors were used, 6 of which are identified in Section (a). These sensors measure the salinity and temperature continuously. Quarterly to the General Secretary refers a report on the recorded data (specifying maximum, minimum, and average) as well as basic information on the state of the sea near the disposal area.

10.4.2 Case of Study 2: Santa Eulalia SWRO Desalination Plant

Santa Eulalia SWRO Desalination Plant is located in northeast of the municipality of *Santa Eulalia* (Ibiza, Spain) (Fig. 10.3). The desalination plant is not located in any area LIC or natural space protected. Designed and constructed by Aqualia, initial operation began in 2011 with an initial capacity of 10,000 m^3/day. The expansion of the desalination plant was completed to its current production of 15,000 m^3/day. Process building and services deposits, etc. are extended and occupy approximately 5400 m^2. General data and technical specification of SWRO Santa Eulalia desalination plant are summarized in Table 10.3.

Unable to provide seawater through beach wells due to the low permeability of the terrain, the seawater intake is conducted by open water intake by submerged tower 18 m deep. The emissary consists of a first underground section executed in tunnel to prevent aggression on *Posidonia oceanica* seagrass until it reaches an existing sandy bed, where a second anchored section begins.

In the pretreatment section, the sulfuric acid dosing (pH adjustment), sodium hypochlorite dosing (chlorination to prevent preventing biological growth), ferric chloride, antiscalant, and Sodium metabisulfite dosing are necessary in order to neutralize any remaining chlorine in the feed water and then the seawater is fed to the RO system. Chemical doses use is shown in Table 10.4. Physical pretreatment is provided by a filtration system con-

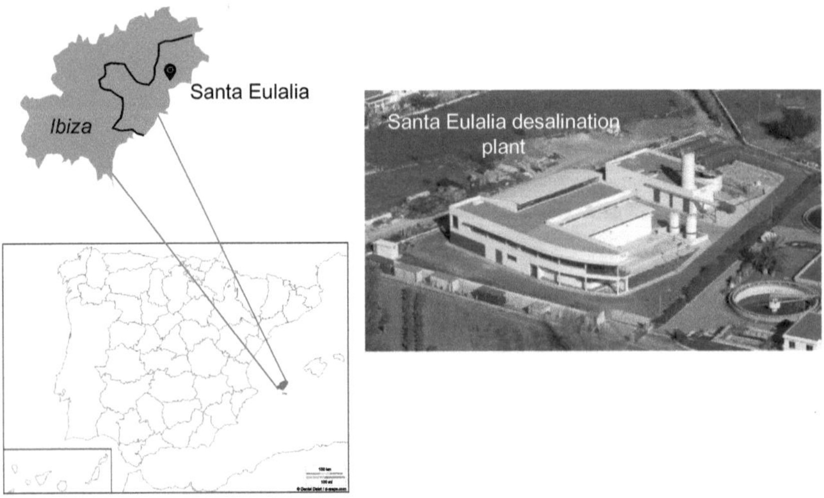

Fig. 10.3 Location of Santa Eulalia SWRO desalination plant (Santa Eulalia, Ibiza, Spain) [33]. *(Personal compilation by the authors.)*

Table 10.3 General data and technical specifications of Santa Eulalia SWRO desalination plant [33,34]

General data

Commissioning	2011
Capacity	15,000 m^3/day
Desalination process	Reverse Osmosis
Raw water type	Seawater
Recovery rate	45%
Product quality (STD)	<400 ppm (domestic consumption)
Technical specifications	
Type of intake	Open intake (3 + 1)
Feed water pumps (N°/unit pot)	3 + 1
Sand/cartridge Filters	6/5
N° High pressure pumps	3 + 1
N° Process line	3
Nª R.O. racks	3
Pressure Vessel	66 × 3
N° of elements per PV	7
RO element type	Polyamide
Energy recovery system	3 Pelton Turbines
Post treatment	Remineralization with calcium hydroxide + CO_2 and hypochlorite sodium disinfection
Product water pumping	3 + 1
Brine discharge	(Dilution) Submarine emissary

(Personal compilation by the authors.)

Table 10.4 Chemical doses used in Santa Eulalia SWRO desalination plant [33]

Chemical doses	ppm
NaClO	2.5
H_2SO_4	25
$NaHSO_3$	20
Antiscalant	5
$FeCl_3$	5

(Personal compilation by the authors.)

sisting of six horizontal filters (anthracite, silica sand, and garnet) and 5 units cartridge filters (150 cartridges per filter).

The RO unit section consisted of three racks (3 × 5000 m^3/day, 45% recovery), which work in a 1 step—1 stage system. To optimize and reduce the final energy consumption, each train has implemented one Pelton Turbine (efficiency 81.5%). These devices allow an energy recovery of 382.43 kW.

For brine discharge, three alternatives were studied in the environmental impact study as it is specified in the RODs of Santa Eulalia SWRO desalination plant [34]. In addition, three different depths were study in the EsIA: 10, 5, and 3 m. These alternatives are the following:

Discharge by outfall with diffusers and after dilution.

Discharge by outfall with diffusers and flow rate.

Discharge by outfall with one mouth and flow rate.Santa Eulalia brine discharge is carried out by one underwater outfall that is 320 m away from the shore and 500 mm diameter with a final section of five diffusers 150 mm diameter, equipped with risers at an angle of 60 degress and at a height above the bottom of 1.5 m. The injection point is 90 m from the *Posidonia oceanica* seagrass and at a depth of −5 or −6 m. Brine concentration in the point of discharge is 68.8 g/L. Calculations of dilution and dispersion the discharge give an interval of 38.0–38.5 g/L in the *Posidonia oceanica* seagrass.

The other possibilities referred here presented a model dilution and dispersion of highest brine concentration values.

10.4.2.1 Analysis of Significant Environmental Impacts and Corrective Measure

In order to comply with the EIS, the Santa Eulalia SWRO desalination plant has been subjected to the following measures [34]:

(a) *Impact on habitats and species listed in the Directive 92/43/CEE.* The project was modified with the appropriate technical requirements for the capacity to dilute and disperse the brine discharge with a dilution factor 2:1. Calculations of brine dilution and dispersion result in very few cases where the concentration of salts in the area of *Posidonia oceanica* seagrass exceeds 38.5 g/L.

(b) *Impact to the land use.* To protect birdlife power lines must be underground. The areas where it has proceeded to removal of vegetation cover will be replanted and 80% the first year, proceeding to the replanting during the first 3 years to ensure complete coverage of the affected area. Autochthonous species it will be used. The promoter is responsible for the maintenance and conservation of plantations during this period.

(c) *Landscape protection.* The existing facilities must be integrated scenically to minimize the visual impact.

(d) *Population protection.* The facilities will have acoustic insulation required to meet specified noise in the national law and the relevant regulations of the Autonomous Community. For this purpose, emission levels and noise emission are controlled.

10.4.2.2 Environmental Monitoring Plan

The monitoring plan requirements for Santa Eulalia SWRO desalination plant are the following [34]:

(a) *Salinity control.* To prevent salinity exceeding the tolerance threshold of *Posidonia oceanica*, a stop protocol of the desalination plant is established. The facility will include a conductivity meter for measuring and recording continuously, which will determine the salinity of the resulting mixture between brine and dilution water before discharge.

For the protocol of progressive stop in the event of overshooting the limit of 38.3 psu in >25% of the observations or 39.5 psu limit at >5% of observations based on monthly data in any of measuring stations, one can proceed to implement the protocol progressive plant stop.

(b) *Evolution of Posidonia oceanica seagrass and benthic organisms.* Three stations of sampling are established in order to ensure that in immediate environment to the limits of the *Posidonia oceanica* seagrass, the initial conditions are maintained. Sediments and benthic organisms are monitored by sampling superficial sediment. In addition, it carried out measurements of salinity, temperature, flow rate, concentration of suspended solids, turbidity, concentration of nutrients (nitrite, nitrate, orthophosphate, and ammonium), pH.

(c) *Land, natural vegetation, and wildlife protection.* Revegetation operations will be monitored.

(d) *Protection of the hydrological system.* Groundwater pollution and surface water is controlled. The structure of the discharge is monitored to prevent or fix possible cracks or pipe mismatches. For chemicals, a holding tank safety will be constructed, waterproofed properly depending on the chemical, which in the case of accidental breakage, prevents spillage. The tightness of tanks and pipes will be reviewed to prevent diffuse pollution of aquifers.

(e) *Protection of the marine environment.* The quality of receiving marine waters is monitored. A meter with continuous data transmission is installed on the discharge points, to control the degree of dilution of the effluent output. In addition, four sampling points are established (two on either side of the discharge driving and one at startup thereof) and another at the output of effluent. The sampling frequency is four times a year and the following steps are performed: thermohaline structure, dissolved oxygen concentration, water transparency, concentration of suspended solids, turbidity and nutrient concentrations (nitrite, nitrate, orthophosphate, and ammonium), and pH. It also should include an annual inspection of the entire length of the stretch submerged driving the discharge and its main elements by using scuba divers or submersible instrumental.

(f) *Heritage protection.* Actions on natural heritage and historical and artistic heritage will be monitored.

(g) *Implementation of an emergency system.* The implementation of an emergency plan will prevent, in the case of accidental leaks or spills of brine or chemical reagents, those that exceed the physical limits of the treatment plant, preventing reach nearby surface waters and channeling. An alarm system coordinated with the sampling network is established. In case of an accidental spill of chemicals produced or brines, it proceeds to washing and restoration of contaminated soil. The emergency plan includes how to proceed in such cases.

10.4.3 Case of Study 3: El Mojón BWRO Desalination Plant

El Mojón BWRO desalination plant and its auxiliary facilities are located in *Campo de Cartagena* in the province of *Murcia (Spain)*, in the municipalities of *San Pedro del Pinatar, San Javier, Torre Pacheco, Los Alcázares*, and *Cartagena*. As shown in Fig. 10.4, there are also two desalination plants nearby, San Pedro I SWRO desalination plant and San Pedro II SWRO desalination plant.

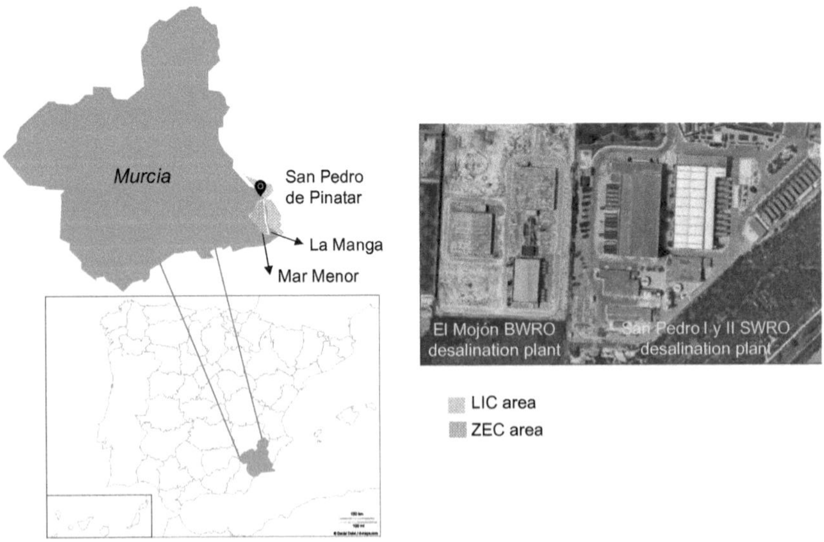

Fig. 10.4 Location of *El Mojón* BWRO desalination plant (Murcia, Spain). *(Personal compilation by the authors.)*

The plants were constructed by Ministry of Public Works and Urbanism, in order to facilitate the reuse of drainage waters from *Campo de Cartagena* providing new resources for later use in irrigation and recovering waters, which due to its excessive salinity, are not suitable for irrigation. Initial operation began in 1996 with an initial capacity of 6000 m^3/day.

The expansion of the desalination plant was completed to its current production of 16,000 m^3/day. General data and technical specification of El Mojón BWRO desalination plant are summarized in Table 10.5. The expansion project promoter is the State Society Waters of the Mediterranean Basin, S.A. (ACUAMED) and the Hydrographic Confederation of Segura.

The marine environment in the area of the *Mar Menor* (see Fig. 10.4) is at space as Natura 2000 network, including the following LIC areas (*Mar Menor*, open areas and *Mar Menor* islands, salt works and sandbanks of *San Pedro del Pinatar*) and ZEC area (*Mar Menor* open areas and Mar Menor islands). The Mediterranean coastline is protected by the following areas LIC (*Franja Litoral Sumergida de la Región de Murcia* y *Cabo Roig*).

The plant design is similar to other studied cases, with chemical and physical pretreatment, and post treatment of product water. 10% of raw water is reserved for mixing with the product water, which is driven to irrigation canal, in order to reach 1000 μS/cm salinity.

After neutralization, brine is discharged by outfall. This outfall part of desalination plant is located on the adjacent plot desalination facilities of *El Mojón* and flows underground until it reaches the sea. From this, it consists

Table 10.5 General data and technical specifications of El Mojón BWRO desalination plant [35]

General data	
Commissioning	1996
Capacity	16,000 m^3/day
Desalination process	Reverse Osmosis
Raw water type	Brackish water
Recovery rate	60%
Product quality	1000 μS/cm (Agriculture)
Technical specifications	
Type of intake	Drainage water (5000–8000 μS/cm)
N° Process line	3
Nª R.O. racks	2 × 5000 m^3/day and 1 × 6000 m^3/day
RO element type	Polyamide
Brine discharge	Underwater outfall

(Personal compilation by the authors.)

of an underwater outfall of about 5 km in length and 32.5 m depth. The discharge contains both brine from El Mojón BWRO desalination plant (10,080 m^3/day with 7.3–18 psu salinity) and brine from *San Pedro I* and *San Pedro II* SWRO desalination plant. The mixture of both discharge improves the quality and increases the degree of dilution of the final discharge.

10.4.3.1 Analysis of Significant Environmental Impacts and Corrective Measure

In order to comply with the EIS, the El Mojón SWRO desalination plant has been subjected to the following measures [35]:

(a) *Protected areas.* In order to guarantee a minimal impact of works on birds and the present terrestrial wildlife in the construction area, performing activities of potential environmental effect should be avoided, such as brush clearing and noisy operations, between the months of March and May (breeding period). For species of fauna interest, the EsIA establishes a monitoring and conducting a census. After completion, a permanent fence around desalination plant will be installed, in order to avoid accidental falls of species or entry into the facilities.

(b) *Marine Environment.* Given the importance of the marine environment in the study area, including mostly in the Natura 2000 network, the EsIA presents a detailed analysis of the characteristics of the final effluent resulting from the mixture, for different operating situations analyses, brine from El Mojón BWRO desalination plant and brine from San Pedro I and San Pedro II SWRO desalination plants. Table 10.6 shows the composition for the most critical parameters.

To assess the potential environmental effects, a simulation of the behavior of the discharge with CORMIX program has been carried out.

Table 10.6 Composition for the most critical parameters of brine discharge *San Pedro de Pinatar* outfall [35]

Operating combination	Nitrogen (mg/L)	Phosphorus (mg/L)	Salinity (g/L)
BWRO	93.4	2.4	15
BWRO + 1 SWRO	19.9	0.9	63.5
BWRO + 2 SWRO	15.3	0.8	66.5
Limit value according to the Directorate General of Quality and Environmental Assessment of the Ministry of Agriculture Water and Environment	15.0	2.0	

(Personal compilation by the authors.)

Considering the whole discharge through the emissary of *San Pedro del Pinatar*, the brine rejection from El Mojón BWRO desalination plant with a salinity (7.3–12.3 psu) much lower than that of seawater (37.5 psu) and a high pollutant load (especially, nutrients and pesticides) would mix with the brine rejected from San Pedro I y II SWRO desalination plants, hypersaline respect to seawater but without contaminants.

The combination of these two effluents into the emissary would have a positive effect on the final discharge, which a lower effluent salinity that rejected brine from SWRO desalination plants and lower pollutant load that the reject brine from BWRO desalination plant would ensure.

(c) *Heritage elements.* The study area is characterized by the presence of numerous deposits and heritage elements. The EsIA has proposed the presence of an archeologist on site in order to prevent the potential environmental effect on not-inventoried archeological sites that may arise during the execution of the works.

(d) *Landscape.* As a corrective measure, the impact on the landscape and in order to contribute to the conservation of the elements of the Natura 2000 network (LIC, ZEPA (bird protection special zone) and species protected under Directive 92/43 y 79/409), the landscape restoration of the affected areas is proposed.

10.4.3.2 Environmental Monitoring Plan

In order to comply with the EIS, the El Mojón BWRO desalination plant has been subjected to the following measures [35]:

(a) *Pollutant load control of the brine discharge.* Quarterly monitoring of the pollution load of effluent from El Mojón BWRO desalination plant is done, with the following measurement parameters: pH, temperature, conductivity, dissolved oxygen, turbidity, nitrates, phosphates, and pesticides. Where these limits are exceeded, harmful effects motivated by the composition of the discharge detected or not reached to the required dispersion, the owner of the plant must correct or purify the effluent prior to discharge. Every time this situation occurs, the holder shall draft and submit a specific report to the General Secretariat stating the reasons therefore and the measures taken to correct them.

(b) *Underwater emissary inspection.* An annual evaluation of the brine underwater emissary by using scuba divers or submersible instrumental is required.

(c) *Protection of Posidonia oceanica seagrass.* During the construction phase, the water quality and benthic populations in the marine environment will be monitored. It is also proposed to establish monitoring stations for measuring parameters of interest during the operational phase of the work, as well as quality control of *Posidonia oceanica* likely to affect seagrass. No additional environmental condition for the protection of seagrasses is established, regarding those established in the corresponding RODs of the underwater outfall San Pedro del Pinatar. In any case, the conditions established by the scientific community must be met, referred to in Table 10.1.

(d) *Remission of periodical reports.* Annual to the General Secretary refers a report containing the results of the structural review of the outfall of San Pedro del Pinatar, and data flow and pollutant load f brine rejection from El Mojón BWRO desalination plant.

10.4.4 Case of Study 4: Maspalomas II SWRO Desalination Plant

This desalination plant is located in the south of the island, on the left side of Barranco del Toro. The desalination facility is located in an area LIC and ZEC as shown in Fig. 10.5. Furthermore, this locality is an area of special conservation interest due to the presence of the most extensive and ecologically relevant significant area of *Cymodocea nodosa* (Ucria) *Aschers* seagrass meadow on the island, as shown in Fig. 10.5.

Designed and constructed by Pridesa, initial operation of the plant began in 1988 with an initial capacity of 14,000 m^3/day in order to supply potable water to the region. The continuous expansion of the Maspalomas II desalination plant was completed in 2001, to its current production of 26,184 m^3/day.

The raw seawater is delivered via an offshore, open, submerged intake. The seawater is filtered through two sets of vertical media filters containing anthracite and sand. The filtered seawater then passes through two sets of cartridge filters sized at 10 and 5 μm. The conventional SWRO plant at Maspalomas II consisted of five trains. The seawater intake capacity is 42,000 m^3/day. Maspalomas II desalination facility has been a pioneering worldwide installation as it was the first plant in the world to set up a brine concentrator as a second stage that allowed raising a 60% water recovery. Technical specifications are shown in Table 10.7.

As illustrated in Fig. 10.5, Maspalomas II is an example of desalination facility, which discharges the brine in the near-shore tidal zone. The brine discharge of this desalination plant is carried out by two underwater concentrate outfalls, which extend around 300 m away from the shore and have 300- and 600-mm diameter, respectively [38,39]. The discharge depth is 7–8 m. The outlet of the discharge outfalls does not have diffusers and the mixing between the reject brine and seawater is mainly driven by the

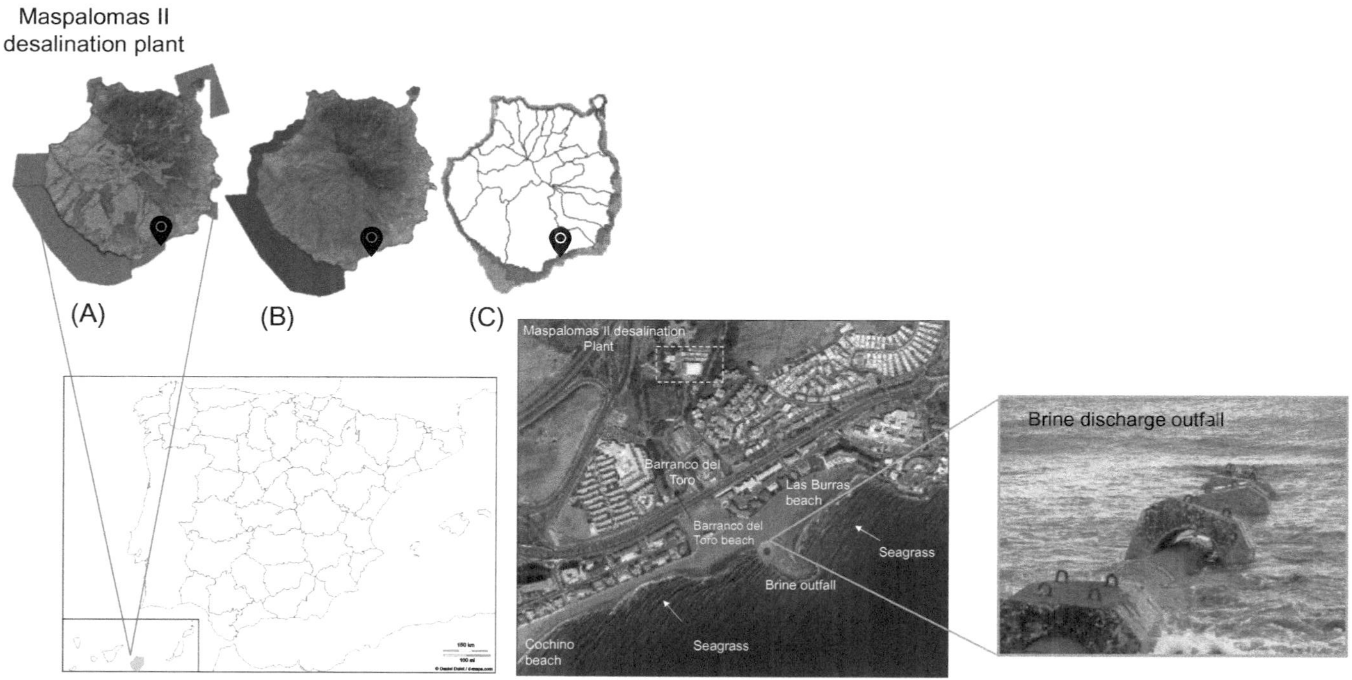

Fig. 10.5 Location of Maspalomas II SWRO desalination plant (Canary Islands, Spain) [36,37]. (A) Area LIC (Place of Community Interest) and area ZEC (Special Area of Conservation), special conservation zones of the UE Nature 2000 network, (B) Biosphere Reserve Area, and (C) Distribution of *Cymodocea nodosa* (Ucria) *Aschers* around de island. *(Personal compilation by the authors.)*

Table 10.7 General data and technical specifications of Maspalomas II SWRO desalination plant

General data	
Commissioning	1988
Capacity	26,184 m^3/day
Desalination process	Reverse Osmosis
Raw water type	Seawater
Recovery rate	60%
Product quality (STD)	<350 ppm
Technical specifications	
Type of intake	Open intake
N° Process line	5
Nª R.O. racks	5
Energy consumption	3.77 kWh/m^3
Brine discharge	Underwater Outfall

(Personal compilation by the authors.)

velocity of the discharge and the fact that the discharge is located in an area with naturally occurring underwater currents of high intensity.

The Maspalomas II brine discharge conditions are challenging for two reasons: (1) very high salinity of the concentrate (TDS ~ 90,000 mg/L), due to the presence of a second stage at 90 bar pressure, raising the recovery to 60%; and (2) seagrass beds habitat for fish and other marine organisms. Due to the naturally occurring near-shore mixing, the salinity of the discharge is dissipated down to 38 ppt (parts per thousand) or 38 PSU (practical salinity units) of salinity concentration within 20 m from the discharge point [39].

Based on environmental studies of the brine discharge area, the area initial dilution of the Maspalomas II desalination plant is a sandy bed with practically no flora [39] [40]. In the first studies of Talavera, the *Cymodocea nodosa* are not significantly affected by the desalination plant discharge [39]. However, recently, it is seen that *Cymodocea nodosa* were not present in most of the area affected by the hypersaline plume associated with the brine effluent of Maspalomas II [40]. Results obtained in this study suggest the need to implement effective dilution systems to minimize the environmental impact of brine effluents from desalination plants and their adverse effects on vulnerable marine habitats [39].

10.4.5 Case of Study 5: Arucas-Moya SWRO Desalination Plant

The Arucas-Moya SWRO desalination plant is located in the Northern coast of the island of Gran Canaria by the shoreline, in a spot called *Punta de Camello*, in Arucas. The desalination plant is not located in any area LIC, ZEC, biosphere reserve, or natural space protected as shown in Fig. 10.6.

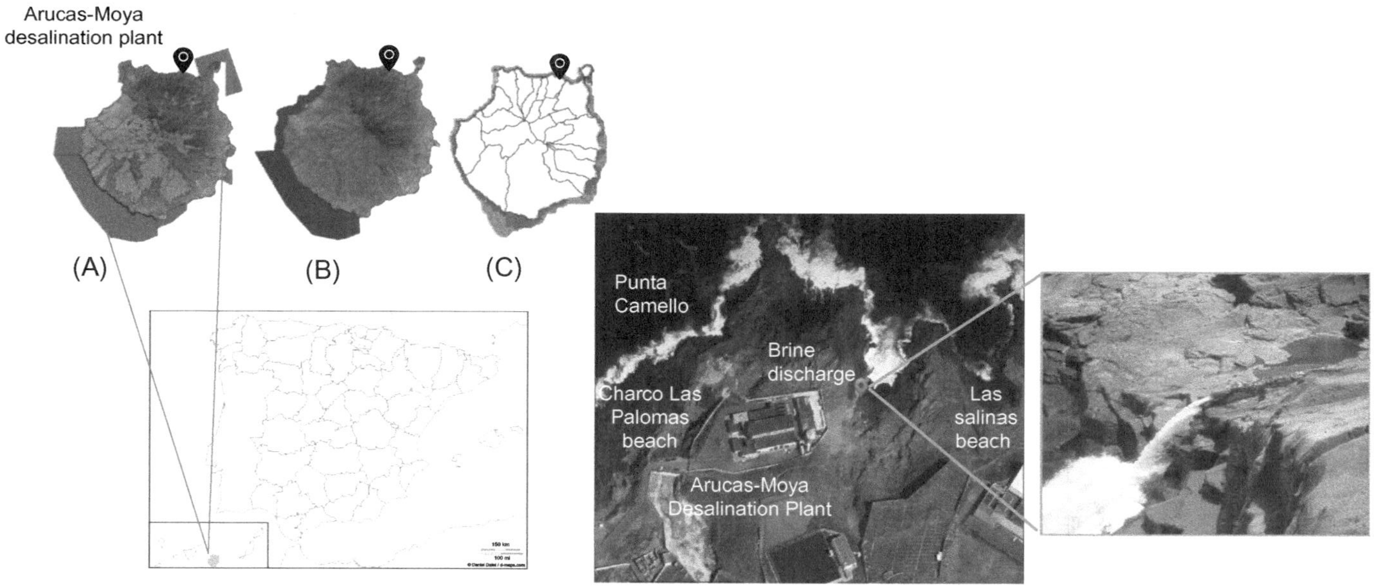

Fig. 10.6 Location of Arucas-Moya SWRO desalination plant (Canary Islands, Spain) [36,37]. (A) Area LIC (Place of Community Interest) and area ZEC (Special Area of Conservation), special conservation zones of the UE Nature 2000 network, (B) Biosphere Reserve Area, and (c) Distribution of Cymodocea nodosa (Ucria) Aschers around de island. *(Personal compilation by the authors.)*

Designed and constructed by Ionics, initial operation of the plant began in 1995 with a capacity of 4000 m^3/day in order to supply potable water to the region. The continuous expansion of the Arucas-Moya desalination plant was completed in 2009, tripling production capacity, actually with a production capacity of 15,000 m^3/day.

Process buildings and services, deposits, etc. are extended occupying approximately 5600 m^2 of which 2200 m^2 contain industrial building. The altitudinal elevation is about 15 m from the sea level. Seawater catchment area occupying it extends approximately 240 m^2 (located in the south of the main building and at a lower level, about 7 m from the sea level).

The RO unit section consisted of two racks (2×7500 m^3/day, 45% recovery), which work in a 1 step—1 stage system, having a total of seven elements per pressure vessel. Technical specifications are shown in Table 10.8. The schematic flow diagram of the desalination plant is shown in Fig. 10.7.

Table 10.8 General data and technical specifications of Arucas-Moya SWRO desalination plant

General data	
Commissioning	1995
Capacity	15,000 m^3/day
Desalination process	Reverse Osmosis
Raw water type	Sea water
Recovery rate	45%
Product quality (STD)	<400 ppm
Technical specifications	
Type of intake	6 coastal wells
Feed water pumps (N°/unit pot)	6+1
Sand/cartridge Filters	4/4
N° High-pressure pumps	2+1
N° Process line	2×7500 m^3/day
Nª R.O. racks	2
Pressure Vessel	83×2
N° of elements per PV	7
RO element type	FILMTEC SW30HR-380
Energy recovery system	Pressure exchangers, 2×10 PX-220
Post treatment	Remineralization with calcium hydroxide, hypochlorite sodium disinfection + CO_2
Energy consumption	3.37 kWh/m^3
Brine discharge	Underwater Outfall

(Personal compilation by the authors.)

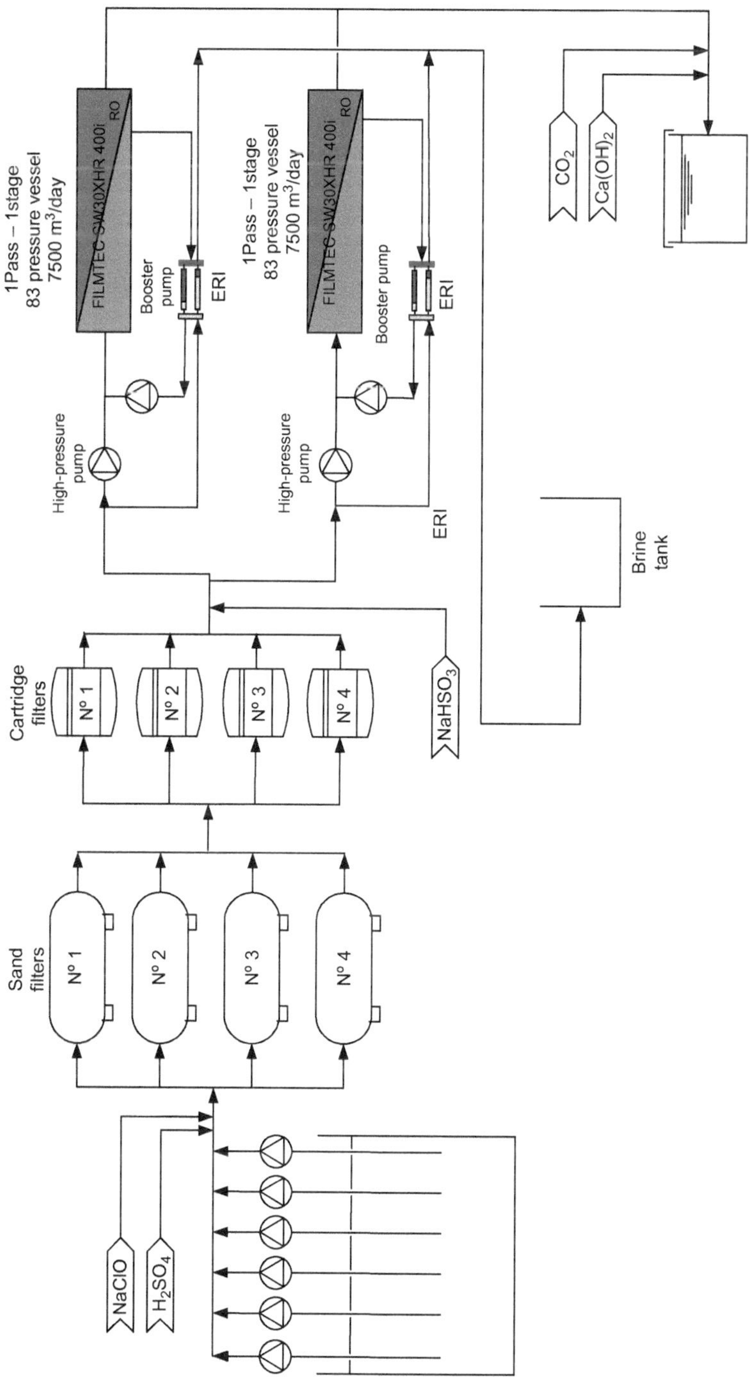

Fig. 10.7 Schematic flow diagram of Arucas-Moya SWRO desalination plant. *(Personal compilation by the authors.)*

The seawater intake is conducted by means of six coastal wells drilled in solid rocks.

In the pretreatment section, it is necessary for the sulfuric acid dosing (pH adjustment), sodium hypochlorite dosing (chlorination to prevent preventing biological growth), and sodium metabisulfite dosing in order to neutralize any remaining chlorine in the feed water and then the seawater is fed to the RO system. In the physical pretreatment, a filtration system consisting of four sand filters and four cartridge filters are provided.

The membranes (1162 units) used in the process were spiral wound RO membranes made of polyamide (FILMTEC SW30HR-380) operating at 68 bar. Usually about 5%–8% of the membranes are annually replaced in order to maintain the targeted product quality as well as the permeate quantity. The membranes are chemically cleaned every 3 months. The replacements of RO elements are required since otherwise the permeate quality would be lower and the operational pressure would increase.

As illustrated in Fig. 10.6, the Arucas-Moya brine discharge system, located on the shoreline, is a free jet discharge via weir (area surface discharges directly at the coastline) with very low mixing capacities.

10.5 CONCLUSIONS

Water desalination presents an effective solution to face the demand for drinking water, preserving existing natural water resources, and contributing to develop irrigated agricultural areas, industries, and tourism infrastructure, where traditional water resources have been scarce. Although water desalination is mature technology, its environmental impact has raised concern and debate.

From the point of view of prevention of potential environmental effects of desalination infrastructure, the qualitative leap in the prevention and treatment of environmental impacts in Spain was the entry into force, at the state level, the Royal Decree 1302/86. Since that time, it has made great progress in the last 30 years in terms of environmental integration of desalination plants but still has a long way to go.

These years of development and growth have resulted in existence of a critical mass of professionals with knowledge and experience to be able to contribute ideas and positive and realistic proposals for the future.

The EIA is presented as a methodological tool of establishing the real impacts of a desalination plant in order to achieve more friendly projects with the environment.

In the different case studies around Spain (Alicante, Balearic Islands, Murcia, and Canary Islands), most of them are located by coastline in areas designated for tourist population. External electric energy is supplied by thermal plants to produce electricity and normally is required to build some electric infrastructure of large capacity. In the case of Alicante II SWRO, the energy consumption is complemented by the use of renewable energy to electricity production through a photovoltaic plant of 1700 MWh present ing a savings of 1.87% of the total energy consumption of the desalination plant.

In all cases, brine discharge on the marine environment represents the greatest environmental drawback and an economic challenge for most desalination plants studies. Current brine management options have failed to solve this environmental problem. For that reason, the RODs establishes a specific monitoring needed to evaluate the long-term impact of discharges on the marine environment or requires that the promoter arrange a plan before construction begins.

Briefly, it can be itemized that desalination and environment are not challengers but complementary to each other. Environmental effects such as global warming by the increased use of energy, adverse effects on the marine environment, noise pollution, land use, among others are to be minimized with the respective preventive and corrective measures set by the EIS to avoid or reduce the impact during execution of works (preoperational phase, construction phase, and operation phase). In addition, the environmental monitoring plan plays an important role as it implies a control and monitoring of the environmental impact and effectiveness of corrective measures set out in the EIS and the specific conditions established in the RODs.

REFERENCES

[1] ICEX. Instituto Español de Comercio Exterior. Available from: http://www.icex.es; 2010.

[2] Sadhwani Alonso JJ, Álvarez Álvarez L, Melián-Martel N, Sadhwani Díaz J. The Canary Islands and its passion for water desalination. Desalin Water Treat 2015;55:2340–50.

[3] European Union. Directive 85/337/EEC of the Council of June 27th 1985, on the assessment of the effects of certain public and private projects on the environment. Off J Eur Union 1985;L175:40–8.

[4] European Union. Directive 97/11/EC of the Council of March 3rd 1997, amending Directive 85/337/EEC of June 27th 1985, on the assessment of the effects of certain public and private projects on the environment. Off J Eur Union 1997;L73:5.

[5] European Union. Directive 2014/52/EU of the European Parliament and of the Council of 16 April 2014 on the assessment of the effects of certain public and private projects on the environment. Off J Eur Union 2014;L124:1–18.

[6] European Union. Directive 2001/42/EC of the European Parliament and of the Council of June 27th 2011, on the assessment of the effects of certain plans and programmes on the environment. Off J Eur Union 2001;L197:30–7.

[7] Spanish Government. Royal Decree 1302/1986, of the 28th of June, on Environmental Impact Assessment. Bol Oficial Estado 1986;155:23733–6.

[8] Spanish Government. Royal Decree 1/2008, on Environmental Impact Assessment—hereinafter, EIA—approving the revised text of the Law on the Environmental Impact Assessment of projects. Bol Oficial Estado 2008;23:1–28.

[9] Spanish Government. Ley 6/2010 of 24th March, Amending the Consolidated Law on Environmental Impact Assessment. Bol Oficial Estado 2010;73:28590–7.

[10] Spanish Government. Law 21/2013 of 9 December on Environmental Assessment. Bol Oficial Estado 2013;296:98151–227.

[11] Sandhwani Alonso JJ, Veza JM, Santana C. Cases studies on environmental impact of seawater desalination. Desalination 2005;185:1–8.

[12] Fuentes-Bargues JL. Analysis of the process of environmental impact assessment for seawater. Desalination 2014;347:166–74.

[13] Procaccini G, Gambi MC, Pérez MC, Pergent G, Pergent-Martini C, Romero J. Seagrasses of the western Mediterranean. In: Green EP, Short FT, editors. World atlas of seagrasses. Los Angeles, CA: University of California Press; 2003. p. 48–52.

[14] European Union. Council Directive 92/43/CEE of 21th May on the conservation of natural habitats and of wild fauna and flora. Off J Eur Union L26 1992;7–50.

[15] Spanish Government. Law 14/2014 of 26th December, on Harmonization and Simplification concerning Protection of Land and Natural Resources. Bol Oficial de Canarias 2015;2:116–240.

[16] Lattemann S, Höper T. Environmental impact and impact assessment of seawater. Desalination 2008;220:1–15.

[17] Tsiourties NX. Desalination and the environment. Desalination 2001;141:223–36.

[18] Einav R, Hamssi K, Periy D. The footprint of the desalination processes on the environment. Desalination 2002;152:141–54.

[19] WHO. Desalination for safe water supply, guidance for the health and environmental aspects applicable to desalination. Cairo, Egypt: World Health Organization (WHO), Eastern Mediterranean Regional Office (EMRO); 2010.

[20] Höpner T, Lattemann S. Chemical impacts from seawater desalination plants—a case study of the northern Red Sea. Desalination 2002;152:133–40.

[21] Roberts DA, Johnston EL, Knott NA. Impacts of desalination plant discharges on the marine environment: a critical review of published studies. Water Res 2010;44:5117–28.

[22] Morton AJ, Callister IK, Wade NM. Environmental impacts of seawater distillation and reverse osmosis processes. Desalination 1996;108:1–10.

[23] Palomar P, Losada IJ. Desalination in Spain: recent developments and recommendations. Desalination 2010;255:97–106.

[24] Sánchez-Lizaso JL, Romero J, Ruiz J, Gacia E, Buceta JL, Invers O, et al. Salinity tolerance of the Mediterranean seagrass Posidonia oceanica: recommendations to minimize the impact of brine discharges from desalination plants. Desalination 2008;221:602–7.

[25] Fernández Torquemada Y, Sánchez Lisazo JL. Effects of salinity on growth and survival of Cymodocea nodosa (Ucria) Aschernon and Zostera noltii Hornemann. Biol Mar Mediterr 2006;13(4):46–7.

[26] Gabelich CJ, Xu P, Cohen Y. Chapter 10 concentrate treatment for inland desalting. Sustain Sci Eng 2010;2:295–326.

[27] Pérez-González A, Urtiaga AM, Ibáñez R, Ortiz I. State of the art and review on the treatment technologies of water reverse osmosis concentrates. Water Res 2012;46:267–83.

[28] MacHarg JP. The evolution of SWRO energy recovery systems. Desalination Water Reuse Q 2001;11(3):49–53.

[29] Raluy G, Serra L, Uche J. Life cycle assessment of MSF, MED and RO desalination technologies. Energy 2006;31:2361–72.
[30] G.S. Inima Environment S.A. Available from: http://www.inima.es/.
[31] Government of Spain. Ministry of Agriculture, Food and Environment. Mancomunidad de los Canales del Taibilla. Available from: https://www.mct.es/alicante-ii.
[32] Spanish Government. Resolución de 7 de junio de 2005, de la Secretaría General para la Prevención de la Contaminación y el Cambio Climático, por la que se formula declaración de impacto ambiental sobre la evaluación del proyecto de «nueva desaladora del Canal de Alicante», promovido por la Mancomunidad de los Canales de Taibilla. Bol Oficial Estado 2005;155:23426–32.
[33] Aqualia Infraestructuras, SWRO Santa Eulalia (Ibiza). Available from: http://www.aqualia-infraestructuras.com/.
[34] Spanish Government. Resolución de 23 de junio de 2005, de la Secretaría General para la Prevención de la Contaminación y el Cambio Climático, por la que se formula declaración de impacto ambiental sobre la evaluación del proyecto «Construcción de la instalación desaladora de agua marina de Santa Eulalia (Ibiza)», promovido por la Dirección General del Agua. Bol Oficial Estado 2005;166:25011–6.
[35] Spanish Government. Resolución de 8 de mayo de 2006, de la Secretaría General para la Prevención de la Contaminación y el Cambio Climático, por la que se formula declaración de impacto ambiental sobre la evaluación del proyecto de «Ampliación de la Estación desaladora de Aguas Salobres de El Mojón y sus colectores (Murcia y Alicante)», promovido por Aguas de las Cuencas Mediterráneas. Bol Oficial Estado 2006;141:22777–81.
[36] Database of biodiversity of the Canary Islands. Available from: http://www.biodiversidadcanarias.es/atlantis/common/index.jsf.
[37] Cartografias Canarias S.A. Available from: http://www.grafcan.es/.
[38] Census discharges into the sea of the Canary Islands. Cartografias Canarias S.A. Available from: http://www.grafcan.es/.
[39] Pérez Talavera JL, Quesada Ruiz JJ. Identification of the mixing processes in brine discharges carried out in Barranco del Toro Beach, south of Gran Canaria carried out in Barranco del Toro Beach, south of Gran Canaria (Canary Islands). Desalination 2001;139:277–86.
[40] Portillo E, Ruiz de la Rosa M, Louzara G, Quesada J, Ruiz JM, Mendoza H. Dispersion of desalination plant brine discharge under varied hydrodynamic conditions in the south of Gran Canaria. Desalin Water Treat 2014;52:164–77.

CHAPTER 11

Impacts of Seawater Desalination on Coastal Environments

Karen L. Petersen*, Hila Frank[†,‡], Adina Paytan*, Edo Bar-Zeev[†]

*University of California, Santa Cruz, CA, United States
[†]Ben-Gurion University of the Negev, The Jacob Blaustein Institutes for Desert Research, Zuckerberg Institute for Water Research, Ben-Gurion, Israel
[‡]National Institute of Oceanography, Israel Oceanographic and Limnological Research, Haifa, Israel

11.1 INTRODUCTION

Arid and semiarid regions of the world are under constant water stress as they attempt to accommodate the growing need for food and potable water [1–3]. Current limitations in freshwater availability have stimulated the use of desalination technologies worldwide, with desalinated water coming online at a rate of ~85 M $m^3 d^{-1}$ [4,5]. Advances in membrane technology during the last decades have favored reverse osmosis, which currently accounts for >80% of the global desalination industry [6,7]. Large-scale desalination facilities that are based on seawater reverse osmosis (SWRO) use coastal water as feed and continuously discharge concentrated brine effluent back into the marine environment. The SWRO brine effluent contains high concentration of salts, as well as various chemicals such as antiscalants and coagulants if used in the facility and discharged with the brine effluent [8,9]. Brine effluent from large-scale desalination facilities is discharged back to the coastal environment via direct surface discharge on the coastline or via diffuser systems away from the shore (up to 2 km) [9–11]. To maximize dilution and increase the buoyancy of the SWRO brine effluent, it is often mixed with cooling water of adjacent power plants resulting in a warm (up to 25% over ambient) and saline (up to 10% over ambient) brine effluent plume. Nevertheless, the brine effluent can be denser than ambient seawater (depending on the volume of cooling water that is used for mixing) [12]. In such cases, the brine effluent tends to sink and flows as a concentrated stream on the seabed [12–14]. It has been previously pointed out that such dense saline plumes can extend further away from the discharge site along the seafloor (Table 11.1) and possibly accumulates within the sediment pore water. The brine effluent plume is dynamic and dispersed along with the

Sustainable Desalination Handbook
https://doi.org/10.1016/B978-0-12-809240-8.00011-3

Table 11.1 Spatial dispersion of the brine-effluent plume at the seabed around desalination plants worldwide

Desalination plant	Brine discharge flux (M m^3 y^{-1})	Discharge technology	2%–5% salinity above ambient (km^2)	5%–10% salinity above ambient (km^2)	References
Alicante (Spain)[a]	66	Beach discharge	~23	9.8	[13]
Maspalomas II (Spain)	22	Beach discharge	~0.2	~0.04	[15]
Ashkelon (Israel)[b]	120	Beach discharge	1.4	0.06	[16,17]
Hadera (Israel)[b]	145	Beach discharge	~2	~0.1	[16,17]
Sorek/Palmachim (Israel)[c]	90–150	Diffuser system	<2	<0.1	[18]
Perth (Australia)	0.14	Diffuser system	~0.11	<0.1	[19]
Monterey Bay regional water project (USA)	30	Diffuser system	~0.3	~0.12	[20]

[a]Brine effluent was not diluted prior to discharge.
[b]The dispersion area is linked to the volume of cooling water from the adjacent power plant that was mixed with the brine effluent.
[c]Two separate discharge points that are close to each other.
(Table was modified after Frank H, Rahav E, Bar-Zeev E. Short-term effects of SWRO desalination brine on benthic heterotrophic microbial communities. Desalination 2017:417;52–59.)

cooling-water outflow from the adjacent power plant (in the case of surface discharge that is diluted) or from the outflow pipe in cases of direct dispersion by diffuser systems. The dispersion of the brine-effluent plume can also be affected by the bathymetry of the seabed at the outfall region. Although the overall footprint area of the brine-effluent plume can be estimated and modeled, the specific location of the brine effluent on the seafloor often vary due to changing currents [13,21]. These dynamic conditions can result in episodic short-term (hours or days) exposure of benthic organisms to the brine effluent at the perimeters of the plume as well as long-term chronic exposure (months or years) at the outfall or the plume center.

The possible impacts on coastal ecosystems related to SWRO desalination can be divided into three main categories [8,10,22–24]:

(i) Impingement and entrainment of marine organisms associated with seawater that is drawn to the desalination facility (typically related to surface intake) [23,25–27]. Impingement refers to adult marine organisms such as fish, crabs, etc., which are large enough to be retained by the intake screens. Entrainment is associated with smaller (often planktonic) organisms including larva and juveniles of different marine species that pass through the intake screens and are transported with feedwater into the desalination facility. The survival rates of entrained organisms are often considered to be nearly zero. Yet, the actual survival rates and significance of impingement and entrainment of organisms are different from site to site and are often not clear.

(ii) Constant release of brine effluent that is a byproduct of the desalination process. The disposal of brine effluent may result in osmotic stress due to elevated salinities compared to those in the receiving environment. Thermal stresses are also associated with the desalination brine effluent if it is mixed with coolant water of adjacent power plants.

(iii) Discharge of different chemicals that are often used in the desalination process into the coastal environment along with the brine effluent. These chemicals include different types of antiscalants and coagulants. Antiscalants may induce local eutrophication of oligotrophic coastal environments by adding organic phosphorus (e.g., polyphosphates), while coagulants can enhance water turbidity and water coloration.

In the past decade a comprehensive, long-term monitoring of large-scale SWRO desalination facilities along the Israeli coastline has reported that antiscalants, coagulants, and heavy metals were not detected at the outfall area [18,29–31]. In addition, it was determined that antiscalants and coagulants did not accumulate around the outfall area of these specific desalination facilities. However, previous studies and surveys around other operational desalination facilities worldwide have shown that exposure to the brine effluent (mainly to elevated salinities) could impact marine organisms, including vertebrates, invertebrates, seagrass, and polychaeta as well as plankton and fish larva, in a diameter of up to a several 100 m from the outfall [8,16–18,25,30,32–35]. Moreover, it has been suggested that these impacts may be more significant in enclosed basins, nature reserves, rocky shores and/or around other sensitive marine environments where water circulation is limited [36–38]. Nevertheless, to date, the effects of brine-effluent discharge on coastal marine ecosystems are poorly understood, thereby merit further research via controlled bioassay experiments and more importantly long-term monitoring campaigns around the outfall of operational desalination facilities.

In this book chapter, we review the latest reports and studies focusing on the impact of SWRO desalination brine-effluent discharge on marine ecosystems, namely, coastal flora and fauna. It should be noted that throughout the chapter we do not report on lifecycle assessments nor define impact as "positive" or "negative" implications but rather specify "impacts" as any deviation from the natural environmental conditions at the site prior to discharge. Finally, we describe the needs for dynamic site-specific monitoring approaches and suggest possible means to minimize the effects of SWRO desalination on coastal marine organisms and ecosystems.

11.2 THE IMPACT OF DESALINATION BRINE EFFLUENT ON ZOOPLANKTON

Plankton are free-floating organisms that depend on ocean currents for movement and consists of diverse groups including bacterioplankton, phytoplankton, and zooplankton. Planktonic organisms range in size from picoplankton (0.2–2 μm) to large zooplankton (up to few mm long). In this chapter, we will focus on zooplankton while phytoplankton and bacterioplankton are covered in greater detail in another chapter (see Chapter by Nurit Kress et al.) of this book.

Zooplankton consists of a large and complex group of animals including crustacean, copepods, marine larvae, and various worms. The life cycle of many zooplankton organisms can be complex and include egg, larvae, and adult stages. The potential impacts of desalination brine effluent may be vary for these different life stages making assessment of the impacts on specific species harder to quantify [39]. Zooplankton can inhabit both the pelagic and the benthic zones of the ocean. In this chapter, we will focus on pelagic zooplankton those that live in the water column. Zooplankton grazes on microplankton, primarily phytoplankton and bacterioplankton, which are in turn dependent on seasonal changes such as temperature [40,41] that affect the growth rates, metabolism, and respiration rates [42,43].

To date, there have only been a handful of studies investigating the effects of desalination brine effluent on pelagic zooplankton. These studies have found that the adult stages of zooplankton have a higher tolerance for salinity changes than larvae stage [44]. For adult zooplankton, mortality is generally not observed until salinity is increased by about 40% over ambient salinity, whereas reproduction and survival of eggs and juvenile can be affected by a salinity increase of about 20% above ambient [44–49]. However, the response varies among species, for example, an incubation experiment exposing rotifers to desalination brine effluent where the salinity was 40% above ambient conditions reported no significant mortality [50]. A summary of reported mortality rates for different zooplankton species at elevated salinities is detailed in Table 11.2.

Table 11.2 A summary of mortality of various zooplankton species in increased salinities

Species	Location	Salinity increase from ambient seawater (%)	Salinity measured	Mortality (%)	References
Rotifer	South Korea	30–40	40	7	[50]
Mysid shrimp	Texas, USA	60	45	40	[49]
Copepod juvenile	Tungkang, Taiwan	25	25	30	[44]
Copepod adult	Tungkang, Taiwan	25	25	0	[44]
Copepod juvenile	Florida, USA	30	45	40	[48]
Copepod adult	Florida, USA	30	45	15	[48]
Copepod juvenile	Wukan Bay, Japan	20	35	70	[47]

It should be noted that the salinity increase over the ambient coastal environment following the brine-effluent discharge of SWRO is often lower than 10%.

Increases in temperature (within an organism's tolerance range) typically enhance growth rates of larvae and increase respiration in adult copepods [39,51]. Increases in sea-surface temperature on a global scale have been connected with decreases in zooplankton abundance and shifts in zooplankton communities [43,52]. Accordingly, local temperature increases within the discharge plume of SWRO plants, at facilities in which the brine effluent is diluted with power-plant cooling water, are expected to show a similar pattern on zooplankton communities.

Chemical additive used in some SWRO desalination facilities such as antiscalants or coagulants may also impact zooplankton if discharged with the brine effluent. To the best of our knowledge, there are currently no studies published on the effects of antiscalants or coagulants on zooplankton. However, a study in the fresh water Lake Tahoe (California, USA) pointed to a significant decrease in reproduction among copepods cultivated with the addition of aluminum-based coagulants [53].

The overall impacts of desalination brine-effluent discharge on zooplankton are vague and not well understood. Moreover, the possible effects of SWRO brine on the zooplankton food web (related to phytoplankton and bacterioplankton) were not studied and are currently unknown. It is possible that SWRO effluent will impact zooplankton species around the outfall, but at this point, there are not sufficient data to arrive at a satisfying conclusion.

11.3 BENTHIC BACTERIA AROUND THE OUTFALL OF DESALINATION FACILITIES

Marine sediments are biodiverse ecosystems playing a key role in different biogeochemical cycles such as nutrient recycling and organic matter decomposition [54,55]. Heterotrophic bacteria have a central role within the benthic fauna community as they regulate various biochemical processes such as decomposition of organic matter and nutrients remineralization in the sediment [56,57]. Organic matter and nutrients can then be released to the water column and/or assimilated by benthic bacteria. This bacterial biomass can then be consumed by bacterivores and utilized by higher trophic levels [57–61]. The growth efficiency and community structure of benthic bacteria is often affected by environmental parameters such as temperature, salinity, oxygen concentrations, organic matter quantity and quality, as well as nutrients availability. These parameters impact the metabolic patterns of bacteria, for example, by changing the rates of anabolic reactions (carbon assimilation) or catabolic processes (respiration) [62,63]. It should be noted that cyanobacteria (photosynthetic prokaryotes) also constitute an important component of the benthic microbial assemblage and they play an important role as primary producers [64].

Currently, studies or technical reports regarding the impact of SWRO brine-effluent discharge on benthic bacteria are highly scarce. To the best of our knowledge, there are also no published data on the effects of desalination brine effluent on benthonic cyanobacteria. In the following section, we will focus on benthic heterotrophic bacteria and review the possible effects of osmotic stress due to SWRO brine-effluent discharge.

11.4 SHORT- AND LONG-TERM IMPACT OF SWRO BRINE-EFFLUENT DISCHARGE ON BENTHIC BACTERIA

Exposure of benthic bacteria to SWRO brine effluent may impose short-term (days) to chronic (years) effects. It was recently shown in controlled microcosm experiments that short-term (2 days) exposure to elevated salinity (>5% above the ambient) resulted in a significant reduction (60%) in bacteria abundance [21]. In addition, the metabolic activity per bacterial cell was found to increase following these short-term (days) exposures to SWRO brine effluent. It has been proposed that under these higher salinity conditions osmotic shock has prompt production of osmoprotectants [21], namely, carbon-rich molecules that are used to adjust the osmotic pressure of the bacterial cell to allow survival [65–67].

Chronic effects of SWRO brine effluent on benthic bacteria are of great importance due to long-term operation of desalination facilities. Environmental measurements were recently conducted near a large-scale operating desalination facility with a surface discharge system (producing ~150 $\mathrm{M\,m^3 Y^{-1}}$ of desalinated water) along the Israeli coastline. Sediments were sampled by the *RV Mediterranean Explorer*. Three locations were selected for each desalination facility: at the brine-effluent outfall, within the brine-effluent plume, and at reference stations that were not affected by the brine effluent. The outfall station was chronically exposed to salinity of up to 9% above ambient, while the "plume" station was exposed to variable conditions between 2% and 5% above ambient salinity. The chronic exposure to brine effluent at the outfall resulted in higher bacterial abundance and production (52% and 60%, respectively) compared to the reference station (Fig. 11.1). Diversity analysis indicated that community structure at the outfall and reference stations were significantly different. The main families at the outfall station comprised primarily *Stramenopiles, Pirellulaceae, Pisckirickettsiaceae*, while the reference station was colonized by *Bacteriodaceae, Prevotelaceae, and Enterobacteriacea* families (Frank et al., unpublished data).

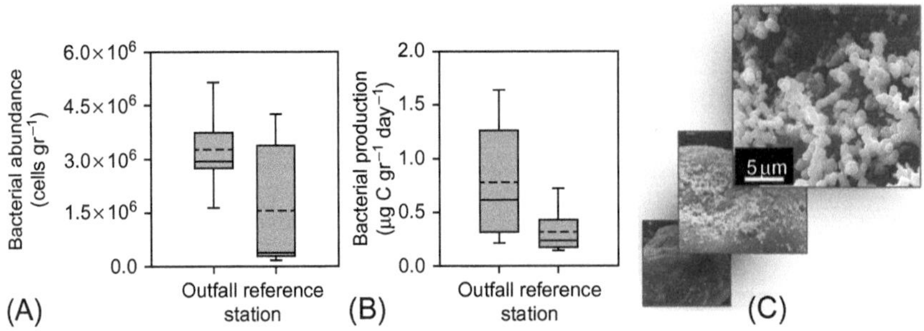

Fig. 11.1 (A) Benthic bacterial abundance and (B) production near a large-scale SWRO desalination facility within the outfall and reference stations (Frank et al., unpublished data). The figures represent the pooled data from four annual cruises. (C) Representative micrograph of benthic heterotrophic bacteria attached to a sand grain.

We stress that additional environmental campaigns from these sites as well as different coastal ecosystems are needed to establish a clear conclusion regarding the chronic effects of SWRO brine effluent on benthic bacteria.

Changes in the community structure and metabolic traits of benthic bacteria can potentially affect ecosystem functionalities on long time scales [68]. The effect of brine effluent on benthic bacteria needs to be further assessed in regards to the size of the outfall area and habitat properties such as sediment type and bathymetry. It should be highlighted that the impact of water temperature and chemical discharge (such as antiscalants) on benthic microbial communities has been overlooked so far and therefore warrants further investigation.

11.5 IMPACT OF OSMOTIC STRESS ON BENTHIC MEIOFAUNA

Soft bottom meiofauna include various microscopic organisms that range from tens of micrometers to one mm. These organisms consist of invertebrates from different phyla such as Crustacea, Echinodermata, Mollusca, Annelida, Nematoda, and Foramenifera [69,70]. Meiofauna could impact microbenthic abundances and shape the macrobenthic populations by grazing on their larvae [71]. In coastal environments, the number of meiofaunal species dwelling within the sediment is high and often exceeds the number of species living on the seabed due to the turbulence conditions [70].

Previous reports have highlighted that benthic Meiofauna (specifically metazoans) are highly sensitive to different anthropogenic effects [33,69]. Therefore, metazoans are often used as bioindicators to assess impacts on the meiofaunal community in coastal environments. Nematodes are the most abundant and diverse phylum within the meiofauna group [71–73]. As permanent dwellers of the interstitial coastal environment, they are constantly exposed to anthropogenic pollutants [72,74], and possibly also to osmotic stress due to brine effluent from desalination discharge. Although Nematodes are used as bioindicators for different pollutants, [71,74] to date no dedicated studies were conducted to assess the impact of desalination on the activity and biodiversity of nematodes.

Crustaceans were also shown to be highly sensitive to many different anthropogenic pollutants [75,76] as well as to salinity increase due to SWRO brine-effluent discharge. For example, amphipods assemblages studied near the San Pedro desalination facility in SE Spain where they were exposed to high salinity levels of up to 53 near the brine-effluent outfall [77] show reduced abundance and diversity (Fig. 11.2, Shannon-Winner index). Moreover, salinity changed the biological traits and functionality of the community selecting for detrivorous and domicolous species, at the expense of carnivorous, omnivorous, and fossorial species. It should be noted that once a diffuser system was deployed at the end of the discharge pipeline (4 years after operation initiation), amphipods abundance rapidly recovered.

Near the same desalination facility, an investigation of brine-effluent effects on Polychaeta assemblages was also conducted [78]. Community parameters including family richness, diversity, and abundance were measured. Polychaeta assemblage was mainly disturbed close (0–300 m) to the discharge

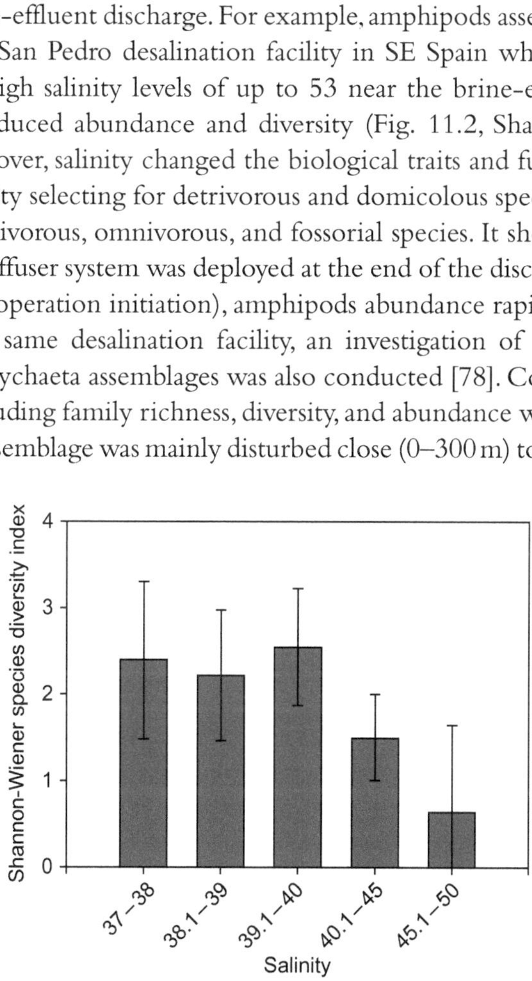

Fig. 11.2 Meiofaunal diversity in different salinities measured along a brine-effluent plume from four operating desalination facilities [77–80].

point, where abundance, richness and diversity were reduced. Following the deployment of the diffuser, nearly full recovery of the Polychaeta diversity and richness was achieved. Similar findings were reported in a study that was conducted in Chabahar bay (Oman Sea, Iran), indicating a decrease in abundance, richness, and diversity closest to the brine-effluent outfall [79]. Polychaeta were also investigated near the outfall of a large desalination facility in Alicante (NE Spain) discharging ~65,000 $m^3 d^{-1}$ of brine effluent at a salinity of 68 (note, brine was not mixed prior to discharge). Polychaeta assemblages were different at the outfall, and composed of a homogeneous group of several families. Measurements at the two stations closest to the brine-effluent outfall (distance of ~660 m between them) showed a decrease in richness, diversity, and abundance at all sampling dates (compared to un-impacted locations). It should be noted that a previous study has indicated that few Polychaeta families (Syllidae and Capitellidae) displayed short-term resistance or tolerance (Paraonodae) to osmotic stress, while the abundance of other families (Mpharetidae, Nephtyidae, and Spionidae) sharply decreased [80].

Echinoderms such as starfish, sea urchins, and sea cucumbers are also ecologically important benthic macro fauna. These organisms function as filter feeders and predators of benthic algae, benthic bacterial consortiums (termed also as biofilms), as well as other meiofauna such as bivalves and snails. It has been shown that this group of organisms can comprise >90% of the benthic biomass [81]. Echinoderms were reported to be strict marine phyla and therefore are expected to be highly sensitive to changes in salinity [46]. A study conducted near the Alicante desalination plant pointed out a significant decrease in Echinoderms abundance (100%) after the desalination plant started operating (2003) [82]. A significant recovery of Echinoderms abundance was noted after initiating mixing of the brine effluent with seawater prior to discharge. A recent study [77] noted that the use of Echinoderms abundance and the benthic opportunistic polychaeta and amphipods (BOPA) indices were an effective and convenient measure for the determination of environmental degradation as a result of brine-effluent discharge. The BOPA (Eq. 11.1) indices measure the proportion of opportunistic Polychaeta families (Capitellidae, Eunicidae, Magelonidae, Nephtydae, and Paraonidae) [78] in relation to the proportion of amphipods, which are considered to be sensitive to environmental changes. It is calculated according to the equation:

$$\mathrm{BOPA} = \log\left[\frac{fp_{op}}{f_a + 1} + 1\right] \qquad (11.1)$$

where fp_{op} is the opportunistic Polychaeta proportion of all fauna (0–1) and f_a is the amphipod proportion of all fauna (0–1) [83]. The maximum value of the index is 0.3013, and it indicates a highly disturbed area with only opportunistic Polychaeta families, while 0 indicates no opportunistic Polychaeta, or higher proportion of Amphipods in the benthic metazoan community.

During the last decade, a series of comprehensive monitoring surveys for benthic meiofauna (abundance and community structure) were conducted adjacent to large-scale SWRO desalination facilities located on the Israeli coastline [16,18,29–31]. These facilities apply two disposal approaches: surface discharge of brine, which is diluted with cooling water of an adjacent power plant (used at Hadera and Ashkelon desalination plants) or a diffuser system discharging the brine away from the shore (used at Sorek\Palmachim desalination plants). Significant changes to the community structure of benthic meiofauna near the brine\cooling-water outfall of Hadera and Ashkelon desalination facilities were reported [17,28]. In some cases, there was also a reduction in the abundance of benthic meiofauna next to the outfall brine\cooling water. It should be stressed that these effects were only observed within a few hundred of meters of the discharge point and could not be linked directly to osmotic/temperature stresses related to brine-effluent discharge [17,28]. Instead, it was suggested that these changes in benthic meiofauna resulted from physical disturbance of the sediment due to the strong water currents of the discharged brine/cooling water [17,28]. Monitoring reports from the Sorek\Palmachim desalination outfall list no significant impacts of brine effluent on benthic meiofauna [18,30,31]. However, a report from 2014 noted that some local effects on the community structure of benthic meiofauna may be linked to the brine-effluent discharge [30]. These reports concluded that additional monitoring campaigns were recommended to determine the significance of these effects.

11.6 THE EFFECTS OF DESALINATION BRINE EFFLUENT ON SEAGRASS

Seagrass meadows are highly productive habitats and form key ecosystems in coastal environments worldwide [84,85]. Seagrass meadows export on average 24% [86] of their net production and serve as important trophic links to nearby ecosystems [84]. The contribution of seagrass meadows includes sediment stabilization and improvement of water clarity, source of food to the coastal and adjacent ecosystems, provision of oxygen to the water and sediment as well as a habitat for various organisms [84,85]. Accordingly, seagrass meadows are estimated as one of the most valuable habitats in terms of ecosystem functionality [87]. SWRO desalination facilities have shown to impact the physiology and growth of seagrass meadows due to osmotic stress around the brine-effluent discharge point (Fig. 11.3) [88–91].

Posidonia oceanica is an endemic specie to the Mediterranean Sea [92]. In a series of mesocosm experiments, the susceptibility of *P. oceanica* to elevated salinities was tested by evaluating the mortality of leaves and their recovery after returning to ambient salinities [91]. It was found that the growth of the leaves was inhibited in salinities above 39 and overall mortality occurred at salinities above 53. A later study examined the effects of elevated salinities [37–43] on photosynthesis by *P. oceanica*. The results indicate that net and gross photosynthetic rates of *P. oceanica* were significantly reduced by 25%–33% and 13%–20%, respectively, following exposure to higher salinities than

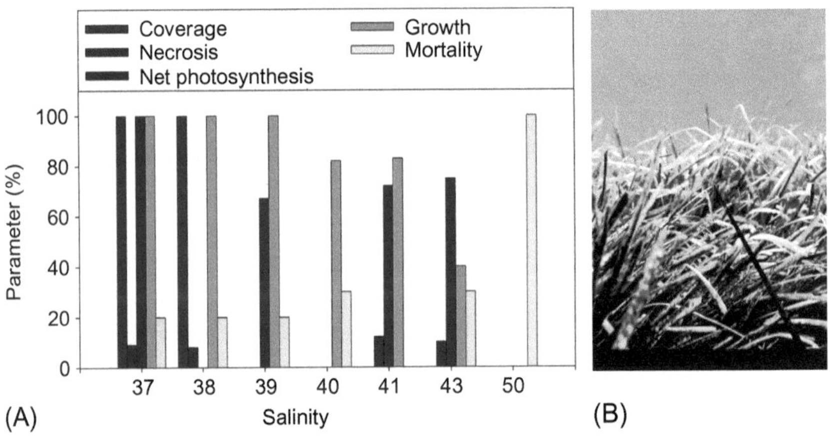

Fig. 11.3 (A) Ecological and physiological parameters of Seagrass that were altered due to elevated salinities. Measurements were taken from mesocosm experiments and different environmental samplings. All data represent percent change from the control in the specific study or experiment the data were taken from. (B) An example of *Posidonia oceanica* meadow, Photo credit (personal permission): Sagi Maayan. *(The data for the figure were adapted from Gacia E, Invers O, Manzanera M, Ballesteros E, Romero J. Impact of the brine from a desalination plant on a shallow seagrass (Posidonia oceanica) meadow. Estuar Coast Shelf Sci 2007;72(4):579–590. Fernández-Torquemada Y, Sánchez-Lizaso J.L. Effects of salinity on leaf growth and survival of the Mediterranean seagrass Posidonia oceanica. J Exp Mar Bio Ecol 2005;320(1):57–63. Marin-Guirao L, Sandoval-Gil J.M, Ruz J.M., Sánchez-Lizaso J.L. Corrigendum to photosynthesis, growth and survival of the Mediterranean seagrass Posidonia oceanica in response to simulated salinity increases in a laboratory mesocosm system. [Estuar Coast Shelf Sci 2011;92:286–296]. Estuar Coast Shelf Sci 2011;93(1);86.)*

the control. In addition, dark respiration of *P. oceanica* increased significantly (by 98%) when exposed to high salinities [41,43]. It was suggested that the respiratory demand for osmoregulation reduces the photosynthetic rates and therefore inhibits growth [89]. An in situ study that was performed close to an operating desalination facility at the Balearic Islands of Spain found that chronic exposure to the brine effluent resulted in *P. oceanica* leaf necrosis [90]. Finally, it has been reported that no *P. oceanica* were observed up to 25 m from the discharge point [90].

Cymodocea nodosa also showed sensitivity to exposure to SWRO brine effluent [93]. *C. nodosa* is a relatively small, fast-growing seagrass that can tolerate a broad range of environmental conditions. Nevertheless, exposure to brine effluent for 1 month (near the Alicante desalination plant, NW Spain) caused reduction in growth rates and higher mortality of shoots closer to the brine-effluent discharge [93]. It has been suggested that the constant salinity increase following brine-effluent discharge resulted in higher energetic costs, due to a need for maintenance of a proper turgor pressure. It has also been speculated that *C. nodosa* plants were losing their inner cellular water content and were accumulating ions from the environment to achieve the proper pressure to cope with the elevated salinities. Changes in water salinities and osmoregulation were possibly the reason for shoots deterioration following the chronic exposure to SWRO brine-effluent.

Hence, it has been previously suggested that seagrass species inhabiting soft bottom habitats may serve as bioindicators for habitat degradation in response to brine-effluent discharge [92,94]. Criteria for seagrass species to be used as a bioindicator for the impact of brine effluent on the coastal environment were established by carrying controlled, laboratory experiments. The Seagrass *Posidonia australis* was incubated for 6 weeks at elevated salinities [37–54] and the following physiological parameters were monitored: survival, growth, photosynthesis, metabolism parameters, carbohydrate, and amino acid concentration [95]. Additional mesocosm experiments were conducted in order to evaluate the growth and survival of other seagrass species including *C. nodosa* and *Zostrea noltii hornemann*. These seagrass species were grown with salinities of up to 72 [91]. It should be noted that these salinities levels are not found next to the outfall of operating desalination facilities. Nonetheless, in these studies, survival and growth of shoots were significantly decreased compared to the control treatments. Additional physiological parameters of *P. australis* were found to be affected by the high salinity concentrations after 6 weeks of incubation [95]. Specifically, it was found that amino acids composition in rhizome and leaf tissues increased, and compatible solutes concentration in leaf tissue raised [95].

It has recently been reported that for large seagrass species necrosis had the shortest response time to hypersalinity, while in small species photosynthetic rate, necrosis, leaf growth, and mortality all had a similarly short response time. In both cases, response time ranged from immediate to 10 weeks [94]. We conclude that it is essential to test what are the proper physiological parameters (e.g., photosynthetic rate, leaf necrosis, and meadow density) to be used in order to evaluate the extent of habitat deterioration. It should be highlighted that the seagrass species and the relevant brine-exposure time should also be considered before using seagrass as brine-effluent bioindicators.

11.7 DESALINATION BRINE-EFFLUENT IMPACTS ON FISH LARVAE

Fish eggs and larvae are small (1 mm to >10 mm, species dependent) plankton that drift in the upper photic zone of the water column. Fish eggs are completely dependent on ocean currents, buoyancy, and water density for their position in the water column, while the larval stage have some limited swimming ability [96–98]. The larvae are hatched with yolk sac still attached to their body, which is consumed within the first few days after hatching (Fig. 11.4). As larvae grow, they start developing swim bladders, which are used by adult fish to maintain buoyancy [98,107]. Temperature increase (within the species' tolerance range) tends to accelerate embryo and larvae

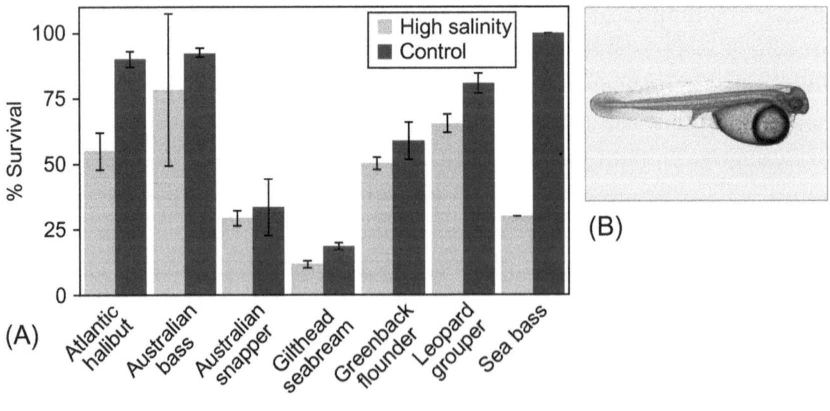

Fig. 11.4 (A) The percent survival of the larvae of seven species of fish. Larvae were hatched in either "optimum salinity" (Control) or in salinity treatments ranging from 5% to 15% over optimum [99–105]. (B) A newly hatched Sea Bass larvae, photocredit (personal permission): David J Ostrach, pH.D [106].

development, which can increase the risk of larvae mortality since the development time is shortened [107,108]. At the early developmental stage, fish larvae undergo significant morphological changes. Therefore, the larvae are more susceptible to environmental changes [108,109]. The development of the egg and larvae can vary from days to weeks depending on the species; therefore, larvae can be affected by changes in their environment even if the exposure is on time scale of hours [98,107,110,111].

Larvae and eggs of different fish species appear to have great resilience to small changes in salinity. However, at salinity increases of ~50% or higher than ambient (which are rarely or ever encountered in modern desalination facilities), larvae and eggs show significant and acute mortality (after 24–48 h of exposure) [112–114]. Increasing salinity to 5%–15% above ambient (which could be encountered sometimes close to the outfall) lowers the larvae's survival rate, but the impact is rarely found to be significant. Fig. 11.4 compares the survival of larvae of 7 species of fish when salinity is increased by 5%–10% over ambient conditions [99–105]. In addition to increased mortality, it has been shown that fish larvae decrease in size and have higher degree of defective and/or inefficient inflation of swim bladders when cultivated in salinity of 10%–15% higher than ambient [99,100,104,105,113].

In laboratory experiments, higher temperatures (~5°C over ambient) lowered the survival of fish eggs and larvae [102,103,109]. Higher temperatures can also shorten the hatching time of eggs resulting in undeveloped larvae or a quicker absorption of the yolk sac of developed larvae [115]. Certain species have shown swim bladder defects at higher temperatures, and one species (Australian Snapper, *Pagrus auratus*) had a 100% mortality rate with a 30% temperature increase above maximum environmental temperatures for 9 days [100,102,116,117].

Larvae of different species of fish respond differently to various stressors. For some species, the combined effect of increasing salinity and temperature increased mortality and swim bladder deficiency [103,116], but for other species this synergy was not found and only temperature changes were found to have a significant impact on the larvae [100,102].

At present, no studies have investigated the effects of chemicals used (antiscalants and coagulants) in SWRO desalination facilities on fish larvae or fish. However, studies on coagulants (aluminum based) in a fresh water setting have shown a negative impact on fish embryo development and fish larvae survival at high concentrations [53,118]. To the best of our knowledge, there are currently also no published studies reporting on experiments with actual brine effluent from desalination plants; hence, the direct effects of SWRO brine-effluent discharge on fish eggs or larva remain unknown. We stress that dedicated research as well as monitoring programs are highly needed to evaluate the impacts of desalination brine-effluent discharge on prominent species of fish eggs and larvae around the outfall of these facilities.

11.8 THE EFFECTS OF DESALINATION BRINE EFFLUENT ON CORALS PHYSIOLOGY

Coral reefs comprise some of the richest habitats in the ocean, supporting great biodiversity, biomass, and productivity [119]. In addition, coral reefs are important for local economies, fisheries, and tourisms [119–121]. Corals are marine invertebrates (phylum: Cnidaria; class: Anthozoa) and the hermatypic coral species (reef building) can consist of large colonies that span many kilometers. These corals have a mutualistic relationship with dinoflagellates of the genus *Symbiodinium* (referred to as Zooxanthellae) that are integrated within the tissue of the coral where they provide organic matter to the coral via photosynthesis in return of nutrients [122,123]. Corals are osmoconformers and adjust their osmotic balance with intercellular osmolytes. The osmotic balance can be affected by many environmental stressor (temperature, nutrient levels, and salinity changes) and an imbalance of osmolytes often leads to coral bleaching [124]. Coral reefs are often found around tropical, subtropical, and even cold-water regions resulting in specific adaptions (e.g., to temperature, salinity, etc.) of species even within the same genus [125,126]. In the last decade, coral reefs have experienced an increase in the extent and frequency of bleaching events due to different environmental stressors such as rising ocean temperatures, seawater acidification as well as anthropogenic impacts such as coastal eutrophication [120,121,127–130]. In line with these, the construction of new desalination facilities around coastal zones with coral reef ecosystems raises concerns over the possible impact on coral physiology and survivability [131,132].

Coral habitats can experience changes in salinity on diurnal and seasonal time scales as tides and rainfall/evaporation can impact local salinity [133]. Short-term exposure studies (5–7 days) with elevated salinities (~15% above ambient) have shown a similar response in different coral species (*Porites furcata, Siderastrea radians, and S. siderea*): (i) decrease in primary production within the first 6–12h following exposure, (ii) retraction of coral polyps, and (iii) discoloring of the tissue [134–136]. In these studies, corals were able to recover to normal primary productivity rates (postincubation) within 36h to 1 week, and made a full recovery of polyps and tissue within 1 week to a month.

Previous reports have indicated that exposure of *Stylophora pistillata* for 3 weeks to increase salinity of ~10% above ambient resulted in 70% mortality [137]. However, in a recent study with *Stylophora pistillata* in 10% above ambient salinity, mortality was not detected, but a significant drop in protein content and zooxanthellae abundance was measured (Fig. 11.5). A similar response was observed for *Acropora tenuis* and *Pocillopora verrucosa* both in salinity of 10% over ambient and in 10% increased salinity with addition of phosphanate (<2 mg/L).

The dramatic decrease in zooxanthellae and protein contents (Fig. 11.5) and the mortality reported by Ferrier-Pagès [137] stresses the need for monitoring of coral reefs in the vicinity of desalintion plants. A recent in situ study (the only one to date) in the Red Sea has positioned the coral *Fungia granulosa* at the discharge channel of an operating SWRO plant and followed different parameters [132]. The corals showed no change in primary production and none of the specimens were significantly affected. Furthermore, pure cultures of the symbiotic zooxanthellae (*S. microadriaticum*) were incubated in salinities between 25 and 55. Cell growth of *S. microadriaticum* was found to decrease only in salinities of 55 (which is not likely to occur around the outfall of SWRO desalination facilities), indicating a high salinity tolerance of the zooxanthellae.

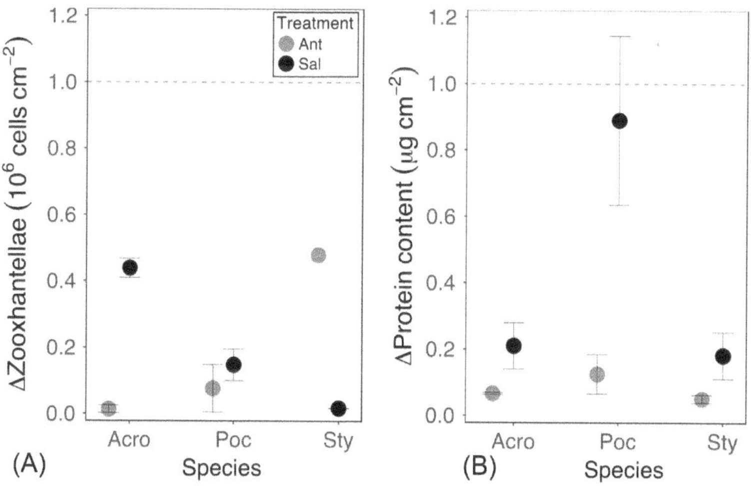

Fig. 11.5 Data from 4-week-long coral incubation with 10% salinity increase over ambient *(solid black)* and 10% salinity increase with 6 μL L^{-}1 phosphanate addition *(solid gray)*. (A) The relative change between treatment and control zooxanthellae abundance for *Acropora* (Acro), *Pocillopora* (Poc) and *Stylophora* (Sty). (B) Relative change in protein content between treatment and control.

In general, antiscalants (phosphonate based) are not harmful in the doses used in desalination plants (~2 mg/L) [138]; however, recent observations pointed that these concentrations may impact the physiology of *S. pistillata, A. tenuis,* and *P. verrucosa* (Petersen et al., unpublished data). To the best of our knowledge, no published studies have reported on the effects of coagulants on corals.

Rising ocean temperatures is one of the main stressors causing coral bleaching [128,139]. The water temperature of desalination brine effluent can be elevated compared to background ambient temperatures (by up to 25% over the ambient), and this results in local plumes of warm seawater around the discharge area ([8,10,140], Frank et al., unpublished data). Exposure to temperature of as little as 3°C above mean maximum levels for several weeks has shown to cause bleaching in *Pocillopora* sp., *Montastrea* sp., *Acropora* sp., and other species spanning reefs in Hawaii, Caribbean, and Australia [125,126,140–142]. Thermal stress was also positively correlated with coral disease outbreaks [125,139]. However, the response varies based on location as *S. pistillata, P. damicornis,* and *A. eurystoma* that were collected in the Gulf of Aqaba (Red Sea) did not show any significant change after 4 weeks of incubation in temperatures of 10% above ambient. Only at a 40% temperature increase (from ~25°C to 34°C), the abundance of zooxanthellae was significantly reduced [143].

Corals exposed to multiple environmental stressors tend to have a higher risk of bleaching and a higher mortality rate [128,134,139]. This is because the efficiency in which the corals sustain energy-demanding processes becomes compromised under multiple stressors. This effect was shown by Lirman et al. [134] who exposed corals to salinity of 25% above ambient and found that they were unable to clear their tissue of sediment [134]. The response to different stress factors, such as increased salinity and temperature, is variable between species of coral even in the same geographic location. This underlines the difficulty of providing globally relevant guidelines for desalination management around coral reefs to minimize possible impacts, and emphasizes the need for local environmental studies prior to the construction of large-scale desalination facilities. At this point, published data on the direct effect of desalination brine-effluent discharge on coral reefs are extremely scarce and virtually nonexisting. Therefore, we stress that coral mesocosm studies (short term) as well as dedicated monitoring schemes (long term), should be carried out to determine the local impact of brine-effluent discharge on coral reefs.

11.9 LOOKING FORWARD: NEXUS OF SWRO DESALINATION AND COASTAL ENVIRONMENTS

SWRO desalination industry is booming with a tight, bidirectional (intake and discharge) interface to the coastal environment. Extensive information has been accumulating during the past decade with respect to the possible impacts imposed by the desalination industry on the marine environment [8,11,17,18,22,24,25,32,36–38,79,138]. These effects were suggested to impact different marine organisms; however, they were more pronounced in sessile organisms such as seagrass or benthic fauna. It should be stressed that the reported impacts of most SWRO desalination were highly local and restricted only to the vicinity of the intake port or the outfall area. Yet, we surmise that due to growing water scarcity, future concerns related to desalination expansion and possible impacts to coastal environments (specifically to enclosed and sensitive ecosystems) should be further evaluated. These impacts may include: (i) Construction of new SWRO desalination facilities at high densities; (ii) Designing increasingly larger SWRO desalination plants with greater production capacities (>300 $M\,m^3\,y^{-1}$) and (iii) Drawing large volumes of feed-water from the ocean, while discharging large amounts of brine effluent to coastal environments with limited water circulation.

We suggest that mitigating the impacts of the SWRO desalination on the coastal environments in years to come could be optimized (together with standard provisions such as near- and far-field modeling) by the following measures:

(i) Carrying dedicated precursory environmental-impacts-assessments that are site specific. Namely, to determine the chemical concentrations at which adverse biological effects are apparent for endemic organisms prior to the contraction of any large-scale desalination facility. These values should be used as a reference point and set the criteria for future, long-term monitoring surveys.

(ii) Conducting long-term monitoring programs that are tailored to each desalination site (e.g., based on the biology and hydrography of the location) or region (such as the oligotrophic environment or semienclosed lagoons). Parameters that will be monitored should be predetermined according to ranges of sensitivity of key organisms that are relevant to the coastal environment, similar to Long et al. [144]. We stress that these monitoring programs should be dynamic and evaluated yearly to minimize oversampling and reduce redundancy in terms of the sampled parameters. Providing such data could contribute to better and more realistic regulation for the interface of the desalination industry and coastal environments.

(iii) Applying best-available technology to reduce any possible effects on the coastal environment that may result from SWRO desalination brine effluent. We propose that using diffuser systems as a favorable discharge approach will reduce the impact of brine on the coastal ecosystems. The advantages of the diffuser systems over surface discharge include minimizing the dependency on power-plant cooling water for dilution and setting the buoyancy of the brine-effluent plume prior to discharge. In addition, using diffuser system will enable to discharge the brine far from the shoreline and/or away from any ecologically sensitive areas. Brine effluent can be allocated via the diffuser system into predetermined locations, thereby maximizing dispersion by coastal currents. Finally, we urge that additional impetus should be allocated toward development of new approaches to minimize the use of chemicals in the desalination process [145] and eventually toward zero liquid discharge solutions.

REFERENCES

[1] Zimmerman JB, Mihelcic JR, Smith J. Global stressors on water quality and quantity. Environ Sci Technol 2008;42:4247–54.

[2] Davies EGR, Simonovic SP. Global water resources modeling with an integrated model of the social-economic-environmental system. Adv Water Resour 2011;34(6):684–700.

[3] Hoekstra AY, Mekonnen MM. The water footprint of humanity. Proc Natl Acad Sci U S A 2012;109(9):3232–7.

[4] Elimelech M, Phillip WA. The future of seawater and the environment: energy, technology, and the environment. Science 2011;333:712–8.

[5] Misdan N, Lau WJ, Ismail AF. Seawater reverse osmosis (SWRO) desalination by thin-film composite membrane-Current development, challenges and future prospects. Desalination 2012;287:228–37.

[6] Greenlee LF, Lawler DF, Freeman BD, Marrot B, Moulin P. Reverse osmosis desalination: water sources, technology, and today's challenges. Water Res 2009;43(9):2317–48.

[7] Shenvi SS, Isloor AM, Ismail AF. A review on RO membrane technology: developments and challenges. Desalination 2015;368:10–26.

[8] Roberts DA, Johnston EL, Knott NA. Impacts of desalination plant discharges on the marine environment: a critical review of published studies. Water Res 2010;44(18):5117–28.

[9] Fritzmann C, Löwenberg J, Wintgens T, Melin T. State-of-the-art of reverse osmosis desalination. Desalination 2007;216(1–3):1–76.

[10] Lattemann S, Höpner T. Environmental impact and impact assessment of seawater desalination. Desalination 2008;220(1–3):1–15.

[11] Voutchkov N. Overview of seawater concentrate disposal alternatives. Desalination 2011;273(1):205–19.
[12] Bleninger, T.; Niepelt, A.; Jirka, G. Desalination plant discharge calculator. Desalin Water Treat 2010, 13, (1–3), 156–173.
[13] Fernández-torquemada Y, Ferrero-vicente L, Gónzalez-correa JM, Loya A, Ferrero LM, Díaz-valdés M, et al. Dispersion of brine discharge from seawater reverse osmosis desalination plants. Desalin Water Treat 2009;5:137–45.
[14] Fernandez-Torquemada Y, Sanchez-Lizaso JL, Gonzalez-Correa JM. Preliminary results of the monitoring of the brine discharge produced by the SWRO desalination plant of Alicante (SE Spain). Desalination 2005;182(May):395–402.
[15] Portillo E, Ruiz de la Rosa M, Louzara G, Quesada J, Ruiz JM, Mendoza H. Dispersion of desalination plant brine discharge under varied hydrodynamic conditions in the south of Gran Canaria. Desalin Water Treat 2013;52(1–3):164–77.
[16] Gefen-Glazer A. Marine environment monitoring around "Orot Rabin" power plant and H_2ID desalination plant. The Israeli Ministry of Environmental Protection; 2015. R- ELP-1-2015. [in Hebrew].
[17] Gefen-Glazer A. Monitoring of the marine environment around the "Orot Rabin" power plant and H_2ID desalination plant. The Israeli Ministry of Environmental Protection; 2014. R- ELP-6-2014. [in Hebrew].
[18] Kress N, Shoham-Frider E, Lubinevsky H. Marine monitoring around the brine outfall of Palmachim and Soreq desalination plants. Israel Oceanographic & Limnological Research Ltd.; 2016. [In Hebrew].
[19] Water Corporation EPA Australia. Perth metropolitan desalination proposal—water quality management, change to implementation conditions; 2009. Report #1327.
[20] Jenkins SA. Brine dilution analysis for DeepWater Desal, Monterey Bay regional water project at moss landing. J Water Supply Res Technol AQUA 2014;1–113.
[21] Frank H, Rahav E, Bar-Zeev E. Short-term effects of SWRO desalination brine on benthic heterotrophic microbial communities. Desalination 2017;417:52–9.
[22] Liu T-K, Sheu H-Y, Tseng C-N. Environmental impact assessment of seawater desalination plant under the framework of integrated coastal management. Desalination 2013;326:10–8.
[23] Lattemann S, Kennedy MD, Amy G. Seawater desalination—a green technology? J Water Supply Res Technol AQUA 2010;59(2–3):134–51.
[24] Dawoud MA, Mulla MMA. Environmental impacts of seawater desalination: Arabian Gulf Case Study. Int J Environ Sustain 2012;1(3):22–37.
[25] Lokiec F. Sustainable desalination: environmental approaches. In Sustainable desalination: environmental approaches, The International Desalination Association World Congress on Desalination and Water Reuse, Tianjin 2013.
[26] Hogan TW. Impingement and entrainment at SWRO desalination facility intakes. In: Intakes and outfalls for seawater reverse-osmosis desalination facilities. Switzerland: Springer International Publishing; 2015. p. 57–78.
[27] Pankratz T. An overview of seawater intake facilities for seawater desalination. The future of desalination in Texas. Texas A&M University; 2004. p. 2.
[28] Abramson K, Shafir V. Marine monitoring of the "Rutenberg"power plant, VID desalination plant, and Mekorot reclamation facility. The Israeli Ministry of Environmental Protection; 2016. RELP-4-2016. [in Hebrew].
[29] Abramson K, Shafir V. Marine monitoring of the "Orot Rabin"power plant and H_2ID desalination plant. The Israeli Ministry of Environmental Protection; 2016. R-ELP-5-2016. [in Hebrew].
[30] Kress N, Shoham-Frider E, Lubinevsky H. Marine monitoring around the outfalls of Palmachim and Soreq desalination facilities. Israel Oceanographic & Limnological Research Ltd.; 2015. [in Hebrew].

[31] Kress N, Shoham-Frider E, Lubinevsky H. Marine monitoring around the outfalls of Palmachim and Soreq desalination facilities. Israel Oceanographic & Limnological Research Ltd.; 2014. [in Hebrew].
[32] Raventos N, Macpherson E, García-Rubiés A. Effect of brine discharge from a desalination plant on macrobenthic communities in the NW Mediterranean. Mar Environ Res 2006;62(1):1–14.
[33] Riera R, Tuya F, Sacramento A, Ramos E, Rodríguez M, Monterroso Ó. The effects of brine disposal on a subtidal meiofauna community. Estuar Coast Shelf Sci 2011;93(4):359–65.
[34] Safrai I, Zask A. Reverse osmosis desalination plants—marine environmentalist regulator point of view. Desalination 2008;220(1–3):72–84.
[35] Ahmad N, Baddour RE. A review of sources, effects, disposal methods, and regulations of brine into marine environments. Ocean Coast Manag 2014;87:1–7.
[36] Höpner T, Windelberg J. Elements of environmental impact studies on coastal desalination plants. Desalination 1997;108:11–8.
[37] Höpner T, Lattemann S. Chemical impacts from seawater desalination plants—a case study of the Northern Red Sea. Desalination 2003;152:133–40.
[38] Einav R, Harussi K, Perry D. The footprint of the desalination processes on the environment. Desalination 2003;152(1–3):141–54.
[39] Chinnery FE, Williams JA. The influence of temperature and salinity on *Acartia* (Copepoda: Calanoida) nauplii survival. Mar Biol 2004;145(4):733–8.
[40] Sterner RW, Schulz KL. Zooplankton nutrition: recent progress and a reality check. Aquat Ecol 1998;32(4):261–79.
[41] Kenitz KM, Visser AW, Mariani P, Andersen KH. Seasonal succession in zooplankton feeding traits reveals trophic trait coupling. Limnol Oceanogr 2017;1184–97.
[42] Huntley M, Boyd C. Food-limited growth of marine zooplankton. Am Soc Nat 1984;124(4):455–78.
[43] Richardson A. In hot water: zooplankton and climate change. ICES J Mar Sci 2008;65:279–95.
[44] Pan YJ, Souissi A, Souissi S, Hwang JS. Effects of salinity on the reproductive performance of *Apocyclops royi* (Copepoda, Cyclopoida). J Exp Mar Biol Ecol 2016;475:108–13.
[45] Yoon SJ, Park GS. Ecotoxicological effects of brine discharge on marine community by seawater desalination. Desalin Water Treat 2011;33:240–7.
[46] De-La-Ossa-Carretero JA, Del-Pilar-Ruso Y, Loya-Fernández A, Ferrero-Vicente LM, Marco-Méndez C, Martinez-Garcia E, et al. Bioindicators as metrics for environmental monitoring of desalination plant discharges. Mar Pollut Bull 2016;103(1–2):313–8.
[47] Chen Q, Sheng J, Lin Q, Gao Y, Lv J. Effect of salinity on reproduction and survival of the copepod *Pseudodiaptomus annandalei*. Aquaculture 2006;258(1–4):575–82.
[48] Ohs CL, Rhyne AL, Grabe SW, DiMaggio MA, Stenn E. Effects of salinity on reproduction and survival of the calanoid copepod *Pseudodiaptomus pelagicus*. Aquaculture 2010;307(3–4):219–24.
[49] Pillard DA, DuFresne DL, Tietge JE, Evans JM. Response of mysid shrimp (*Mysidopsis bahia*), Sheepshead Minnow (*Cyprinodon variegatus*), and inland silverside minnow (*Menidia beryllina*) to changes in artificial seawater salinity. Environ Toxicol Chem 1999;18(3):430–53.
[50] Park GS, Yoon S-M, Park K-S. Impact of desalination byproducts on marine organisms: a case study at Chuja Island desalination plant in Korea. Desalin Water Treat 2011;33(1–3):267–72.
[51] Gaudy R, Cervetto G, Pagano M. Comparison of the metabolism of *Acartia clausi* and *A. tonsa*: influence of temperature and salinity. J Exp Mar Biol Ecol 2000;247(1):51–65.
[52] Roemmich D, McGowan J. Climatic warming and the decline of zooplankton in the california current. Science 1995;267(5202):1324–6.

[53] Bachand PAM, Bachand SM, Lopus SE, Heyvaert A, Werner I. Treatment with chemical coagulants at different dosing levels changes ecotoxicity of stormwater from the Tahoe basin, California, USA. J Environ Sci Health A 2010;45(2):137–54.
[54] Köster M, Meyer-Reil LA. Characterization of carbon and microbial biomass pools in shallow water coastal sediments of the southern Baltic Sea (Nordrügensche Bodden). Mar Ecol Prog Ser 2001;214:25–41.
[55] Snelgrove PVR. Getting to the bottom of marine biodiversity: sedimentary Habitats. Bioscience 1999;49(2):129–38.
[56] Forehead H, Thomson P, Kendrick Ga. Shifts in composition of microbial communities of subtidal sandy sediments maximise retention of nutrients. FEMS Microbiol Ecol 2013;83(2):279–98.
[57] Nealson KH. Sediment bacteria: who's there, what are they doing, and what's new? Annu Rev Earth Planet Sci 1997;25:403–34.
[58] Snelgrove PVR. The importance of marine sediment biodiversity in ecosystem processes. Ambio 1997;26(8):578–83.
[59] Manini E, Fiordelmondo C, Gambi C, Pusceddu A, Danovaro R. Benthic microbial loop functioning in coastal lagoons: a comparative approach. Oceanol Acta 2003;26:27–38.
[60] Pusceddu A, Fiordelmondo C, Danovaro R. Sediment resuspension effects on the benthic microbial loop in experimental microcosms. Microb Ecol 2005;50:602–13.
[61] Zheng B, Wang L, Liu L. Bacterial community structure and its regulating factors in the intertidal sediment along the Liaodong Bay of Bohai Sea, China. Microbiol Res 2014;169(7–8):585–92.
[62] del Giorgio PA, Cole JJ. Bacterial growth efficiency in natural aquatic systems. Annu Rev Ecol Syst 1998;29:503–41.
[63] del Giorgio PA, Williams PJB. Respiration in aquatic ecosystems. New York: Oxford University Press; 2005.
[64] Carreira C, Staal M, Middelboe M, Brussaard CPD. Autofluorescence imaging system to discriminate and quantify the distribution of benthic cyanobacteria and diatoms. Limnol Oceanogr Methods 2015;13(4):169–77.
[65] Sévin DC, Sauer U. Ubiquinone accumulation improves osmotic-stress tolerance in *Escherichia coli*. Nat Chem Biol 2014;10(4):266–72.
[66] Csonka LN. Physiological and genetic responses of bacteria to osmotic stress. Microbiol Rev 1989;53(1):121–47.
[67] Ventosa A, Nieto JJ, Oren A. Biology of moderately halophilic aerobic bacteria. Microbiol Mol Biol Rev 1998;62(2):504–44.
[68] Morin PJ, Mcgrady-steed J. Biodiversity and ecosystem functioning in aquatic microbial systems: a new analysis of temporal variation and species richness- predictability relations. Oikos 2004;(3)458–66.
[69] Balsamo M, Semprucci F, Frontalini F, Coccioni R. Meiofauna as a tool for marine ecosystem biomonitoring. In: Cruzado DA, editor. Marine ecosystems. Rijeka, Croatia: InTech; 2012. p. 77–104.
[70] McLachlan A, Brown A. The ecology of sandy shores. London, UK: Elsevier; 2006.
[71] Balsamo M, Albertelli G, Ceccherelli VU, Coccioni R, Colangelo MA, Curini-Galletti M, et al. Meiofauna of the Adriatic Sea: present knowledge and future perspectives. Chem Ecol 2010;26:45–63.
[72] Semprucci F, Losi V, Moreno M. A review of Italien research on free-living marine nematodes and the future perspectives on their use as ecological indicators. Mediterr Mar Sci 2015;16(2):352–65.
[73] Appeltans W, et al. The magnitude of global marine species diversity. Curr Biol 2012;22(23):2189–202.
[74] Moreno M, Semprucci F, Vezzulli L, Balsamo M, Fabiano M, Albertelli G. The use of nematodes in assessing ecological quality status in the Mediterranean coastal ecosystems. Ecol Indic 2011;11(2):328–36.

[75] Chang JS. Understanding the role of ecological indicator use in assessing the effects of desalination plants. Desalination 2015;365:416–33.

[76] Gesteira JLG, Dauvin J-C. Amphipods are good bioindicators of the impact of oil spills on soft-bottom macrobenthic communities. Mar Pollut Bull 2000;40(11):1017–27.

[77] De-La-Ossa-Carretero JA, Del-Pilar-Ruso Y, Loya-Fernández A, Ferrero-Vicente LM, Marco-Méndez C, Martinez-Garcia E, et al. Response of amphipod assemblages to desalination brine discharge: impact and recovery. Estuar Coast Shelf Sci 2016;172:13–23.

[78] Del-Pilar-Ruso Y, Martinez-Garcia E, Giménez-Casalduero F, Loya-Fernández A, Ferrero-Vicente LM, Marco-Méndez C, et al. Benthic community recovery from brine impact after the implementation of mitigation measures. Water Res 2015;70:325–36.

[79] Begher Nabavi SM, Miri M, Doustshenas B, Safahieh AR. Effects of a brine discharge over bottom polychaeta community structure in Chabahar bay. J Life Sci 2013;7(3):302–7.

[80] Del-Pilar-Ruso Y, De-la-Ossa-Carretero JA, Giménez-Casalduero F, Sánchez-Lizaso JL. Effects of a brine discharge over soft bottom polychaeta assemblage. Environ Pollut 2008;156(2):240–50.

[81] Sugni M, Mozzi D, Barbaglio A, Bonasoro F, Candia Carnevali MD. Endocrine disrupting compounds and echinoderms: new ecotoxicological sentinels for the marine ecosystem. Ecotoxicology 2007;16(1):95–108.

[82] Fernández-Torquemada, Y.; González-Correa, J. M.; Sánchez-Lizaso, J. L. Echinoderms as indicators of brine discharge impacts. Desalin Water Treat 2013, 51 (October), 567–573.

[83] Dauvin JC, Ruellet T. Polychaete/amphipod ratio revisited. Mar Pollut Bull 2007;55(1–6):215–24.

[84] Duarte CM. The future of seagrass meadows. Environ Conserv 2002;29(2):192–206.

[85] Unsworth RKF, van Keulen M, Coles RG. Seagrass meadows in a globally changing environment. Mar Pollut Bull 2014;83(2):383–6.

[86] Duarte CM, Cebrián J. The fate of marine autotrophic production. Limnol Oceanogr 1996;41(8):1758–66.

[87] Duarte CM, Dennison WC, Orth RJW, Carruthers TJB. The charisma of coastal ecosystems: addressing the imbalance. Estuar Coasts 2008;31(2):233–8.

[88] Sánchez-Lizaso JL, Romero J, Ruiz J, Gacia E, Buceta JL, Invers O, et al. Salinity tolerance of the Mediterranean seagrass *Posidonia oceanica*: recommendations to minimize the impact of brine discharges from desalination plants. Desalination 2008;221(1–3):602–7.

[89] Marín-Guirao L, Sandoval-Gil JM, Ruíz JM, Sánchez-Lizaso JL. Photosynthesis, growth and survival of the Mediterranean seagrass *Posidonia oceanica* in response to simulated salinity increases in a laboratory mesocosm system. Estuar Coast Shelf Sci 2011;92(2):286–96.

[90] Gacia E, Invers O, Manzanera M, Ballesteros E, Romero J. Impact of the brine from a desalination plant on a shallow seagrass (*Posidonia oceanica*) meadow. Estuar Coast Shelf Sci 2007;72(4):579–90.

[91] Fernández-Torquemada Y, Sánchez-Lizaso JL. Effects of salinity on leaf growth and survival of the Mediterranean seagrass *Posidonia oceanica*. J Exp Mar Biol Ecol 2005;320(1):57–63.

[92] Torquemada YF, Lizaso JLS. Monitoring of brine discharges from seawater desalination plants in the Mediterranean. Int J Environ Heal 2007;1(3):449–61.

[93] Garrote-Moreno A, Fernández-Torquemada Y, Sánchez-Lizaso JL. Salinity fluctuation of the brine discharge affects growth and survival of the seagrass *Cymodocea nodosa*. Mar Pollut Bull 2014;81(1):61–8.

[94] Roca G, Alcoverro T, Krause-Jensen D, Balsby TJS, Van Katwijk MM, Marbá N, et al. Response of seagrass indicators to shifts in environmental stressors: a global review and management synthesis. Ecol Indic 2016;63:310–23.

[95] Cambridge ML, Zavala-Perez A, Cawthray GR, Mondon J, Kendrick GA. Effects of high salinity from desalination brine on growth, photosynthesis, water relations and osmolyte concentrations of seagrass *Posidonia australis*. Mar Pollut Bull 2017;115(1):252–60.

[96] Norcross BL, Shaw RF. Oceanic and estuarine transport of fish eggs and larvae: a review. Trans Am Fish Soc 1984;113:153–65.

[97] Petereit C, Hinrichsen HH, Franke A, Köster FW. Floating along buoyancy levels: dispersal and survival of western Baltic fish eggs. Prog Oceanogr 2014;122:131–52.

[98] Osse JWM, Van Den Boogaart JGM, Van Snik GMJ, Van Der Sluys L. Priorities during early growth of fish larvae. Aquaculture 1997;155(1–4):249–58.

[99] Battaglene SC, Talbot RB. Effects of salinity and aeration on survival of and initial swim bladder inflation in larval Australian bass. Progress Fish Cult 1993;55(1):35–9.

[100] Fielder DS, Bardsley WJ, Allan GL, Pankhurst PM. The effects of salinity and temperature on growth and survival of Australian snapper, *Pagrus auratus*, larvae. Aquaculture 2005;250(1–2):201–14.

[101] Gracia-López V, Kiewek-Martínez M, Maldonado-García M. Effects of temperature and salinity on artificially reproduced eggs and larvae of the leopard grouper *Mycteroperca rosacea*. Aquaculture 2004;237(1–4):485–98.

[102] Hart PR, Purser GJ. Effects of salinity and temperature on eggs and yolk sac larvae of the greenback flounder *Rhombosolea tapirina*. Aquaculture 1995;136(3–4):221–30.

[103] Johnson DW, Katavic I. Mortality, growth and swim bladder stress syndrome of sea bass (*Dicentrarchus labrax*) larvae under varied environmental-conditions. Aquaculture 1984;38(1):67–78.

[104] Lein I, Tveite S, Gjerde B, Holmefjord I. Effects of salinity on yolk sac larvae of Atlantic halibut (*Hippoglossus hippoglossus* L.). Aquaculture 1997;156(3–4):291–303.

[105] Tandler A, Anav FA, Choshniak I. The effect of salinity on growth rate, survival and swimbladder inflation in gilthead seabream, *Sparus aurata*, larvae. Aquaculture 1995;135(4):343–53.

[106] Ostrach, D.J., Bass Larvae, Available from: http://www.ostrachconsulting.com.

[107] Blaxter JHS. Development: eggs and Larvae. In: Hoar WS, Randall DJ, editors. Fish physiology, 11. San Diego, California, USA: Elsevier; 1988. p. 177–252.

[108] Kestemont P, Jourdan S, Houbart M, Mélard C, Paspatis M, Fontaine P, et al. Size heterogeneity, cannibalism and competition in cultured predatory fish larvae: biotic and abiotic influences. Aquaculture 2003;227(1–4):333–56.

[109] Pepin P. Effect of temperature and size on development, mortality, and survival rates of the pelagic early life history stages of marine fish. Can J Fish Aquat Sci 1991;48(3):503–18.

[110] Vanderplancke G, Claireaux G, Quazuguel P, Madec L, Ferraresso S, Sévère A, et al. Hypoxic episode during the larval period has long-term effects on European sea bass juveniles (*Dicentrarchus labrax*). Mar Biol 2014;162(2):367–76.

[111] Tal TL, McCollum CW, Harris PS, Olin J, Kleinstreuer N, Wood CE, et al. Immediate and long-term consequences of vascular toxicity during zebrafish development. Reprod Toxicol 2014;48:51–61.

[112] Iso S, Suizu S, Maejima A. The lethal effect of hypertonic solutions and avoidance of marine organisms in relation to discharged brine from a destination plant. Desalination 1994;97(1–3):389–99.

[113] Swanson C. Early development of milkfish: effects of salinity on embryonic and larval metabolism, yolk absorption and growth. J Fish Biol 1996;48:405–21.

[114] Mihelakakis A, Kitajima C. Effects of salinity and temperature on incubation period, hatching rate, and morphogenesis of the silver sea bream, *Sparus sarba*. Aquaculture 1994;126(3–4):361–71.

[115] Gibson S, Johnston IA. Temperature and development in larvae of the turbot *Scophthalmus maximus*. Mar Biol 1995;124(1):17–25.

[116] Bagarinao T, Kungvankij P. An incidence of swimbladder stress syndrome in hatchery-reared sea bass. Aquaculture 1986;51:181–8.

[117] Trotter AJ, Battaglene SC, Pankhurst PM. Effects of temperature on initial swim bladder inflation and related development in cultured striped trumpeter (*Latris lineata*) larvae. Aquaculture 2003;224(1–4):141–58.
[118] Bonisławska M, Tański A, Nędzarek A, Tórz A. Effect of the coagulants PAX and PIX on the embryonic development of pike (*Esox lucius*). Limnol Rev 2012;12(3):125–32.
[119] Birkeland C. Coral reefs in the anthropocene. In: Birkeland C, editor. Dordrecht, Netherlands: Springer; 2015.
[120] Moberg FF, Folke C. Ecological goods and services of coral reef ecosystems. Ecol Econ 1999;29(2):215–33.
[121] Graham NAJ, Wilson SK, Jennings S, Polunin NVC, Bijoux JP, Robinson J. Dynamic fragility of oceanic coral reef ecosystems. Proc Natl Acad Sci 2006;103(22):8425–9.
[122] Roth MS. The engine of the reef: photobiology of the coral-algal symbiosis. Front Microbiol 2014;5:1–22.
[123] Muller-Parker G, D'Elia CF, Cook CB. Interactions between corals and their symbiotic algae. In: Birkeland C, editor. Coral reefs in the anthropocene. Dordrecht, Netherlands: Springer; 2015. p. 1–271.
[124] Mayfield AB, Gates RD. Osmoregulation in anthozoan-dinoflagellate symbiosis. Comp Biochem Physiol 2007;147(1):1–10.
[125] Jokiel PL, Coles SL. Response of Hawaiian and other Indo-Pacific reef corals to elevated temperature. Coral Reefs 1990;8(4):155–62.
[126] Coles SL, Riegl BM. Thermal tolerances of reef corals in the Gulf: a review of the potential for increasing coral survival and adaptation to climate change through assisted translocation. Mar Pollut Bull 2013;72(2):323–32.
[127] Hughes TP, Baird A, Bellwood M, Card M, Connolly S, Folke C, et al. Climate change, human impacts and the resilience of coral reefs. Science 2003;301:929–33.
[128] Brown BE. Coral bleaching: causes and consequences. Coral Reefs 1997;16:129–38.
[129] Barshis DJ, Ladner JT, Oliver TA, Seneca FO, Traylor-Knowles N, Palumbi SR. Genomic basis for coral resilience to climate change. Proc Natl Acad Sci U S A 2013;110(4):1387–92.
[130] Sebens KP. Biodiversity of coral reefs: what are we losing and why? Am Zool 1994;34(1):115–33.
[131] Mabrook B. Environmental impact of waste brine disposal of desalination plants, Red Sea, Egypt. Desalination 1994;97(1–3):453–65.
[132] van der Merwe R, RÖthig T, Voolstra CR, OchsenkÜhn MA, Lattemann S, Amy GL. High salinity tolerance of the Red Sea coral *Fungia granulosa* under desalination concentrate discharge conditions: an in situ photophysiology experiment. Front Mar Sci 2014;1:1–8.
[133] Drushka K, Gille ST, Sprintall J. The diurnal salinity in the tropics. J Geophys Res Ocean 2014;119:5874–90.
[134] Lirman D, Manzello D. Patterns of resistance and resilience of the stress-tolerant coral *Siderastrea radians* (Pallas) to sub-optimal salinity and sediment burial. J Exp Mar Biol Ecol 2009;369(1):72–7.
[135] Manzello D, Lirman D. The photosynthetic resilience of *Porites furcata* to salinity disturbance. Coral Reefs 2003;22(4):537–40.
[136] Muthiga NA, Szmant AM. The effects of salinity stress on the rates of aerobic respiration and photosynthesis in the hermatypic coral *Siderastrea siderea*. Biol Bull 1987;173:539–51.
[137] Ferrier-Pagès C, Gattuso JP, Jaubert J. Effect of small variations in salinity on the rates of photosynthesis and respiration of the zooxanthellate coral *Stylophora pistillata*. Mar Ecol Prog Ser 1999;181:309–14.
[138] Lattemann S. Protecting the marine environment. In: Micale G, Rizzuti L, Cipollina A, editors. Seawater desalination, 46. Berlin, Germany: Springer; 2009. p. 273–99.

[139] Bruno JF, Selig ER, Casey KS, Page CA, Willis BL, Harvell CD, et al. Thermal stress and coral cover as drivers of coral disease outbreaks. PLoS Biol 2007;5(6):1220–7.
[140] Porter JW, Lewis SK, Porter KG. The effect of multiple stressors on the Florida Keys coral reef ecosystem: a landscape hypothesis and a physiological test. Limnol Oceanogr 1999;44:941–9.
[141] Schoepf V, Grottoli AG, Levas SJ, Aschaffenburg MD, Baumann JH, Matsui Y, et al. Annual coral bleaching and the long-term recovery capacity of coral. Proc R Soc B Biol Sci 2015;282:20151887.
[142] Hoegh-Guldberg O, Smith GJ. The effect of sudden changes in temperature, light and salinity on the population density and export of zooxanthellae from the reef corals *Stylophora pistillata* Esper and *Seriatopora hystrix* Dana. J Exp Mar Biol Ecol 1989;129:279–303.
[143] Fine M, Gildor H, Genin A. A coral reef refuge in the Red Sea. Glob Chang Biol 2013;19(12):3640–7.
[144] Long ER, Smith SL, Fred D, Der CAL. Incidence of adverse biological effects within ranges of chemical concentrations in marine and estuarine sediments. Environ Manag 1995;19:81–97.
[145] Bar-Zeev E, Belkin N, Liberman B, Berman-Frank I, Berman T. Bioflocculation: chemical free, pre-treatment technology for the desalination industry. Water Res 2013;47(9):3093–102.

FURTHER READING

[1] Marin-Guirao L, Sandoval-Gil JM, Ruz JM, Sánchez-Lizaso JL. Corrigendum to photosynthesis, growth and survival of the Mediterranean seagrass *Posidonia oceanica* in response to simulated salinity increases in a laboratory mesocosm system [Estuarine, Coastal and Shelf Science 92 (2011) 286–296]. Estuar Coast Shelf Sci 2011;93(1):86.
[2] Drami D, Yacobi YZ, Stambler N, Kress N. Seawater quality and microbial communities at a desalination plant marine outfall: a field study at the Israeli Mediterranean coast. Water Res 2011;45(17):5449–62.

CHAPTER 12

Microbial Communities in the Process and Effluents of Seawater Desalination Plants

Natalia Belkin*, Nurit Kress†, Ilana Berman-Frank*
*Bar Ilan University, Ramat Gan, Israel
†The National Institute of Oceanography, Haifa, Israel

12.1 INTRODUCTION

Scarcity of potable water due to the combined effects of climate change (driving changes in precipitation patterns and draughts) and population growth [1] has increased the global demand for desalination. At the end of 2015,[1] global desalination capacity was estimated at 86.6 $Mm^3 day^{-1}$ with seawater (SW) desalination accounting for ca. 60% of the global desalination effort [2,3]. From 1997 to 2008, the compound annual growth rate of desalination was 17%, and increased exponentially at a rate of 14% $year^{-1}$ from 2007 to 2012. Although the yearly rate of increase then declined to 3% $year^{-1}$ from 2012 to 2015 [3,4], desalination and in particular construction of large-scale seawater desalination plants are expected to continue expanding worldwide. At present, two desalination technologies dominate the desalination industry: the thermal process Multistage Flash (MSF), and the membrane-based Reverse Osmosis (RO), the latter accounting for more than 60% of the market share [4].

Regardless of the technology used in seawater desalination, SW (feed water) is pumped into the desalination plant and brine is discharged back to sea—impacting the marine environment [5]. Intake of seawater entrains and impinges marine organisms [6,7]. Yet, its impacts on populations and communities are rarely characterized. Brine discharge effects are associated with increased salinity, accompanied by higher temperatures for thermal processes: brine from SWRO is discharged with about twice the seawater salinity, while brine from MSF has about 1.5 seawater salinity and a temperature higher by ca. 5°C. Environmental effects can be associated also with the

[1]http://www.iwa-network.org/desalination-past-present-future/

Sustainable Desalination Handbook
https://doi.org/10.1016/B978-0-12-809240-8.00012-5

chemicals used in the desalination process that are then discharged at sea. Those may include coagulants in the pretreatment stage (iron or aluminum salts, polymers—RO only); antifoaming agents (MSF only), corrosion inhibitors, biocides (such as chlorine), and neutralizers (sodium sulfite); antiscalants to prevent fouling of the membranes (such as polyphosphates, polyphosphonates, polyacrylic acid, polymaleic acid); cleaning solutions for RO membranes (acidic and alkaline solutions and detergents); pH and hardness adjustors for the product water (limestone). Thermal desalination plants are also subject to corrosion and subsequent discharge of metals (such as copper) with the brine [7].

In contrast to a continuously expanding number of papers published on desalination technologies, a relatively small number of studies have been published quantifying the environmental impacts of discharging brine in situ or in laboratory experiments [7a] (Fig. 12.1). Although the number of publications on environmental impacts is gaining in the last few years, there still exists a large gap between the two perspectives reflecting the high investment in technologies and the lag in funds and regulations for environmental research and impact studies.

The published studies are site specific, demonstrating variability that is also influenced by the desalination process, the size of the plant, the discharge composition, and the sensitivity of the receiving environment.

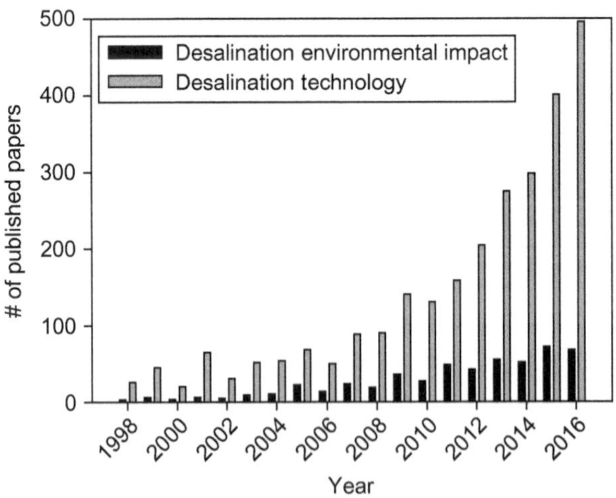

Fig. 12.1 The number of published peer-reviewed scientific papers from 1998 to 2016 on the subjects: Desalination technology; Desalination environmental impact. *Source: Web of Science, 3/5/2017.*

Thus, while brine discharge typically increases salinity (and sometimes temperature) at discharge sites, the areas affected are highly variable. In SE Spain, high salinity (45, compared with the ambient 37–38) was observed up to 2 km from the SWRO outfall in winter, while in summer it was confined to the vicinity of the outfall [8]. In Israel, brine co-discharged with cooling waters of a power plant at the Mediterranean shoreline increased salinity (by up to 1.04) and temperature (up to 7.8°C) with spatial impacts reaching 2 km from the discharge point [9]. In contrast, brine discharged through a marine outfall at 10-m depth increased salinities (45, ambient average 39) of the bottom 1–2 m water layer with a spatial effect of up to1 km from the outfall [10]. In Western Australia, brine discharged through an outfall increased salinity by 1 in bottom layer waters, yet remained within the range of natural variability (ambient salinity of 33 to 37, [11]). More widespread increases in offshore salinity could also occur as reported from Kuwait [12]. The salinity at the discharge area increased from 36–41.5 to 41–43 from 1993 to 2002, while the natural seasonal variation decreased, with no concurrent change in temperature [12]. Following the expansion of the plant, dispersion of higher salinities increased to several kilometers from the outfall. Other chemical characteristics of water quality may be affected by the discharges, yet data are extremely limited. Whether this is valid for all other sites, and at different spatial and temporal scales is currently unknown.

Brine discharges also affect the coastal biota—both benthic and pelagic. Short-term (15 days) mesocosm experiments examining the impacts of salinity increase (by 1 above the natural range) on the seagrasss *Posidonia oceanica* resulted in significant reduction in leaf growth, increased necrosis, enhanced senescence, and an increase in mortality rates [13]. In an in situ salinity manipulation experiment, *P. oceanica* exposed to desalination brine with a salinity of 1 and 2.5 above ambient for a 3-month period showed a decrease in the number of leaves, the maximum leaf length, growth rates, biomass, and plant survival when compared with the control ambient populations [14]. A field survey of a shallow *P. oceanica* meadow exposed to RO brine discharge for more than 6 years revealed low shoot abundance, significant reduction in leaf size, and overload of epiphytes in the area nearest the outfall (salinity of 38.4–39.8) [15]. In contrast, three tropical seagrasses (*Thalassia testudinum, Halodule wrightii,* and *Ruppia maritima*) were highly tolerant of hypersaline conditions. The stress indicators (shoot decline, growth rates, and quantum yields of photosynthesis) demonstrated tolerance up to salinities of 55 to 65 [16]. Moreover, total shoot mortality was not affected for any of the three species across the range of salinities examined (35–70 salinity over 30-day exposures) [16]. Discharges at a site on the southeastern Mediterranean coast of Spain increased coastal salinity and changed the benthic community from a mixed polychaeta, crustacea, and mollusca to one dominated by nematodes [17], while at the same site the amphipod populations decreased in abundance and diversity [18]. Discharges did not show any impacts on benthic communities from the northwest Mediterranean [19], examined for 12 months prior and 12 months following start of operations nor in Antigua [20], following 6 months of exposure to higher salinity and temperature or in Australia (Perth) after 15 months of operations. In the latter, however, benthic communities changed in the whole Cockburn Sound as a result of altered sediment composition, unrelated to the brine discharge [21]. These results emphasize the importance and lack of long-term quantitative studies as well as on the different trophic levels of the marine food web.

Currently, only a few studies investigating the impacts of desalination intakes and discharges on marine microorganisms (heterotrophic bacteria, cyanobacteria and unicellular eukaryotes) have been published [9,22–24]. Marine microorganisms dominate the abundance, diversity, and metabolic activity of the oceans, representing the "unseen majority" of the marine environment [25]. They sustain marine organic productivity (both autotrophic and heterotrophic), comprise a vast and dynamic reservoir of genetic variability, and are a fundamental link within the oceanic biogeochemical cycles of carbon and other nutrients [26,27]. Microbial communities, which are the basis of the marine food web (Fig. 12.2), constitute the primary and direct path of energy and biomass to the higher trophic levels in the ecosystem, and thus can influence the response and vulnerability of the whole ecosystem [28]. The interaction between primary (autotrophic) and heterotrophic

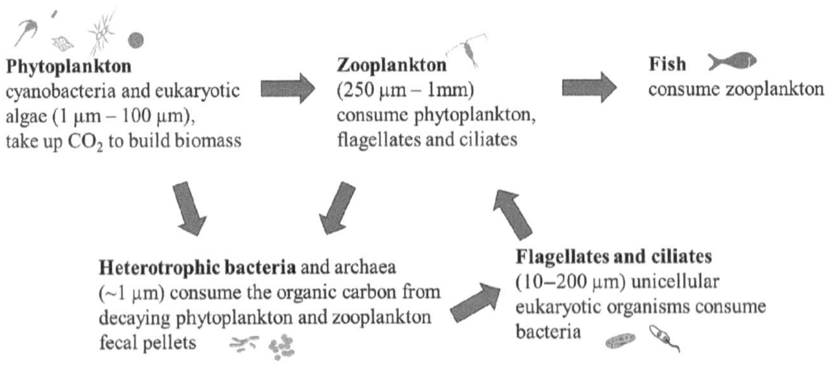

Fig. 12.2 Simplified marine food web and the basic components and functions of the microbial loop.

production within the microbial community is highly variable, and this variability impacts ecosystem function [29]. Moreover, a change in the physical and/or chemical environment due to the combined pressures of climate change and human impacts is often followed by a shift in structure and/or metabolic function change of the planktonic microbial community [30]. This may have potentially deleterious effects on biodiversity, abundance, productivity, and the ecological health of the marine ecosystem.

In this chapter, we will focus on the effects of desalination on the microbial community in seawater. This review complements and expands on Chapter 11. Section 12.2.1 discusses alterations to the microbial communities throughout the seawater desalination process from intake to discharge, while Section 12.2.2 assesses the effects of desalination discharges on coastal microbial communities using data from laboratory and in situ experiments. We conclude (Section 12.3) by addressing gaps in knowledge and with recommendations for plant operations that could mitigate the effects of desalination on the marine microbial community.

12.2 DESALINATION IMPACT ON MICROBIAL COMMUNITIES

Assessment of possible desalination impacts should begin with an environmental impact assessment (EIA) at the planning stages of a desalination plant and continue during construction. Although most of the impacts are assumed to be localized and cease after the construction phase, they could significantly affect populations during the process [7,31]. Seawater quality and microbial communities will most probably be affected by coastal construction that induces sediment resuspension, increases turbidity, and can release nutrients and pollutants to the water column. To the best of our knowledge, no data have been published comparing microbial community structure and function before, during, and postconstruction stages of large-scale desalination plants.

The assessment of possible impacts of desalination on the microbial community can be followed using the parameters and functions of the microbial community summarized in Table 12.1. During plant operations, the first stage affecting microbial populations is the intake of seawater (feed water), which entrains and transports the microbial biota with the flow of seawater into the desalination plant. We know of no published studies that have quantified the loss of entrained biota at either population or regional levels. The common assumption is that the organisms entering the

Table 12.1 Suggested parameters and required methodologies to assess microbial community structure and function

	Parameter/function	Indicator	Methodology	Reference
Autotrophs Phytoplankton	Biomass	Chlorophyll *a*	– Acetone extraction Fluorometric or spectrophotometric estimation – Chlorophyll *a* in situ fluorescence	[32], [33]
	Community composition	Microphytoplankton composition	– Microscopic determination of the microphytoplankton species – Taxonomic discrimination based on pigment combinations and ratios	[34]
		Picophytoplankton composition	Taxonomic discrimination based on the natural fluorescence of photosynthetic pigments, side-scatter and forward-scatter of the cells, enumerated by flow cytometry	[35]
		OTUs assignment	Molecular analysis—16S (cyanobacteria) and 18S (eukaryotic algae) rRNA gene extraction, sequencing, and Operational Taxonomic Units (OTU) assignment	[36]
	Diversity within community	Effective number of species	Calculated from the OTUs count	[37]
	Community composition shift	Diversity shared between communities	UNIFRAC metric for beta diversity measurements based on sequences from molecular analysis	[38]
	Productivity	Primary productivity	^{14}C or ^{13}C incorporation method	[39], [40]

Heterotrophs Bacteria	Biomass	Bacterial cell abundance	Nucleic acid staining, enumerated by flow cytometry	[35]
	Community composition	OTUs assignment	Molecular analysis—16S (cyanobacteria) rRNA gene extraction, sequencing, and Operational Taxonomic Units (OTU) assignment	[36]
	Diversity within community	Effective number of species	Calculated from the OTUs count	[37]
	Community composition shift	Diversity shared between communities	UNIFRAC metric for beta diversity measurements based on sequences from molecular analysis	[38]
	Productivity	Bacterial productivity	^{3}H leucine incorporation method	[41]

desalination plant will perish during the different stages of the desalination process, including when biocides are applied. However, this assumption is not always valid (see Section 12.2.1) and resistant microbial populations may remain in the water during the desalination process.

12.2.1 Entrainment of Microbial Communities and Their Fate Along the Desalination Process

In the desalination literature, microbial populations are mostly addressed in the context of membrane biofouling [42–51]. Several other investigations examined group-specific contributions such as harmful algae blooms (HABs) and their effects from the perspectives of plant operations and product water [52–54]. However, the main impact of entrainment on the marine environment, the quantification of the biomass lost during the desalination process, is usually mentioned only qualitatively. To the best of our knowledge, there are no publications of quantitative data excluding several studies on power plants, mainly concerning fish larvae [55]. Yet, the membrane biofouling literature provides the basis for future research on entrainment effects on the microbial community: loss during the desalination process and the communities, if any, returned to the sea with the brine.

The studies on membrane biofouling usually address the diverse bacterial communities in various compartments of the desalination process: feed and product water (free-living) and biofilm-forming communities in pretreatment filters and compartments and in membranes (usually RO). Following are some results taken from studies performed at operational RO desalination plants. *Alphaproteobacteria* and *Planctomycetaeota* dominated the longest operated SWRO membrane (330 days), while in the less used membrane (10 days) *Betaproteobacteria* predominated [49], hint of succession and temporal changes of the biofilm community. In parallel, the microbial composition of the RO membranes in Southern California was mainly *Alphaproteobacteria* (60%) followed by *Gammaproteobacteria* [50]. Similarly, in Israel, the bacterial community on the RO membrane was dominated by *Alphaproteobacteria* in spring and winter and *Gammaproteobacteria* in the summer, reflecting the feed water free-living communities that included mainly *Alphaproteobacteria* followed by *Gammaproteobacteria* [47]. All studies emphasized the dynamic interaction between the feed water planktonic communities and the biofouling on the membranes. These groups of planktonic heterotrophic bacteria, prevalent in the surface seawater, change their relative abundance seasonally according to the environmental conditions such as temperature, salinity, carbon source, and nutrients levels.

The biofouling on the membranes and the possible effects on the environment are probably also process dependent, in addition to being site and season specific as demonstrated in a bench scale (4-day) laboratory study [51]. The prevalent class in the feed water was the *Alphaproteobacteria*, which was superseded by *Betaproteobacteria* (*Burkholderiales*) in the membrane distillation (MD) filter and *Alphaproteobacteria* in the RO filters. The altered microbial populations for the two different membranes were attributed to the different processes: the community under RO is exposed to high pressure at low (room) temperatures while under MD the temperatures are high (50–80°C) with no hydraulic pressure [51].

Currently, we know of a sole study that addressed the communities discharged with the brine [47]. The general findings demonstrated that free-living communities at the feed waters did not highly differ from other free-living communities along the desalination process. The biofilm-forming communities did differ however and were related to the original seawater communities and therefore were site specific and characteristically seasonally dependent. Importantly, no significant differences were found between bacterial communities from the feed water to those discharged in the brine, showing that SWRO did not promote or stimulate proliferation of nonambient or exotic populations [47].

Free-living bacteria and eukaryotes were quantified in the water along the desalination process in Saudi Arabia, but not in the brine [43,44]. The resulting product water was not sterile and contained both bacteria (i.e., Firmicutes) and eukaryotes (i.e., Fungi). The source of these organisms in the product water was suggested to be bacteria (or only DNA) transferred at the time of installation, or/and intrusion through the membrane during normal plant operation.

12.2.2 Impact of Seawater Desalination Discharges on the Coastal Microbial Communities: Results from Laboratory, Mesocosm, and In Situ Studies

12.2.2.1 Salinity Effects: Mesocosm Study

In addition to impacts from brine discharges and the increasing salinities, the receiving environments of thermal desalination plants are often exposed to higher temperatures that may also affect the microbial populations. Halophilic and thermophilic species may thrive while other species, less tolerant, may dwindle and disappear. The relative contribution of the different species to the microbial communities is expected to change, and in turn, influence the higher trophic levels.

The main impact driver of the desalination effluent is the elevated salinity of the brine. Hypersalinity effects on the structure and function of natural assemblages of planktonic microbial populations were studied in large-volume (1 m^3) mesocosm experiments in the Eastern Mediterranean Sea (EMS) [56]. Excess salinities of 5% and 15% above ambient, simulating the natural salinity increases near desalination plant outfalls were examined. Rapid (within 2h upon treatment application) community physiological responses were detected at the 15% salinity treatment: primary productivity decreased (by 25%–60%) while heterotrophic bacterial productivity increased (by 36%–180%; Table 12.2) [56]. Similar physiological responses may occur in situ when the residence time of plankton at the discharge site is relatively short. Longer-scale responses (11–12 days) occurred mostly in the composition and structure of bacterial and eukaryotic communities accompanied by functional changes of enhanced primary and heterotrophic bacterial productivity (by 100%–150% and by 0%–200%, respectively; Table 12.3). The response was seasonally dependent, with a more distinct difference in the summer communities. The relative abundance of large cyanobacteria (*Oscillatoriophycideae*) and the heterotrophic *Alphaproteobacteria* (*Rhodobacterales*) increased. Concurrently, decrease was measured in the relative abundances of the small cyanobacteria (*Prochlorococcus*), heterotrophic *Alphaproteobacteria* (*Pelagibacteraceae*), and *Gammaproteobacteria* (*Altermondales*). The experiment simulated the in situ continuous brine discharge to the coastal areas with low turbulence that may retain the phytoplankton and bacterial communities under salinity stress. This may in turn lead to altered community composition caused by selection for salinity-tolerant microorganisms that may then reduce subsequent community diversity.

12.2.2.2 Chemical Discharges Along With Salinity Effects: Mesocosm Study

Microbial communities may be affected also by the chemicals discharged with the brine. These chemicals could have both negative (toxic or inhibiting) and positive (increasing bioavailability) impacts on the microbial populations. Coagulants and antiscalants are typically added during the desalination process to aid in the precipitation of large particles to facilitate their removal by prefiltration (coagulants) and to prevent membrane blockage by inorganic scaling (antiscalants). Iron hydroxides serve as common coagulants while antiscalants are characteristically comprised from different compounds including phosphonates. As in other topics covered in this review, only limited published data exist examining the effects of chemical additions to desalination discharges on the microbial populations.

Table 12.2 Immediate (within 2 h) biomass, productivity, and diversity changes (in percent from control treatments) in microbial communities during mesocosm experiments (compiled from Refs. [24,56])

	Parameter	Treatment: Salinity (Sal)	Treatment: Fe [% changes from control treatment]	Treatment: Phosphonate (Pn) [% changes from control treatment]	Treatment: Fe + Pn + Sal
Phytoplankton	Biomass	30 ↓	40–70 ↑	10–40 ↑	20–30 ↑
	Primary productivity (per chl *a*)	25–60 ↓	30–45 ↓	10–25 ↓	80–125 ↑
Heterotrophic bacteria	Cell abundance	No change	45–50 ↓	0–25 ↓	35–45 ↓
	Bacterial productivity (per cell)	40–180 ↑	125–185 ↑	0–25 ↑	30–85 ↑
Bacteria (autotrophs and heterotrophs)	Diversity	No change	45 ↓	No change	50 ↓
Eukaryotic algae	Diversity	No change	80 ↓	No change	60 ↓

Table 12.3 Biomass, productivity and diversity changes (calculated as a percent from the control values) after 10 days of incubation of microbial communities during mesocosm experiments (compiled from Refs. [24, 56])

		Treatment			
	Parameter	Salinity (Sal)	Fe [% changes from control treatment]	Phosphonate (Pn)	Fe + Pn + Sal
Phytoplankton	Biomass	100–400 ↑	0–50 ↓	No change	No change
	Primary productivity (per chl *a*)	100–150 ↑	30–50 ↓	No change	No change
Heterotrophic bacteria	Cell abundance	No change	No change	0–25 ↓	No change
	Bacterial productivity (per cell)	0–200 ↑	0–100 ↑	0–300 ↑	50–200 ↑
Bacteria (autotrophs and heterotrophs)	Diversity	85 ↓	20 ↑	75 ↓	40 ↑
Eukaryotic algae	Diversity	70 ↓	70 ↓	250 ↑	75 ↑

Simulation of such discharges was performed by examining the impacts of iron hydroxide and polyphosphonate additions on coastal microbial populations of the EMS under ambient and elevated salinities [24]. Immediately upon addition, the coagulant (iron hydroxide) changed the composition of the bacterial communities, increased the heterotrophic productivity (by 150%), and reduced primary productivity (by 40%) (Table 12.2). At the same timescale, adding only the antiscalant relieved the phosphorus stress of the community, while the combination of coagulant with antiscalant and elevated salinity (15% above ambient) resulted in synergistic effects reflected by the increased productivity of both primary and bacterial producers (by 100% and 50%, respectively) [24]. These effects may be found in situ during high turbulence and short residence time of the microbial communities within the discharge-affected areas.

Longer-term effect (10 days) revealed significant compositional shift only in the treatments in which coagulants, antiscalants, and high salinity were combined. This treatment reduced both function (photosynthetic rates) and biomass of primary producers by 50%, while the heterotroph bacterial activity and communities increased by 50% (Table 12.3) [24]. These effects may be found in situ during low turbulence and long residence times of the microbial communities within the discharge-affected areas.

12.2.2.3 In Situ Studies

While simulated experiments can isolate the specific factors impacting the microbial populations, in situ studies are necessary to assess the response of communities to the combined effects of these parameters in the natural setting. Changes in microbial communities were investigated from two different brine-discharge sites along the EMS coastline: (1) a submerged marine outfall equipped with diffusers, installed capacity of 90 $Mm^3 year^{-1}$ and (2) open outfall at the shoreline that was mixed with power plant cooling waters and well amelioration brine [24] with an installed capacity of 127 $Mm^3 year^{-1}$. Polyphosphonate-based coagulants were discharged by both plants, while iron-hydroxide coagulants were discharged at the open-discharge plant [57,58].

Seasonality governed the bacterial communities' structure at the nonaffected, ambient salinity stations [24]. Overall, the community shifts were weaker at the submerged outfall compared to an open outfall site, and occurred for the prokaryotic microorganisms (autotrophic and heterotrophic bacteria) at both sites, parallel to diversity changes (Table 12.4). The highest contribution to the shifts in community composition was attributed mostly to the interplay in the

Table 12.4 Biomass, productivity, and diversity changes (calculated as percent changes from the ambient station values) in microbial communities at the submerged outfall and open outfall sites of SWRO desalination brine discharges (compiled from Ref. [24, 56])

		Submerged outfall	Open outfall
		[% changes from control stations]	
Salinity elevation above ambient		10–38	4–5
Phytoplankton	Biomass	30–40 ↑ Summer 20–30 ↓ Winter	50–70 ↓
	Primary productivity (per chl *a*)	30 ↓ Summer	50 ↑ Summer 30 ↓ Winter
Heterotrophic bacteria	Cell abundance	No change	30 ↓ Winter
	Bacterial productivity (per cell)	No change	200 ↑ Summer 30 ↓ Winter
Bacteria (autotrophs and heterotrophs)	Diversity	50 ↓ Summer 27 ↓ Spring 21 ↓ Fall 60 ↑ Winter	5–40 ↓
Eukaryotic algae	Diversity	70 ↑ Summer	70 ↓ Summer 70 ↑ Winter

relative abundance of proteobacteria, cyanobacteria, and diatoms. Physiological changes accompanied the compositional ones. The community physiological responses to elevated salinity (and codischarged chemicals) was seasonal and site dependent (Table 12.4). At the submerged outfall site, only autotrophs were affected, mostly in summer, when reduced productivity (by 30%) was measured. At the open outfall site, the physiological changes were more prominent and included both phytoplankton and heterotrophic bacteria. The biomass was reduced, while productivities fluctuated between seasons. Although at the open outfall site the salinity elevation reached only 5% above the ambient, SWRO discharge included iron hydroxides (not present at the submerged outfall). Thus, the responses of the microbial community were similar to the iron-hydroxide treatment in the mesocosm approach [24].

An earlier preliminary study at the open outfall site was performed when backwash from the sand filters containing iron hydroxide was discharged in pulses with the continuous brine discharge at the shoreline [9]. This pulsed discharge increased turbidity, caused water discoloration, enhanced the concentrations of suspended particulate matter near the discharge point, and reduced phytoplankton densities and their productivity per chlorophyll *a* [9].

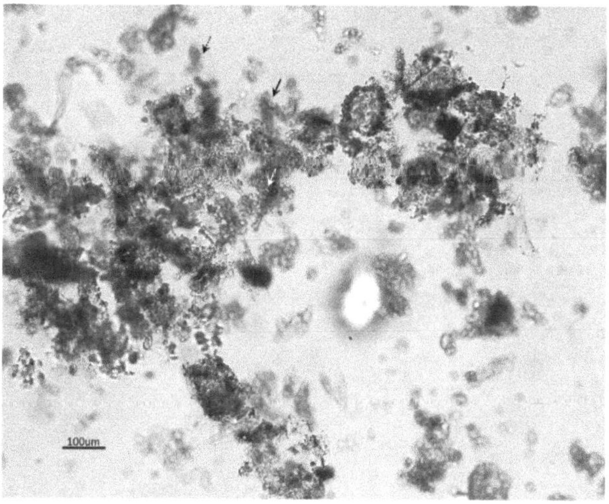

Fig. 12.3 Microscopic image of aggregates collected at the open outfall site on August 2012. Black arrow: phytoplankton, white arrow: bacteria, dashed black arrow: transparent exopolymer particles (TEP).

Mitigation measures stopped the pulsed discharge of the backwash by mixing it with the brine prior to its continuous discharge. This was expected to minimize the impact. Still, changes in community composition and function were detected in the vicinity of the discharge location [24], while aggregates (from the backwash discharge) containing bacteria and algae were present in the surface water (Fig. 12.3).

At other geographical sites, only minor changes were observed in the abundance of in situ microbial populations sampled at different distances from the submerged outfall site. Small variations in species abundance were attributed to a dilution effect, rather than a direct impact of the brine discharged, at the marine outfall of the King Abdullah University of Science and Technology (KAUST) SWRO Plant (Saudi Arabia, Red Sea) [23]. However, the plant is small (installed capacity of 15 $Mm^3 year^{-1}$) compared to the large capacity of the plants along the Mediterranean coastline discussed earlier. An early study from the coastal waters near Al-Jubail desalination plant (Saudi Arabia, Gulf) [22] also examined the plankton community. Higher primary production was evaluated in the coastal area that receives the water from the outfall bay, while the phytoplankton species composition did not change significantly. Concurrently, decreased abundances of diatoms, dinoflagellates, and cyanobacteria were measured at the outfall bay in comparison to those quantified from the open sea and the intake stations.

A summary of the measured effects of desalination discharges on microbial communities is compiled in Table 12.5. The composition of aquatic

Table 12.5 Summary of the effects of desalination brine and codischarged chemicals on the microbial communities based on in situ studies and on mesocosm and laboratory experiments

In situ studies

Area	Desalination technology	Stressor	Parameters	Observed effect (and stressor, when attributed)	Reference
Arabian Gulf	MSF	Salinity (S), Temperature (T)	Chl *a*, phytoplankton, zooplankton	*Increase:* PP *No change*: phytoplankton species composition *Decrease*: abundance of diatoms, dinoflagellates and cyanobacteria	[22]
Eastern Mediterranean Sea	RO with power plant cooling waters	Salinity (S), Temperature (T), Fe coagulant	Chl *a*, picoplankton number, PP, BP	*Inconclusive changes:* PP and BP *Decrease*: chl *a* and cell numbers (T and S) and phytoplankton growth efficiency (Fe)	[9]
Red Sea	RO	Salinity (S),	picoplankton number	*Reduced numbers* (Dilution of effluent)	[23]
Eastern Mediterranean Sea	RO with power plant cooling waters	Salinity (S), Temperature (T), Fe coagulant, Phosphonate (P) antiscalant	Chl *a*, picoplankton and bacteria numbers, PP, BP, 16S and 18S rDNA gene sequencing	*Increase*: *Gammaproteobacteria*, BP (Fe+P) and cell specific BP (Fe+P) *Decrease*: *Alphaproteobacteria* and cyanobacteria, chl *a* (Fe) and PP, diversity of bacteria and eukaryotes	[24]

Mesocosm experiments

Area	Experiment time	Stressor	Parameters	Observed effect (and stressor, when attributed)	Reference
Eastern Mediterranean Sea	Mesocosm, 10 d	Salinity (S)	Chl *a*, picoplankton and bacteria numbers, PP, BP, 16S and 18S rDNA gene sequencing	*Increase*: chl *a* (S), PP (S) *Inconclusive changes*: BP *Decrease*: biodiversity (S, seasonal dependent)	[56]
Eastern Mediterranean Sea	Mesocosm, 10 d	Salinity (S), Fe coagulant, Phosphonate (P) antiscalant	Chl *a*, picoplankton and bacteria numbers, PP, BP, 16S and 18S rDNA gene sequencing	*Increase*: PP (combined S, Fe, P), BP (combined S, Fe, P), eukaryotic diversity (F, combined S, Fe, P) and bacterial diversity (combined S, Fe, P) *Decrease*: Chl *a* (Fe), PP (Fe), eukaryotic diversity (Fe) and bacterial diversity (P)	[24]

Laboratory experiments

Area	Experiment type and time	Stressor	Parameters	Observed effect (and stressor, when attributed)	Reference
Pacific Ocean	Laboratory bioassay, 72 h	Salinity	EC_{50} of phytoplankton growth inhibition	Salinity range of 42–62 ppt	[64a]
Pacific Ocean (Australia)	Laboratory diatoms cultures, 14 d+14 d	Salinity	^{29}Si CP-MAS NMR	Changes in silica structure in one species (S)	[59]
Pacific Ocean (Australia)	Laboratory bioassay, 72 h	Brine (S and P)	NOEC of phytoplankton growth inhibition	At 11% brine in seawater (S and P)	[64b]

EC_{50}—Concentrations that caused an effect on 50% of the population, NOEC—No observed effect concentrations.

bacterial communities varies as a result of both natural (site specific and seasonal) variability and due to desalination impact. Compositional shifts in microbial community can reflect the chronic impacts of the outfall site. Such shifts may compensate for the metabolic stress encountered by some species at the site due to the salinity elevation or change in chemical composition of the local coastal water [60]. Yet, the altered microbial communities may also reduce the ecosystem's resilience to additional stressors such as climate change [61–64].

12.3 GAPS IN KNOWLEDGE AND OUTLOOK

Here we reviewed the potential impacts of seawater desalination on the marine planktonic microbial communities and described the scarce quantitative data available on the different components of the community's composition and functions. Accordingly, we believe the following gaps in knowledge should be addressed in future studies:

- Lack of baseline information on the ambient microbial communities (composition, abundance, metabolic, and growth rates) and their functions within the food web.
- No quantification of losses due to entrainment (including zooplankton, not addressed here).
- Lack of baseline information on the microbial communities discharged back to the coastal environment with the brine.
- Lack of short- and long-term in situ data at outfall areas.
- Lack of laboratory and mesocosm studies at relevant temporal scales.

These gaps may be filled by

- Incorporating the study of microbial communities to environmental impacts studies prior to plant construction and during long-term monitoring studies.
- Performing dedicated laboratory and mesocosm studies to assign a cause-response relationship among the different desalination stressors and their combined effects.
- Employing interdisciplinary in situ surveys and installing in situ automatic measuring devices to monitor changes in environmental parameters along the brine plumes (e.g., conductivity, temperature, density profilers (CTDs)).
  - Deploying new in situ instruments for continuous monitoring of biological parameters such as flow cytometry buoys and other imaging equipment for microbial and zooplankton populations.

- Utilizing newly available techniques such as hydrodynamic (e.g., Ref. [65]) and ecological (e.g., Ref. [66]) modeling.
- Determining eco-region specific ecological indicators, to be continuously monitored [67], including several indices for microbial species and communities.

In the meantime, based on the available knowledge and technology and on precautionary principles, several mitigation measures may be applied to reduce the possible effects of desalination on the marine microbial communities. Impacts can be reduced during the construction of the plants by more environmental sustainable construction measures. For example, pipeline installation by pipe jacking as far as possible from shore with controlled dredging beyond. Entrainment can be reduced by locating the intakes away from biologically productive areas, such as in deeper waters farther offshore, or by using underground beach wells, although the latter may be difficult to implement for large-scale desalination plants [6,68,69]. Brine management should strive to reduce brine discharge, improve its quality [70,71], and optimize its dilution. For direct discharge, the marine outfall should be equipped with a diffuser system, away from the shore so that exposure of the microbial populations to concentrated discharges is reduced. For open-channel discharges at the shoreline, dilution can be achieved by co-disposal, for example, with cooling waters from power plants. Alternative pretreatments for the desalination feed water, such as biofiltration and bioflocculation can be utilized instead of chemicals such iron hydroxide coagulants [72]. Other more environmentally friendly chemicals could replace more traditional antiscalants and coagulants. Alternatively, instead of discharging directly to the sea, the brine may be reused or a zero-liquid discharge policy promoted [31,73].

ACKNOWLEDGMENT

Funding for this study was granted by the Israel Ministry of Science, Technology, and Space—grant number 3-12373 to IBF and NK.

REFERENCES

[1] FAO. Coping with water scarcity. An action framework for agriculture and food security FAO Water Report 2012. p. 38.
[2] Khordagui, H. Assessment of potential cumulative environmental impacts of desalination plants around the Mediterranean Sea. SWIM Final report, Activity 1.3.2.1, 2013, 84.
[3] Lior N. Sustainability as the quantitative norm for water desalination impacts. Desalination 2017;401:99–111.
[4] Gude VG. Desalination and sustainability—an appraisal and current perspective. Water Res 2016;89:87–106.

[5] Kress N, Galil B. Impact of seawater desalination by reverse osmosis on the marine environment. In: Burn S, Gray S, editors. Efficient desalination by reverse osmosis. London: IWA; 2015. p. 177–202.
[6] NRC. Desalination, a national perspective. National Research Council of the National Academies. Washington, DC: The National Academies press; 2008.
[7] UNEP. In: Lattemann S, editor. Desalination resource and guidance manual for environmental impact assessments. United Nations Environment Programme, Regional Office for West Asia, Manama, and World Health Organization, Regional Office for the Eastern Mediterranean, Cairo; 2008. p. 168.
[7a] Roberts DA, Johnston EL, Knott NA. Impacts of desalination plant discharges on the marine environment: a critical review of published studies. Water Res 2010;44:5117–28.
[8] Fernandez-Torquemada Y, Gonzalez-Correa JM, Loya A, Ferrero LM, Diaz-Valdes M, Sanchez-Lizaso JL. Dispersion of brine discharge from seawater reverse osmosis desalination plants. Desalin Water Treat 2009;5:137–45.
[9] Drami D, Yacobi YZ, Stambler N, Kress N. Seawater quality and microbial communities at a desalination plant marine outfall. A field study at the Israeli Mediterranean coast. Water Res 2011;45:5449–62.
[10] Kress N, Galil BS. Seawater desalination in Israel and its environmental impact. Desal Water Reuse 2012;26–9.
[11] Holloway, K. Perth seawater desalination plant water quality monitoring programme. Final Programme summary Report 2005-2008. Report No. 445_001/3. Prepared by Oceanica Consulting Pty LTD for the Water Corporation of Western Australia, 2009.
[12] Uddin S, Al Ghadban AN, Khabbaz A. Localized hyper saline waters in Arabian Gulf from desalination activity-an example from South Kuwait. Environ Monit Assess 2011;181:587–94.
[13] Fernández-Torquemada Y, Sánchez-Lizaso JL, González-Correa JM. Preliminary results of the monitoring of the brine discharge produced by the SWRO desalination plant of Alicante (SE Spain). Desalination 2005;182:395–402.
[14] Ruiz JM, Marin-Guirao L, Sandoval-Gil JM. Responses of the Mediterranean seagrass *Posidonia oceanica* to *in situ* simulated salinity increase. Botanica Marina 2009;52:459–70.
[15] Sánchez-Lizaso JL, Romero J, Ruiz J, Gacia E, Buceta JL, Invers O, et al. Salinity tolerance of the Mediterranean seagrass *Posidonia oceanica*: recommendations to minimize the impact of brine discharges from desalination plants. Desalination 2008;221:602–7.
[16] Koch MS, Schopmeyer SA, Kyhn-Hansen C, Madden CJ, Peters JS. Tropical seagrass species tolerance to hypersalinity stress. Aquatic Botany 2007;86:14–24.
[17] Del Pilar Ruso Y, la Ossa Carretero JAD, Casalduero FG, Lizaso JLS. Spatial and temporal changes in infaunal communities inhabiting soft-bottoms affected by brine discharge. Mar Environ Res 2007;64:492–503.
[18] De-la-Ossa-Carretero JA, Del-Pilar-Ruso Y, Loya-Fernández A, Ferrero-Vicente LM, Marco-Méndez C, Martinez-Garcia E, et al. Response of amphipod assemblages to desalination brine discharge: Impact and recovery. Estuar Coast Shelf Sci 2016;172:13–23.
[19] Raventos N, Macpherson E, García-Rubiés A. Effect of brine discharge from a desalination plant on macrobenthic communities in the NW Mediterranean. Mar Environ Res 2006;62:1–14.
[20] Hammond, M., N. Blake, P. Hallock-Muller, M. Luther, D. Tomasko, and G. Vargo. Effects of disposal of seawater desalination discharges on near shore benthic communities. Phase Two Report. Report of Southwest Florida Water Management District. Electric Power Research Institute, University of South Florida, 1998.
[21] Shute, S. Perth desalination plant-cockburn Sound benthic macrofauna community and sediment habitat, Repeat Macrobenthic survey. Oceanica Consulting. Report No. 604-011/1, 2009, 202.

[22] Azis PKA, Al-tis IA, Daili MA, Green TN, Dalvi AGI, Javeed MA. Chlorophyll and plankton of the Gulf coastal waters of Saudi Arabia bordering a desalination plant. Desalination 2003;154:291–302.
[23] van der Merwe R, Hammes F, Lattemann S, Amy G. Flow cytometric assessment of microbial abundance in the near-field area of seawater reverse osmosis concentrate discharge. Desalination 2014;343:208–16.
[24] Belkin N, Rahav E, Elifantz H, Kress N, Berman-Frank I. The effect of coagulants and antiscalants discharged with seawater desalination brines on coastal microbial communities: a laboratory and in situ study from the southeastern Mediterranean. Water Res 2017;110:321–31.
[25] Whitman WB, Coleman DC, Wiebe WJ. Prokaryotes: the unseen majority. Proc Natl Acad Sci 1998;95:6578–83.
[26] Ignatiades L, Gotsis-Skretas O, Pagou K, Krasakopoulou E. Diversification of phytoplankton community structure and related parameters along a large-scale longitudinal east-west transect of the Mediterranean Sea. J Plankton Res 2009;31:411–28.
[27] Karl DM. Microbial oceanography: paradigms, processes and promise. Nat Rev Microbiol 2007;5:759–69.
[28] Bell T, Newman JA, Silverman BW, Turner SL, Lilley AK. The contribution of species richness and composition to bacterial services. Nature 2005;436:1157–60.
[29] Azam F, Malfatti F. Microbial structuring of marine ecosystems. Nat Rev Microbiol 2007;5:782–91.
[30] Bianchi CN, Morri C. Marine biodiversity of the Mediterranean Sea: situation, problems and prospects for future research. Mar Pollut Bull 2000;40:367–76.
[31] Lokiec F. Sustainable desalination: environmental approaches. In: Sustainable desalination: environmental approaches. The International Desalination Association World Congress on Desalination and Water Reuse, Tianjin, China; 2013.
[32] Lorenzen CJ. Determination of chlorophyll and pheo-pigments: spectrophotometric equations. Limnol Oceanogr 1967;12:343–6.
[33] Welschmeyer NA. Fluorometric analysis of chlorophyll *a* in the presence of chlorophyll *b* and pheopigments. Limnol Oceanogr 1994;39:1985–92.
[34] Knap, A.H., Michaels, A., Close, A.R., Ducklow, H., Dickson, A.G., Protocols for the joint global ocean flux study (JGOFS) core measurements, 1994, 180.
[35] Marie D, Simon N, Vaulot D. Phytoplankton cell counting by flow cytometry. In: Andersen RAE, editor. Algal culturing techniques. San Diego, CA: Academic Press; 2005. p. 1–17.
[36] Caporaso JG, Kuczynski J, Stombaugh J, Bittinger K, Bushman FD, Costello EK, et al. QIIME allows analysis of high- throughput community sequencing data. Nat Methods 2010;7:335–6.
[37] Jost L. Partitioning diversity into independent alpha and beta components. Ecology 2007;88:2427–39.
[38] Lozupone C, Hamady M, Knight R. UniFrac—an online tool for comparing microbial community diversity in a phylogenetic context. BMC Bioinform 2006;7:371–85.
[39] Mulholland MR, Bernhardt PW. The effect of growth rate, phosphorus concentration, and temperature on N_2 fixation, carbon fixation, and nitrogen release in continuous cultures of *Trichodesmium IMS101*. Limnol Oceanogr 2005;50:839–49.
[40] Steemann-Nielsen E. The use of radioactive carbon (^{14}C) for measuring organic production in the sea. J des Cons Int Pour Explor la Mer 1952;18:117–40.
[41] Kirchman D, Knees E, Hodson R. Leucine incorporation and its potential as a measure of protein-synthesis by bacteria in natural aquatic systems. Appl Environ Microbiol 1985;49:599–607.
[42] Ang WL, Mohammad AW, Hilal N, Leo CP. A review on the applicability of integrated/hybrid membrane processes in water treatment and desalination plants. Desalination 2015;363:2–18.

[43] Belila A, El-Chakhtoura J, Otaibi N, Muyzer G, Gonzalez-Gil G, Saikaly PE, et al. Bacterial community structure and variation in a full-scale seawater desalination plant for drinking water production. Water Res 2016;94:62–72.
[44] Belila A, El-Chakhtoura J, Saikaly P, van Loosdrecht M, Vrouwenvelder J. Eukaryotic community diversity and spatial variation during drinking water production (by seawater desalination) and distribution in a full-scale network. Environ Sci Water Res Technol 2017;3:92–105.
[45] Jiang S, Li Y, Ladewig BP. A review of reverse osmosis membrane fouling and control strategies. Sci Total Environ 2017;595:567–83.
[46] Khan MT, Manes C-LdO, Aubry C, Croué J-P. Source water quality shaping different fouling scenarios in a full-scale desalination plant at the Red Sea. Water Res 2013;47:558–68.
[47] Levi A, Bar-Zeev E, Elifantz H, Berman T, Berman-Frank I. Characterization of microbial communities in water and biofilms along a large scale SWRO desalination facility: site-specific prerequisite for biofouling treatments. Desalination 2016;378:44–52.
[48] Liyanaarachchi S, Shu L, Muthukumaran S, Jegatheesan V, Baskaran K. Problems in seawater industrial desalination processes and potential sustainable solutions: a review. Rev Environ Sci Biotechnol 2014;13:203–14.
[49] Manes CLD, West N, Rapenne S, Lebaron P. Dynamic bacterial communities on reverse-osmosis membranes in a full-scale desalination plant. Biofouling 2011;27:47–58.
[50] Zhang M, Jiang S, Tanuwidjaja D, Voutchkov N, Hoek EMV, Cai B. Composition and variability of biofouling organisms in seawater reverse osmosis desalination plants. Appl Environ Microbiol 2011;77:4390–8.
[51] Zodrow KR, Bar-Zeev E, Giannetto MJ, Elimelech M. Biofouling and microbial communities in membrane distillation and reverse osmosis. Environ Sci Technol 2014;48:13155–64.
[52] Anderson DM, McCarthy S. Red tides and harmful algal blooms: impacts on desalination operations. Muscat Oman: Middle East Desalination Research Center; 2012. p. 46.
[53] Caron DA, Garneau M-È, Seubert E, Howard MDA, Darjany L, Schnetzer A, et al. Harmful algae and their potential impacts on desalination operations off southern California. Water Res 2010;44:385–416.
[54] Seubert EL, Trussell S, Eagleton J, Schnetzer A, Cetinić I, Lauri P, et al. Algal toxins and reverse osmosis desalination operations: Laboratory bench testing and field monitoring of domoic acid, saxitoxin, brevetoxin and okadaic acid. Water Res 2012;46:6563–73.
[55] Barnthouse LW. Impacts of entrainment and impingement on fish populations: a review of the scientific evidence. Environ Sci Policy 2013;31:149–56.
[56] Belkin N, Rahav E, Elifantz H, Kress N, Berman-Frank I. Enhanced salinities, as a proxy of seawater desalination discharges, impact coastal microbial communities of the eastern Mediterranean Sea. Environ Microbiol 2015;17:4105–20.
[57] Glazer, A., Monitoring the coastal and marine environment at the discharge site of the Rutenberg power plant, VID desalination plant and Mekorot's well amelioration plant, results from 2014. RELP-4-2015, 2015, 115.
[58] Kress, N., Shoham-Frider, E., Lubinevsky, H., Monitoring the marine environment at the brine discharge site of the Palmahim/Soreq desalination plants, results from 2014, IOLR Report H18/2015, 2015, 151.
[59] Vars S, Johnston M, Hayles J, Gascooke J, Brown M, Leterme S, et al. 29Si{1H} CP-MAS NMR comparison and ATR-FTIR spectroscopic analysis of the diatoms *Chaetoceros muelleri* and *Thalassiosira pseudonana* grown at different salinities. Anal Bioanal Chem 2013;405:3359–65.
[60] Saikaly PE, Oerther DB. Diversity of dominant bacterial taxa in activated sludge promotes functional resistance following toxic shock loading. Microb Ecol 2011;61:557–67.

[61] Balvanera P, Pfisterer AB, Buchmann N, He J-S, Nakashizuka T, Raffaelli D, et al. Quantifying the evidence for biodiversity effects on ecosystem functioning and services. Ecol Lett 2006;9:1146–56.
[62] Bernhardt JR, Leslie HM. Resilience to climate change in coastal marine ecosystems. Ann Rev Mar Sci 2013;5:371–92.
[63] Rapport DJ, Costanza R, McMichael AJ. Assessing ecosystem health. Trends Ecol Evol 1998;13:397–402.
[64] Solan M, Cardinale BJ, Downing AL, Engelhardt KAM, Ruesink JL, Srivastava DS. Extinction and ecosystem function in the marine benthos. Science 2004;306:1177–80.
[64a] Yoon SJ, Park GS. Ecotoxicological effects of brine discharge on marine community by seawater desalination. Desalin Water Treat 2011;33:240–7.
[64b] Hobbs D, Stauber J, Kumar A, Smith R. Ecotoxicity of effluent from the proposed Olympic Dam Desalination Plant. Final Report. Hydrobiology Pty Ltd. Aquatic Environmental Services; 2008.
[65] Blumberg A, O'Neil S, Kim B. Environmental impacts of a desalination facility proposed for Tampa. Florida J Mar Environ Eng 2003;7:25–45.
[66] Heymans J, Coll M, Libralato S, Christensen V. Ecopath theory, modeling, and application to coastal ecosystems. In: Wolanski E, McLusky DS, editors. Treatise on estuarine and coastal science. 2012. p. 93–113.
[67] Chang J-S. Understanding the role of ecological indicator use in assessing the effects of desalination plants. Desalination 2015;365:416–33.
[68] Elimelech M, Phillip WA. The future of seawater desalination: energy, technology, and the environment. Science 2011;333:712–7.
[69] Rachman RM, Li S, Missimer TM. SWRO feed water quality improvement using subsurface intakes in Oman, Spain, Turks and Caicos Islands, and Saudi Arabia. Desalination 2014;351:88–100.
[70] Giwa A, Dufour V, Al Marzooqi F, Al Kaabi M, Hasan SW. Brine management methods: recent innovations and current status. Desalination 2017;407:1–23.
[71] Voutchkov N. Overview of seawater concentrate disposal alternatives. Desalination 2011;273:205–19.
[72] Bar-Zeev E, Belkin N, Liberman B, Berman-Frank I, Berman T. Bioflocculation: chemical free, pre-treatment technology for the desalination industry. Water Res 2013;47:3093–3102.
[73] Tong T, Elimelech M. The global rise of zero liquid discharge for wastewater management: drivers, technologies, and future directions. Environ Sci Technol 2016;50:6846–55.

FURTHER READING

[1] Del Pilar-Ruso Y, De-la-Ossa-Carretero JA, Gimenez-Casalduero F, Sanchez-Lizaso JL. Effects of a brine discharge over soft bottom Polychaeta assemblage. Environ Pollut 2008;156:240–50.
[2] Dupavillon JL, Gillanders BM. Impacts of seawater desalination on the giant Australian cuttlefish *Sepia apama* in the upper Spencer Gulf, South Australia. Mar Environ Res 2009;67:207–18.
[3] Fernández-Torquemada Y, Sánchez-Lizaso J. Responses of two Mediterranean seagrasses to experimental changes in salinity. Hydrobiologia 2011;669:21–33.
[4] Lattemann S, Hopner T. Impacts of seawater desalination plants on the marine environment of the Gulf. In: Abuzinada AH, Barth HJ, Krupp F, Böer B, Al Abdessalaam TZ, editors. Protecting the Gulf's marine ecosystems from pollution. Switzerland: Birkhäuser Verlag; 2008. p. 191–205.

[5] Le Page, S. Salinity Tolerance Investigations: a Supplemental report for the Carlsbad, CA Desalination project. Report presented to Poseidon Resources, 2005, 8.
[6] Marín-Guirao L, Sandoval-Gil JM, Bernardeau-Esteller J, Ruíz JM, Sánchez-Lizaso JL. Responses of the Mediterranean seagrass *Posidonia oceanica* to hypersaline stress duration and recovery. Mar Environ Res 2013;84:60–75.
[7] McConnell, R. Tampa Bay Seawater Desalination Facility—environmental impact monitoring. Proceedings of 2009 Annual WateReuse Conference, Seattle, WA, 2009.
[8] Nasrolahi A, Pansch C, Lenz M, Wahl M. Being young in a changing world: how temperature and salinity changes interactively modify the performance of larval stages of the barnacle *Amphibalanus improvisus*. Mar Biol 2012;159:331–40.
[9] Purnama A, Al-Barwani HH, Smith R. Calculating the environmental cost of seawater desalination in the Arabian marginal seas. Desalination 2005;185:79–86.
[10] Riera R, Tuya F, Ramos E, Rodríguez M, Monterroso Ó. Variability of macrofaunal assemblages on the surroundings of a brine disposal. Desalination 2012;291:94–100.
[11] Riera R, Tuya F, Sacramento A, Ramos E, Rodriguez M, Monterroso O. The effects of brine disposal on a subtidal meiofauna community. Estuar Coast Shelf Sci 2011;93:359–65.
[12] Sandoval-Gil JM, Marin-Guirao L, Ruiz JM. Tolerance of Mediterranean seagrasses (*Posidonia oceanica* and *Cymodocea nodosa*) to hypersaline stress: water relations and osmolyte concentrations. Mar Biol 2012;159:1129–41.
[13] Walker DI, McComb AJ. Salinity response of the seagrass *Amphibolis antarctica* (Labill.) Sonder et Aschers. an experimental validation of field results. Aquatic Botany 1990;36:359–66.

CHAPTER 13

Impact of Algal Blooms and Their Toxins on Reverse Osmosis Desalination Plant Operations

Erica L. Seubert[1], David A. Caron
University of Southern California, Los Angeles, CA, United States

The combination of a swelling global population along with the pressures of climate change has brought potable water availability and safety to the forefront of international concern. While desert regions and island nations have long employed desalination to augment local water supplies, semiarid regions have recently begun turning to the technology to supplement supplies during periods of prolonged drought. One example of this trend is southern California, USA, which has experienced periodic shortages of freshwater availability since the early 1900s. A period of sustained and severe drought in California persisted from 2012 to 2016, and during that time the state experienced the worst drought conditions it has seen in the last 1000 years [1]. Climate change models predict that droughts in the southwestern United States will only increase in incidence and severity [2], making the need to expand seawater desalination efforts in the area even more pressing. This demand has resulted in an expansion of desalination methodologies, and testing of these approaches in laboratory studies and pilot plants [3].

At present, reverse osmosis (RO) is the dominant technology used for desalination worldwide [4,5]. It surpassed the installed capacity of thermal desalination in the late 1990s, and its usage continues to rise in importance as cost decreases and the demand for potable water increases. A primary limiting factor behind RO desalination is the price of the membranes, and any action that a plant can take to extend the life of their membranes potentially reduces expenditures. Relevant to this point, the microalgal content of source water used in a plant can have a tremendous impact on RO desalination

[1]Present address: Department of Life Sciences, Los Angeles Mission College, 13,356 Eldridge Avenue, Sylmar, CA 91342, United States.

Sustainable Desalination Handbook
https://doi.org/10.1016/B978-0-12-809240-8.00013-7

operation and cost, including prefiltration processes, RO membranes, and/or amount of chemical consumption [6–10].

A well-documented increase in the global incidence of harmful algal blooms (HABs) in recent decades foreshadows an increased risk to seawater desalination plants [11–13]. Addressing this threat requires knowledge and expertise regarding the causes, duration, and type of HAB events. Broadly defined, HABs include any instance in which a microalgal species has a negative impact on the local ecosystem, although they range tremendously in nature, magnitude, and potential for toxicity. The most common HAB event is typified by a dramatic increase in microalgal abundance, but can also include production of compounds that alter the concentration of nonliving particulate and dissolved organic matter, as well as toxic and/or noxious compounds. Over 65 countries have reported experiencing at least one HAB event, and that number will undoubtedly continue to rise as HAB awareness and monitoring improves and HAB events escalate due to anthropogenic influences (Fig. 13.1; [14–17]). The massive abundances of microalgal cells that can exemplify HAB events may cause desalination plants to suspend operations until the event has ended. In 2008–2009, an HAB event involving large abundances of the dinoflagellate *Cochlodinium polykrikoides* caused desalination plants in the United Arab Emirates to halt operation during the several month-long bloom [18–20]. The risk for large microalgal abundances to impact desalination activities has prompted many investigations to focus on pretreatment strategies to, among other concerns, reduce the impact on blooms on the desalination process [10,21–28]. Various intake methodologies have also received attention including

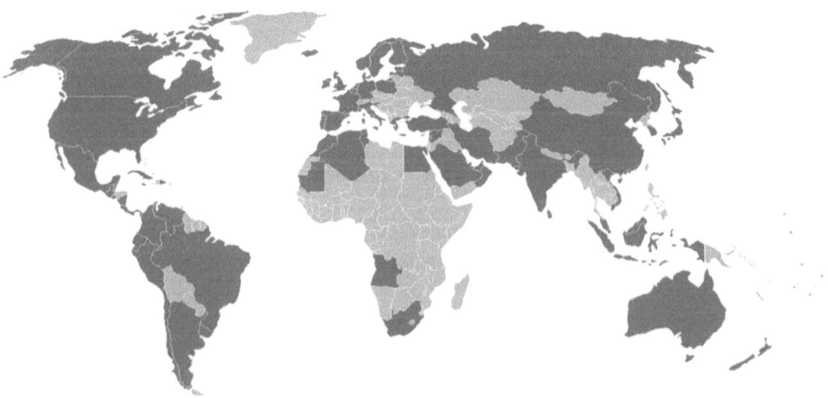

Fig. 13.1 Countries in which an HAB event has been reported in the primary literature is colored red, while countries without a documented HAB event are colored light gray. (For interpretation of the references to color in this figure legend, the reader is referred to the web version of this article.)

installation of a subsurface intake designed to reduce microalgal abundances prior to pretreatment versus an open water intake [29,30].

Large standing stocks of microalgae can cause immediate and direct problems for desalination operations, as discussed earlier; however, organic and inorganic materials resulting from their growth can also impact RO membrane performance. Transparent exopolymer particles (TEP) are gel-like polysaccharide particles found in both fresh and marine waters [31,32]. Their sticky nature causes them to form aggregates and provide a surface on which bacterial biofilms can develop [32–34]. Abundances of TEP have been recently recognized as being tightly coupled with microalgal abundances [31,32], and any period of high microalgal biomass should be expected to include high amounts of TEP. For desalination plants, large abundances of TEP in source waters can cause membrane fouling and therefore impact plant performance [35–37]. Various pretreatment strategies for TEP removal have been investigated; however, those investigations have also included findings that TEP removal was not always complete [38–41]. Monitoring of source water for TEP abundances and adjusting pretreatment strategies in response are the best suggestion for plant operators currently available [42]. TEP can also increase the amount of marine snow present in a given area, by assisting the aggregation of dead microalgal material as it sinks from the euphotic zone [43,44]. Marine snow is ubiquitously distributed throughout the oceans, but can be concentrated in "thin layers" and at subsurface depths [45,46]. Significant concentrations of marine snow in SWRO plant intake waters have been identified as a potential problem for membrane performance and should be considered for monitoring [47].

Newly realized contributors to HAB events are surfactant-like proteins produced by red-tide forming dinoflagellates, such as *Akashiwo sanguinea* [48,49]. These compounds have been implicated in mass mortalities of seabirds due to hypothermia stemming from wetting of the birds' feathers. The potential role of these surfactants in desalination operations is poorly known; however, they would be expected to adversely affect RO membrane performance. While surfactant production by *A. sanguinea* was unknown prior to the large 2009 bloom on the west coast of the United States, it has been implicated in HAB events in many regions of the world [50–54]. The worldwide distribution of this species highlights the need for desalination plants to be aware of the potential for surfactants in the intake water.

HAB events without large increases in microalgal abundance can also present potentially detrimental situations to desalination plant operations if toxic or noxious compounds are produced by the microalgae [11,13,55]. There are a wide variety of exceptionally toxic compounds produced naturally by microalgae, and it is possible for more than one of these compounds to be present in the water at any given time [56]. In southern California coastal waters, for example, there are microalgal species in the microalgal community capable of producing the neurotoxins domoic acid (DA), saxitoxin (STX), and brevetoxin (PbTx) [11,54,57,58]. These toxins are well known around the world, with many health departments and other consumer regulatory agencies dedicated to monitoring for the presence of these microalgal toxins in seafood [59–62]. RO membranes can be effective in removing these compounds [57,63], but the predicted molecule rejection of an RO membrane via molecular weight alone may not necessarily determine whether a particular compound will be successfully excluded from the product water [64]. The rejection of molecules can be influenced by the surface charge of the molecule and its hydrophobicity, duration of the toxic algal bloom, as well as other plant-operating conditions including specific flux and composition of the intake water. For these reasons, knowledge of the local potentially toxic species and their seasonality, and pilot testing of methods are highly recommended prior to installation of a large scale RO desalination plant.

Pseudo-nitzschia spp., a known producer of DA, is the microalgal genus most frequently involved in HAB events in the southern California region [11,57,58,65–68], although the global distribution of these species presents a concern to desalination operations in temperate regions worldwide [69]. Records of this diatom occurring in California waters have existed since the early 1900s [70–72]; however, the ability by local species to produce DA was not understood until the 1990s when there was an unusual mortality event involving brown pelicans (*Pelecanus occidentalis*) and Brandt's cormorants (*Phalacrocorax*) in Santa Cruz, CA [73,74]. The risk for unusual mortality events involving marine mammals and birds caused by DA in the coastal food web is a continuing concern for the US west coast [58,75–78]. The first and only human deaths due to DA poisoning occurred in 1987, when three people died after consuming DA contaminated blue mussels (*Mytilus edulis*) in Prince Edward Island, Canada, during a *P. multiseries* bloom [79,80]. Over 150 people became ill during the Canadian HAB event, with symptoms ranging from gastroenteritis (i.e., vomiting, diarrhea, abdominal cramping) to memory loss, disorientation, seizures, and coma [81–83]. The vigilance of present-day monitoring programs of this toxin in shellfish by state health departments, such as the Marine Biotoxin Monitoring Program (MBMP) by the California Department of Health [84], has held human illness due to DA poisoning largely in check since the Canadian event. However, new threats to human health due to DA recently identified, such as recreational fishing activities [85,86], demonstrated presence of DA in trophic levels not known to directly consume microalgae [85,87–89], and seawater desalination [11,13,55,57,63,90]. RO membranes typically employed in seawater desalination are thought to reject molecules with molecular weights larger than 250 g/mol, and DA has a molecular weight of 311.14 g/mol. However, the nearness of the molecular weight of DA relative to the molecular weight target of the average RO membrane, coupled with the concern for human health, has precipitated a few empirical tests to verify the effectiveness of RO to remove this toxin [57,63].

STX is the parent compound of a group of neurotoxins identified as paralytic shellfish toxins (PSTs), and it is the most potent of the PSTs [91]. STX is also the most powerful marine toxin currently known and is listed as a Schedule 1 chemical weapon [91,92]. PST poisoning in humans produces symptoms that may include tingling of the lips and/or face, headache, dizziness, increased salivation, thirst and perspiration, vomiting, diarrhea, and stomach cramps. Death, caused by muscular paralysis and respiratory difficulties culminating in complete respiratory arrest, can occur within hours of consuming a lethal dose. HAB events involving PSTs are responsible for over 2000 illnesses worldwide every year and have a 5%–10% mortality rate [59]. Relatively low abundances of any STX-producing microalga can constitute a threat to human health [84,93,94]. STX production is known to be accomplished by the dinoflagellate *Alexandrium* spp., with the exact species involved and the characteristics of the HAB event they can produce differing between regions. For example, in the United States, there is a dramatic difference in the HAB events resulting from *Alexandrium* spp. on the northeast versus the west coast of the country. In the northeast United States, *Alexandrium* spp. produces toxic events that are accompanied by large microalgal abundances [95–97]. The HAB events have been shown to be seasonal, and the seasonality of the events has facilitated bloom forecast efforts, researchers are able to release a prediction as to the magnitude and spatial extent of the bloom expected for that year [98,99]. In southern California, *Alexandrium catanella* has been identified as the species responsible for STX production; however, toxic events are not accompanied by large microalgal biomass [11,54,57,100]. Though it is generally not as dominant along the coast of southern California as *Pseudo-nitzschia* spp., the toxicity of STX makes it a year-round concern in the region. Recent research has expanded the known distribution of STX-producing organisms along the California coast [94,100–102]. STX may be a concern for RO desalination operations as the molecular weight of the hydrosoluble STX is 299.3 g/mol, close to the hypothetical RO membrane rejection of 250 g/mol.

PbTxs encompass a suite of neurotoxins, of which 13 derivatives have been identified. The most common derivatives known to occur in the marine environment are PbTx-2, PbTx-3, and PbTx-9, with PbTx-2 and PbTx-3 known to be the most potent [103]. PbTx poisoning via consumption of contaminated shellfish by humans produces symptoms of nausea, diarrhea, abdominal pain, paresthesia, and respiratory irritation and/or failure [104]. Historically in the United States, PbTxs have presented the greatest threat to human health in the Gulf of Mexico, where the PbTx-producing dinoflagellate *Karenia brevis* forms nearly annual blooms [104–106]. The fragile *K. brevis* cells release PbTxs into the air when breaking waves cause the cells to rupture. The aerosolized toxins can have a tremendous impact on human health, for not only individuals at the beach [107–110], but also up to 6.4 km inland [111]. The lung functions of asthmatics are significantly affected by aerosolized PbTx presence, and they experience not only more intensive symptoms but also experience symptoms for longer periods than otherwise healthy individuals [112–115]. PbTxs are a worldwide concern, with multiple species of *Karenia* identified as PbTx producers during HAB events in many different countries [116–118]. Further complicating our complete understanding of PbTx risk in any given region is the fact that these toxins can also be produced by raphidophyte microalgae such as *Chattonella* spp., *Heterosigma akashiwo*, and *Fibrocapsa japonica* [54,119–122]. These raphidophytes are known to exploit brackish waters, forming blooms when coastal waters are influenced by significant freshwater flows. Raphidophyte HAB events are often accompanied by fish kill events, due to the combination of PbTx production and the production of other toxic compounds such as reactive oxygen species [123–128]. The worldwide distribution and multiple microalgal groups capable of PbTx production make it a microalgal toxin for concern. PbTxs are liposoluble, with large molecular masses around 899 g/mol, well above the theoretical RO membrane rejection capability of 250 g/mol. Effective rejection of these toxins by RO membranes can be inferred from their size, but unique solute-membrane interactions may occur due to the hydrophobicity of PbTxs.

13.1 ESTABLISHING THE PRESENCE OF MICROALGAL TOXINS IN DESALINATION INTAKE WATERS

Microalgal shellfish toxin data collected by health departments can be used by desalination plants in order to understand the risk for microalgal toxins to be present in their intake waters. For example, desalination plants planned for the southern California region have benefited from the decades of shellfish toxin data generated by the state's MBMP; however, information on shellfish toxins may not directly correlate with toxin concentration in the intake water. Shellfish toxin concentration can be influenced by many factors including, but not limited to, the spatial and temporal extent of a bloom as well as the shellfish toxin depuration rate and tissues selected for sampling [129–131]. In addition, plankton monitoring by health departments and research groups typically focuses on toxins contained in phytoplankton cells because particulate toxins pose the greatest risk to human health. Extracellular toxin concentrations, those dissolved in the surrounding seawater, are not routinely monitored; however, they are important when considering potential threats to desalination operations.

Confident determination of microalgal toxin risk to any given desalination plant can only be established by monitoring the water being used for desalination for both particulate and dissolved toxin concentrations. For instance, microalgal toxin monitoring was conducted at a southern California pilot RO desalination plant in El Segundo, CA, with the intent of understanding the impact that microalgal toxins may have on plant operations [57]. The plant was operated between 2005 and 2009 by the West Basin Municipal Water District (Carson, CA), and was colocated at the El Segundo Generating Station (33°55′49.88″N, 118°26′8.31″W). Sampling of the microprefiltrated intake and resulting desalinated water (permeate) was conducted on a weekly basis during months of full plant operation, and was examined for the presence and concentrations of the microalgal toxins DA, STX, and PbTx. Although microalgal toxins were detected frequently in the intake waters, they were not detected in the permeate at any point of the monitoring effort, suggesting the RO membranes were effective in removing the toxins. The surprising result from this monitoring effort was the pervasive presence of dissolved DA and STX (Fig. 13.2). This type of information on pilot plant operation can help generate best management practices for full-scale SWRO plants in regions impacted by toxic HAB events.

Particulate DA concentrations in the intake water of the pilot plant were most commonly detected in the spring and early summer months [57], consistent with previous work conducted in the southern California region [11,65–67,84,85,132–134]. Dissolved DA was measured in the intake water during periods concomitant with high particulate DA concentrations, but it was also infrequently detected in winter months. The maximum concentration of dissolved DA (10 μg/L) detected during the study period was higher

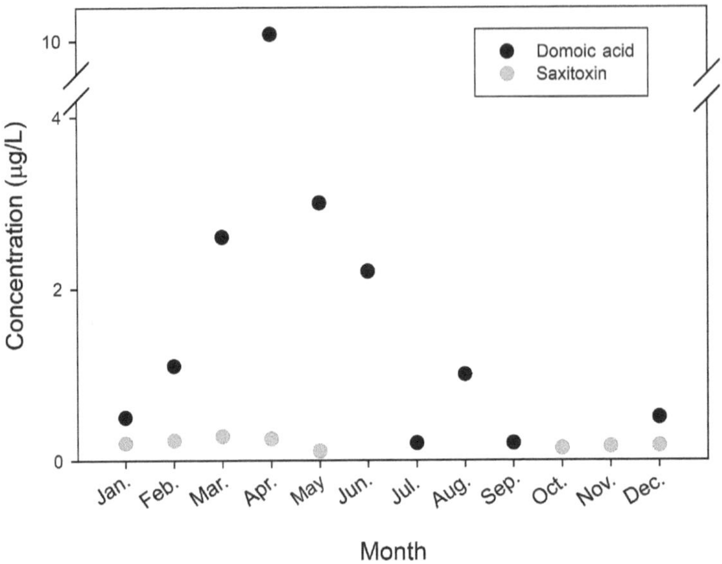

Fig. 13.2 Maximum concentrations of dissolved DA (black) and STX (gray) detected per month in the intake waters of a pilot SWRO desalination plant in El Segundo, CA.

than the maximum concentration of particulate DA (4.0 μg/L). Dissolved microalgal toxins are not commonly included in monitoring programs for reasons noted earlier, but the emergence of new technologies such as solid-phase adsorption toxin tracking (SPATT) has dramatically increased understanding of their importance. SPATT uses small mesh bags containing chromatography resins to adsorb dissolved toxins present in the water during their deployment. A diversity of resins are commercially available to allow researchers around the world to employ SPATT methodologies tailored to the chemical properties of the toxins known to be present in their specific locales. When employed in Monterey Bay, CA, dissolved DA concentrations detected using SPATT indicated an impending *Pseudo-nitzschia* spp. bloom weeks before "traditional" shore-based sampling signaled the presence of a bloom [135].

The El Segundo monitoring effort focused on dissolved STX sampling during 2008 and 2009, and STX was detected in nearly every month sampled (Fig. 13.2; [57]). Prior to this work, measurements of dissolved STX were not a routine component of any southern California monitoring effort, and STX was considered to be an infrequent contributor to toxic HAB events in the region [11,100]. An analysis of PST concentrations in shellfish along the California coast by the MBMP indicates that, in recent years, STX has been more prevalent in northern California than in southern California [11,84]. Accordingly, most previous studies of this class of toxins on the US west coast have focused primarily on occurrence in central and northern California as well as Oregon and Washington [101,102,136]. Even minute concentrations of STX in oceanic waters can raise human health concerns due to the extreme toxicity of this class of compounds [59]. SPATT monitoring in the area conducted in the years since has validated the El Segundo monitoring results, demonstrating that dissolved STX can be present throughout the year, albeit at low concentrations [137]. These findings of the pervasiveness of dissolved STX in southern California highlight the need for a clearer understanding of the distribution and concentration of dissolved STX in any region known to have STX-producing species of microalgae in their waters.

Monitoring at the El Segundo pilot plant failed to yield detectable levels of either particulate or dissolved PbTx, which is in keeping with the fact that PbTx has not been previously demonstrated in the region [11,57]. Raphidophyte species capable of producing PbTx have been documented in these coastal waters [11,54], making it prudent for any desalination plant for the region to investigate potential presence of the toxin near any planned intake pipes.

13.2 EXPERIMENTAL INVESTIGATIONS INTO MICROALGAL TOXIN REMOVAL BY RO MEMBRANES

An understanding of the potential impact microalgal species may have on RO desalination efforts in a given area is necessary in the planning and implementation effort of any desalination plant. Knowledge of microalgal species of concern common to the intake area and an awareness of the impact microalgae may have on the product water can allow plant managers to plan for times of unsuitable production more effectively. The combination of monitoring efforts and bench-scale experiments demonstrating the same result of microalgal toxin rejection by RO membranes allows confident determination of the efficacy of RO removal of microalgal toxins.

For example, the monitoring efforts at the El Segundo pilot plant discussed earlier was accompanied by bench-scale RO experiments modeling worst-case scenario toxin concentrations for southern California [57]. The bench-scale RO setup was designed to represent normal operating conditions (i.e., pressure, specific flux, and recovery) expected for the pilot plant. Concentrations were chosen to reflect worst-case scenario concentrations, 50 μg/L DA, 2 μg/L STX, and 20 μg/L PbTx-2, well above what has been reported to naturally occur in the region [Reviewed in Ref. [11]]. No toxins were detected in the permeate water, suggesting RO membranes are capable of completely removing DA, STX, and PbTx from seawater under normal operating conditions, without the need for posttreatment.

Similar toxin rejection results were demonstrated in a study conducted with a manually operated RO desalination apparatus [63]. The authors reported >99% removal of toxin by the RO membranes, although the microalgal concentrations employed were significantly higher than the experiments conducted by Seubert et al. [57]. Initial toxin concentrations significantly exceeded levels observed in natural algal blooms [Reviewed in Ref. [11]], using dissolved concentration of 17,000 μg/L DA, 10,300 μg/L STX, and 900 μg/L PbTx. The similarity in results between the two sets of RO desalination experiments, despite employing different methods, further support the supposition that RO membranes are capable of preventing microalgal toxins from passing into product water under routine RO operations.

Microalgal toxins constitute a significant human health risk when contaminated seafood is consumed. While it has been demonstrated that these toxins can be present in SWRO desalination plant intake waters [57], the removal of DA, STX, and PbTx by RO membranes has been observed in both pilot plant investigations and laboratory experiments [57,63]. These results suggest microalgal toxin presence does not impact the desalinated product water; however, additional research should investigate whether or not the presence of other microalgal compounds (i.e., TEP, marine snow, surfactants) influence the efficiency of toxin removal by RO membranes.

REFERENCES

[1] Griffin D, Anchukaitis K. How unusual is the 2012–2014 California drought? Geophys Res Lett 2014;41(24):9017–23.
[2] Cayan D, et al. Future dryness in the southwest US and the hydrology of the early 21st century drought. Proc Natl Acad Sci U S A 2010;107(50):21271–6.
[3] Goh P, Ismail A, Hilal N. Nano-enabled membranes technology: sustainable and revolutionary solutions for membrane desalination? Desalination 2016;380:100–4.
[4] Ghaffour N, Missimer T, Amy G. Technical review and evaluation of the economics of water desalination: current and future challenges for better water supply sustainability. Desalination 2013;309:197–207.
[5] Greenlee L, et al. Reverse osmosis desalination: water sources, technology, and today's challenges. Water Res 2009;43:2317–48.
[6] Abdul Azis P, et al. Effects of environment on source water for desalination plants on the eastern coast of Saudi Arabia. Desalination 2000;132:29–40.
[7] Ladner D, Vardon D, Clark M. Effects of shear on microfiltration and ultrafiltration fouling by marine bloom-forming algae. J Membr Sci 2010;356(1–2):33–43.

[8] Zhang M, et al. Composition and variability of biofouling organisms in seawater reverse osmosis desalination plants. Appl Environ Microbiol 2011;77(13):4390–8.
[9] Peleka E, Matis K. Application of flotation as a pretreatment process during desalination. Desalination 2008;222:1–8.
[10] Penate B, Garcia-Rodriguez L. Current trends and future prospects in the design of seawater reverse osmosis desalination technology. Desalination 2012;284:1–8.
[11] Caron D, et al. Harmful algae and their potential impacts on desalination operations off southern California. Water Res 2010;44:385–416.
[12] Gobler C, et al. Ocean warming since 1982 has expanded the niche of toxic algal blooms in the North Atlantic and North Pacific oceans. Proc Natl Acad Sci U S A 2017;114(19):4975–80.
[13] Villacorte L, et al. Algal blooms: an emerging threat to seawater reverse osmosis desalination. Desalin Water Treat 2014;55(10):2601–11.
[14] Berdalet E, et al. Marine harmful algal blooms, human health and wellbeing: challenges and opportunities in the 21st century. J Mar Biol Assoc U K 2016;96(1):61–91.
[15] Glibert P, et al. The role of eutrophication in the global proliferation of harmful algal blooms. Oceanography 2005;18:198–209.
[16] Glibert P, Burford M. Globally changing nutrient loads and harmful algal blooms: recent advances, new paradigms, and continuing challenges. Oceanography 2017;30(1):58–69.
[17] Smayda T. Novel and nuisance phytoplankton blooms in the sea: evidence for a global epidemic. In: Graneli E, et al., editors. Toxic Marine Phytoplankton. New York: Elsevier; 1990. p. 29–40.
[18] Pankrantz T. Red tides close desal plants. Water Desalin Rep 2008;44.
[19] Richlen M, et al. The catastrophic 2008-2009 red tide in the Arabian gulf region, with observations on the identification and phylogeny of the fish-killing dinoflagellate *Cochlodinium polykrikoides*. Harmful Algae 2010;9:163–72.
[20] Nazzal N. 'Red tide' shuts desalination plant. Gulf News 2009.
[21] Pearce G, et al. Pretreatment options for large scale SWRO plants: case studies of UF trials at Kindasa, Saudi Arabia, and conventional pretreatment in Spain. Desalination 2004;167:175–89.
[22] Schurer R, et al. Three years operational experience with ultrafiltration as SWRO pre-treatment during algal bloom. Desalin Water Treat 2013;51(4–6):1034–42.
[23] Park K, et al. Algal bloom—critical to designing SWRO pretreatment and pretreatment as built in Shuwaikh, Kuwait SWRO by Doosan. Desalin Water Treat 2013;51(31–33):6317–28.
[24] Schurer R, et al. Performance of ultrafiltration and coagulation in an UF-RO seawater desalination demonstration plant. Desalin Water Treat 2012;42(1–3):57–64.
[25] Castaing J, et al. Investigating submerged ultrafiltration (UF) and microfiltration (MF) membranes for seawater pre-treatment dedicated to removal of undesirable micro-algae. Desalination 2010;253:71–7.
[26] Kim S-H, Yoon J-S. Optimization of microfiltration for seawater suffering from red tide contamination. Desalination 2005;182:315–21.
[27] Li L, Pan G. A universal method for flocculating harmful algal blooms in marine and fresh waters using modified sand. Environ Sci Technol 2013;47:4555–62.
[28] Gaid K, Treal Y. Le dessalement des eaux par osmose inverse: l'expérience de Véolia Water. Desalination 2007;203:1–14.
[29] Missimer T, et al. Subsurface intakes for seawater reverse osmosis facilities: capacity limitation, water quality improvement, and economics. Desalination 2013;322:37–51.
[30] Rachman R, Li S, Missimer T. SWRO feed water quality improvement using subsurface intakes in Oman, Spain, Turks and Caicos Islands, and Saudi Arabia. Desalination 2014;351:88–100.

[31] Alldredge A, Passow U, Logan B. The abundance and significance of a class of large, transparent organic particles in the ocean. Deep-Sea Res I 1993;40(6):1131–40.
[32] Passow U. Transparent exopolymer particles (TEP) in aquatic environments. Prog Oceanogr 2002;55:287–333.
[33] Azam F, Malfatti F. Microbial structuring of marine ecosystems. Nat Rev Microbiol 2007;5:782–91.
[34] Passow U, Alldredge A. Distribution, size and bacterial-colonization of transparent exopolymer particles (TEP) in the ocean. Mar Ecol Prog Ser 1994;113:185–98.
[35] Berman T, Mizrahi R, Dosoretz C. Transparent exopolymer particles (TEP): a critical factor in aquatic biofilm initiation and fouling on filtration membranes. Desalination 2011;276:184–90.
[36] Bar-Zeev E, et al. Transparent exopolymer particles: potential agents for organic fouling and biofilm formation in desalination and water treatment plants. Desalin Water Treat 2009;3:136–42.
[37] Meng S, et al. Transparent exopolymer particles (TEP) and their potential effect on membrane biofouling. Appl Microbiol Biotechnol 2013;97(13):5705–10.
[38] Villacorte L, et al. Measuring transparent exopolymer particles (TEP) as indicator of the (bio)fouling potential of RO feed water. Desalin Water Treat 2009;5:207–12.
[39] Villacorte, L., et al., The fate of transparent exopolymer particles (TEP) in seawater UF-RO system: a pilot plant study in Zeeland, the Netherlands. Desalin Water Treat, 2009. 13: p. 109–119.
[40] Kennedy M, et al. Transparent exopolymer particle (TEP) fouling of ultrafiltration membrane systems. Desalin Water Treat 2009;6:169–76.
[41] Villacorte L, et al. MF/UF rejection and fouling potential of algal organic matter from bloom-forming marine and freshwater algae. Desalination 2015;367:1–10.
[42] Berman T, Parparova R. Visualization of transparent exopolymer particles (TEP) in various source waters. Desalin Water Treat 2010;21:382–9.
[43] Turner J. Zooplankton fecal pellets, marine snow and sinking phytoplankton blooms. Aquat Microb Ecol 2002;27:57–102.
[44] Passow U, et al. Origin of transport exopolymer particles (TEP) and their role in the sedimentation of particulate matter. Cont Shelf Res 2001;21:327–46.
[45] Alldredge A, et al. Occurrence and mechanisms of formation of a dramatic thin layer of marine snow in a shallow Pacific fjord. Mar Ecol Prog Ser 2002;233:1–12.
[46] McManus M, et al. Characteristics, distribution and persistence of thin layers over a 48 hour period. Mar Ecol Prog Ser 2003;261:1–19.
[47] Ning R. Reverse osmosis process chemistry relevant to the Gulf. Desalination 1999;123:157–64.
[48] Jessup D, et al. Mass stranding of marine birds caused by a surfactant-producing red tide. PLoS One 2009;4(2):e4550. https://doi.org/10.1371/journal.pone.0004550.
[49] Phillips E, et al. Summary of birds killed by a harmful algal bloom along the south Washington and north Oregon coasts during October 2009. Northwest Nat 2011;92:120–6.
[50] Robinson M, Brown L. A recurrent red tide in a British Columbia coastal lagoon. Can J Fish Aquat Sci 1983;40:2135–43.
[51] Votolina D. The origin of recurrent blooms of *Gymnodinium sanguineum* (Hirasaka) in a shallow lagoon. J Exp Mar Biol Ecol 1993;168:217–22.
[52] Lu S, Hodgkiss I. Harmful algal bloom causative collected from Hong Kong waters. Hydrobiologia 2004;512:231–8.
[53] Badylak S, Philips E, Mathews A. *Akashiwo sanguinea* (Dinophyceae) blooms in a sub-tropical estuary: an alga for all seasons. Plankton Benthos Res 2014;9(3):147–55.
[54] Lewitus A, et al. Harmful algal blooms along the North American west coast region: history, trends, causes and impacts. Harmful Algae 2012;19:133–59.

[55] Villacorte L, et al. Seawater reverse osmosis desalination and (harmful) algal blooms. Desalination 2015;360:61–80.
[56] Van Dolah F. Marine algal toxins: origins, health effects, and their increased occurrence. Environ Health Perspect 2000;108(Suppl. 1):133–41.
[57] Seubert E, et al. Algal toxins and reverse osmosis desalination operations: laboratory bench testing and field monitoring of domoic acid, saxitoxin, brevetoxin and okadaic acid. Water Res 2012;46:6563–73.
[58] Smith J, et al. A decade and a half of Pseudo-nitzschia spp. and domoic acid along the coast of southern California. Harmful Algae submitted for publication.
[59] Hallegraeff G. Harmful algal blooms: a global overview. In: Hallegraeff G, Anderson D, Cembella A, editors. Manual on Harmful Marine Microalgae. Monographs on Oceanographic MethodologyParis: UNESCO Publishing; 2003. p. 25–49.
[60] Lelong A, et al. Pseudo-nitzschia (Bacillariophyceae) species, domoic acid and amnesic shellfish poisoning: revisiting previous paradigms. Phycologia 2012;51(2):168–216.
[61] Lilly E, Halanych K, Anderson D. Species boundaries and global biogeography of the *Alexandrium tamarense* complex (Dinophyceae). J Phycol 2007;43:1329–38.
[62] Wang D. Neurotoxins from marine dinoflagellates: a brief review. Mar Drugs 2008;6(2):349–71.
[63] Laycock M, et al. Laboratory desalination experiments with some algal toxins. Desalination 2012;293:1–6.
[64] Verliefde A, et al. A semi-quantitative method for prediction of the rejection of uncharged organic micropollutants with nanofiltration. Water Practice Technol 2006;1(4):https://doi.org/10.2166/WPT.2006084.
[65] Schnetzer A, et al. Coastal upwelling linked to toxic *Pseudo-nitzschia australis* blooms in Los Angeles coastal waters, 2005-2007. J Plankton Res 2013;35(5):1080–92.
[66] Seubert E, et al. Seasonal and annual dynamics of harmful algae and algal toxins revealed through weekly monitoring at two coastal ocean sites off southern California, USA. Environ Sci Pollut Res 2013;20(10):6878–95.
[67] Schnetzer A, et al. Blooms of Pseudo-nitzschia and domoic acid in the San Pedro Channel and Los Angeles harbor areas of the Southern California Bight, 2003-2004. Harmful Algae 2007;6(3):372–87.
[68] Seegers B, et al. Subsurface seeding of surface harmful algal blooms observed through the integration of autonomous gliders, moored environmental sample processors, and satellite remote sensing in Southern California. Limnol Oceanogr 2015;60:754–64.
[69] Trainer V, et al. Pseudo-nitzschia physiological ecology, phylogeny, toxicity, monitoring and impacts on ecosystem health. Harmful Algae 2012;14:271–300.
[70] Allen W. Marine plankton diatoms of lower California in 1931. Bot Gaz 1934; 95(3):485–92.
[71] Allen W. Occurrence of marine plankton diatoms in a ten-year series of daily catches in southern California. Am J Bot 1936;23(1):60–3.
[72] Fryxell G, Villace M, Shapiro L. The occurrence of the toxic diatom genus Pseudo-nitzschia (Bacillariophyceae) on the West Coast of the USA, 1920-1996: a review. Phycologia 1997;36(6):419–37.
[73] Fritz L, et al. An outbreak of domoic acid poisoning attributed to the pennate diatom *Pseudo-nitzschia australis*. J Phycol 1992;28:439–42.
[74] Work T, et al. Epidemiology of domoic acid poisoning in brown pelicans (*Pelecanus occidentalis*) and Brandt's cormorants (*Phalacrocorax penicillatus*) in California. J Zoo Wild Med 1993;24(1):54–62.
[75] Lefebvre K, et al. Detection of domoic acid in northern anchovies and California sea lions associated with an unusual mortality event. Nat Toxins 1999;7:85–92.
[76] Scholin C, et al. Mortality of sea lions along the central California coast linked to a toxic diatom bloom. Nature 2000;403:80–4.

[77] de la Riva G, et al. Association of an unusual marine mammal mortality event with Pseudo-nitzschia spp. blooms along the southern California coastline. J Wildl Dis 2009;45(1):109–21.
[78] Lefebvre K, et al. Clinical signs and histopathology associated with domoic acid poisoning in northern fur seals (*Callorhinus ursinus*) and comparison of toxin detection methods. Harmful Algae 2010;9:374–83.
[79] Bates S, et al. Pennate diatom *Nitzschia pungens* as the primary source of domoic acid, a toxin in shellfish from eastern Prince Edward Island, Canada. Can J Fish Aquat Sci 1989;46:1203–15.
[80] Wright J, et al. Identification of domoic acid, a neuroexcitatory amino acid, in toxic mussles from eastern Prince Edward Island. Can J Chem 1989;67:481–90.
[81] Perl T, et al. Amnesic shellfish poisoning: a new clinical syndrome due to domoic acid. Can Dis Weekly Rep 1990;16:7–8.
[82] Perl T, et al. An outbreak of toxic encephalopathy caused by eating mussels contaminated with domoic acid. N Engl J Med 1990;322:1775–80.
[83] Wright J, et al. Chemistry, biology and toxicology of domoic acid and its isomers. Can Dis Weekly Rep 1990;16:21–6.
[84] Langlois G. Marine biotoxin monitoring program annual report. California Department of Public Health for California Department of Fish and Game, Sacramento; 2007
[85] Busse L, et al. Domoic acid in phytoplankton and fish in San Diego, CA, USA. Harmful Algae 2006;5(1):91–101.
[86] Mazzillo F, et al. Angler exposure to domoic acid via consumption of contaminate fishes. Aquat Biol 2010;9:1–12.
[87] Vigilant V, Silver M. Domoic acid in benthic flatfish on the continental shelf of Monterey Bay, California, USA. Mar Biol 2007;151:2053–62.
[88] Kvitek R, et al. Domoic acid contamination within eight representative species from the benthic food web of Monterey Bay, California, USA. Mar Ecol Prog Ser 2008;367:35–47.
[89] Mazzillo F, et al. A note on the detection of the neurotoxin domoic acid in beach-stranded *Dosidicus gigas* in the southern California bight. CalCOFI Rep 2011;52:109–15.
[90] Boerlage S, Nada N. Algal toxin removal in seawater desalination processes. Desalin Water Treat 2014;55(10):2575–93.
[91] Llewellyn L. Saxitoxin, a toxic marine natural product that targets a multitude of receptors. Nat Prod Rep 2006;23:200–22.
[92] Schantz E, et al. Paralytic shellfish poison. VI. A procedure for the isolation and purification of the poison from toxic clam and mussel tissues. J Am Chem Soc 1957;79(19):5230–5.
[93] Burkholder J, Azanza R, Sako Y. The ecology of harmful dinoflagellates. In: Graneli E, Turner J, editors. Ecology of Harmful Algae. Berlin: Springer Heidelberg; 2006. p. 53–66.
[94] Jester R, et al. A shift in the dominant toxin-producing algal species in central California alters phycotoxins in food webs. Harmful Algae 2009;8:291–8.
[95] Gribble K, et al. Distribution and toxicity of *Alexandrium ostenfeldii* (Dinophyceae) in the Gulf of Maine, USA. Deep-Sea Res II 2005;52:2745–63.
[96] Townsend D, Pettigrew N, Thomas A. Offshore blooms of the red tide dinoflagellate, *Alexandrium* sp., in the Gulf of Maine. Cont Shelf Res 2001;21:347–69.
[97] Anderson D, et al. Alexandrium fundyense cysts in the Gulf of Maine: long-term time series of abundance and distribution, and linkages to past and future blooms. Deep-Sea Res II 2014;103:6–26.
[98] McGillicuddy Jr. D, et al. Mechanisms regulating large-scale seasonal fluctuations in *Alexandrium fundyense* populations in the Gulf of Maine: results from a physical-biological model. Deep-Sea Res II 2005;52:2698–714.

[99] Stock C, et al. Evaluating hypotheses for the initiation and development of *Alexandrium fundyense* blooms in the western Gulf of Maine using a coupled physical-biological model. Deep-Sea Res II 2005;52:2715–44.
[100] Garneau M-E, et al. Examination of the seasonal dynamics of the toxic dinoflagellate *Alexandrium catenella* at Redondo Beach, California, by quantitative PCR. Appl Environ Microbiol 2011;77(21):7669–80.
[101] Jester R, Baugh K, Lefebvre K. Presence of *Alexandrium catenella* and paralytic shellfish toxins in finfish, shellfish and rock crabs in Monterey Bay, California, USA. Mar Biol 2009;156(3):493–504.
[102] Jester R. An investigation into the prevalence of alexandrium derived toxins in marine food webs. Santa Cruz: University of California, Santa Cruz; 2008116.
[103] Baden D, et al. Natural and derivative brevetoxins: historical background, multiplicity and effects. Environ Health Perspect 2005;113:621–5.
[104] Kirkpatrick L, et al. Literature review of Florida red tide: implications for human health effects. Harmful Algae 2004;3:99–115.
[105] Kusek K, Vargo G, Steidinger K. *Gymnodinium breve* in the field in the lab and in the newspaper—scientific and journalistic analysis of Florida red tides. Contrib Mar Sci 1999;31(1–229).
[106] Fleming L, et al. Review of Florida red tide and human health effects. Harmful Algae 2011;10:225–33.
[107] Backer L, et al. Recreational exposure to aerosolized brevetoxins during Florida red tide events. Harmful Algae 2003;2:19–28.
[108] Pierce R, et al. Brevetoxin composition in water and marine aerosol along a Florida beach: assessing potential human exposure to marine biotoxins. Harmful Algae 2005;4:965–72.
[109] Pierce R, et al. Brevetoxin concentrations in marine aerosol: human exposure levels during a Karenia brevis harmful algal bloom. Bull Environ Contam Toxicol 2003;70:161–5.
[110] Backer L, et al. Occupational exposure to aerosolized brevetoxins during Florida red tide events: effects on a healthy worker population. Environ Health Perspect 2005;113:644–9.
[111] Kirkpatrick B, et al. Inland transport of aerosolized Florida red tide toxins. Harmful Algae 2010;9:186–9.
[112] Fleming L, et al. Aerosolized red-tide toxins (brevetoxins) and asthma. Chest 2007;131(1):187–94.
[113] Fleming L, et al. Initial evaluation of the effects of aerosolized Florida red tide toxins (brevetoxins) in persons with asthma. Environ Health Perspect 2005;113:650–7.
[114] Fleming L, et al. Exposure and effect assessment of aerosolized red tide toxins (brevetoxins) and asthma. Environ Health Perspect 2009;117:1095–100.
[115] Kirkpatrick B, et al. Aerosolized red tide toxins (brevetoxins) and asthma: continued health effects after 1 h beach exposure. Harmful Algae 2011;10:138–43.
[116] Truman P, et al. Lipophilic toxicity from the marine dinoflagellate *Karenia brevisulcata*: use of the brevetoxin neuroblastoma assay to assess toxin presence and concentration. Toxicon 2005;46:441–5.
[117] Silke J, O'Beirn F, Cronin M. *Karenia mikimotoi*: an exceptional dinoflagellate bloom in western Irish waters. Mar Environ Health Ser 2005;21:44pp. http://hdl.handle.net/10793/240.
[118] Brand L, Campbell L, Bresnan E. Karenia: the biology and ecology of a toxic genus. Harmful Algae 2012;14:156–78.
[119] Bourdelais A, et al. New fish-killing alga in coastal Delaware produces neurotoxins. Environ Health Perspect 2002;110:465–70.

[120] Khan S, Arakawa O, Onoue Y. Neurotoxin production by a Chloromonad *Fibrocapsa japonica* (Raphidophyceae). J World Aquacult Soc 1996;27(3):254–63.
[121] Khan S, Arakawa O, Onoue Y. A toxicological study of the marine phytoflagellate, *Chattonella antiqua* (Raphidophyceae). Phycologia 1996;35(3):239–44.
[122] Khan S, Arakawa O, Onoue Y. Neurotoxins in a toxic red tide of *Heterosigma akashiwo* (Raphiphyceae) in Kagoshima Bay, Japan. Aquac Res 1997;28:9–14.
[123] Keppler C, et al. Sublethal cellular effects of short-term raphidophyte and brevetoxin exposures on the eastern oyster *Crassotrea virginica*. Mar Ecol Prog Ser 2006;312:141–7.
[124] Chang F, Anderson C, Bousted N. First record of a Heterosigma (Raphidophyceae) bloom with associated mortality of cage-reared salmon in Big Glory Bay, New Zealand. N Z J Mar Freshw Res 1990;24:461–9.
[125] Hiroishi S, et al. High toxicity of the novel bloom-forming species of *Chattonella ovata* (Raphidophyceae) to cultured fish. Harmful Algae 2005;4:783–7.
[126] Marshall J, et al. Ichthyotoxicity of *Chattonella marina* (Raphidophyceae) to damselfish (*Acanthochromis polycanthus*): the synergistic role of reactive oxygen species and free fatty acids. Harmful Algae 2003;2:273–81.
[127] Imai I, Yamaguchi M. Life cycle, physiology, ecology and red tide occurrences of the fish-killing raphidophyte Chattonella. Harmful Algae 2012;14:46–70.
[128] Karin de Boer M, et al. Haemolytic activity within the species *Fibrocapsa japonica* (Raphidophyceae). Harmful Algae 2009;8:699–705.
[129] Blanco J, et al. Depuration of domoic acid from different body compartments of the king scallop *Pecten maximus* grown in raft culture and natural bed. Aquat Living Resour 2006;19:257–65.
[130] Novaczek I, et al. Depuration of domoic acid from live blue mussels (*Mytilus edulis*). Can J Fish Aquat Sci 1992;49:312–8.
[131] Blanco J, et al. Depuration of mussels (*Mytilus galloprovinicialis*) contaminated with domoic acid. Aquat Living Resour 2002;15(1):53–60.
[132] Sekula-Wood E, et al. Rapid downward transport of the neurotoxin domoic acid in coastal waters. Nat Geosci 2009;2:272–5.
[133] Anderson C, et al. Circulation and environmental conditions during a toxigenic *Pseudo-nitzschia australis* bloom in the Santa Barbara Channel, California. Mar Ecol Prog Ser 2006;327:119–33.
[134] Mengelt C. Ultravioloet photoecology, dark survival and seasonal abundance of *Pseudo-nitzschia australis* and *P. multiseries* in coastal waters of central California. Santa Barbara: University of California; 2006.
[135] Lane J, et al. Application of solid phase adsorption toxin tracking (SPATT) for field detection of the hydrophilic phycotoxins domoic acid and saxitoxin in coastal California. Limnol Oceanogr Methods 2010;8:645–60.
[136] Lefebvre K, et al. Characterization of intracellular and extracellular saxitoxin levels in both field and cultured *Alexandrium* spp. samples from Sequim Bay, Washington. Mar Drugs 2008;6:103–16.
[137] Seubert E, et al. In: Implementation of solid phase adsorption toxin tracking (SPATT) for the monitoring of harmful algal blooms in southern California. Seventh Symposium on Harmful Algae in the U.S., Sarasota, FL; 2013.

CHAPTER 14

Social Issues and Public Acceptance of Seawater Desalination Plants

Brent Haddad, Nadine Heck, Adina Paytan, Donald Potts
University of California, Santa Cruz, CA, United States

14.1 INTRODUCTION

In a prescient 2002 article, Means et al. write:

> *Utilities will likely turn increasingly to recycling, ocean desalination, and use of other "marginal" supplies in order to meet growing water needs. These supplies will bring unique water quality and perception issues to the forefront.*

The authors identified three trends: growing/aging populations with concerns about health impacts of drinking water; improved pollutant detection technologies that are outpacing the ability to equate the presence of small amounts of pollutants with health impacts; and global warming, which could disrupt historical surface and groundwater supplies. The combination of growing interest in water quality and supply unreliability is pushing urban regions toward new supplies, including desalination and urban water reuse. A new water world is upon us.

Public perception matters because infrastructure projects exist to serve the public good. The public also influences policy decisions through participation and voting. Attitudes can inform whether a project gets built, influence facility operational and expansion decisions, and determine whether to pursue similar projects elsewhere. Public opposition to essential infrastructure projects implies a failure of past policy to communicate the needs to the general public and could lead to future underfunding, weak oversight, or neglect of existing systems, threatening an essential service.

Means et al. [1] focused on what could be called the old rationality of public health and water reliability concerns. Historically, water agencies interacted with the public by presenting scientific and economic evidence to support recommendations on water infrastructure planning and

Sustainable Desalination Handbook
https://doi.org/10.1016/B978-0-12-809240-8.00014-9

management. Today's agencies recognize a new form of rationality that includes psychological aspects of perception, such as fear of contamination. Scholars note that "there are many examples where people's feelings or intuitions run counter to their own rationally considered self-interest, and sometimes the public interest" ([2], citations omitted). Psychological impacts of water supply are just one part of the water dialog; the public is also interested in water's connection to environmental impacts, urban congestion, economic growth, and regional independence. All of these topics contribute to understanding how the public views proposed and existing water infrastructure projects. Political questions concern whether and how public perception matters for policy making, and how public perception can be informed, influenced, or even manipulated.

Public perception and water resources management is a more complex and nuanced field in 2017 than it was even two decades earlier. The search for communications paradigms continues, including understanding what members of the public know about water supply, what they feel is important, what role the public wants to play in decision making and management, and how they want to be informed. These questions represent an ongoing research agenda and in this chapter we address one segment: public perception of desalination facilities.

Describing a treatment method such as desalination as a *source* of potable water is somewhat arbitrary. Nearly all water is part of the global hydrological cycle but, designating one intervention in the cycle as a distinguishing characteristic or origin of a water supply, we can begin the conversation on public perception. The origin of a water supply is commonly understood to be the location in nature from which water is captured for human use. A groundwater basin can be a source, as can a river, lake, spring, or the ocean. An actual use becomes a new origin for the same flow of water (e.g., urban water), and when the water is returned to a natural sink or conveyance, the perceived origin is updated again. When water from one origin is mixed with water with another, such as river water mixed with treated urban wastewater, the length of time the water is in the river before it is again captured for human use influences perceptions of the water's origin and its suitability for use. Water of natural origin is preferred to water of urban origin [3].

Throughout human history, knowledge of the origins of a water supply has been a useful proxy for water quality, and we should not be surprised that people feel revulsion as a defense against perceived contamination in water. As water treatment technologies improve, a struggle between perception/response and science-based assurances of water quality has emerged. Treatment technologies can turn any quality of water into potable quality. The ability to tap into previously unused source waters, such as the ocean, grows. Humanity is less dependent on natural biological and physical processes to deliver high-quality source water for human use. But, as Means et al. [1] explain, the technology and science of water treatment have advanced far beyond our understanding of how to communicate about water.

Technological advances in water treatment are timely and necessary because many of the natural and physical systems currently relied upon, such as groundwater percolation, have either been compromised by human activities or do not have the capacity to provide the amount of water needed by today's cities. Different approaches to potable water supply are needed.

Sources of potable water now emerging as crucial to urban water supply planning in the 21st century include saline water from either inland saline aquifers or coastal bays and shorelines. Inland saline aquifers present the fewest barriers to implementation, but they generally do not occur where large concentrations of people live. Historically, the existence of saline groundwater rather than fresh groundwater has been a hindrance to human settlement. In recent decades, human migration to the world's coastal regions has accelerated, creating both pressure and opportunity to reuse existing urban supplies and to develop seawater desalination. Urban reclaimed water and seawater desalination share the similarities of being located in growing coastal cities and of utilizing generally the same treatment technologies. However, public perception and policy implementation processes are different. Urban reclaimed water has undergone a great deal of scrutiny with regard to public perception and policy in recent years (for a summary, see Ref. [4]), but public perception of desalination has received less attention. Although desalination does not carry the negative baggage of prior human use, its public perception is more strongly associated with coastal environmental impacts and high energy demand. A long-standing preference for desalinated water for potable use over other alternative supplies was confirmed in a recent international survey [5]. The combination of continuing improvement in the economics of desalination technology, growing coastal populations, multiple demands on surface water and coastal groundwater supplies, reticence to utilize urban wastewater, and unlimited availability of seawater calls for greater scrutiny of desalination [6]. Coastal desalination projects are likely to be more numerous in the coming decades and have the potential for wide swings in positive and negative impacts depending on how they are implemented.

Because seawater desalination is a relatively new source of urban water in the United States, much of the social sciences literature has focused on processes for establishing new facilities and for accepting desalinated water as a safe drinking source. A more recent literature is now emerging that examines the effects of operating desalination facilities on the regions in which they are located. The benefits of an operating seawater desalination facility include overall water supply augmentation, a year-round reliable water supply that reduces drought risks, and a robust technology that reduces system failure risk and increases system resilience. These benefits are experienced by the region taking the desalinated water as well as nearby regions and uses competing for the same inland water supplies. Impacts of desalination are concentrated near the coastal location of the facility, both on shore and in the near-shore ocean. Other impacts, including the costs of constructing and operating desalination facilities, are additional considerations that may be experienced by the region consuming the water as these costs may be included in the cost for the water supply. A classic environmental context of disbursed benefits and concentrated costs/impacts calls for careful scrutiny of how significant are the impacts, who is bearing them, and are they being considered in the regional governance of desalination.

Public perception research can occur in many contexts. For complex engineering/natural systems, the intervening layers of technological/scientific complexity and interest group politics can influence the public's understanding of and support for a project. Researchers examine whether the public has been given adequate opportunity to express collective or individual opinions in ways that could influence a decision or action. Another issue is whether the public has been informed about potential impacts and benefits of a proposed system. Infrastructure projects have impacts in the short term and long term, both positive and negative, and it is the duty of proponents to provide baseline information on these impacts, after which robust debate contributes to an informed decision. Further research can examine the extent to which the public understands potential impacts and is therefore able to make an informed decision about a proposed project. In this chapter, we first list potential impacts of desalination, then explore how they are perceived by the public, and whether these perceptions influence public acceptance or opposition to a desalination facility. We also examine psychological and demographic aspects of acceptance or opposition to desalination facilities, before discussing future research needs.

14.2 POTENTIAL IMPACTS OF SEAWATER DESALINATION

For some individuals, any adverse impact on the ocean is unacceptable. This perspective could arise from recognition of the ongoing decline of ocean biota, growing ocean pollution, and changes in ocean chemistry, or from a spiritual or personal attachment to the ocean that does not countenance its use as a source of fresh water for onshore uses. This perspective could also arise as an opening salvo in a public policy process in which opponents state their strongest possible opposition to a proposed desalination facility about which they have other more specific reservations. A categorical rejection of terrestrial uses of ocean water means that one doesn't have to consider any categories of use.

Categories of potential physical and environmental impacts include piping, coastal water, coastal land areas including the atmosphere and construction. Coastal seawater desalination facilities require three separate pipelines. The first is the seawater intake, and its design influences its impact on nearby ocean ecology. In general, the greater the flow rates at the point of intake, the greater the potential for ongoing ecological damage through entrainment (capture) of marine organisms. With more widely dispersed intakes, less disturbance is likely. Intakes installed underneath sand have also been implemented at some locations and are considered to have lower entrainment rates. The second pipeline delivers concentrated brine from the facility back to the ocean. Numerous factors influence the potential impact of the brine on sea life, including the rate at which the brine mixes with ocean water thereby reducing the potential harm from localized concentrated salts. Although location, physical structure, and flow rates play crucial roles in determining the level of ocean impacts, it is impossible to have no impacts at all from the operation of a seawater desalination facility. The third pipeline is the onshore link between the treatment facility and the rest of the water system. With occasional exceptions, such as Riyadh's 400-km pipeline from the coast to the inland city, the proximity of treatment to end use means extensive inland piping is not needed. Urban wastewater treatment systems typically have unused capacity, so it is unlikely that an expansion to an existing treatment infrastructure will be needed with the addition of desalinated water. Potable water originating from seawater does not require special treatment after use.

In the onshore coastal area, a desalination facility takes the form of a small-to-moderate-sized factory. Inside the building are the water pumps, water treatment devices, piping, a water sampling/testing area, and other necessities of industrial facilities, such as offices. There is also a parking lot for the relatively small crew of facility operators. Sound is produced by the operation of water pumps and the occasional arrival and departure of vehicles. Desalination facilities do not generate noxious odors. Overall, a desalination plant is an unobtrusive light industrial facility except that it is located next to the ocean. For many coastal cities, near-shore real estate is valued both for private residences and for public spaces. The presence of a desalination facility can be detrimental to the alternative nearby uses, and a new facility might replace a previously valued public use where it is constructed. In addition, the facility uses electricity as its power source and is thus a net generator of atmospheric carbon dioxide, which is of concern because it exacerbates climate change impacts.

Construction involves laying the pipelines, often crossing both the surf zone and existing shoreline roads, building the facility, and attaching it to the existing potable water system through inland pipelines. The construction process could be preceded by a multi-year smaller scale pilot project utilizing temporary systems. Construction of inland piping is likely to interrupt transportation as streets are dug up and pipes laid. The coastal ocean impacts described above all occur in the zones adjoining the intakes and outfalls.

Another broad category of costs involves economic costs and benefits. In the United States, coastal desalination facilities can cost in the range of $100 million to design, permit, and build. Once built, operating costs are likely to be higher than for other existing water supplies because of the energy cost of removing salt from the water. These costs are distributed to the customers of the facility. The capital (construction) costs are typically bond financed and paid off over a few decades. Overall, the region's cost of water will increase with the addition of desalinated water. Economic benefits include the wages paid and purchases made to build and operate the facility, and the ongoing beneficial impact of the additional more reliable water supply.

14.3 PUBLIC PERCEPTION OF SEAWATER DESALINATION

Studies of attitudes toward desalination fall into three contexts. The first includes studies on general attitudes toward desalination as a water supply option by asking the public wide-ranging questions that are not attached to a specific project. The second are studies focusing on public opinions about actual forthcoming water projects, and the third examines attitudes in locations where a desalination system has already been established [6a], such as in Spain [7] and Australia [8]. To determine whose perceptions matter, one can generate such categories as all people, people who take water from the desalination-connected system, and people who live near the ocean and/or the desalination facility and infrastructure. For example, local residents might share regional feelings about the cost of monthly water bills while holding differing views about coastal impacts related to the desalination process that provides this water.

Most studies suggest that residents look favorably upon desalination solutions for potable water supply. In Carlsbad, a small coastal community in southern California with no previous experience with desalination, Heck et al. [9,10] found more than 70% of coastal residents supported the local seawater desalination plant. Gibson et al. [11] reported similar findings for desalination plants in Perth, Australia, where 74% of residents supported the expansion of desalination plants in the city. Perth, a large urban region, already receives about 47% of its water from two desalination plants [11a]. While there are only a few such studies, the similar high support in Perth and Carlsbad suggests that the size of a community and degree of experience with desalination may have little influence on support for specific desalination projects.

14.3.1 Perception of Environmental Impacts

Coastal residents may oppose desalination plants due to beliefs about biological, ecological, and environmental impacts from desalination plants on personal values, including effects on benthic organisms and small marine organisms that are entrained in the water intake [8–10]. While there is society-wide positive association with and support for ocean stewardship, coastal residents appear to value the local marine ecosystem very highly, including aspects that are less visible such as small marine organisms (e.g., fish larvae, phytoplankton) and marine life on the ocean floor. The public places importance on ecological functions of many species, not just charismatic flagship species such as marine mammals)[11b]. Evidence confirming coastal residents' concerns about new and complex impacts on marine ecosystems comes from a study on public perceptions of impacts on marine environments by Gelcich et al. [11c]. These findings suggest that even though people engage with marine areas in different ways compared to terrestrial areas, and impacts are commonly less visible in the marine realm [11d,e], local residents are aware of less visible ecosystem features and are concerned about impacts that affect ecosystem functions and lead to declines in the health of marine ecosystems.

14.3.2 Perception of Social Impacts of Desalination

Ocean access and scenic impacts are another onshore concern of coastal residents. Some coastal regions are industrial in character, such as harbors with commercial/industrial facilities, and a desalination facility would be an unobtrusive addition to a coastal industrial area. However, the best location for a coastal desalination facility is determined by other factors such as linkage to existing water infrastructure and offshore factors. It is also the nature of desalination, as a new water supply to be introduced into regions where the human population is growing, that results in competition for coastal locations to provide housing, commercial areas, recreation, and environmental benefits. A factory-like building of any kind would be an unwelcome addition to such coastal neighborhoods.

In the Carlsbad region of southern California, coastal neighborhoods had grown around an older power plant that was the proposed location for the desalination facility. The potential impacts of the facility on coastal scenery and coastal access significantly reduced support for the facility, even though the site was adjacent to and smaller than the existing power plant, and located further away from the beach (Fig. 14.1). Residents apparently

Fig. 14.1 Carlsbad, California desalination facility, operated by Poseidon Resources, Inc. Colocated with and using inflow/outflow infrastructure of an existing power plant. Residential neighborhoods are located in upper left across the Agua Hedionda Lagoon. The public has access to the beach located across the coastal highway. *Source: Google Earth, 33°08′21.04″ N and 117°20′22.19 W (11/8/16).*

still did not appreciate the seemingly minimal additional impacts on coastal scenery. In Santa Cruz in central California, one resident speaking out in a public hearing on a proposed desalination plant stated that having paid over 1 million dollars for his ocean-front home, he did not want to be disturbed by the sound of water pumps. In Tampa, Florida, the state's largest desalination facility is also colocated with a power plant close to the community of Apollo Beach. This power plant's visual and ocean impacts are so much larger than those of the desalination facility that the latter has not been considered an issue of local scenic or sound concern. In general, coastal residents appear to feel protective about the remaining coastal scenery and natural quietude, and they do not welcome additional development close to the shore.

Residents sometimes oppose natural resource projects in their close proximity even though they would otherwise generally support them. The not-in-my-back-yard (NIMBY) framework has been widely used to explore whether proximity to a new development influences local residents' attitudes. NIMBY has been studied since the early 1980s in multiple natural resource contexts, including energy plants, natural gas extraction, wind farms, and offshore oil drilling platforms [12–18]. For desalination, NIMBY could provide insights to the question of whether coastal residents close to a facility oppose it more than those living further away. In these instances, the public may support desalination as a water supply option in general, but not the construction of such a facility in their immediate vicinity.

The Tampa Bay facility, colocated with the Big Bend Power Station, did not elicit NIMBY reactions from the community of Apollo Beach, and the Carlsbad facility also showed no significant effect of distance from the plant on support. Residents living closer to the plant did not support the new facility more or less than residents living further away. Limited explanatory power of proximity has also been found in multiple studies on public attitudes in other natural resource sectors [12,15,19] and desalination in Australia [19a]. The Carlsbad desalination plant is located in a mainly industrial area next to the Encina power plant and separated from residential areas by the Agua Hedionda Lagoon. Because this facility is small compared to the power plant and does not produce high levels of noise or visible point source air emissions, residents living closer to the plant may not be affected by the plant more than other residents, which may explain the absence of the NIMBY effect [10]. In contrast, the proposed desalination plant in Santa Cruz, California, would have been located close to or in an affluent residential neighborhood. It was put on hold after strong community resistance, including a number of complaints about expected impacts on quality of life and property prices [20]. The Santa Cruz resident mentioned earlier made a NIMBY-consistent point about the potential sound of pumps near his coastal residence and hence the details of coastal location seem to matter. Siting a plant in an existing residential area was hotly opposed, while operating facilities in nearby-yet-isolated locations such as the Carlsbad and Tampa Bay locations may have contributed to the absence of the NIMBY reaction.

14.3.3 The Influence of Social and Psychological Variables on Attitudes

The sociodemographic profile of a community may predict local support or opposition to some extent. Previous studies on public support in Australia and the United States investigated the influence of sociodemographic variables such as gender, age, education, and race on public acceptance for desalination plants. These studies found acceptance was correlated with all four of these variables. In Carlsbad, race and age were the only sociodemographic variables that predicted support for a coastal desalination facility, with white and older residents more likely to be supportive of the plant. Education and gender did not predict support in a study by this chapter's authors [9,10], which suggests that sociodemographic variables may have different, location-specific effects on the acceptance of desalination. Gibson et al. [11] also reported almost no influence of sociodemographic variables on the level of support for proposed desalination plants in Perth. Previous studies reported that sociodemographic factors shape acceptance to use desalination water. For example, in Australia, individuals receptive to using desalinated water were older, male, educated, and had previously used desalinated water [21]. These mixed findings suggest that the sociodemographic profile of a community in general might not be a decisive factor determining local support or rejection of a desalination facility.

Based on one study in Carlsbad, residents who frequently use local marine areas for fishing, surfing, swimming, and beach walking seem less supportive of seawater desalination. The Carlsbad plant is situated at a very popular beach with a high volume of recreational use. One potential explanation for less support by surfers and swimmers may be that these activities require physical contact with ocean water in the vicinity of the plant, and people may have concerns about the quality of ocean water in areas close to the point of discharge. Monitoring water quality may help reduce conflicts with recreational users close to a desalination facility, and with fishers who may worry about declining fish resources due to impingement and entrainment at the open-ocean water intake, or effects of brine discharge on food webs.

Psychological factors including beliefs and perceptions about seawater desalination outcomes have been found to influence attitudes toward the technology. Heck et al. [10] found that beliefs about negative impacts from the Carlsbad desalination facility are the strongest predictors of reduced support for seawater desalination. This facility is the largest desalination plant operating on the U.S. west coast, and the study spanned the construction/preoperation period of the facility and its initial operation. These findings are similar to the results of Gibson et al. [11] who also found that perceptions about outcomes predicted support for proposed desalination plants.

Another psychological variable that influences acceptance of local desalination facilities is place attachment to near-shore marine areas. While sense of place is concerned with factors creating bonds to a place (e.g., Ref. [22]), place attachment refers to the strength of bonds between humans and locations [23]. Place attachment can be assessed in many ways, but the two main dimensions with strong foundations in the academic literature are place identity and place dependence [24,25]. Place identity, or emotional ties to a place, usually develop over time and refers to the symbolic meanings of an area. Place dependence is the functionality associated with an area and is represented by its tangible physical characteristics and attributes [25]. As coastal residents and marine users often place emotional and functional importance on marine areas (e.g., Refs. [26–28]), place attachment can shape their attitudes toward the use or protection of these areas. Heck et al. [10] found that place attachment correlated significantly and negatively with support for a local desalination plant, indicating that strong emotional and functional ties to local marine areas shaped local views on the new facility. This finding is similar to a study by Devine-Wright and Howes [29] that found a perceived threat to place identity led to negative attitudes toward an offshore wind farm. In general, residents who attach high emotional and functional values to marine areas appeared to perceive the desalination plants as a threat to those values based on the expected negative impacts on local marine ecosystems. Place dependence on marine areas may also explain why marine users were less supportive of such facilities compared to other members of the community, since functional ties to an area and actual use are related.

Another consideration for public support of water supply systems is the notion of threat perception [30]. The public is more likely to support a proposed new water supply if there is a severe water crisis that threatens the existing system [31]. Studies on public support in the United States found that the strongest predictor for public support was concern about local water supply after a prolonged drought lasting multiple years that led to mandated emergency conservation [10,32]. Support therefore may decline in periods with higher precipitation and less concern about drinking water supply.

Research informed by theoretical approaches in social psychology and psychological risk perception (e.g., [33,34]) explore the influence of beliefs and attitudes toward desalinated water on the acceptance of using this water for different purposes. These studies found that concerns about the low quality of desalinated water and associated health concerns reduce acceptance of using desalinated water. Positive attitudes toward conservation and the environment, and social norms were also predictive of behavioral responses to using desalinated water [21,35].

14.3.4 Institutional Factors Shaping Attitudes Toward Desalination

Support or opposition to desalination or any other public project does not occur in a vacuum. Communities are aware, to some extent, of the institutional context in which decisions are made, and they have formed opinions about the institutional context. For example, in Carlsbad, trust was relatively low for the desalination plant operator and for the State Water Resources Control Board, both of which are heavily involved in operating and regulating the specific plant and seawater desalination facilities in general. Low levels of public trust in agencies to manage marine areas have been reported in other studies (e.g., see Refs. [11c,35a,b]). In California, agencies typically engage the public in open hearings, town halls, and public comments on draft proposals and Environmental Impact Reports related to seawater desalination. These approaches are widely used in numerous natural resource settings, and agencies face the dilemma of developing a plan with enough detail that meaningful discussion can ensue while not delivering a *fait accompli*. The process has been criticized as after-the-fact and lacking in early public engagement that helps shape the project under consideration. Lack of meaningful early and continuing public engagement has been found to reduce trust in management and oversight agencies [35c,d].

14.3.5 Management Strategies to Increase Public Support for Desalination

Since beliefs about negative environmental impacts erode support for desalination, it seems advantageous for gaining public support to design and build desalination facilities that minimize such impacts. One example is colocating a facility with an existing power plant and/or wastewater treatment plant and sharing the plant's ocean intake and/or outfall infrastructure. In such cases, the desalination facility does not require construction of additional infrastructure with its associated environmental disruptions. As with Florida's Tampa Bay facility, colocation with a power plant also reduces visual concerns. Adjacent marine ecosystems are already being affected by water intake and return flows from the existing infrastructures using seawater as a coolant in power production or releasing treated urban wastewater to the ocean. Even with these design elements, perceived negative impacts from the plant on the marine ecosystem may still be high and significantly reduce support for desalination plants [9,10]. Residents may be especially concerned about desalination-specific impacts (e.g., discharge of brine) that add additional stress to local marine ecosystems.

In California, the logic of combining desalination infrastructure with power plant infrastructure became enmeshed with an effort to end once-through cooling of coastal power plants. Once-through cooling uses orders of magnitude more ocean water than desalination. The water returned to the ocean is warmer than intake water, and tidal flow characteristics near intake and outfall are altered. Once-through cooling opponents feared that colocated desalination facilities would provide additional justification to maintain the operation of existing cooling systems through shared infrastructure. In California, the political skill and influence of the opponents, honed in hard-fought policy battles over coastal energy production, came as a shock to desalination proponents. The antidesalination effort ended a number of proposals while delaying others by several years. The rationale from the water production side for colocation remains, so the issue could emerge in other states in the future.

Onshore, expected CO_2 emissions from the desalination facility, driven by the power demands of high-pressure membrane separation of salt from water, also significantly reduced public support. Residents are aware that the means for generating electricity are predominantly fossil fuels, so operation of a desalination facility will increase greenhouse gas loading into the atmosphere. Energy demand and greenhouse gas emissions caused by desalination plants could further reduce public support if concerns about climate change and its impacts on marine ecosystems (e.g., ocean acidification) remain or increase.

The plant operator in Carlsbad developed a mitigation plan to offset impacts of greenhouse gas emission. Yet, residents may not be aware of the mitigation plan, or they might prefer that the plant directly reduce rather than offset emission through market-based methods. This would be consistent with a more general negative public attitude toward market-based approaches to greenhouse gas mitigation. The public has reacted ambivalently to this approach because it prefers to see polluters take individual action on site to reduce pollution, even if the actions are more expensive to the polluter [36]. Taking action at one's own facility makes it easier to hold the polluter accountable, since evidence of greenhouse gas reductions can be easily demonstrated. A plan to mitigate greenhouse gas pollution from a proposed desalination facility in Santa Cruz, California, encountered criticism for its proposal to mitigate greenhouse gas emissions via least-cost means because of the nonlocal market-based nature of the proposed means. The proposed approach involved purchasing greenhouse gas credits that represented greenhouse gas reductions in other sectors and parts of the world. Neither management strategy to address greenhouse gas impacts was particularly successful.

14.4 RESEARCH NEEDS IN PUBLIC PERCEPTION AND DESALINATION

Desalination has the potential to be a vital water supply source to millions of coastal residents. The small but growing body of research on public perception and attitudes toward desalination will be an important complement to ongoing research on technological and ecological impacts. Based on our review of the existing literature and our own investigations, we have identified the following research needs and opportunities. The focus is on public perception and social impacts, but there is an implied need in nearly all identified issues for ongoing and expanded technical and scientific research as well.

- *Public attitude dynamics.* While public attitudes toward desalination plants are important, they are not stable and statistically demonstrated predictors may change over time. Gibson et al. [11] explored public acceptance of expanding desalination in Perth, Australia, in 2007 and 2012 and found that support for the expansion of desalination remained high in both years (74.5% in 2007 and 73.4% in 2012), but the drivers of acceptance changed over time. In addition, public support may change between periods of sufficient water supply and drought. Since threat perceptions about local water supply appear to be a significant predictor for public support, support may wane once the threat recedes. One should not expect public perception to be stable since underlying environmental conditions themselves are not stable. Rather, one should expect perception to track and correlate with actual conditions (e.g., ocean impacts, onshore impacts) and how individuals evaluate them. The dynamic nature of public perception and public support/opposition is an important, underexplored issue. Dynamic public opinion contrasts with multidecadal or permanent impacts of a proposed infrastructure facility, and focuses our attention on what the appropriate role of public attitudes should be in infrastructure planning.

- *Use of existing offshore piping infrastructure.* One aspect of desalination environmental impacts is the construction of ocean intakes and outfalls, including piping to and from the facility. Colocation with coastal power plants that already have intake/outfall infrastructure and the utilization of existing outfalls from wastewater treatment plants present opportunities to avoid additional construction of ocean infrastructure. To determine whether colocation has any effect on local perceptions and support for desalination facilities, future studies could investigate whether perceived impacts on marine ecosystems are different for desalination plants that utilize pre-existing infrastructure and those that do not.
- *Energy and water supply.* Surveys have identified the importance to the public of reducing the energy demand of desalination and delinking water treatment from greenhouse gas generation. Technical research on improved membrane processes and integration of novel pre-reverse osmosis steps could help reduce this cost and negative perception barriers. Another relationship concerns the global warming impact of desalination. This can be reduced by the water utility or facility owner/operator investing in renewable energy, or by the regional power utility reducing the fossil fuel share of its primary energy. Further public perception research is needed to delve deeper into the connection between greenhouse gas emissions and water utility installations. It appears that the public focuses more on water utility emissions than on private industry emissions, even when the private industry product is less essential to public wellbeing than potable water. A deeper understanding into how the public evaluates the performance of natural resource-related utilities versus other private sector actors could influence how society understands and implements greenhouse gas regulation.
- *Users of marine ecosystems.* As desalination expands, the incidence and impacts of increased offshore physical infrastructure, additional inflows, and additional brine discharges will grow. Marine resource users, such as fishers, kayakers, and divers, directly engage the regions of impact and may have specific concerns. Understanding their uses and how new

facilities could impact them could be an important part of avoiding and/or managing conflicts.

- *Place attachment.* Individuals living near coastlines have personal and community connections to the ocean. If this connection is to be honored and integrated into coastal policy decision making, it needs to be better understood. Studies on place attachment to marine areas, and its related concept NIMBY, would help clarify the meaning and potential importance of place attachment.
- *Immediate vs. long-term planning.* A desalination facility takes several years to design, approve, build, and operate. A study of public perception captures a snapshot in time in this long process. In doing so, it can reveal what long-term elements are influencing opinion, versus ephemera. Some issues that initially do not emerge as influential, such as long-term projections of water supply and demand, could be emphasized more in the public debate regarding new water infrastructure. It could also help set agendas for early discussions between utilities and the public, avoiding the critique of ex post facto consultation. Further studies and meta-studies can identify short-term and long-term drivers of public opinion

14.5 CONCLUSION

Even in the highly technical world of potable water supply, the public plays an important role in decisions to build and manage new systems. Public policy processes should arrive at the best option on the table, or identify a better one through dialog. Processes should also avoid the outcomes of failure to build a needed facility or the construction of an unneeded facility. An informed and engaged public can influence technical and managerial choices. An informed public will not only grasp and evaluate valid points in the policy discourse, but it will also be able to dismiss misleading or inaccurate representations. Water supply broadly, and seawater desalination in particular require further investigation of public knowledge and perception of desalination processes and impacts.

REFERENCES

[1] Means E, Brueck T, Dixon L, Manning A, Miles J, Patrick R. Drinking water quality in the new millennium: The risk of underestimating public perception. J AWWA 2002;94(8):28. pp. 30–32, 34.
[2] Rozin P, Haddad B, Nemeroff C, Slovic P. Psychological aspects of the rejection of recycled water: contamination, purification and disgust. Judgem Decis Mak 2015;10(1):50–63.

[3] Haddad B, Rozin P, Nemeroff C, Slovic P. The psychology of water reclamation and reuse: survey findings and research roadmap. Alexandria, VA: WateReuse Foundation; 2009.
[4] National Research Council. Water reuse: potential for expanding the Nation's water supply through reuse of Municipal wastewater. National Academies Press; 2012.
[5] Hurlimann A, Dolnicar S. Public acceptance and perceptions of alternative water sources: a comparative study in nine locations. Int J Water Resourc Develop 2016;32(4):650–73.
[6] Haddad B. A case for an ecological-economic research program for desalination. Desalination 2013;324:72–8.
[6a] Friedler E, Lahav O, Jizhaki H, Lahav T. Study of urban population attitudes towards various wastewater reuse options: Israel as a case study. J Environ Manage 2006;81(4):360–70.
[7] Domènech L, March H, Saurí D. Degrowth initiatives in the urban water sector? A social multi-criteria evaluation of non-conventional water alternatives in Metropolitan Barcelona. J Clean Prod 2013;38:44–55.
[8] King TJ, Ooi D, Cary J, Fisher A, Schibeci R, Murphy K, et al. Public perceptions of, and responses to, desalination in Australia. Alfred Deakin Research Institute Working Paper Series Deakin University; 2012.
[9] Heck N, Paytan A, Potts DC, Haddad B. Coastal residents' literacy about seawater desalination and its impacts on marine ecosystems in California. Mar Policy 2016;68:178–86.
[10] Heck N, Paytan A, Potts DC, Haddad B. Predictors of local support for a seawater desalination plant in a small coastal community. Environ Sci Policy 2016;66:101–11. https://doi.org/10.1016/j.envsci.2016.08.009.
[11] Gibson FL, Tapsuwan S, Walker I, Randrema E. Drivers of an urban community's acceptance of a large desalination scheme for drinking water. J Hydrol 2015;528:38–44.
[11a] Water Corporation. Desalination 2014. Available online at: https://www.watercorporation.com.au/water-supply/our-water-sources/desalination?pid=res-wss-spw-np-des (accessed 09-27-2017).
[11b] Jefferson R, et al. Understanding audiences: Making public perceptions research matter to marine conservation. Ocean Coastal Manage 2015;115:61–70.
[11c] Gelcich Crossley S. Public awareness, concerns, and priorities about anthropogenic impacts on marine environments. PNAS; 2014.
[11d] Patel A, Rapport DJ, Vanderlinden L, Eyles J. Forests and societal values: comparing scientific and public perception of forest health. Environmentalist 1999;19(3):239–49.
[11e] Shackeroff JM, Hazen EL, Crowder LB. The oceans as peopled seascapes. In: Ecosystem-Based Management for the Oceans 2009;33–54.
[12] Jacquet JB. Landowner attitudes toward natural gas and wind farm development in northern Pennsylvania. Energy Policy 2012;50:677–88.
[13] Johansson M, Laike T. Intention to respond to local wind turbines: the role of attitudes and visual perception. Wind Energy 2007;10:435–51.
[14] Kahn RD. Siting struggles: the unique challenge of permitting renewable energy power plants. Electricity J 2000;13:21–33.
[15] Michaud K, Carlisle JE, Smith ERAN. Nimbyism vs. environmentalism in attitudes toward energy development. Environ Politics 2008;17:20–39.
[16] Swofford J, Slattery M. Public attitudes of wind energy in Texas: local communities in close proximity to wind farms and their effect on decision-making. Energy Policy 2010;38:2508–19.
[17] Wolsink M. Entanglement of interests and motives: assumptions behind the NIMBY-theory on facility siting. Urban Studies 1994;31:851–66.
[18] Wolsink M. Wind power and the NIMBY-myth: institutional capacity and the limited significance of public support. Renew Energy 2000;21:49–64.

[19] Ek K. Public and private attitudes towards "green" electricity: the case of Swedish wind power. Energy Policy 2005;33:1677–89.

[19a] King TJ, Ooi D, Cary J, Fisher A, Schibeci R, Murphy K, et al. Public perceptions of, and responses to, desalination in Australia: A report on findings. Alfred Deakin Research Institute; 2012.

[20] City of Santa Cruz. Draft environmental impact report—public comments. Santa Cruz CA, 2013.

[21] Dolnicar S, Schäfer AI. Desalinated versus recycled water: public perceptions and profiles of the accepters. J Environ Manage 2009;90:888–900.

[22] Stedman RC. Toward a social psychology of place predicting behavior from place-based cognitions, attitude, and identity. Environ Behav 2002;34(5):561–81.

[23] Wynveen CJ, Kyle GT, Absher JD, Theodori GL. The meanings associated with varying degrees of attachment to a natural landscape. J Leisure Res 2011;43:290.

[24] Manning R. Studies in outdoor recreation: search and research for satisfaction. Corvallis, Oregon, USA: Oregon State University Press; 2011.

[25] Williams DR, Vaske JJ. The measurement of place attachment: validity and generalizability of a psychometric approach. Forest Sci 2003;49:830–40.

[26] Gee K, Burkhard B. Cultural ecosystem services in the context of offshore wind farming: a case study from the west coast of Schleswig-Holstein. Ecol Complex 2010;7:349–58.

[27] Gray DL, Canessa RR, Rollins RB, Dearden P, Keller CP. Understanding recreational boater attitudes to zoning in a proposed marine protected area. Coastal Manag 2010;38:575–97.

[28] Ressurreição A, Simas A, Santos RS, Porteiro F. Resident and expert opinions on marine related issues: implications for the ecosystem approach. Ocean Coastal Manag 2012;69:243–54.

[29] Devine-Wright P, Howes Y. Disruption to place attachment and the protection of restorative environments: a wind energy case study. J Environ Psychol 2010;30(3):271–80.

[30] Marks JS. Advancing community acceptance of reclaimed water. Australian Water and Wastewater Association; 2004.

[31] Mankad A, Tapsuwan S. Review of socio-economic drivers of community acceptance and adoption of decentralised water systems. J Environ Manage 2011;92(3):380–91.

[32] State Water Resources Control Board, (2015) Emergency Conservation Regulations. Implementing 25% Conservation Statewide.

[33] Ajzen I, Fishbein M. Theory of reasoned action—theory of planned behavior. University of South Florida; 1988.

[34] Pidgeon NF, Beattie J. The psychology of risk and uncertainty. London: Blackwell Science; 1998.

[35] Dolnicar S, Hurlimann A, Grün B. What affects public acceptance of recycled and desalinated water? Water Res 2011;45:933–43.

[35a] Hynes S, Norton D, Corless R. Investigating societal attitudes towards the marine environment of Ireland. Marine Policy 2014;47:57–65.

[35b] Potts T, Pita C, O'Higgins T, Mee L. Who cares? European attitudes towards marine and coastal environments. Marine Policy 2016;72:59–66.

[35c] Innes JE, Booher DE. Reframing public participation: strategies for the 21st century. Plan Theory Pract 2004;5(4):419–36.

[35d] Gray S, Shwom R, Jordan R. Understanding factors that influence stakeholder trust of natural resource science and institutions. Environ Manage 2012;49(3):663–74.

[36] Mufson, S., and Agiesta, J. (2009) "Majority of poll respondents say U.S. should limit greenhouse gases." Washington Post. Downloaded July 24, 2017 from http://www.washingtonpost.com/wp-dyn/content/article/2009/06/24/AR2009062403648.html?hpid=topnews.

FURTHER READING

[1] Ambrose RF. Mitigating the effects of a coastal power plant on a kelp forest community: rationale and requirements for an artificial reef. Bull Marine Sci 1994;55:694–708.
[2] Anderson E. Drinking water starts flowing from Carlsbad desalination plant. KPBS; 2015.
[3] Bridgeman J. Public perception towards water recycling in California. Water Environ J 2004;18(3):150–4.
[4] Bourne G. California desalination planning handbook. Sacramento: Center for Collaborative Policy, California State University Sacramento; 2008.
[5] California Coastal Commission. Seawater desalination and the California Coastal Act. San Francisco, CA: California Coastal Commission; 2004.
[6] California Demographics, (2014) Carlsbad demographics.
[7] California Department of Water Resources. Desalination. Sacramento: California Department of Water Resources; 2003.
[8] California Water Boards. Desalination facility intakes, brine discharges, and the incorporation of other nonsubstantive changes. Sacramento: State of California, Environmental Protection Agency; 2014.
[9] Ching L. A lived-experience investigation of narratives: recycled drinking water. Int J Water Resourc Develop 2016;32(4):637–49.
[10] Connelly NA, Brown TL, Decker DJ. Factors affecting response rates to natural resource-focused mail surveys: empirical evidence of declining rates over time. Soc Nat Resour 2003;16:541–9.
[11] Cooley H, Ajam N, Heberger M. Key issues in seawater desalination in California: marine impacts. Oakland, CA: Pacific Institute; 2013.
[12] Cooley H, Heberger M. Key issues in seawater desalination in California: energy and greenhouse gas emission. Oakland, CA: Pacific Institute; 2013.
[13] Côté P, Siverns S, Monti S. Comparison of membrane-based solutions for water reclamation and desalination. Desalination 2005;182:251–7.
[14] Dawoud MA. The role of desalination in augmentation of water supply in GCC countries. Desalination 2005;186:187–98.
[15] Alternatives D. Desals' problems. Santa Cruz, CA: Desal Alternatives; 2013.
[16] Devine-Wright P. Place attachment and public acceptance of renewable energy: a tidal energy case study. J Environ Psychol 2011;31:336–43.
[17] Dillman DA, Smyth JD, Christian LM. Internet, phone, mail, and mixed-mode surveys: the tailored design method. John Wiley & Sons; 2014.
[18] Dolnicar S, Hurlimann A, Nghiem LD. The effect of information on public acceptance—the case of water from alternative sources. J Environ Manage 2010;91:1288–93.
[19] Elimelech M, Phillip WA. The future of seawater desalination: energy, technology, and the environment. Science 2011;333:712–7.
[20] Friedler E, Lahav O, Jizhaki H, Lahav T. Study of urban population attitudes towards various wastewater reuse options: Israel as a case study. J Environ Manage 2006;81:360–70.
[21] Fuentes-Bargues JL. Analysis of the process of environmental impact assessment for seawater desalination plants in Spain. Desalination 2014;347:166–74.
[22] Halpern BS, Walbridge S, Selkoe KA, Kappel CV, Micheli F, D'Agrosa C, et al. A global map of human impact on marine ecosystems. Science 2008;319:948–52.
[23] Hartley TW. Public perception and participation in water reuse. Desalination 2006;187(1):115–26.
[24] Heberger M, Cooley H, Herrera P, Gleick P, Morre E. The impacts of sea-level rise on the California Coast. Oakland, CA: Pacific Institute; 2009.
[25] Hinkebein TE, Price MK. Progress with the desalination and water purification technologies US roadmap. Desalination 2005;182:19–28.

[26] Hurlimann A, Dolnicar S. When public opposition defeats alternative water projects—the case of Toowoomba Australia. Water Res 2010;44:287–97.
[27] International Desalination Association, (2014) Desalination by numbers.
[28] Lattemann S, Höpner T. Environmental impact and impact assessment of seawater desalination. Desalination 2008;220:1–15.
[29] Lejano RP, Leong C. A hermeneutic approach to explaining and understanding public controversies. J Public Adm Res Theory 2012;22(4):793–814.
[30] Liu T-K, Sheu H-Y, Tseng C-N. Environmental impact assessment of seawater desalination plant under the framework of integrated coastal management. Desalination 2013;326:10–8.
[31] Mainali B, Ngo H, Guo W, Pham T, Wang X, Johnston A. SWOT analysis to assist identification of the critical factors for the successful implementation of water reuse schemes. Desalination Water Treat 2011;32:297–306.
[32] Martin CA. Landscape water use in Phoenix, Arizona. Desert Plants 2015;.
[33] Miller S, Shemer H, Semiat R. Energy and environmental issues in desalination. Desalination 2015;366:2–8.
[34] Mirchi A, Madani K, Roos M, Watkins D. Climate change impacts on California's water resources. In: Schwabe K, Albiac J, Connor JD, Hassan RM, Meza González L, editors. Drought in arid and semi-arid regions. Netherlands: Springer; 2013. p. 301–19.
[35] Miri R, Chouikhi A. Ecotoxicological marine impacts from seawater desalination plants. Desalination 2005;182:403–10.
[36] Monterey Bay National Marine Sanctuary, National Marine Fisheries Service, (2010) Guidelines for desalination plants in the Monterey Bay National Marine Sanctuary.
[37] Natural Resources Defense Council. Proceed with Caution: California's Drought and Seawater Desalination, NRDC Issue Brief May 2014. New York: Natural Resources Defense Council; 2014.
[38] Oki T, Kanae S. Global hydrological cycles and world water resources. Science 2006;313:1068–72.
[39] Pasko, J., (2013) A change of plans. Good times.
[40] Peel J, Choy J. Water governance and climate change. In: Drought in California as a lens on our climate future. Stanford: Stanford Woods Institute for the Environment; 2014.
[41] Perry EE, Needham MD, Cramer LA, Rosenberger RS. Coastal resident knowledge of new marine reserves in Oregon: The impact of proximity and attachment. Ocean Coastal Manag 2014;95:107–16.
[42] Pierce, J., (2013) A history of Santa Cruz's Desal fight.
[43] Poseidon Water. Poseidon Resources Marine Life Mitigation Plan Poseidon Water, Carlsbad, 2008.
[44] Poseidon Water. Carlsbad desalination plant. Available at http://carlsbaddesal.com/faqs (accessed 08/01/2016), 2016.
[45] Roberts DA, Johnston EL, Knott NA. Impacts of desalination plant discharges on the marine environment: a critical review of published studies. Water Res 2010;44:5117–28.
[46] San Diego County Water Authority. Seawater Desalination. San Diego: San Diego County Water Authority; 2012.
[47] San Diego County Water Authority. Seawater desalination: the carlsbad desalination project. San Diego: San Diego County Water Authority; 2014.
[48] Sanchez-Lizaso JL, Romero J, Ruiz J, Gacia E, Buceta JL, Invers O, et al. Salinity tolerance of the Mediterranean seagrass *Posidonia oceanica*: recommendations to minimize the impact of brine discharges from desalination plants. Desalination 2008;221:602–7.
[49] Schiffler M. Perspectives and challenges for desalination in the 21st century. Desalination 2004;165:1–9.
[50] Schively C. Understanding the NIMBY and LULU phenomena: reassessing our knowledge base and informing future research. J Planning Literature 2007;21:255–66.

[51] Seller J. Desalination policy in a multilevel regulatory state. In: Escuder C, Bert G, editors. Desalaction de agua con energias renovables. Mexico City: National Autonomous University of Mexico Press; 2008. p. 173–88.
[52] Sellers J. Desalination policy in a multilevel regulatory state. In: Escuder CN, Bert GHL, editors. Desalaction de agua con energias renovables. Mexico City: National Autonomous University of Mexico Press; 2008. p. 173–88.
[53] State of California Public Utilities Commission. Am Monterey Peninsula Water Supply Project. California Public Utilities Commission, 2015.
[54] State Water Resources Control Board, (2014) Amendment to the water quality control plan for ocean waters of California. Draft Staff Report Including the Draft Substitute Environmental Documentation. In CEP Agency (Ed.). State Water Resources Control Board, Sacramento.
[55] Stratus Consulting. Research on estimating the environmental benefits of restoration to mitigate or avoid environmental impacts caused by California power plant cooling water intake structures. Prepared for the California Energy Commission's Public Interest Energy Research Program Boulder, CO: California Energy Commission; 2004.
[56] Theodori GL, Wynveen BJ, Fox WE, Burnett DB. Public perception of desalinated water from oil and gas field operations: data from Texas. Soc Nat Resour 2009;22:674–85.
[57] Tortajada C, Joshi YK. Water demand management in Singapore: involving the public. Water Resourc Manag 2013;27(8):2729–46.
[58] United States Census Bureau, Carlsbad, California, 2014. Online available at https://www.census.gov/quickfacts/fact/table/carlsbadcitycalifornia/PST045216 (accessed 27 September, 2017).
[59] Vaske J. Survey research and analysis: applications in parks, recreation, and human dimensions. State College, PA: Venture Publishing; 2008.
[60] Vorkinn M, Riese H. Environmental concern in a local context: the significance of place attachment. Environ Behav 2001;33:249–63.
[61] World Water Assessment Programme. The United Nations World Water Development Report 3: water in a changing world. Paris: UNESCO; 2009.

CHAPTER 15

Environmental Life Cycle Analysis of Water Desalination Processes

Habib Cherif*, **Jamel Belhadj***,†
*Université de Tunis El Manar, Tunis, Tunisia
†Université de Tunis, Tunis, Tunisia

15.1 INTRODUCTION

The increasing fresh water demand is one of the world's major problems associated with the growth of the world population and the climate change. Fresh water accounts for only 3% of the planet's water sources contained in the poles, ground water, lakes, and rivers, whereas the remaining 97% is salty seawater [1]. Water desalination can be a promising and viable solution to provide fresh water from seawater or brackish water in rural or urban areas. In 2014, about 1% of all fresh water consumed globally was derived from desalination, and there are about 17,000 desalination plants in 150 countries serving >300 million people worldwide [2]. The installed capacity of desalination plants has increased in recent decades and reached in 2014 over 80 million m^3/day with the biggest increase in the Middle East and should reach by 2018 about 128 million m^3/day [3]. USA and especially Middle East paid much attention to create many desalination systems to satisfy freshwater demand for the people.

Due to a combination of renewable energy integration, improvement in energy recovery devices and improvement in membrane technology, the cost of desalinated fresh water has declined significantly over the past three decades. In terms of desalination technology, RO has the highest market share, accounting for about 60% of global desalination capacity.

15.1.1 Desalination—Energy Nexus

Desalination of seawater or brackish water requires a great amount of energy, which often based on fossil fuels as the energy input. The electrical energy consumption for RO desalination with unit size of 24,000 m^3/day is nearly 5 kWh/m^3 for seawater (salinity is around 41,500 ppm) and is 2.10 kWh/m^3 for brackish water (salinity is around 5000 ppm) [4]. The use of conventional

Sustainable Desalination Handbook
https://doi.org/10.1016/B978-0-12-809240-8.00015-0

fossil fuels in desalination is increasingly raising concerns over climate change and environmental impacts. Also, many issues of desalination are the discharge of brine and the embodied energy associated to industrial desalination. Renewable energies can reduce the environmental impacts and reduce the dependency on fossil fuels of produced water by desalination plants especially in remote regions where climatologically characteristics are favorable [5]. Life Cycle Assessment (LCA) method is applied in order to evaluate environmental impacts and relevant emissions during the life cycle of desalination technologies [6]. LCA of a product life cycle includes acquisition of raw materials, design, production, transportation, use, and the end-of-life phase. Most system components of a typical desalination plant are manufactured using fossil fuel, and this consumed energy is commonly referred to Embodied Energy (EE). Embodied energy expressed in MJ or kWh is a concept that simplifies the assessment of the environmental impact for materials and products used.

In addition, because of the shortage of freshwater, the opportunity for the use of renewable energy to power desalination can be a solution in many regions of the world, including southern Spain, the Maghreb, the Middle East, Central Asia, Pakistan, Southern India, and Northern China. Due to population and economic growth along with water shortages, China and India are high potential markets for desalination. Various renewable energy sources like solar, wind, and wave energy can be used to drive desalination plants and produce potable water. Also, there are various possible combinations to couple renewable energy with desalination technologies [4]. So, water desalination and energy are intrinsically linked together and depend on one another [7]. Both water and energy resources should be considered and dealt in a single and efficient system in order to reach a more sustainable fresh water production.

This chapter presents a description of an environmental life cycle analysis of water desalination processes. Desalination technologies are presented and special attention is given to the use of membrane techniques as reverse osmosis desalination process, which is industrially matured. Various renewable-energy-sources-driven reverse osmosis desalination plants are reviewed. Assessment of desalination environmental impact includes GHG emission and embodied energy is presented. Special attention is given to life cycle assessment method.

15.2 DESALINATION TECHNOLOGIES

15.2.1 Technologies and Performance

Desalination technologies are divided into two broad classes: the thermal techniques (traditional technology) such as Multistage Flash Distillation

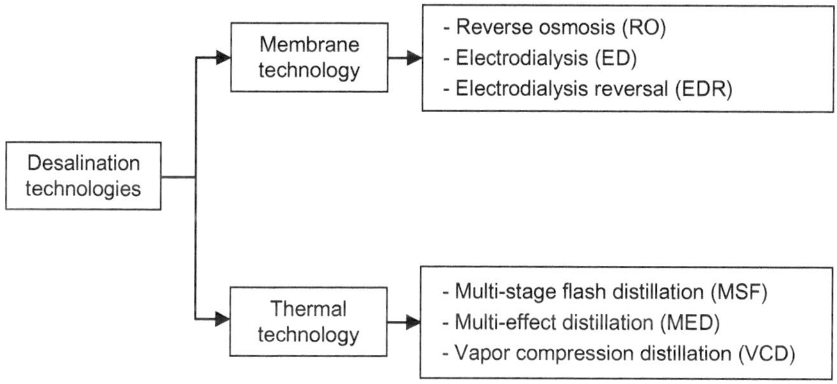

Fig. 15.1 Water desalination technologies and processes.

(MSF), Multieffect Distillation (MED), Vapor Compression Distillation (VCD), and the membrane techniques (modern technology) such as Reverse Osmosis (RO) and Electrodialysis/Electrodialysis Reversal (ED/EDR) [8]. Fig. 15.1 presents the most applied water desalination technologies. Membrane desalination uses high pressure from motor pumps to separate permeates water from brackish water or seawater based on membrane, while thermal desalination uses heat to vaporize permeate water. Due to the considerable amount of energy required for desalination, the amounts of thermal and electrical energy required for a desalination plant in function of the applied technology are shown in Table 15.1 [9–11]. Thermal desalination requires both thermal and electrical energy, while membrane desalination requires only electrical energy [12]. Energy consumption for RO technology is between 3.5 and 5 kWh/m^3.

Membrane techniques tend to replace thermal techniques in desalination plants (seawater or brackish water) due to the recent development of membrane technology. As shown in Fig.15.2, in 2013, the worldwide installed desalination capacity was 65% for RO, 21% for MSF, and only 9% for MED [13,14].

15.2.2 Reverse Osmosis Desalination

Most modern desalination plants use a reverse osmosis technique due to the lower energy demand, lower cost, and improved membrane durability compared to other desalination technologies.

A typical scheme of an RO desalination process is displayed in Fig. 15.3. As seen, the desalination unit includes Low-Pressure (LP) motor pump, High-Pressure (HP) motor pump, stage of pressure vessels (membrane elements), pretreatment of the feed water, and a post-treatment phase. In this

Table 15.1 Performance assessment of desalination technologies

Desalination technology	Water type	Energy consumption (kWh/m^3)	Thermal energy (kWh/m^3)	Operation temperature (°C)
MSF	Seawater, Brackish water	2.5–3.5	12	90–110
MED	Seawater, Brackish water	1.5–2.5	6	70
TVC	Seawater, Brackish water	1.6–1.8	14.6	63–70
MD	Seawater, Brackish water	0.6–1.8	54–350	80
RO	Seawater, Brackish water	3.5–5	–	Ambient
ED	Brackish water	1.5–4	–	Ambient
NF	Brackish water	2.54–3.35	–	Ambient
MVC	Seawater, Brackish water	7–12	–	–

TVC, thermovapor compression; *MD*, membrane distillation; *NF*, nanofiltration; *MVC*, mechanical vapor compression.

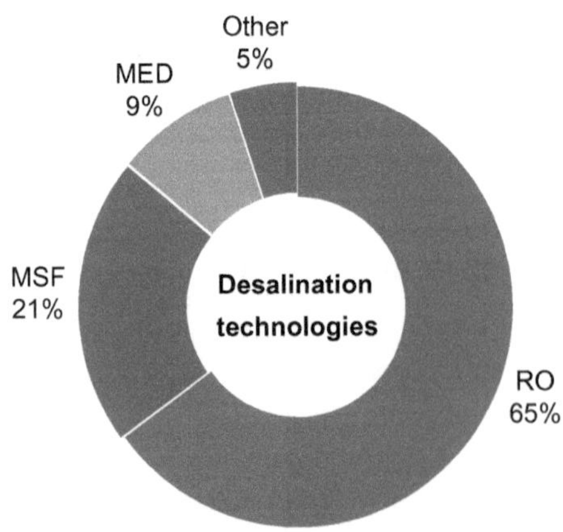

Fig. 15.2 Installed desalination capacity by technology.

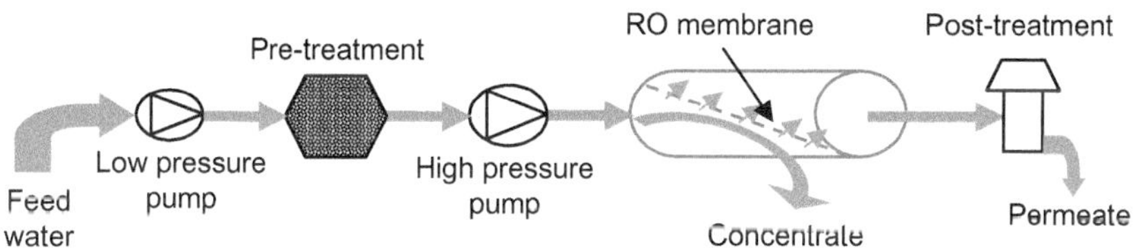

Fig. 15.3 Principle of operation of a reverse osmosis system.

way, an experimental test bench of LSE lab of ENIT, in Tunis, Tunisia, consists in a small-scale RO desalination unit as shown in Fig.15.4.

RO is a pressure-driven membrane separation process in which the water from a pressurized saline solution is separated from the solutes via diffusion across a semipermeable membrane under the influence of osmotic pressure [15,16]. We note that brackish water has a much lower osmotic pressure than seawater; therefore, its desalination requires less energy. High salinity

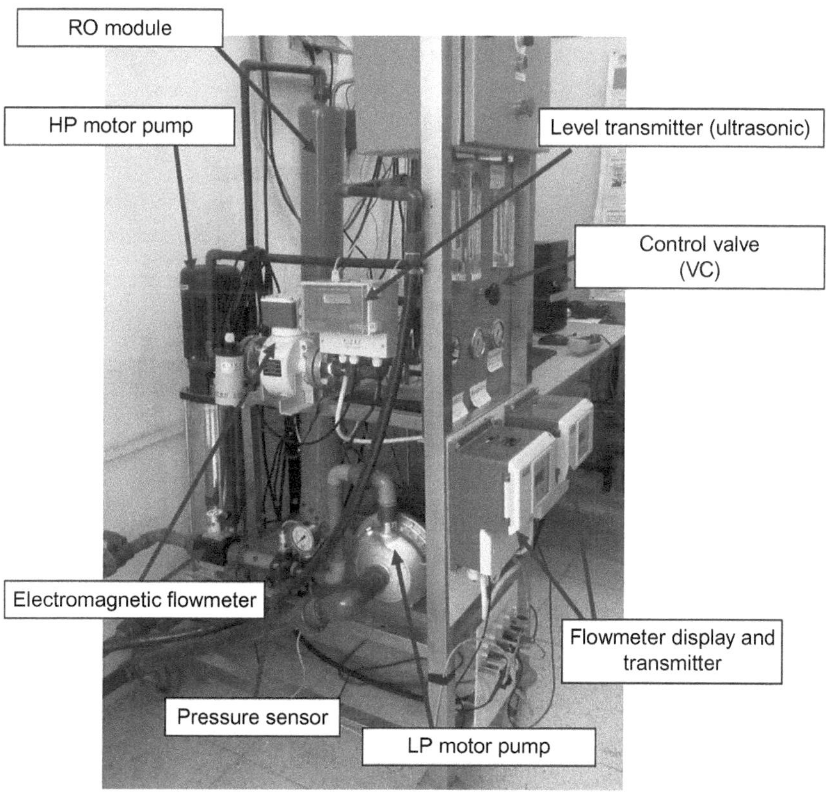

Fig. 15.4 Test bench of RO water desalination_LSE lab, Tunisia.

of seawater can limit the osmosis process as it affects the osmosis pressure, requiring more energy (see Table 15.1). To take an idea on the water quality, three raw water having different qualities from the Mediterranean Sea, the Black Sea, brackish water from Southern Tunisia and WHO (World Health Organization) requirement are listed in Table 15.2 [17]. It's clear that the water of the Mediterranean Sea is very salty compared to the Black Sea or brackish water.

The RO plants are very sensitive to feed pressure, feed water composition, recovery rate, and temperature compared to other distillation technologies that are not so demanding in this respect [18]. Also, the feed water quality has an impact on the architecture and design of the reverse osmosis unit. As shown in Figs. 15.5 and 15.6, according to the feed water quality, the system configuration of the RO unit is designed with the permeate feed in the second stage in order to desalinate the water again or with the concentrate feed in the second stage and the permeate water in each stage is combined together in order to maximize the fresh water production [19].

Different industrial RO membranes (FILMTEC membrane) for seawater and brackish water are listed in Table 15.3. Single RO elements are operated with a recovery of 4%–15%. The high-pressure pump supplies the pressure needed to enable the water to pass through the reverse osmosis membrane in order to product permeate water and reject the salts. Both pretreatment and post-treatment phase are necessary in water desalination in order to reduce fouling and stabilize fresh water. In fact, the seawater or brackish water needs to pass the pretreatment process to neglect the big material, the microbacteria, other organic material, solid material, metal, silica, etc. After main treatment, the post treatment is also necessary to introduce carbonate alkalinity and calcium hardness to the mineral-free desalinated water so as to make the water palatable and noncorrosive as well as to meet the WHO (World Health Organization) requirement.

In reverse osmosis process, the recovery rate is calculated by the following relation:

$$\tau = \frac{Q_p}{Q_f} \tag{15.1}$$

where Qp is the permeate flow and Q_f is the feed flow.

The feed power can be calculated by Eq. (15.2) and the concentrate flow can be deducted by Eq. (15.3):

$$P(W) = Q_f \times P_f \tag{15.2}$$

Table 15.2 Quality parameters of seawater, brackish water and WHO requirement

Component	Mediterranean Sea	Black Sea	Brackish water (Southern Tunisia)	WHO requirements
PH	8.3	8.2	7.8	≥6.5 and ≤9.5
TDS (mg/L)	37,100	19,100	5401	<1000
Sodium (mg/L)	11,544	5930	1430	200
Potassium (mg/L)	388	220	25.6	12
Calcium (mg/L)	1302	226	320	<100
Magnesium (mg/L)	415	646	106	50
Ammonium (mg/L)	0	0	–	–
Strontium (mg/L)	6.3	4	–	–
Barium (mg/L)	0.01	0	–	–
Bicarbonate (mg/L)	146	157	167	–
Chloride (mg/L)	20,500	10,420	1900	250
Sulfate (mg/L)	2790	1490	1450	500
Fluoride (mg/L)	<1	<1	0	1.5
Nitrate (mg/L)	<1	<1	–	–
Silica (mg/L)	5	<1	0	–
Boron (mg/L)	–	–	0	0.5
Iron & Manganese (mg/L)	<0.1	<0.1	–	–

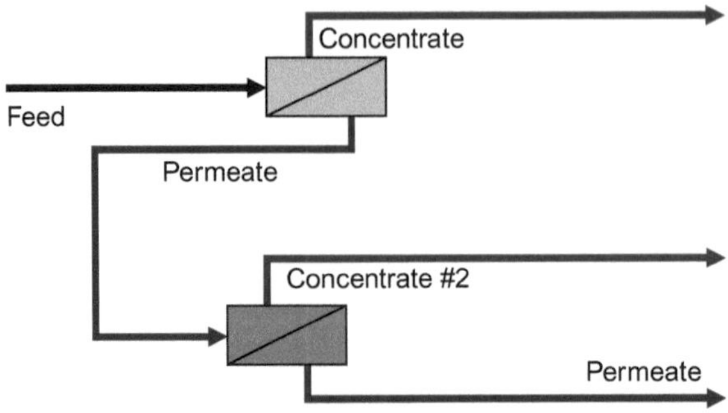

Fig. 15.5 Two-stage system configuration of an RO unit with permeate feed in the second stage.

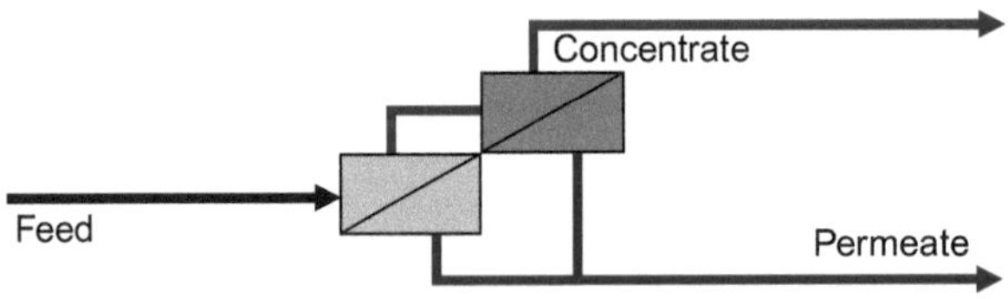

Fig. 15.6 Two-stage system configuration of an RO unit with concentrate feed in the second stage.

$$Q_c = Q_f - Q_p \tag{15.3}$$

where P_f is the feed pressure.

15.2.2.1 RO Desalination Costs

The cost of reverse osmosis membrane is very sensitive to many parameters as RO plant capacity, climatic conditions (feed water temperature, etc.), associated site-specific factors, the energy requirement of the RO desalination plant, the properties of the membranes, and the economic and financial parameters. The current price of desalted water produced by large plants is about \$5 [20]. In Ref. [21], the costs for brackish water RO desalination are in the range of 0.2–0.3 $\$/m^3$. Water distribution cost of various factors for an RO desalination plant of seawater and brackish water is given in Fig. 15.7 and Fig. 15.8, respectively. It can be observed that the energy consumption is the major difference between desalination of seawater and brackish water while the fixed costs are a major factor for both seawater and brackish water. Membrane exchange, maintenance, labor, and consumables represent small costs [10,21–23].

Table 15.3 Different RO membranes for seawater and brackish water

Elements	Active Area (m^2)	Flow (m^3/day)	Recovery (%)
BW30–2540	2.6	3.2	15
BW30–330	30.7	28.4	15
BW30–365	33.9	36	15
BW30–400	37	40	15
BW30LE-440	40.9	43.5	15
TW30–2514	0.7	0.76	5
TW30–2521	1.2	1.23	5
TW30–4040	7.2	9.1	15
XLE-2540	2.6	3.2	15
XLE-4021	3.3	3.9	8
XLE-4040	8.1	9.8	15
LP-2540	2.6	3.8	15
LP-4040	7.2	11	15
NF200–4040	7.6	5.1	15
NF200–400	37.2	25.7	15
SW30–2540	2.6	2.6	8
SW30–4021	3.1	3	4
SW30HRLE-400	37.2	28.4	8
SW30–380	35.3	34.1	8

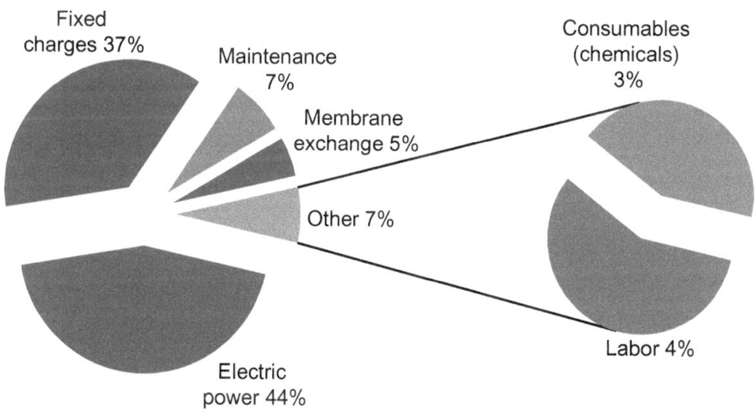

Fig. 15.7 Water distribution cost of a seawater RO plant.

15.2.2.2 Operational Window of RO Membrane and Technologies

The RO spiral membrane module, as shown in Fig. 15.9, consists of several flat sheet membranes, permeate spacer, and feed spacers. Typically, the recovery of an RO element is operated between 5% and 15% [24].

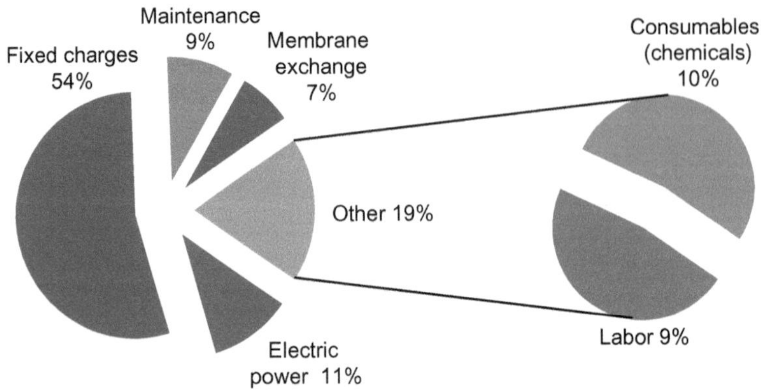

Fig. 15.8 Water distribution cost of a brackish water RO plant.

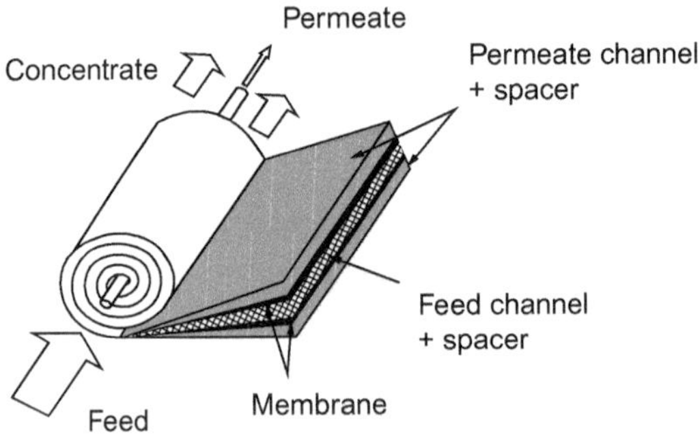

Fig. 15.9 Reverse osmosis membrane composition [24].

The operational parameter variation of RO membranes must create the operating point, which is in the operational window as shown in Fig. 15.10. The four limits that define this window are [25]:

- Maximum feed pressure—determined by the membrane mechanical resistance;
- Maximum brine flow rate—should not be exceeded to avoid membrane deterioration;
- Minimum brine flow rate—it should be observed to avoid precipitation and consequent membrane fouling;
- Maximum product concentration—if the applied pressure is less than a determined value, permeate concentration will be too high.

The number of RO elements needed in desalination plant can be calculated by

$$N_E = \frac{Q_P\left(l/h\right)}{\Phi_d\left(l/h/m^2\right)\cdot S\left(m^2\right)} \tag{15.4}$$

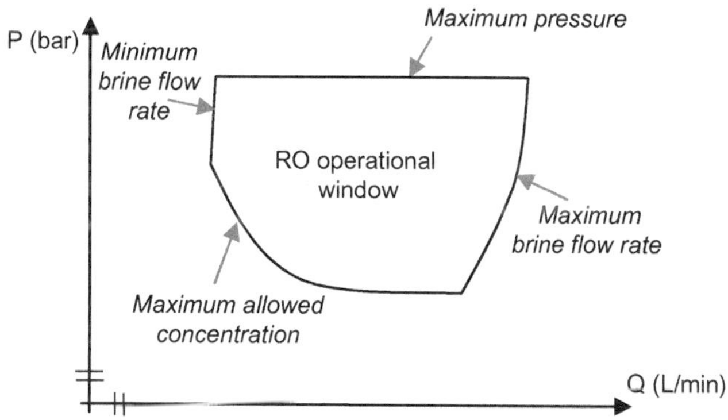

Fig. 15.10 Reverse osmosis membrane operational window (Limits and domains) [25].

where Qp is permeate flow rate, Φ_d is the design flux, and S is the membrane surface area of the selected element.

The number of pressure vessels needed can be calculated by

$$N_{TP} = \frac{N_E}{N_{E/TP}} \tag{15.5}$$

where N_E is the number of elements and $N_{E/TP}$ is the number of elements per pressure vessel. Note that for large systems, six-element vessels are standard, but vessels with up to eight elements are available.

The most leading RO membrane producers for brackish water and seawater elements are [26]:

- Dow Filmtec membranes;
- Hydranautics membranes;
- Toray membranes;
- GE-Osmonics (Desal) membranes;
- Koch Membrane Systems (Fluid Systems);
- CSM/Saehan membranes;
- Inge membranes;
- Parker membranes;
- Ropur membranes;
- TriSep membranes;
- Axeon membranes;
- LewaBrane membranes;
- Pentair RO membranes;
- Microdyn Nadir;

Table 15.4 Different RO system design software

Membrane brands	Software
Dow Filmtec membranes	Rosa
GE-Osmonics (Desal) membranes	Winflows
Hydranautics membranes	IMS Design
Koch Membrane Systems	ROPRO
CSM/Saehan membranes	CSMPRO
Toray membranes	CAROL

- LG RO Elements;
- Mann + Hummel;
- Membranium

Software tools offered by the membrane suppliers only evaluate RO plant design using their own membrane elements. The most commonly used software tools for plant design of RO systems are available from the major RO membrane suppliers are summarized in Table 15.4. These programs offer the possibility to design, sizing, and simulation of the desalination plant equipped with RO membranes [21,26].

15.2.3 Thermal Desalination

In this section, different thermal desalination technologies that have been utilized to produce fresh water will be addressed. The main thermal distillation technologies are Multistage flash (MSF), Multieffect distillation (MED), Membrane distillation (MD), vapor compression (Mechanical Vapor Compression (MVC), and Thermovapor Compression (TVC)):

- MSF units are currently the second most commonly installed desalination process worldwide after the RO process and are widely used in the Middle East. MSF is an energy-intensive process based on the generation of vapor from seawater or brine that requires both thermal and electrical energy. This process requires an external steam supply at temperature around 100°C and a reduced pressure in order to moves the feed seawater in heat exchangers through the stages and gains some heat that helps to reduce the external thermal energy needed for hot brine and also to condense the water vapor for collection as permeate water in each stage [10].
- MED is a distillation process that consists of multiple stages. The performance of this process is proportional to the number of stages. In each stage, the feed water is heated by steam in tubes. Some of the water evaporates, and this steam flows into the tubes of the next stage.

The tubes of the first stage are heated using an external source of steam (at temperature around 70°C)·There are many benefits of the MED process such as the low-energy consumption compared to other thermal processes, highly reliable and simple to operate, and low maintenance cost with minimum supervision [27].

- Vapor compression desalination refers to a distillation process where the evaporation of seawater or saline water is obtained using vapor compression. The feed water penetrates the Vapor Compression (VC) process through a heat exchanger, and vapor is generated in the evaporator. The effect of compressing water vapor can be done by mechanical vapor compression (MVC) or thermovapor compression (TVC) [10,27].
- MD is a temperature-driven separation process in which separation through a microporous hydrophobic membrane is enabled due to phase change. A hydrophobic membrane displays a barrier for the liquid phase, allowing the vapor phase to pass through the membrane's pores. The driving force of MD process is the vapor pressure difference induced by the temperature difference across the hydrophobic membrane. There are several MD configurations that have been utilized to separate aqueous feed solution such as Direct Contact Membrane Distillation (DCMD), Air Gap Membrane Distillation (AGMD), Sweeping Gas Membrane Distillation (SGMD), and Vacuum Membrane Distillation (VMD) [9,28].

15.3 DESALINATION PROCESSES POWERED BY RENEWABLE ENERGY SOURCES

Since desalination is highly energy consuming and usually uses fossil fuels, environmental impacts can be a real issue [29,30]. Fortunately, in order to limit the use of fossil fuels and provide a sustainable way to produce fresh water, renewable energies appear as a good alternative especially in remote rural areas and for less developed countries with both water and energy shortage problems. According to the rise of fossil fuel prices and decreasing costs of renewable technologies as solar panels and wind turbines, desalination based on the use of renewable energy sources has become a viable solution especially in rural areas. Therefore, desalination coupled to renewable energy systems falls into two classes:

- Distillation processes driven by heat produced by the renewable energy systems.
- Membrane and distillation processes driven by electricity or mechanical energy produced by renewable energy systems.

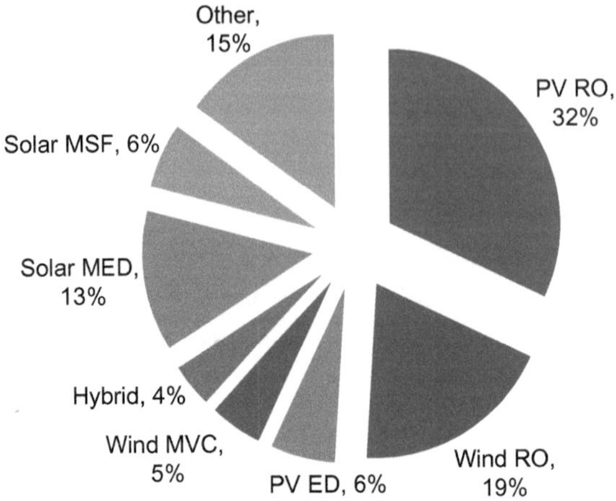

Fig. 15.11 Desalination techniques fed by renewable energy sources.

The distribution of renewable energy coupled to desalination technologies is presented in Fig. 15.11 [31]. Solar and wind energy are regarded as the most suitable options to power desalination at present with which the coupling of PV/RO represent 32% and Wind/RO represents 19%.

Membrane desalination technologies are the most popular in the desalination system coupled to renewable energy sources especially for electrical power conversion. Potential renewable energy resources are solar, wind, geothermal, while the membrane desalination technologies are reverse osmosis, membrane distillation, and electrodialysis.

Desalination based on an RO technique can operate using renewable energy technologies such as Wind Turbines (WT) and solar Photovoltaic (PV) due to the lower energy demand and improved membrane durability [32–34]. Therefore, energy generation systems based on renewable energy sources, such as wind energy and solar photovoltaic, are playing a major role in the clean energy production, especially when the generated electricity will be used to product the fresh water to consumers in remote rural areas or urban areas [35,36]. On the whole life cycle (life cycle analysis), renewable-energy-sources-driven RO desalination plants contribute to reduce environmental impacts and CO_2 emissions [37,38]. RO desalination based on many combinations of renewable energy sources have been reported in Table 15.5. Thus, it can be seen that RO desalination units have various energy supply combinations including PV/RO, WT/RO, PV/WT/RO, PV/diesel/RO, PV/FC/RO, and WT/PV/FC/RO. The combination can include RO with single renewable energy source or with hybrid renewable

Table 15.5 RO desalination powered by renewable energy sources

References	Combinations of renewable sources/RO desalination
[39–42]	PV/RO
[1,43–45]	WT/RO
[46–49]	PV/WT/RO
[50,51]	PV/diesel/RO
[32,52,53]	PV/FC[a]/RO & WT/PV/FC/RO

[a]FC: Fuel Cell.

energy sources. Renewable energy sources (solar and wind) are combined with each other to power an RO desalination in order to provide more continuous electrical power compared with single renewable energy systems [54–56]. According to the intermittency of weather conditions as solar radiation, wind speed, and temperature, battery storage bank is necessary in various cases [57].

15.4 ASSESSMENT OF DESALINATION ENVIRONMENTAL IMPACT

Major environmental impacts of water desalination plants include Greenhouse Gas (GHG) emission generated by the required energy, which is based on fossil fuels, the discharge of brine, and the embodied energy of consumed for components manufacturing of desalination plants including the acquisition of raw materials, processing, transportation to site, and construction. To evaluate the environmental impacts potential of the water desalination plants, Life Cycle Assessment (LCA) tool is applied across the whole life cycle.

15.4.1 Environmental Evaluation Tool

In order to measure the environmental impacts, sustainable development requires methods and tools. Life Cycle Assessment (LCA) is a useful tool that can be used to measure and calculate the environmental impacts of systems or products while taking the entire life cycle into consideration [17,58,59]. As seen in Fig. 15.12, LCA takes into account all the phases of a product's life cycle, starting from the acquisition of raw materials to the end-of-life phase (collection/sorting, reuse, recycling, waste disposal).

LCA technique consists of four stages, as shown in Fig. 15.13, according to ISO 14040–14044 standards [60]:

- Goal and scope definition
- Inventory analysis

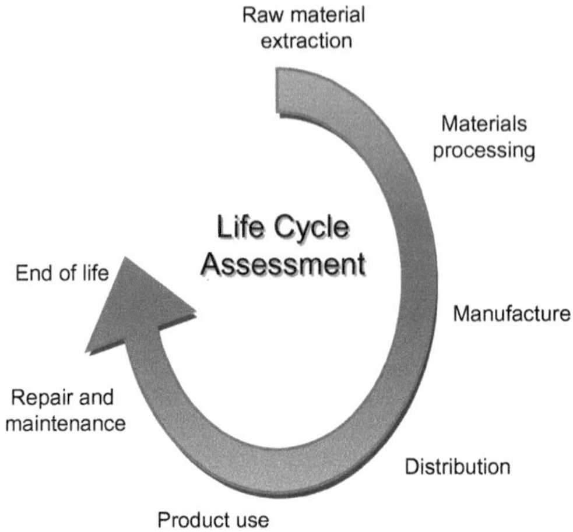

Fig. 15.12 Product life cycle.

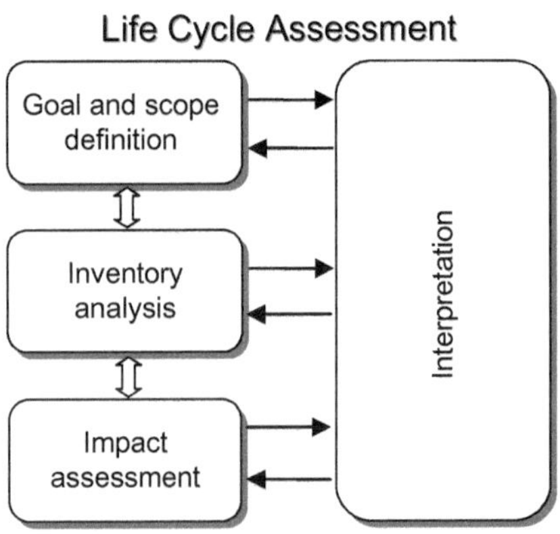

Fig. 15.13 Phases of life cycle assessment method.

- Impact assessment
- Interpretation.

In fact, the description of a studied system or product in terms of the system boundaries and the functional unit is provided by the goal and scope definition. Inventory analysis during life cycle is a method of identification and quantification of the matter/energy, raw material, and environmental releases. During Life Cycle Impact Assessment (LCIA), indicators are presented for analyzing the potential contribution of resource extraction and waste/emissions in an inventory of potential impacts. To evaluate the results of the inventory analysis and impact assessment, the interpretation phase of the LCA is applied in order to select the preferred system, product, or process. Analysis and impact assessment are summarized during the life cycle interpretation phase.

Many researchers have studied LCA method to investigate the total life cycle environmental impact for water desalination. A life cycle Greenhouse Gas (GHG) emissions of seawater RO desalination plant located in Perth, Western Australia, has been evaluated in Ref. [61]. The Simapro Australian and Ecoinvent databases are used for operational phase Life Cycle Inventory (LCI). Based on LCA, the overall environmental impact associated with renewable energy powered desalination plants is carried out. Based on different LCIA methods, the environmental impacts of RO desalination plant in the United States have been examined by Zhou et al. [62]. LCI used was adopted from the Ecoinvent Database with specific US datasets. The authors showed that choosing an inappropriate LCIA method might only have minimal impacts in the large-scale issues such as global warming potential and ozone depletion potential. Authors of Ref. [63] have evaluated the environmental impact of desalination technologies such as MSF, MED, and RO by applying life cycle analysis. The software SimaPro 6.0 has been used and different evaluation methods such as CML 2 baseline 2000, Ecopoints 97, and EI 99 are applied. The results showed that RO technology has an environmental load associated significantly lower than thermal desalination processes (MSF and MED). This result is further reinforced if the energy production is oriented to renewable energies. Ref. [64] investigated a comparative LCA methodology of three functionally equivalent seawater desalination and wastewater reclamation processes. SimaPro LCA software was used to perform interpretation of the LCIA data and for aggregate the chemical and material requirements. The LCIA of the inventories was performed with the CML 2001 method.

Methodology, databases, and software have been continuously developed to improve the scientific use of LCA. Commonly used LCA software are EIMEV5, Cycle IT System V1.1, e-LICCO, Open LCA 1.2, GaBi, SimaPro Analyst 7.3.3, Umberto 5.6, etc.

15.4.2 GHG Emissions

CO_2 emissions from consumed energy account for the largest share of global anthropogenic Greenhouse Gas (GHG) emissions, representing about 60% of global emissions [65]. In 2014, global CO_2 emissions from fuel combustion were 32 $GtCO_2$ and between 2013 and 2014, CO_2 emissions from electricity and heat increased by 0.2%. In fact, due to the considerable

Table 15.6 Greenhouse gases for different desalination technologies

References		RO	MSF	MED
[21,66,67]	$kgCO_2/m^3$	2	20	20
	gNO_x/m^3	3	25	25
	gSO_x/m^3	12	27	27
	$gNMVOC/m^3$	1.5	7	7
[63]	$kgCO_2/m^3$	1.78	23.41	18.05
	gNO_x/m^3	3.87	28.3	21.41
	gSO_x/m^3	10.68	27.91	26.48
	$gNMVOC/m^3$	1.10	7.90	5.85

Table 15.7 Greenhouse gases for RO desalination integrated with renewable energies and fossil fuels

Process	$kg\,CO_2/m^3$	$g\,NO_x/m^3$	$g\,SO_x/m^3$	$g\,NMVOC/m^3$
Fossil fuels /RO	2.79	3.38	3.25	0.93
WT/RO	0.11–0.17	0.42	1.80–2	0.08–0.22
PV/RO	0.34–0.9	1–2.1	4.73–16.15	0.36–0.72
Hydro/RO	0.082	0.24	1.68	0.05

amount of energy required for water desalination, desalination plants emit greenhouse gases into the atmosphere.

The amount of greenhouse gases for the production of 1 m^3 of fresh water along all life cycle in seawater desalination is shown in Table 15.6. Compared to thermal processes (MSF and MED), RO desalination is considered the least pollutant to the atmosphere. RO desalination process remains a pollutant for the atmosphere because it requires fossil fuels as energy resource to produce fresh water. Also, in Table 15.7, Raluy et al. [63] showed that RO integrated with renewable energies (PV, WT, and hydroenergy) have the lowest pollutant technology when compared to fossil fuels as primary power source.

Therefore, to reduce greenhouse gases of desalination and to reduce the dependency on fossil fuels, the application of renewable energies should be considered and applied in desalination. Renewable powered desalination technologies can provide substantial benefits in the next future by supplying water needs and reducing the environmental emissions (including CO_2 emissions).

15.4.3 Embodied Energy

Embodied Energy (EE) represents the nonrenewable energy consumed in the acquisition of raw materials, their processing, manufacturing, transportation to site, and construction throughout the whole life cycle. LCA tool is

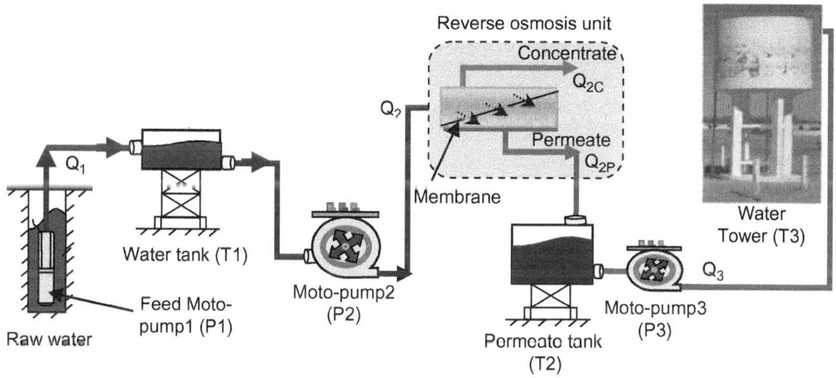

Fig. 15.14 Typical reverse osmosis desalination plant.

applied during a life cycle period of 20 years in order to investigate the embodied energy of a typical reverse osmosis desalination plants (water pumping, water storage and an RO desalination unit) as shown in Fig. 15.14 [68].

The typical reverse osmosis desalination plant includes three motor pumps (submerged pump P1, desalination pump P2, and storage pump P3) with different functions, RO desalination unit, and three water tanks (T1, T2, and T3). Water tower T3 is used as hydraulic storage [69].

The amount of materials used in the typical RO desalination plant for 1 m^3 of permeate flow as the components of an RO module, piping, tanks, motor pumps are listed in Table 15.8. The types of materials used in RO module are fiberglass-reinforced plastic, Cotton fabric, Polypropylene, and Cellulose acetate. Also, many types of materials used in motor pumps as Stainless steel, Copper, Iron, Cast iron, etc. Therefore, each material used in the typical RO desalination plant has its embodied energy, which depends on the amount of the materials.

Global embodied energy of the typical RO desalination plant during a life cycle period of 20 years for 1 m^3/day of permeate flow is summarized in Table 15.9. Global embodied energy integrates balance-of-system (BOS) such as cables, connectors, and protections, the exchange of each component as RO membrane and Static Converter (SCV).

Embodied energy breakdown for a typical reverse osmosis desalination plant during a life cycle of 20 years is presented in Fig. 15.15. It can be observed that RO membrane and water tower represent the major parts of embodied energy consumed due to regular exchange of RO membrane (every 5 years) and the amount of materials used in water tower (90% concrete, 10% Iron). In Ref. [68], the authors show that the embodied energy

Table 15.8 Amount of materials used in the hydraulic process

Components	Quantity	Materials
Pressure vessels (RO Module) 1 m3/d		
RO module casing	11.28 kg	100% fiberglass-reinforced plastic
Permeate spacer	3.36 kg	100% Cotton fabric
Feed spacer	3.68 kg	100% Polypropylene
Membrane	1.12 kg	100% Cellulose acetate
Piping (PEHD) kg/m		
Piping for HP applications	6.71 kg/m (d = 160 mm, NP16)	10% Stainless steel, 90% polyethylene resins
Piping for LP applications	3.05 kg/m (d = 160 mm, NP6)	100% polyethylene resins
Permeate piping	3.05 kg/m (d = 160 mm, NP6)	100% polyethylene resins
Reservoirs kg/m3		
Brackish water tank	4.2 kg /m^3	80% PVC, 20% fiber reinforced plastic
Permeate tank	4.2 kg/m^3	80% PVC, 20% fiber reinforced plastic
Water tower	350 kg/m^3	90% concrete, 10% Iron
Motor pumps		
Immersed motor pump1 (*3.5 kW*)	28.4 kg	90% Stainless steel, 5% Copper, 5% Iron
Motor pump2 (*7.5 kW*)	130 kg	85% Stainless steel, 5% Cast iron, 5% Copper, 5% Iron
Motor pump3 (*2 kW*)	32 kg	75% Stainless steel, 15% Cast iron, 5% Copper, 5% Iron

for a 20-year lifetime of 1 m^3 of fresh water (embodied energy of only hydraulic process) is around 2.2 MJ/m^3 (0.61 kWh/m^3).

From the LCA investigation of a typical RO desalination plant, embodied energy for various industrial motor pumps from Grundfos manufacturer data as CRN 3–10, CRN 20–16 SF, SP 8A-5, SP 30–5, CRTE 2–7 is presented in Fig. 15.16, while Fig. 15.17 presents the evaluation of embodied energy for different industrial RO membranes.

Global embodied energy model of motor pump is given by

$$EE_{MP} = 5.52 \times W_{MP}^{1.47} \quad (15.6)$$

Table 15.9 Embodied energy for a typical RO desalination plant components

Materials RO Module (MJ/m^3)	1536.7
Materials RO Module after 5 years (MJ/m^3)	1383
Materials RO Module after 10 years (MJ/m^3)	1229.3
Materials RO Module after 15 years (MJ/m^3)	1075.7
Piping (MJ/m)	1375.8
Tanks (MJ/m^3)	2263.23
Motor pump P1 (MJ/kW)	283.35
Motor pump P2 (MJ/kW)	684.87
Motor pump P3 (MJ/kW)	679.20
BOS (MJ/m^3)	100
SCV (MJ/kW)	1260
SCV after 10 years (MJ/kW)	940

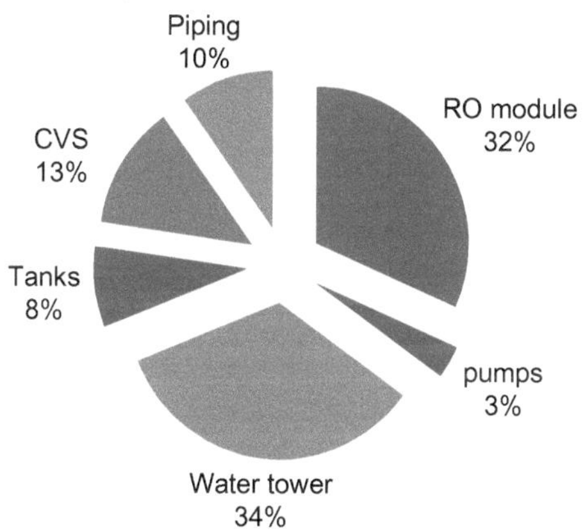

Fig. 15.15 Embodied energy distribution of a typical RO desalination plant.

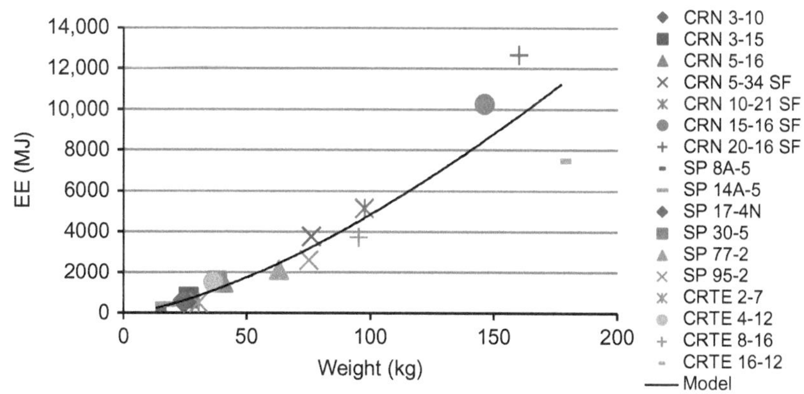

Fig. 15.16 Global EE of various motor pumps.

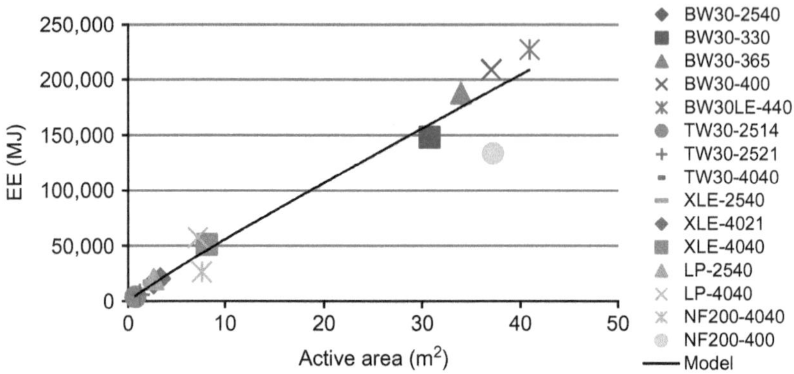

Fig. 15.17 Global EE of various RO membranes.

where EE_{MP} is the embodied energy model of motor pump expressed as a function of weight motor pump (*WMP*).

Global embodied energy model of desalination RO membrane is given by

$$EE_{RO} = 6443.5 \times A_c^{0.94} \tag{15.7}$$

where EE_{RO} is the embodied energy model of an RO membrane expressed as a function of Active area (Ac).

15.4.4 GHG Emissions and Embodied Energy of Renewable PV/Wind Source

It can be noted that in the whole life cycle, renewable-energy-sources (solar and wind)-driven RO desalination plants contribute to reduce environmental impacts and CO_2 emissions. In fact, in the manufacture process, these renewable sources, as shown in Fig. 15.18, emit many greenhouse gases and consume many embodied energy. So, it is important to compute and integrate the environmental impacts of the renewable sources in the environmental balance sheet.

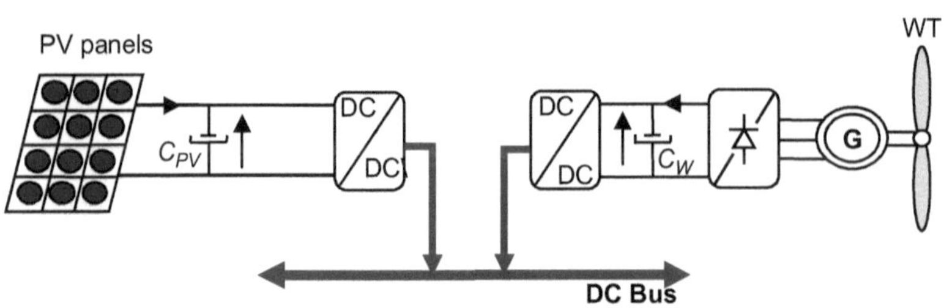

Fig. 15.18 Renewable solar and wind source.

Table 15.10 Embodied energy for various solar panels technologies

References	Technologies				
	Mono-Si (MJ/m^2)	Poly-Si (MJ/m^2)	a-Si (MJ/m^2)	[a]CdTe (MJ/m^2)	[b]CIS (MJ/m^2)
[70–74]	2860–5253	2699–5150	710–1990	790–1803	1069–1684

[a]CdTe: Cadmium Telluride.
[b]CIS: Cuivre Indium Selenium.

Over 25 years, the embodied energies of 1 m^2 solar panels according panel technology are summarized in Table 15.10. Embodied energy of manufacturing process for poly-Si feedstock production (purification, crystallization), wafer and cell production are about 1562, 617, and 300 MJ/m^2, respectively [75]. Embodied energy of the frame is between 100 and 300 MJ/m^2 while embodied energy of Balance of System (BOS) components is 300 MJ/m^2 [76].

For wind turbines, embodied energy of manufacturing process for blades, hub, gearbox generator, and nacelle is about 61.8, 36.8, 36, 86.2, and 61.8 MJ/kg, respectively [76].

The EE evolution of various industrial solar panels (poly-Si) versus PV area (*APV*) and the EE evolution of various industrial wind turbines versus WT swept area (*AWT*) are presented in Fig. 15.19 and Fig. 15.20.

Therefore, global EE model of PV system (poly-Si) expressed as a function of PV surface area (*APV*) is given by

$$EE_{PV} = 3863 \times A_{PV} - 47.26 \tag{15.8}$$

Global EE model of wind turbine system expressed as a function of wind turbine swept area (*AWT*) is given by

$$EE_{WT} = 2359.7 \times A_{WT} + 49.43 \tag{15.9}$$

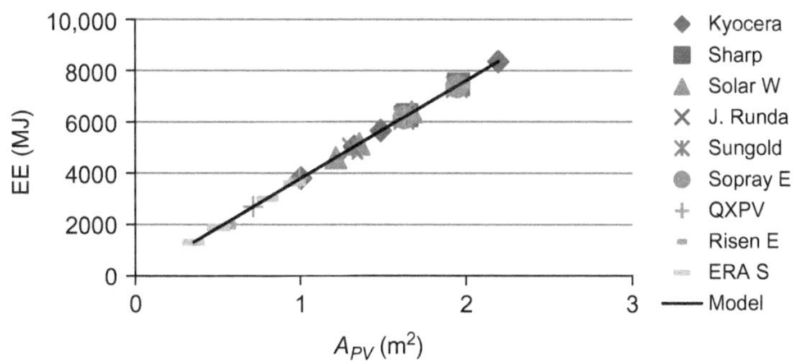

Fig. 15.19 Embodied energy of various poly-Si PV modules.

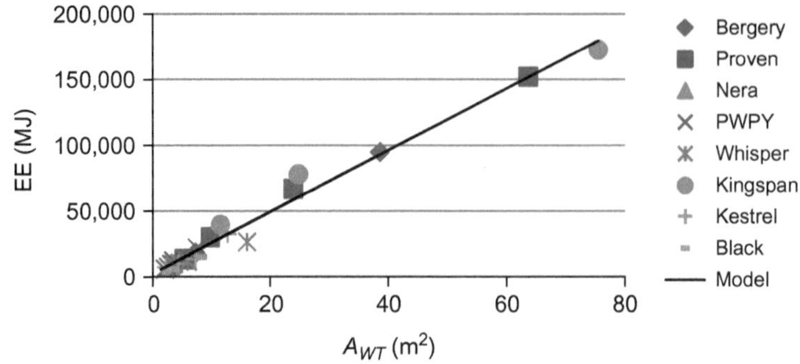

Fig. 15.20 Embodied energy of various wind turbines.

In addition, solar panels and wind turbines emit greenhouse gases into the atmosphere during the manufacturing process [77–82]. Greenhouse Gas (GHG) emissions are measured in carbon dioxide equivalent (CO_2eq). GHG emissions evolution of various industrial poly-Si solar panels versus PV area (*APV*) is presented in Fig. 15.21, while GHG emissions evolution of various industrial wind turbines versus WT swept area (*AWT*) is shown in Fig. 15.22.

Global GHG emission model of poly-Si PV system expressed as a function of PV surface area (*APV*) is given by

$$GHG_{PV} = 138.3 \times A_{PV} - 2.54 \tag{15.10}$$

Global GHG emission model of wind turbine system expressed as a function of wind turbine swept area (*AWT*) is given by the following relation:

$$GHG_{WT} = 156 \times A_{WT} \tag{15.11}$$

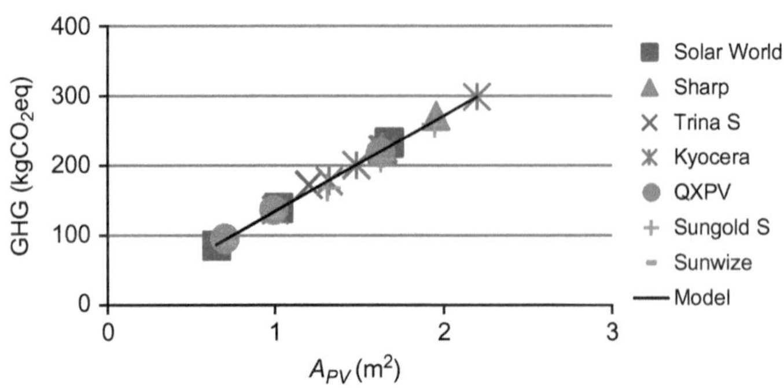

Fig. 15.21 GHG emission of various poly-Si PV modules.

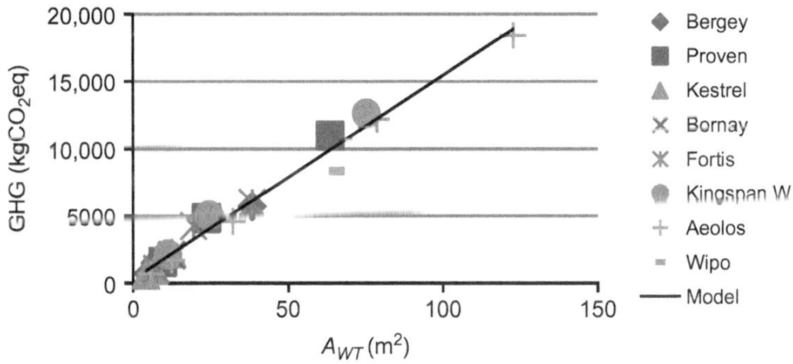

Fig. 15.22 GHG emission of various wind turbines.

15.4.5 Energy, Environmental, and Cost Payback Periods of Renewable PV/Wind Source

Estimating energy, environmental, and cost payback periods for each of the renewable energy technologies in a specific region is very difficult, because it is affected by so many parameters, such as life cycle energy requirement, local weather conditions, and systems' life time. Payback time is defined as the number of years required to recover during the life time the energy, the cost, and associated generation of pollution and CO_2 that went into making the system or the product, in the first place. Energy payback time is a ratio between primary energy consumed during LCA and electricity production per year. Cost payback time is defined as the number of years required to recover the cost invested during the life time of the product or system. GHG payback time is calculated as the ratio between LCA emissions and grid emission to produce the same amount of electricity than the renewable source.

Using process-based LCA methods (over a period of 30 years), Ref. [83] showed that the Energy Payback Time (EPBT) for mono-Si, multi-Si, a-Si, CdTe, and CIS PV systems were in the range of 1.7–2.7, 1.5–2.6, 1.8–3.5, 0.75–2.1, and 1.45–2.2 years, respectively. In addition, the total GHG emissions rates were in the order of 29–45, 23–44, 18–50, 14–35, and 10.5–46 g CO_2-eq/kWh, respectively. It is found that a large amount of energy and GHG emissions has been consumed and emitted during the whole life time for the mono-Si and a-Si PV systems. Wild-Scholten [84] investigated the energy payback time and carbon footprint of commercial roof-top photovoltaic systems. The results showed that the energy payback times and carbon footprints are 1.96, 1.24, 1.39, 0.92, 0.68, and 1.02 years and 38.1, 27.2, 34.8, 22.8, 15.8, and 21.4 g CO_2-eq/kWh for mono-Si, multi-Si, a-Si, μm-Si (micromorph silicon), CdTe and CIGS roof-top photovoltaic systems, respectively.

Mazzanti et al. [85] estimated the payback time (over the 25 years) of PV plants of different sizes and destinations in Italy (in northern, central, and southern Italy), considering the geographical area and the market economy of the plants. As results, the payback periods are equal to 4.4–6.8, 4.5–6.8, and 4.3–7.3 years for residential, tertiary, and industrial users, respectively.

Concerning wind turbines, the energy and environmental performance of an Italian wind farm for the generation of 1-kWh electricity using the average European data have been evaluated by Ardente et al. [86]. In this study, energy intensity varies from 0.04 to 0.07 kWhprim/kWhel and CO_2 intensity index varies from 8.8 to 18.5 g/kWh. The payback indexes were found to be lower than 1 year. Three configurations of wind turbines to produce a nameplate power of 100 kW (20 Endurance (EN) 5 kW, or 5 Jacobs (JA) 20 kW, or 1 Northern Power (NP) 100 kW turbines) applying LCA methodologies over a lifetime of 25 years have been studied in Ref. [79]. As a result, energy payback time for the turbine configurations is 1.4, 0.8, and 0.6 years, respectively, for EN, JA, and NP and the GHG emission payback periods were found 1.4, 0.8, and 0.5 years, respectively, for EN, JA, and NP. For economic analysis, simple payback period is in the range of 6.89–9.68, 6.85–9.53, and 5.92–7.82 years, respectively, for EN, JA, and NP. Tremeac and Meunier [87] obtained for 4.5-MW and 250-W wind turbines have a primary energy payback time of 0.58 and 2.29 years, respectively. GHG payback time is about 0.7 year and 2 years for the 4.5-MW wind turbine and 250-W wind turbine, respectively.

15.5 ECO-DESIGN AND ECO-OPTIMIZATION OF AN RO WATER DESALINATION

The designers integrate the economic aspects into desalination systems design in order to reduce the total annual cost. The integration of environmental impact considerations (embodied energy and GHG emissions) into systems design and into systems optimization is a very important issue. In fact, today it is necessary to design and optimize desalination systems based on an environmental, economic, and technical viewpoint taking into account LCA indicators in order to continuously satisfy the fresh water load demand with the minimum of economic and environmental requirements. As shown in Fig. 15.23, the LCA indicators are:

- Environmental indicator: Embodied Energy (EE) and Greenhouse Gas (GHG) emission.
- Economic indicator: Life Cycle Cost (LCC)

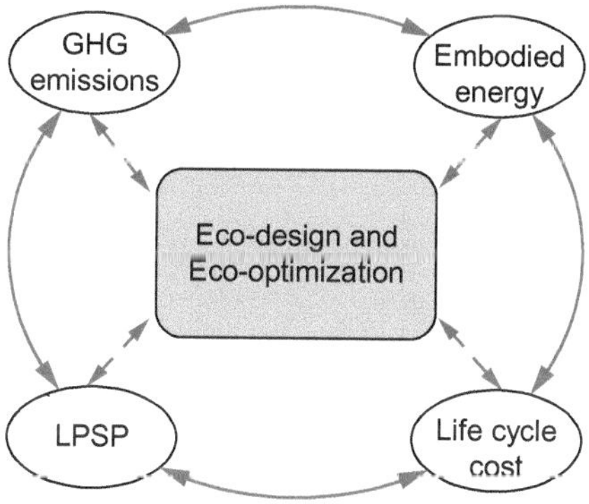

Fig. 15.23 LCA indicators.

- Reliability indicator: Loss of Power Supply Probability (LPSP)

By this way, sizing and optimizing of desalination based on renewable energy systems have been carried out by a number of researchers. The optimal configuration of a standalone hybrid photovoltaic/wind/hydrogen system supplying a desalination unit was predicted on the basis of the minimum initial investment cost [88]. Maleki et al. [46] proposed an optimization, sizing, and technoeconomic assessment of standalone renewable energy systems to supply electricity and potable water where needed. Particle swarm optimization is compared to HOMER for the simultaneous optimization of size and power management in standalone hybrid systems. The multiobjective functions are: minimize total NPC (Net Present Cost) and lifetime CO_2 emissions. In Ref. [53], the authors presented a novel optimizer approach for optimization of a hybrid solar-wind-powered reverse osmosis water desalination system located in Davarzan, Khorasan, Iran. The decision variables are optimized using a harmony-search-based chaotic search for the most cost-effective system. The use of harmony search with chaotic search is proposed to determine the optimal values of parameters for the hybrid system that satisfy the load: minimizing the life cycle cost and not exceeding the maximum allowable loss of power supply probability.

The sizing, modeling, and design for complex systems like desalination based on renewable energy systems are complex tasks, in particular when the generated electricity by the renewable sources will be used to product the fresh water to consumers in remote rural areas or urban areas. Therefore, it is essential to find an optimal configuration of different sources of energy and system components, which depend on consumers demand (water/electricity) and intermittent weather conditions (wind speed, solar radiation, ambient temperature, etc.).

15.6 TECHNOLOGICAL CHALLENGES AND THE FUTURE OF DESALINATION

In the future, the need exists for better knowledge of how to operate desalination plants with the lowest possible inputs of energy in order to reduce the life cycle cost and the environmental impacts. To achieve this vision, the following general approaches and objectives will be pursued:

- The application of renewable energies should be considered and applied in desalination to reduce the environmental impact on the atmosphere and to reduce the dependency on fossil fuels. Also, desalination based on cogeneration can be used to desalinate water, without further major demands for energy.
- The integration of environmental impact considerations into systems design (eco-design) and into systems optimization (eco-optimization) is very important to exploit a new energy and water supplies that meet sustainability and environmental requirements.
- The reduction of the total embodied energy of desalination processes may be achieved by construction of plants near populated areas. This eliminates the need for a transportation of the product waters for long distances.

In the near future, both urban and rural environments will manage water and energy as an integrated system. Therefore, engineers and scientists will contribute to integrate the water and energy systems in a single and efficient system with (1) the identification of renewable energy sources inside urban and rural areas, (2) Research new methods for water and energy caption/storage (3) Stimulate the intelligent use of the available water and energetic resources, and (4) develop new and better ways to desalt seawater, and purify low-quality water, including brackish groundwater, storm water, gray water, and wastewater.

15.7 CONCLUSION

This chapter presents an environmental life cycle analysis of water desalination processes based on life cycle assessment method. Water desalination technologies and processes include membrane and thermal technologies are presented. Special attention is given to the use of RO desalination technique due to the lower energy demand, lower cost, and improved membrane durability compared to other desalination technologies. The review shows that the energy requirement for desalination process to get permeate from brackish water is very less compared to seawater desalination technique. Different industrial RO membranes for seawater and brackish water and most commonly used software tools for plant design of RO systems are presented. Various renewable-energy-sources-driven reverse osmosis desalination plant are reviewed. The review shows that renewable-energy-sources-driven RO desalination plants contribute to reduce environmental impacts and CO_2 emissions. Assessment of desalination environmental impact is presented. To evaluate the environmental impacts potential of the water desalination plants, LCA tool is applied across the whole life cycle. The results showed that RO integrated with renewable energies sources have the lowest pollutant technology (including CO_2 emissions) when compared to fossil fuels. Greenhouse gases includes CO_2 emissions, nitrogen oxides (NO_x), sulfur oxides (SO_x), and nonmethane volatile organic compounds (NMVOC). Embodied energy for a typical RO desalination plant is estimated during a life cycle period of 20 years. It can be observed that RO membrane and water tower represent the major parts of embodied energy consumed due to regular exchange of RO membrane and the amount of materials used in water tower (90% concrete, 10% Iron). For a developing country, renewable-energy-driven desalination technology is a competitive alternative for generating freshwater to meet the future fresh water requirement based on an environmental(embodied energy and greenhouse gases), economic (life cycle cost), and technical (loss of power supply probability) viewpoint.

ACKNOWLEDGMENT

This work was supported by the Tunisian Ministry of High Education and Research under the ERANETMED-NEXUS "Energy and Water Systems Integration and Management" ID Number 14-044 and the UTIQUE program CMCU-12G/1103 for the financial support.

REFERENCES

[1] Gökçek M, Gökçek ÖB. Technical and economic evaluation of freshwater production from a wind-powered small-scale seawater reverse osmosis system (WP-SWRO). Desalination 2016;381:47–57.
[2] International Desalination Association. http://idadesal.org; 2014.
[3] RobecoSAM. Water: the market of the future. 2015. 1–40. www.robecosam.com.
[4] Manju S, Sagar N. Renewable energy integrated desalination: a sustainable solution to overcome future fresh-water scarcity in India. Renew Sustain Energy Rev 2017;73:594–609.
[5] Gilau A, Small MJ. Designing cost-effective seawater reverse osmosis system under optimal energy options. Renew Energy 2008;33:617–30.

[6] Raluy RG, Serra L, Uche J. Life cycle assessment of desalination technologies integrated with renewable energies. Desalination 2005;183:81–93.
[7] Hamiche AM, Stambouli AB, Flazi S. A review of the water-energy nexus. Renew Sustain Energy Rev 2016;65:319–31.
[8] Miller S, Shemer H, Semiat R. Energy and environmental issues in desalination. Desalination 2015;366:2–8.
[9] González D, Amigo J, Suárez F. Membrane distillation: perspectives for sustainable and improved desalination. Renew Sustain Energy Rev 2017;80:238–59.
[10] Triki Z, Bouaziz MN, Boumaza M. Techno-economic feasibility of wind powered reverse osmosis brackish water desalination systems in southern Algeria. Desalin Water Treat 2014;52:1745–60.
[11] Zhou D, Zhu L, Fu Y, Zhu M, Xue L. Development of lower cost seawater desalination processes using nanofiltration technologies—a review. Desalination 2015; 376:109–16.
[12] IEA-ETSAP and IRENA Technology Brief I12. Water desalination using renewable energy. 2012. 1–24. http://www.irena.org.
[13] International Desalination Association (IDA), Desalination yearbook, Global Data Report DesalData, 2013–2014.
[14] Balfaqih H, Al-Nory MT, Nopiah ZM, Saibani N. Environmental and economic performance assessment of desalination supply chain. Desalination 2017;406:2–9.
[15] Greenlee LF, Lawler DF, Freeman DB, Marrot B, Moulin P. Reverse osmosis desalination: water sources, technology, and today's challenges. Water Res 2009;43:2317–48.
[16] Mohamed ES, Papadakis G, Mathioulakis E, Belessiotis V. An experimental comparative study of the technical and economic performance of a small reverse osmosis desalination system equipped with a hydraulic energy recovery unit. Desalination 2006;194:239–50.
[17] Rebitzer G, et al. Life cycle assessment part 1: framework, goal and scope definition, inventory analysis, and applications. Environ Int 2004;30:701–20.
[18] Cherif H, Belhadj J. Large-scale time evaluation for energy estimation of stand-alone hybrid photovoltaic-wind system feeding a reverse osmosis desalination unit. Energy 2011;36:6058–67.
[19] ROSA, Reverse Osmosis System Analysis, Version 6.1.4, 2007.
[20] Lior N. Sustainability as the quantitative norm for water desalination impacts. Desalination 2017;401:99–111.
[21] Fritzmann C, Löwenberg J, Wintgens T, Melin T. State-of-the-art of reverse osmosis desalination. Desalination 2007;216:1–76.
[22] Semiat R. Desalination: present and future. Water Int 2000;25:54–65.
[23] Eltawil MA, Zhengming Z, Yuan L. A review of renewable energy technologies integrated with desalination systems. Renew Sustain Energy Rev 2009;13:2245–62.
[24] Schwinge J, Neal PR, Wiley DE, Fletcher DF, Fane AG. Spiral wound modules and spacers. Rev Anal J Membr Sci 2004;242:129–53.
[25] Miranda MS, Infield DA. Wind-powered seawater reverse-osmosis system without batteries. Desalination 2002;153:9–16.
[26] Reverse Osmosis Membranes, Lenntech, Available at: http://www.lenntech.com/products/membranes.
[27] Al-Karaghouli A, Kazmerski L. Energy consumption and water production cost of conventional and renewable-energy-powered desalination processes. Renew Sustain Energy Rev 2013;24:343–56.
[28] Alkhudhiri A, Darwish N, Hilal N. Membrane distillation: a comprehensive review. Desalination 2012;287:2–18.
[29] He W, Wang Y, Shaheed MH. Stand-alone seawater RO (reverse osmosis) desalination powered by PV (photovoltaic) and PRO (pressure retarded osmosis). Energy 2015;86:423–35.

[30] Subramani A, Badruzzaman M, Oppenheimer J, Jacangelo JG. Energy minimization strategies and renewable energy utilization for desalination: a review. Water Res 2011;45:1907–20.
[31] Jijakli K, Arafat H, Kennedy S, Mande P, Theeyattuparampil VV. How green solar desalination really is? Environmental assessment using life-cycle analysis (LCA) approach. Desalination 2012;287:123–31.
[32] Maleki A, Pourfayaz F, Ahmadi MH. Design of a cost-effective wind/photovoltaic/hydrogen energy system for supplying a desalination unit by a heuristic approach. Sol Energy 2016;139:666–75.
[33] Koutroulis E, Kolokotsa D. Design optimization of desalination systems power-supplied by PV and W/G energy sources. Desalination 2010;258:171–81.
[34] Novosel T, et al. The influence of reverse osmosis desalination in a combination with pump storage on the penetration of wind and PV energy: a case study for Jordan. Energy 2014;76:73–81.
[35] Segurado R, Madeira JFA, Costa M, Duić N, Carvalho MG. Optimization of a wind powered desalination and pumped hydro storage system. Appl Energy 2016;177:487–99.
[36] Tsai Y, Chiu C, Ko F, Chen T, Yang J. Desalination plants and renewables combined to solve power and water issues. Energy 2016;113:1018–30.
[37] Bilgili F, Koçak E, Bulut Ü. The dynamic impact of renewable energy consumption on CO2 emissions: a revisited environmental Kuznets curve approach. Renew Sustain Energy Rev 2016;54:838–45.
[38] Dogan E, Seker F. Determinants of CO2 emissions in the European Union: the role of renewable and non-renewable energy. Renew Energy 2016;94:429–39.
[39] Bilton AM, Wiesman R, Arif AEM, Zubair S, Dubowsky S. On the feasibility of communutiy-scale photovoltaic-powered reverse osmosis desalination systems for remote locations. Renew Energy 2011;36:3246–56.
[40] Joyce A, Loureiro D, Rodrigues C, Castro S. Small reverse osmosis units using PV systems for water purification in rural places. Desalination 2001;137:39–44.
[41] Kyriakarakos G, Dounis AI, Arvanitis KG, Papadakis G. Design of a Fuzzy Cognitive Maps variable-load energy management system for autonomous PV-reverse osmosis desalination systems: a simulation survey. App Energy 2017;187:575–84.
[42] Alghoul MA, Poovanaesvaran P, Mohammed MH, Fadhil AM, Muftah AF, Alkilani MM, et al. Design and experimental performance of brackish water reverse osmosis desalination unit powered by 2 kW photovoltaic system. Renew Energy 2016;93:101–14.
[43] Koklas PA, Papathanassiou SA. Component sizing for an autonomous wind driven desalination plant. Renew Energy 2006;31:2122–39.
[44] Loutatidou S, Liosis N, Pohl R, Ouarda TBMJ, Arafat HA. Wind-powered desalination for strategic water storage: techno-economic assessment of concept. Desalination 2017;408:36–51.
[45] Lai W, Ma Q, Lu H, Weng S, Fan J, Fang H. Effects of wind intermittence and fluctuation on reverse osmosis desalination process and solution strategies. Desalination 2016;395:17–27.
[46] Maleki A, Khajeh MG, Rosen MA. Weather forecasting for optimization of a hybrid solar-wind–powered reverse osmosis water desalination system using a novel optimizer approach. Energy 2016;114:1120–34.
[47] Sinha S, Chandel S. Review of recent trends in optimization techniques for solar photovoltaic–wind based hybrid energy systems. Renew Sustain Energy Rev 2015; 50:755–69.
[48] Ekren O, Ekren BY. Size optimization of a PV/wind hybrid energy conversion system with battery storage using simulated annealing. Appl Energy 2010;87:592–8.
[49] Mohamed ES, Papadakis G. Design, simulation and economic analysis of a stand-alone reverse osmosis desalination unit powered by wind turbines and photovoltaics. Desalination 2004;164:87–97.

[50] Scrivani A. Energy management and DSM techniques for a PV-diesel powered sea water reverse osmosis desalination plant in Ginostra, Sicily. Desalination 2005;183:63–72.

[51] Helal A, Al-Malek S, Al-Katheeri E. Economic feasibility of alternative designs of a PV-RO desalination unit for remote areas in the United Arab Emirates. Desalination 2008;221:1–16.

[52] Touati S, Belkaid A, Benabid R, Halbaoui K, Chelali M. Pre-feasibility design and simulation of hybrid PV/fuel cell energy system for application to desalination plants loads. Proc Eng 2012;33:366–76.

[53] Clarke DP, Al-Abdeli YM, Kothapalli G. Multi-objective optimization of renewable hybrid energy systems with desalination. Energy 2015;88:457–68.

[54] Asensio M, Contreras J. Stochastic unit commitment in isolated systems with renewable penetration under CVaR assessment. IEEE Trans Smart Grid 2016;7:1356–67.

[55] Tani A, Camara MB, Dakyo B. Energy management in the decentralized generation systems based on renewable energy—ultracapacitors and battery to compensate the wind/load power fluctuations. IEEE Trans Indus Appl 2015;51:1817–27.

[56] Marinelli B, et al. Wind and photovoltaic large-scale regional models for hourly production evaluation. IEEE Trans Sustain Energy 2015;6:916–23.

[57] Alami HA, Nojavan S. Energy storage system and demand response program effects on stochastic energy procurement of large consumers considering renewable generation. IET Gener Transm Distrib 2016;10:107–14.

[58] Crawford RH. Validation of a hybrid life-cycle inventory analysis method. J Environ Manage 2008;88:496–506.

[59] Suh S, Huppes G. Methods for life cycle inventory of a product. J Clean Prod 2005;13:687–97.

[60] Mälkki H, Alanne K. An overview of life cycle assessment (LCA) and research based teaching in renewable and sustainable energy education. Renew Sustain Energy Rev 2017;69:218–31.

[61] Shahabi M, McHugh A, Anda M, Ho G. Environmental life cycle assessment of seawater reverse osmosis desalination plant powered by renewable energy. Renew Energy 2014;67:53–8.

[62] Zhou J, Chang V, Fane A. Environmental life cycle assessment of reverse osmosis desalination: the influence of different life cycle impact assessment methods on the characterization results. Desalination 2011;283:227–36.

[63] Raluy G, Serra L, Uche J. Life cycle assessment of MSF, MED and RO desalination technologies. Energy 2006;31:2361–72.

[64] Hancock N, Black N, Cath T. A comparative life cycle assessment of hybrid osmotic dilution desalination and established seawater desalination and wastewater reclamation processes. Water Res 2012;46:1145–54.

[65] International Energy Agency, 2016. Available: https://www.iea.org/.

[66] Raluy RG, Serra L, Uche J, Valero A. Life-cycle assessment of desalination technologies integrated with energy production systems. Desalination 2004;167:445–58.

[67] Meerganz GL. "Direct" and socially induced environmental impacts of desalination. Desalination 2005;185:57–70.

[68] Cherif H, Champenois G, Belhadj J. Environmental life cycle analysis of a water pumping and desalination process powered by intermittent renewable energy sources. Renew Sustain Energy Rev 2016;59:1504–13.

[69] Cherif H, Champenois G, Belhadj J. Methodology for life cycle assessment of brackish water desalination unit powered by hybrid photovoltaic-wind power system. Intern Conf on green energy, Tunisia, March 25–27; 2014, p. 38–43.

[70] Abbes D, Martinez A, Champenois G. Eco-design optimisation of an autonomous hybrid wind–photovoltaic system with battery storage. IET Renew Power Gener 2012;6:358–71.

[71] Alsema EA, Nieuwlaar E. Energy viability of photovoltaic systems. Energy Policy 2000;28:999–1010.
[72] Sumper A, Garcıa MR, Robles RV, Jané JB, Pieiro JA. Life-cycle assessment of a photovoltaic system n Catalonia (Spain). Renew Sustain Energy Rev 2011;15:3888–96.
[73] Fthenakis VM, Kim HC. Environmental impacts of photovoltaic life cycles. Compr Renew Energy 2012;1:143–59.
[74] Fthenakis VM, Kim HC. Photovoltaics: life-cycle analyses. Solar Energy 2011;85:1609–28.
[75] Kaldellis JK, Zafirakis D, Kondil E. Energy pay-back period analysis of stand- alone photovoltaic systems. Renew Energy 2010;35:1444–54.
[76] Lenzen M, Munksgaard J. Energy and CO2 life-cycle analyses of wind turbines—review and applications. Renew Energy 2002;26:339–62.
[77] Fleck B, Huot M. Comparative life-cycle assessment of a small wind turbine for residential off-grid use. Renew Energy 2009;34:2688–96.
[78] Kabir MR, Rooke B, Dassanayake GDM, Fleck BA. Comparative life cycle energy, emission, and economic analysis of 100 kW nameplate wind power generation. Renew Energy 2012;37:133–41.
[79] Uddin MS, Kumar S. Energy, emissions and environmental impact analysis of wind turbine using life cycle assessment technique. J Clean Prod 2014;69:153–64.
[80] Kristjansdottir TF, Good CS, Inman MR, Schlanbusch RD, Andresen I. Embodied greenhouse gas emissions from PV systems in Norwegian residential zero emission pilot buildings. Sol Energy 2016;133:155–71.
[81] Nugent D, Sovacool BK. Assessing the lifecycle greenhouse gas emissions from solar PV and wind energy: a critical meta-survey. Energy Policy 2014;65:229–44.
[82] Alsema E.V-2—Energypay-back time and CO2 emissions of PV systems. In: Practical Handbook of photovoltaics, fundamentals and applications. 2003. p. 869–86.
[83] Peng J, Lu L, Yang H. Review on life cycle assessment of energy payback and greenhouse gas emission of solar photovoltaic systems. Renew Sustain Energy Rev 2013;19:255–74.
[84] Wild-Scholten MJ. Energy payback time and carbon footprint of commercial photovoltaic systems. Sol Energy Mater Sol Cells 2013;119:296–305.
[85] Mazzanti G, Santini E, Zaccagnini Romito D. Towards grid parity of solar energy in Italy: The payback time trend of photovoltaic plants during the last years. IEEE Power and Energy Society General Meeting 2012; pp. 1–8.
[86] Ardente F, Beccali M, Cellura M, Brano VL. Energy performances and life cycle assessment of an Italian wind farm. Renew Sustain Energy Rev 2008;12:200–17.
[87] Tremeac B, Meunier F. Life cycle analysis of 4.5MW and 250W wind turbines. Renew Sustain Energy Rev 2009;13:2104–10.
[88] Smaoui M, Abdelkafi A, Krichen L. Optimal sizing of stand-alone photovoltaic/wind/hydrogen hybrid system supplying a desalination unit. Sol Energy 2015;120:263–76.

INDEX

Note: Page numbers followed by *f* indicate figures, and *t* indicate tables.

E

F

G

H

I

J

K

L

M

N

O

FSC
www.fsc.org
MIX
Papier | Fördert gute Waldnutzung
FSC® C083411

Kolibri 360 GmbH
Ferdinand-Jühlke-Straße 7
99095 Erfurt, Deutschland
produktsicherheit@kolibri360.de